# Communications in Computer and Information Science 1455

Hector Florez ·
Ma Florencia Pollo-Cattaneo (Eds.)

# Applied Informatics

Fourth International Conference, ICAI 2021
Buenos Aires, Argentina, October 28–30, 2021
Proceedings

*Editors*
Hector Florez
Universidad Distrital Francisco Jose de Caldas
Bogotá, Colombia

Ma Florencia Pollo-Cattaneo
Universidad Tecnológica Nacional Facultad Regional Buenos Aires
Buenos Aires, Argentina

ISSN 1865-0929 ISSN 1865-0937 (electronic)
Communications in Computer and Information Science
ISBN 978-3-030-89653-9 ISBN 978-3-030-89654-6 (eBook)
https://doi.org/10.1007/978-3-030-89654-6



This Springer imprint is published by the registered company Springer Nature Switzerland AG
The registered company address is: Gewerbestrasse 11, 6330 Cham, Switzerland

# Preface

The Fourth International Conference on Applied Informatics (ICAI 2021) aimed to bring together researchers and practitioners working in different domains in the field of informatics in order to exchange their expertise and to discuss the perspectives of development and collaboration.

ICAI 2021 held (virtually) at the Universidad Tecnológica Nacional Facultad Regional Buenos Aires located in Buenos Aires, Argentina, during October 28-30, 2021. It was organized by the Information Technologies Innovation (ITI) research group that belongs to the Universidad Distrital Francisco José de Caldas and the GEMIS research group (Grupo de Estudio de Metodologías para Ingeniería en Software y Sistemas de Información) that belongs to the Universidad Tecnológica Nacional Facultad Regional Buenos Aires. In addition, ICAI 2021 was proudly sponsored by Springer and Science Based Platforms.

ICAI 2021 received 89 submissions on informatics topics. Accepted papers covered artificial intelligence, data analysis, decision systems, health care information systems, image processing, security services, simulation and emulation, smart cities, software and systems modeling, and software design engineering. Authors of accepted submissions came from the following countries: Argentina, China, Colombia, Ecuador, France, Latvia, Lithuania, Mexico, Morocco, the Netherlands, Nigeria, Norway, Poland, Tunisia, the UK, and the USA.

All submissions were reviewed through a double-blind peer-review process. Each paper was reviewed by at least three experts. To achieve this, ICAI 2021 was supported by 54 program committee members, who hold Ph.D. degree. PC members come from the following countries: Argentina, Brazil, China, Colombia, Cyprus, Ecuador, France, Germany, India, Latvia, Lithuania, Luxembourg, Mexico, Netherlands, Nigeria, Portugal, Spain, Switzerland, Ukraine, the United Arab Emirates, the USA, and Uruguay. Based on the double-blind review process, 35 full papers were accepted to be included in this volume of Communications in Computer and Information Sciences (CCIS) proceedings published by Springer.

We would like to thank Jorge Nakahara, Guido Zosimo-Landolfo, Alla Freund, and Ramvijay Subramani from Springer for their helpful advice, guidance, and support in publishing the proceedings.

We trust that the ICAI 2021 proceedings open you to new vistas of discovery and knowledge.

October 2021

Hector Florez
Ma Florencia Pollo-Cattaneo

# Organization

## General Chairs

| | |
|---|---|
| Hector Florez | Universidad Distrital Francisco José de Caldas, Colombia |
| Ma Florencia Pollo-Cattaneo | Universidad Tecnológica Nacional Facultad Regional Buenos Aires, Argentina |

## Steering Committee

| | |
|---|---|
| Jaime Chavarriaga | Universidad de los Andes, Colombia |
| Cesar Diaz | OCOX AI, Colombia |
| Hector Florez | Universidad Distrital Francisco José de Caldas, Colombia |
| Ixent Galpin | Universidad de Bogotá Jorge Tadeo Lozano, Colombia |
| Olmer García | Universidad de Bogotá Jorge Tadeo Lozano, Colombia |
| Christian Grévisse | Université du Luxembourg, Luxembourg |
| Sanjay Misra | Covenant University, Nigeria |
| Fernando Yepes-Calderon | Children's Hospital Los Angeles, USA |

## Organizing Committee

| | |
|---|---|
| Ignacio Bengoechea | Universidad Tecnológica Nacional Facultad Regional Buenos Aires, Argentina |
| Paola Britos | Universidad Nacional de Río Negro, Argentina |
| Andres Bursztyn | Universidad Tecnológica Nacional Facultad Regional Buenos Aires, Argentina |
| Ines Casanovas | Universidad Tecnológica Nacional Facultad Regional Buenos Aires, Argentina |
| Alejandro Hossian | Universidad Tecnológica Nacional Facultad Regional Neuquén, Argentina |
| Hernan Merlino | Universidad Tecnológica Nacional Facultad Regional Buenos Aires, Argentina |
| Ma Florencia Pollo-Cattaneo | Universidad Tecnológica Nacional Facultad Regional Buenos Aires, Argentina |
| Giovanni Rottoli | Universidad Tecnológica Nacional Facultad Regional Concepción del Uruguay, Argentina |
| Cinthia Vegega | Universidad Tecnológica Nacional Facultad Regional Buenos Aires, Argentina |

## Workshops Committee

| | |
|---|---|
| Hector Florez | Universidad Distrital Francisco José de Caldas, Colombia |
| Ixent Galpin | Universidad de Bogotá Jorge Tadeo Lozano, Colombia |
| Christian Grévisse | Université du Luxembourg, Luxembourg |

## Program Committee Chairs

| | |
|---|---|
| Ixent Galpin | Universidad de Bogotá Jorge Tadeo Lozano, Colombia |
| Christian Grévisse | Université du Luxembourg, Luxembourg |
| Hector Florez | Universidad Distrital Francisco José de Caldas, Colombia |
| Ma Florencia Pollo-Cattaneo | Universidad Tecnológica Nacional Facultad Regional Buenos Aires, Argentina |
| Fernando Yepes-Calderon | Children's Hospital Los Angeles, USA |

## Program Committee

| | |
|---|---|
| Francisco Alvarez | Universidad Autónoma de Aguascalientes, Mexico |
| Jorge Bacca | Fundacion Universitaria Konrad Lorenz, Colombia |
| Carlos Balsa | Instituto Politécnico de Bragança, Portugal |
| Hüseyin Bicen | Yakin Dogu Üniversitesi, Cyprus |
| Alexander Boch | Universität Duisburg Essen, Germany |
| Paola Britos | Universidad Nacional de Río Negro, Argentina |
| Raymundo Buenrostro | Universidad de Colima, Mexico |
| Santiago Caballero | Universidad Popular Autónoma del Estado de Puebla, Mexico |
| Patricia Cano-Olivos | Universidad Popular Autónoma del Estado de Puebla, Mexico |
| Jaime Chavarriaga | Universidad de los Andes, Colombia |
| Robertas Damasevicius | Kauno Technologijos Universitetas, Lithuania |
| Victor Darriba | Universidade de Vigo, Spain |
| Cesar Diaz | OCOX AI, Colombia |
| Helga Duarte | Universidad Nacional de Colombia, Colombia |
| Mauri Ferrandin | Universidade Federal de Santa Catarina, Brazil |
| Hans-Georg Fill | Université de Fribourg, Switzerland |
| Hector Florez | Universidad Distrital Francisco José de Caldas, Colombia |
| Ixent Galpin | Universidad de Bogotá Jorge Tadeo Lozano, Colombia |
| Olmer Garcia | Universidad de Bogotá Jorge Tadeo Lozano, Colombia |
| Raphael Gomes | Instituto Federal de Goiás, Brazil |
| Daniel Görlich | SRH Hochschule Heidelberg, Germany |
| Jānis Grabis | Rīgas Tehniskā Universitāte, Latvia |
| Christian Grévisse | Université du Luxembourg, Luxembourg |

# Contents

**Health Care Information Systems**



**Image Processing**

## Security Services



## Simulation and Emulation



## Smart Cities



## Software and Systems Modeling

## Software Design Engineering

# Artificial Intelligence

# A Chatterbot Based on Genetic Algorithm: Preliminary Results

Cristian Orellana[1], Martín Tobar[1], Jeremy Yazán[1], D. Peluffo-Ordóñez[2,3,4], and Lorena Guachi-Guachi[1,3](✉)

[1] Department of Mechatronics, Universidad Internacional del Ecuador, Av. Simon Bolivar, Quito 170411, Ecuador
loguachigu@uide.edu.ec

[2] Mohammed VI Polytechnic University, Ben Guerir, Morocco

[3] SDAS Research Group, Ibarra, Ecuador

[4] Faculty of Engineering, Department of Electronics, Corporación Universitaria Autónoma de Nariño, Pasto 520001, Colombia
http://www.sdas-group.com

**Abstract.** Chatterbots are programs that simulate an intelligent conversation with people. They are commonly used in customer service, product suggestions, e-commerce, travel and vacations, queries, and complaints. Although some works have presented valuable studies by using several technologies including evolutionary computing, artificial intelligence, machine learning, and natural language processing, creating chatterbots with a low rate of grammatical errors and good user satisfaction is still a challenging task. Therefore, this work introduces a preliminary study for the development of a GA-based chatterbot that generates intelligent dialogues with a low rate of grammatical errors and a strong sense of responsiveness, so boosting the personals satisfaction of individuals who interact with it. Preliminary results show that the proposed GA-based chatterbot yields 69% of "Good" responses for typical conversations regarding orders and receipts in a cafeteria.

**Keywords:** Chatbot · Evolutionary programming · Natural language processing

## 1 Introduction

Chatterbots are computer programs that simulate intelligent human conversations. The usual chatterbot execution involves a user providing natural language input, to which the chatterbot responds with a reasonable and presumably intelligent response to the input sentence. Chatterbots are commonly used in customer service, product suggestions, e-commerce, travel and vacations, queries and complaints [1,2]. Since the 1990s, several technologies have been developed for the development of chatterbot-based solutions, including evolutionary computing, artificial intelligence, machine learning, natural language processing [2].


H. Florez and M. F. Pollo-Cattaneo (Eds.): ICAI 2021, CCIS 1455, pp. 3–12, 2021.
https://doi.org/10.1007/978-3-030-89654-6_1

In recent research, Genetic Algorithms (GA), a technique from evolutionary computing, have been used to create chatterbots that retrieve dialog responses to user input [3–6]. GA-based chatterbots reflects the process of natural selection and attempts to improve the fitness function by fitting patterns and rules to human inputs. This technique has been used primarily for trivial conversations, and it, like other techniques, such as deep learning, is prone to producing meaningless or inappropriate responses [7].

In this work, we present a GA-based chatterbot aiming at providing satisfactory and grammatically correct responses. For purposes of this work, we create a dataset of 120 phrases appropriate for conversations about food orders or recipes in a cafeteria. To reduce the risk of grammatical errors and to maintain structure, the population is created with randomly selected words from the database (adverbs, nouns, verbs, etc.). The mutation operator determines the population's diversity in each generation replacing a random word from a phrase with a new word of the same type (adverbs, nouns, verbs, etc.) from our dataset.

This paper is structured as follows: Sect. 2 presents a brief review of state-of-the-art related works. The database is described in Sect. 3. The use of GA for building a chatterbot is described in Sect. 4. Section 5 describes the experiments carried out over the database suitable for ordering food or recipes in a cafeteria. Section 6 presents the preliminary results and discussion across the experiments. Finally, Sect. 7 gathers the concluding remarks.

## 2 Related Works

On one hand, GAs have proven to be effective in the development of chatterbots and the identification of appropriate responses. For instance, [8], proposes a chatterbot model built on a natural language-adapted algorithm that combines indexing and pattern matching to generate new phrases from preloaded words. A GA-based chatterbot to learn through samples of conversations between the user and the system is proposed in [9]. It generates the response evaluating the type of expressions used by the user based on rules built from the last samples of conversation. In [10], an inductive chatterbot model based on a GA with sexual selection (SeGa-ILSD) is developed in order to handle different languages: English and Japanese.

On the other hand, some hybrid techniques based on GA have been also introduced. For instance, the work carried out in [11] addresses a user-adaptive communication robot using interactive evolutionary computation (IEC) and machine-learning approaches for robot's movements. In particular, IEC genes are initialized at random. Then, one of them is translated into the parameters of the robot's value system, and the robot interacts with the user. In this case, fitness is provided by users at the end of each interaction. This sequence continues until all genes are evaluated. Finally, more appropriate genes are generated.

The study in [12] determines that others relate better to people who have similar personality characteristics to them. In this sense, this research introduces a GA capable of changing their responses behavior in real-time. The change in GA behavior depends on the fitness function, which is calculated based on how comfortable/satisfactory the response has been from the user's perception. The comfortable scale ranges from 1 to 10. The best responses (comfortable/satisfactory) are passed on to the next generation, while the uncomfortable ones are combined with a different random generation.

Furthermore, more specialized research in [3] address the creation of a chatterbot with AIML, which is similar to an XML database that employs specific tags for each instance. The new chatterbot is being used for an interactive university chatterbot that will answer related FAQs. Users submit their questions, which are subsequently processed to match the specified answer and returned. In [4] a worth-method for creating new sentences in Spanish by combining prior sentences is presented. Each phrase is treated as an individual, with the genes representing the words that make up the phrase. The mutation coefficient is inversely proportional to the number of words, and there is no crossover. Mutation, which is carried out depending on the feature of the word, provides population diversity. The amount of results found for a specific phrase on the WWW is used to assess the fitness function.

Although there are valuable research on GA-based chatterbots in the literature, the usage of GA with a low rate of grammatical errors and good user satisfaction is a challenging task yet. Therefore, we introduce a chatterbot based on GA that generates intelligent dialogues with a low rate of grammatical errors and a high sense of responsiveness, increasing the personal pleasure of those who interact with it.

## 3 Database

The database consists of 120 statements suitable for ordering food or recipes in a cafeteria. All sentences are written in a .JSON file and more sentences can be added depending on the target application. Samples of the entries in the database are as follows:

- **"Greeting"**
  - *Patterns:* "Hi", "How are you", "Is anyone there?", "Hello", "Good day", "Whats up"
  - *Responses:* "I can help you", "You have a good day", "I am happy to see you"
- **"Goodbye"**
  - *Patterns:* "See you later", "Goodbye", "I am Leaving", "Have a Good day", "Bye"
  - *Responses:* "I am sad", "I see you later", "It was a pleasure"

– **"Age"**
  – *Patterns:* "how old", "what is your age", "how old are you", "age?"
  – *Responses:* "I am 18", "I have no idea", "It does not matter", "It is a secret"
– **"Name"**
  – *Patterns:* "what is your name", "what should I call you", "whats your name?"
  – *Responses:* "My name is Tim", "I am Tim", "It is Tim"
– **"Shop"**
  – *Patterns:* "Id like to buy something", "whats on the menu", "what do you recommend?", "could i get something to eat"
  – *Responses:* "It costs 20", "We sell cookies"
– **"Hours"**
  – *Patterns:* "when are you guys open", "what are your hours", "hours of operation"
  – *Responses:* "We are open at 8", "We are closed at 20"

## 4 The Use of Genetic Algorithm

This work uses a GA to generate a suitable sentence based on the combination of a set of words. The general workflow of the proposed approach is schematized in Fig. 1. The population of $s$ sentences is made up of $w$ words randomly selected from dataset of words to reduce the risk of grammatical errors. Every sentence is mutated by substituting a random word with a new one from word dataset. The "fitness" of each sentence is determined by its frequency on the World Wide Web (WWW).

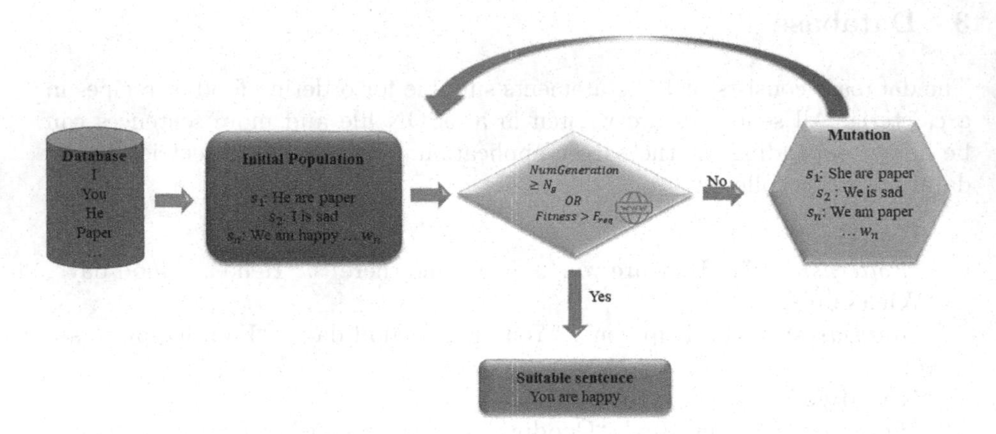


**Fig. 1.** Workflow of genetic algorithm to generate the most appropriate sentence.

### 4.1 Initial Population Generation

This step selects random phrases from database pool which contains sentences that frequently used in human conversations. Like "How was your day?", "How are you?", "I'm doing great and you"?, etc. Then, initial population of sentences is created extracting words from selected phrases and keeping the structure (adverbs, nouns, verbs, etc.).

### 4.2 Mutation

For purposes of this work, the mutation probability ($P$) is the inverse of the total amount of words in the phrase.

The mutation probability ($P$) is given by:

$$P(N) = \frac{1}{N} \tag{1}$$

where $N$ is the number of words.

The algorithm identifies the word to be replaced (adverbs, nouns, verbs, etc.) and substitutes it with the same kind of word in order to keep the structure as good as possible.

### 4.3 Evaluation

The evaluation is carried out by searching the WWW and finding out how frequently the phrase is found. To accomplish this, each phrase is typed into the Google search engine, and the frequency is determined by the number of times the engine finds the same phrase. The greater the frequency of phrase, the better the individual's fitness. When a complete generated phrase can not be found literally on the searching engine, a phrase partition strategy is proposed, since the shorter the strings entered into the engine, the greater the number of results are obtained. Furthermore, $x$ threshold values will be assigned to classify whether a generated sentence is good or bad.

### 4.4 Pseudo-code

Genes (individuals) are composed of three essential parts: subject, verb, and predicate. Each one is represented by a char string with a distinct meaning. The initial population is created with phrases chosen randomly. Then, the fitness is calculated by comparing each phrase to a corresponding sentence in the database. If the sentence is consistently logical, it is Google-searched to determine the search frequency. This implementation is extended in Algorithm 1.

**Algorithm 1.** Used Algorithm's Pseudo Code

```
1: Select random phrases from database                          ▷ Create Initial Population
2: Generations = 0
3: function EVALUATE                                            ▷ Evaluate new population
4:     Compare generated phrase similarity with data base phrases.
5:     Fitness = Amount of Google results [4]     ▷ Check the amount of times the
   phrase is found on google in a literal way.
6: end function
7: while Generations < 5 or Fitness < 2 000 000 do
8:     Mutate Population
9:     Probability = 1 / AmountOfWords
10:    for N do                                                 ▷ N is the amount of words
11:        R = Random Number between 0 and 1
12:        if Probability > R then
13:            Substitute word for a new one
14:        end if
15:        function EVALUATE                                    ▷ Call Evaluate function
16:        end function
17:    end for
18:    Generations = Generations + 1
19: end while
```

## 5 Experimental Setup

For experimental purposes, Python routines have been implemented using Google Colab[1]. This environment has 12 GB of RAM and 100 GB of storage space. To work with human language data, the Natural Language Toolkit (NLTK) was also used. Arrays were also manipulated using Numpy and Scikit-learn to obtain the linear regression function. Python reads json files using JSON. BS4 was used to connect to the Internet. Difflib was used to compare two char strings, and the Requests library was used to search a website for a specific query.

The quality of the responses generated by the proposed chatterbot (fitness function) are evaluated using the database presented in this work together with an Internet search algorithm; if these answers do not meet a 90 % similarity percentage with the dictionary along with a minimum number of 2000000 internet searches, the algorithm generates a new response until these conditions are satisfied, otherwise, the program ends at the end of four iterations. It is worth noting that the minimum search value was obtained as the most optimal from experimental tests. The program was run on a total of 20 times and subsequently, the resulting conversations are evaluated by the users according to the degree of satisfaction they had with the responses made by the chatterbot

[1] Source: https://colab.research.google.com/drive/1Y7Cd5xTVZ0TtGIXDGNmBBEQe7u_Lic3b?usp=sharing.

## 6 Experimental Results

Samples of answer obtained are depicted in Fig. 2 and Fig. 3.

(a) First example of a conversation test user-Chatbot

(b) Second example of a conversation test user-Chatbot

**Fig. 2.** Conversation sample with answers obtained by using the proposed genetic algorithm. The numbers under the user's sentence correspond to the number of searches found on the internet of the answers generated by the Chatbot. (Color figure online)

The fifty-eight responses obtained by the Chatbot to experimental questions have been evaluated based on an application metric in which users assigned a

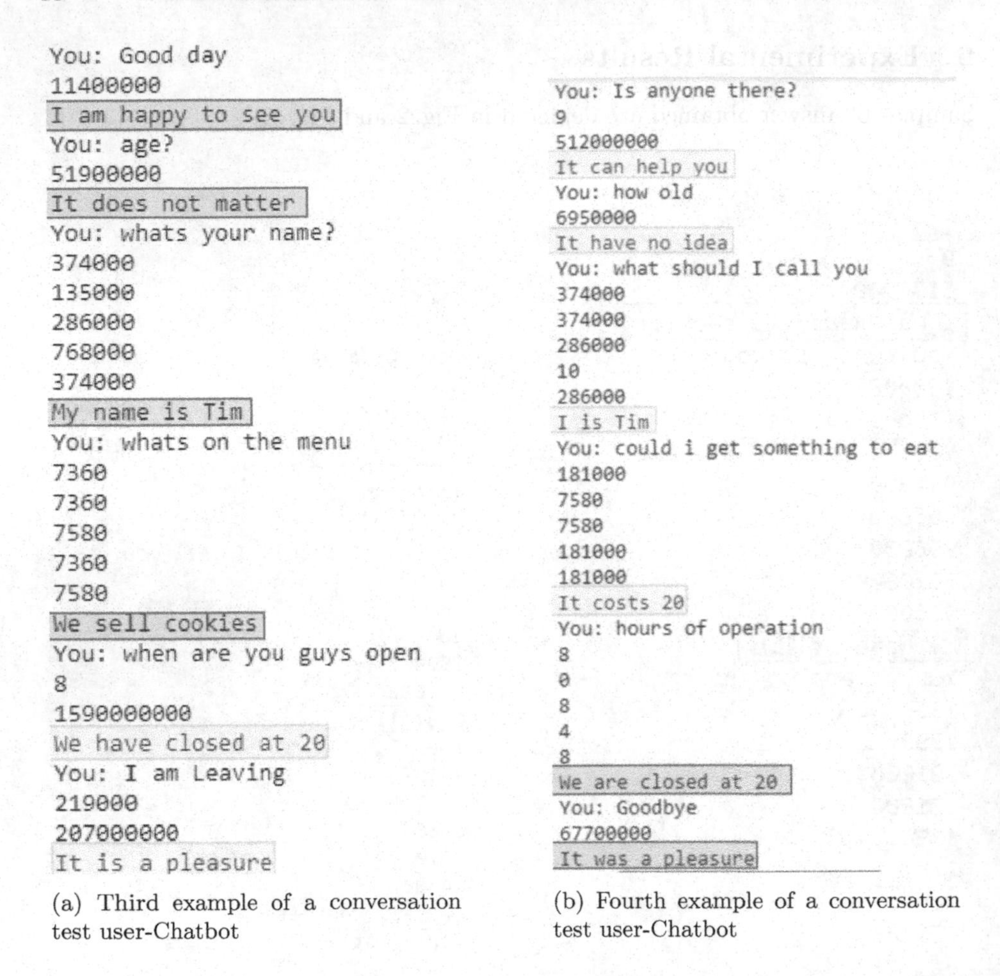


(a) Third example of a conversation test user-Chatbot

(b) Fourth example of a conversation test user-Chatbot

**Fig. 3.** Conversation sample with answers obtained by using the proposed genetic algorithm. The numbers under the user's sentence correspond to the number of searches found on the internet of the answers generated by the Chatbot. (Color figure online)

rating based on the level of personal satisfaction, which is structured in three categories:

- Good: Green square.
- Regular: Yellow square.
- Bad: Red square

The users' evaluation is based on the orthography and coherence in the structure of the ChatBot's answers, as well as their similarity to the words in the database. In this way, if the answer meets the aforementioned criteria, it will be rated "good". If it contains spelling errors that do not significantly alter the meaning of the answer, it will be rated "regular." Finally, if the response is grammatically incorrect, it will receive a "bad" grade. The number of answers grouped by category is illustrated in Figure 4.

From Fig. 4, it can be seen that the proposed chatterbot by using genetic algorithm generates 69% Good phrases, 31% regular, and 0% Bad phrases. Therefore, the proposed chatterbot meets to a great extent the need for a cafeteria application since the answers classified as regular did not present errors that mean a loss of direction in the user-chattbot conversation.

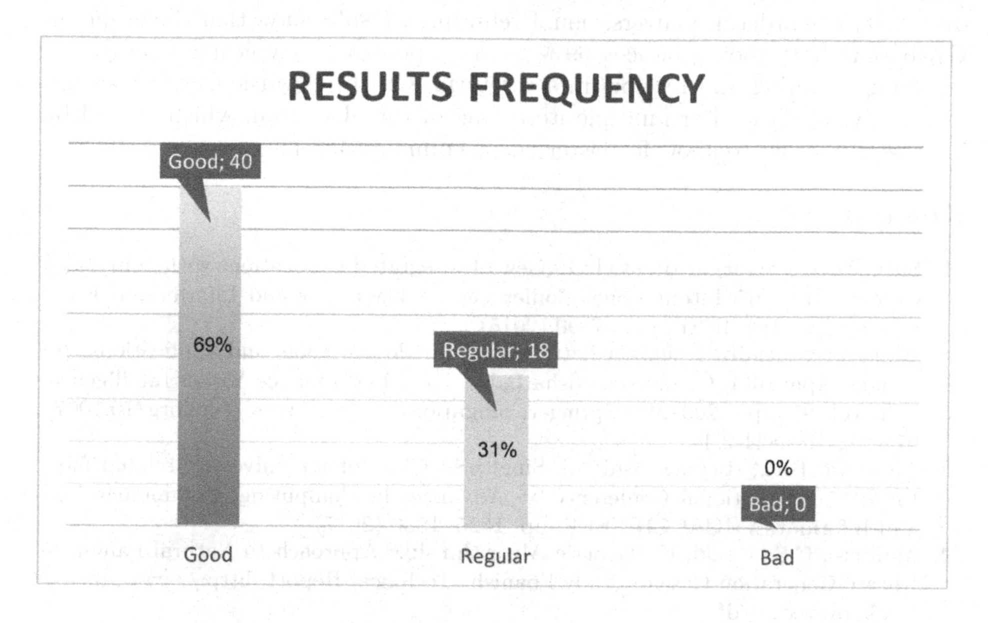


**Fig. 4.** Personal satisfaction

The processing time of the proposed approach has been also evaluated. Table 1 shows the average of the time consumed by the operations included in the genetic algorithm, besides, we can see the significant consumption of the RAM memory. Since the execution of the algorithm involves a reduced time, the proposed chatterbot can be an appropriated tool for providing an immediate support.

**Table 1.** Average of the time and RAM consumed for each execution of the algorithm.

| Metric | Average value |
|---|---|
| Time (sec) | 0.112 |
| RAM used (MB) | 840 |

## 7 Conclusions

This work presents a GA-based chatterbot that can be used to improve customer experience in situations related to ordering in a cafeteria. It generates responses with a low rate of grammatical errors and a strong sense of responsiveness, so boosting the personals satisfaction of individuals who interact with it. The algorithm is still in a preliminary stage. Genetic evolution is able to generate new responses and improve the diversity of the chatterbot while maintaining the overall style of ordering conversation. Preliminary results show that the proposed GA-based chatterbot produces 69% good responses for typical conversations regarding ordering and receipts in a cafeteria. On the downside, responses are frequently supplied after multiple iterations of the algorithm, which should be improved in order to allow for faster client communication.

## References


1. Satu, M.S., Parvez, M.H., et al.: Review of integrated applications with aiml based chatbot. In: 2015 International Conference on Computer and Information Engineering (ICCIE). IEEE, pp. 87–90 (2015)
2. Ahirwar, Gajendra Kumar: Chatterbot: technologies, tools and applications. In: Nanda, Aparajita, Chaurasia, Nisha (eds.) High Performance Vision Intelligence. SCI, vol. 913, pp. 203–213. Springer, Singapore (2020). https://doi.org/10.1007/978-981-15-6844-2_14
3. Ranoliya, B.R., Raghuwanshi, N., Singh, S.: Chatbot for university related faqs. In: 2017 International Conference on Advances in Computing, Communications and Informatics (ICACCI). IEEE, pp. 1525–1530 (2017)
4. Montero, C.S., Araki, K.: Genetic Algorithm-like Approach to Natural Language Phrase Generation-Case of Study Spanish. Technical Report. http://www.um.es/lacell/proyectos/dfe/
5. Pham, X.L., Pham, T., Nguyen, Q.M., Nguyen, T.H., Cao, T.T.H.: Chatbot as an intelligent personal assistant for mobile language learning. In: Proceedings of the 2018 2nd International Conference on Education and E-Learning, pp. 16–21 (2018)
6. Daniel, G., Cabot, J., Deruelle, L., Derras, M.: Xatkit: a multimodal low-code chatbot development framework. IEEE Access **8**, 15 332-15 346 (2020)
7. Hadjiev, A.M., Araki, K.: Chatbot system for open ended topics based on multiple response generation methods
8. Vrajitoru, D.: Evolutionary sentence building for chatterbots. In: Genetic and Evolutionary Computation Conference (GECCO). Citeseer (2003)
9. Kimura, Y., Araki, K., Momouchi, Y., Tochinai, K.: Spoken dialogue processing method using inductive learning with genetic algorithm. Syst. Comput. Japan **35**(12), 67–82 (2004)
10. Araki, K., Kuroda, M.: Generality of spoken dialogue system using sega-il for different languages. In: Computational Intelligence, pp. 72–77 (2006)
11. Suga, Y., Ikuma, Y., Ogata, T., Sugano, S.: Development of user-adaptive value system of learning function using interactive ec. IFAC ProC. Vol. **41**(2), 9156–9161 (2008). 17th IFAC World Congress. https://www.sciencedirect.com/science/article/pii/S1474667016404258
12. Yorita, A., Egerton, S., Oakman, J., Chan, C., Kubota, N.: Self-adapting chatbot personalities for better peer support. In: 2019 IEEE International Conference on Systems, Man and Cybernetics (SMC). IEEE, pp. 4094–4100 (2019)

# A Supervised Approach to Credit Card Fraud Detection Using an Artificial Neural Network

Oluwatobi Noah Akande[1], Sanjay Misra[2(✉)], Hakeem Babalola Akande[3], Jonathan Oluranti[4], and Robertas Damasevicius[5]

[1] Computer Science Department, Landmark University, Omu-Aran, Kwara State, Nigeria
akande.noah@lmu.edu.ng

[2] Department of Computer Science and Communication, Ostfold University College, Halden, Norway

[3] Department of Telecommunication, University of Ilorin, Ilorin, Kwara State, Nigeria
akande.hb@unilorin.edu.ng

[4] Center of ICT/ICE, Covenant University, Ota, Nigeria
jonathan.oluranti@covenantuniversity.edu.ng

[5] Kaunas University of Technology, Kaunas, Lithuania

**Abstract.** The wide acceptability and usage of credit card-based transactions can be attributed to improved technological availability and increased demand due to ease of use. As a result of the increased adoption levels, this domain has become profitable and one of the most popular targets for fraudsters who use it to conduct regular exploitations or assaults. Merchants and financial processing providers that sell credit cards suffer substantial financial damages as a result of credit card theft. Because of the possibility of large casualties, it is one of the most serious risks to these organizations and individuals. Credit card fraudulent transaction can be viewed as a binary classification task in which a supervised machine learning technique could be used to analyze and classify a credit card transaction dataset into genuine or fraudulent cases. Therefore, this study explored the use of Artificial Neural Network (ANN) for credit card fraud detection. ULB Machine Learning Group dataset that has 284, 315 legitimate and 492 fraudulent transaction were used to validate the proposed model. Performance evaluation results revealed that model achieved a 100% and 99.95% classification accuracy during training and testing respectively. This affirmed the fact that ANN model could be efficiently used to predict credit card fraudulent transactions.

**Keywords:** Sales forecasting · XGBoost algorithm · Machine learning · Walmart dataset

## 1 Introduction

The advancement in Information and Communication Technology (ICT) has made it possible for buying and selling to happen via the Internet without any need for face-to-face interaction. Financial institutions have increased the flexibility of transacting businesses online with the aid of innovative solutions supported by credit cards and


H. Florez and M. F. Pollo-Cattaneo (Eds.): ICAI 2021, CCIS 1455, pp. 13–25, 2021.
https://doi.org/10.1007/978-3-030-89654-6_2

mobile banking applications. These solutions have made transacting businesses online easier, faster and have also eradicated long queue waiting time in banks. Thanks to the widespread use of credit cards and the exponential growth of e-services, the volume of credit card purchases has increased significantly [1]. However, the continuous reliance and usage of credit cards and mobile banking applications without strict oversight and verification have opened up many customers to diverse kinds of financial frauds and attacks. The increased credit card transactions during the COVID-19 lockdown gave fraudsters the more opportunities to perpetrate their illicit acts. According to a US based Fidelity National Information Services, the dollar volume of attempted illegal transactions in dollars increased by 35% in April 2020 alone. With US as the largest country of credit card fraud cases, credit card fraud cost the world $24.2 billion in 2018, with credit card fraud transactions estimated to hit $40 billion by 2027 [2]. According to the Unisys protection index, credit and debit card frauds are Americans' top concern, far surpassing terrorist concerns [3].

Credit-card fraud occurs when an individual uses a credit card for personal purposes without the owner's permission and with no intention of repaying the payment. Furthermore, the person who uses the card has no ties to the cardholder or issuer, and has no intention of approaching the card's owner or repaying the transactions received [4]. Credit card fraud can occur when an unauthorized cardholder uses a fake identity to gain the confidence of a bank official, or when stolen credit cards are used [5]. It is an unfair or criminal deceit with the goal of gaining personal benefit [5]. Contrary to popular belief, when a fraudster steals with the use of a credit card, the bill is the responsibility of the retailers [6]. Also, when a customer claims that he did not receive the goods, he ordered for, there will be a need for the retailers to pay back if the claim can be proved by the customer. If the corporation is unable to refute this argument, the money will be returned to the customer's account, and the goods will be discarded (if it has been shipped). Furthermore, if the chargeback rate exceeds the card associations' limits, retailers can be liable to chargeback penalties and penalties [7]. Ten different types of credit card frauds were reported in [26]. Application fraud occurs when a fraudster gains control of an application, obtains the customer's information, creates a phony account, and then conducts transactions. Electronic or manual card imprints: In electronic imprint fraud attempt, the fraudster retrieves the needed information from the card's magnetic strip. this information is then utilized to carry out fraud transactions. in card not present fraud attempt, the hacker does not make use of the actual physical card at the time of the transaction while in counterfeit card fraud attempt, the hacker replicates data available on the magnetic strip of the original card to create a fake card that will be used in the transaction. Lost/stolen fraudulent card attack occurs when the actual owner of the card lost the card and is found by the fraudster or when the fraudster deliberately steals the card from the owner. In card id theft instances, the cardholder's id or credentials is stolen and the stolen credentials are used in perpetrating fraud. Mail non-received card fraud attack occurs when the hacker intercepts a mail sent by the bank to inform the cardholder that his/her card is ready for collection. Here, the true recipient may not receive the mail or the content of the mail may be manipulated such that the intended recipient will receive the mail after the card details have been retrieved. Similarly, in account takeover: attempt, the hacker hijacks the account of the cardholder during the illegal transaction period. So, the owner of the card won't be aware of the fraudulent transactions going on with his/her account. In fake fraud on website attack, a hacker inserts malicious code into the website of the ATM producing company or bank and uses the information retrieved for fraudulent activities. Conflict between merchants' attack occurs when there is a leak of card information between the card manufacturer, the financial institution and a third party. Generally, credit card frauds can be categorized into two: online fraud and offline fraud; the first is committed by using a stolen credit card for transactions while the second is committed by manipulating victim identification such as credit card numbers, credit card holders' names, expiration dates, and passwords [10].

However, authors in [1] classified credit card fraud into application fraud [8] and behaviour fraud [9]. Application fraud occurs when a fraudster requests for a new card using another person's identity. Behavior fraud occurs when a fraudster steals or forges a card or carries out illegitimate transactions with or without the credit card. The most popular form of behaviour fraud occurs when a stolen credit card is used for unauthorized transaction. The method of determining whether a transaction carried out using credit card is legitimate or fake is known as credit card fraud identification [4, 10]. To reduce fraud losses, a sophisticated fraud detection system with a cutting-edge fraud detection model is considered important [1]. Regardless of the fraud identification model adopted, fraudulent transactions which are always the minority-class samples must be distinguished from the legitimates transactions which are always the majority-class samples [11, 12]. However, instances of normal transactions are always more than the suspicious transactions in the fraudulent transactions class; this makes the classification task a delicate but surmountable task [10, 13–15]. Nevertheless, several innovative solutions such as the Address Verification System (AVS), Chip and Pin identification, and Card Verification Code (CVV) have been explored to deter credit card fraud [6]. However, most of these solutions have been compromised by the fast fingers of hackers. Therefore, the advancement of fraud detecting approaches is critical and fraud detection techniques must continue to grow at a quicker rate than fraudsters. Supervised and unsupervised machine learning approach have been adopted in the literature to detect credit card frauds. In the supervised approach, transactional data records are grouped into fraudulent and non-fraudulent transactions while in unsupervised approach, secret trends in non-labeled transactional data are identified with the help of machine learning algorithms. Account numbers, credit card and payment forms, transaction location and time, customer name, merchant code, transaction size, and so on are some of the transaction information that can be found in these transactional data records. This information can be used as pointers to decide whether a transaction is illegitimate or legal, as well as to investigate outliers that might indicate a suspicious event. A supervised approach to detecting credit card fraud using Artificial Neural Networks is presented in this study. The dataset employed to train and test the resulting ANN model contains credit card transactions made by European cardholders in September 2013. This dataset has 284,807 transactions that occurred in two days; 284, 315 of these are legit credit card transactions while the remaining 492 are fraudulent credit card transactions.

## 2 Related Works

Machine Learning (ML) models are used by card issuers and network providers to detect credit card fraud. Despite the fact that much research has been done in both industry and academia to develop machine learning models, finding effective solutions remains a challenge. Studied on security in banking [16, 17], mobile applications [18, 19] can be found in various papers. ML techniques have been greatly explored to provide solutions to several security threats [20, 21]. A supervised and unsupervised approach for improving credit card fraud detection accuracy was proposed in [22]. Unsupervised outlier scores were computed at various levels of granularity from an annotated credit card fraud detection dataset. The results obtained revealed that combining techniques from both supervised and unsupervised techniques could improve the accuracy of credit card fraud detection. RIBIB, a cost-sensitive Risk Induced Bayesian Inference Bagging model for credit card fraud detection was proposed in [15]. The proposed model is made up of a cost-sensitive weighted voting combiner, a constrained bag formation process, and a Risk Induced Bayesian Inference method as a base learner. Brazilian bank data was used to validate the proposed technique and resulted to a cost reduction of 1.04–1.5 times. Experiments on UCSD-FICO data show that the model is capable of processing unknown data without the need for fine-tuning of domain-specific parameters. Furthermore, authors in [23] suggested a Bayesian Network Classifier (BNC) algorithm for a credit card fraud detection. The model was created automatically using a dataset from an online payment system. When the results achieved was compared to seven different algorithms, the proposed technique achieved a better classification performance. To detect fraudulent credit card transaction behavior, authors in [24] proposed an ensemble model focused on sequential data processing using deep RNN and a voting system based on ANN. The proposed model was more effective in terms of classification time. Moreover, a new hybrid approach built on the divide-and-conquer concept to address the issue of class imbalance was proposed in [25]. The proposed model attempts to exclude minority class outliers as well as a large number of majorities so as to achieve a better classification accuracy. After that, a non-linear classifier was used to deal with this complicated overlapping subset in order to separate them well. The results obtained was better than similar works. Furthermore, authors in [6] investigated the use of both manual and automated classification, as well as providing insights into the whole implementation process and comparing various machine learning processes. As a result, the paper will assist researchers and professionals in the creation and implementation of data mining-based systems for fraud detection and other issues. This project provided the fraud analysts with not only an automated method, but also insights into how to improve their manual revision process, resulting in overall superior results. This study explored the classification prowess of ANN for credit card fraud detection. ULB credit card transactions dataset was used to validate the proposed model. The dataset has 284,807 transactions with 284, 315 being legit credit card transactions while the remaining 492 are fraudulent credit card transactions.

## 3 Methodology

An ANN based credit card fraud detection model is presented in this study. The model is expected to be able to analyze credit card transactions and determine whether the transaction is legit or that the transaction is a fraudulent one. This section outlines the experiment used for creating the detection model; this involves dataset collection and exploration, feature scaling, model training and testing as well as the performance evaluation.

### 3.1 Data Collection and Exploration

The dataset employed in this study was retrieved from ULB Machine Learning Group. The dataset contains credit card transactions made by European cardholders in September 2013. It contains record of 284,807 transactions that occurred in two days; 284, 315 of these are legit credit card transactions while the remaining 492 are fraudulent credit card transactions. As a result, the positive class (fraud cases) accounts for 0.172 percent of all transactions. The evidence was somewhat unbalanced and biased toward the optimistic side. It only has numerical (continuous) input variables, which are the result of a feature selection transformation using Principal Component Analysis (PCA) that yielded 28 principal components. In this analysis, a total of 30 input features are used. Owing to confidentiality concerns, the specifications and context information for the features cannot be shared. The seconds elapsed between each transaction and the first transaction in the dataset are stored in the time function. The transaction sum is represented by the 'amount' function. The 'class' takes a value 1 for positive fraudulent cases and 0 for non-fraudulent cases. The data exploration was carried out to understand the various features of the dataset better. These were visualized so as to further examine the related features that will be used for the fraud detection.

### 3.2 Feature Scaling

One of the most important stages in the pre-processing of data prior to constructing a machine learning model is feature scaling. Scaling will make the difference between a bad and a good machine learning model. Machine learning algorithms that calculate distances between data include feature scaling. When measuring distances, if the function with the higher value range is not scaled, the function with the higher value range takes precedence. Scaling is needed in many algorithms that need faster convergence, such as Neural Networks. The feature scaling technique adopted in this study is the Standard Scaler. This is available in Phython's scikit-learn or sklearn library. Scikit-learn is perhaps Python's most useful machine learning library. Classification, regression, clustering, and dimensionality reduction are only a few of the useful methods in the sklearn library for machine learning and statistical modelling. Each column of the dataset is rescaled to have a 0 mean and 1 Standard Deviation. By subtracting the mean and dividing by the standard deviation, it standardizes a function. If the original distribution is not normally distributed, the relative space between the features will be distorted. The data is scaled when dividing the dataset into the training data and the testing data. By calculating the necessary statistics on the samples in the training set, each function is individually centered and scaled.

### 3.3 Model Development

Artificial Neural Networks were used to create the proposed credit card fraud detection model (ANN). As seen in Fig. 1, ANNs are multi-layer fully connected neural networks. An input layer, several hidden layers, and an output layer make up all layers. Each node in one layer is connected to the next layer's nodes. The network gets stronger as the number of hidden layers increases.

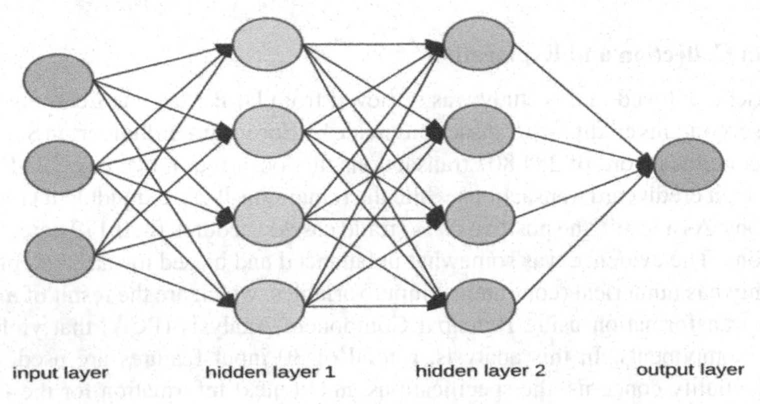


**Fig. 1.** ANN structure

There is an input layer, a hidden layer (there can be more than 1) and an output layer in the network architecture. Because of the multiple layers, it is also called Multi-Layer Perceptron (MLP). The concealed layer functions as a "distillation layer," extracting valuable patterns from the inputs and passing them on to the next layer to be revealed. It makes the network quicker and more efficient by distinguishing only critical information from the inputs and discarding the redundant data. A given node uses a non-linear activation function to process the weighted number of its inputs. This is the node's output, which is then used as an entry by another node in the next layer. The signal travels from left to right, with the final output calculated by repeating the procedure for all nodes. The model was built using Keras and TensorFlow library. TensorFlow is an open-source machine learning framework that runs from start to finish. It is a simple concept. It is a vast and adaptable ecosystem of software, databases, and other resources that provide high-level APIs for workflows. Keras, on the other hand, is a sophisticated neural network library that uses TensorFlow, CNTK, and Theano as its foundation.

### 3.4 Training and Testing of the Model

After building the ANN model, the next stage in the report is to train the model. The training process will go through the dataset for a specified number of iterations called epochs, which was defined from the onset with the epochs statement. The batch size was also set using the batch size argument. The number of epochs used in training the model was 300 epochs and the batch size was 2048. The dataset was further divided into 70% for training and 30% for testing. The training dataset was further divided into

80% for training and 20% for dataset validation. The validation dataset was used to provide an unbiased evaluation of a model fit on the training dataset while tuning model hyper parameters. It was used during the training of the dataset. In summary, training this deep neural network involves learning the weights associated with all the edges. So, the training task is aimed at teaching the model how to learn the weights. The training procedure works as follows:

- Initialize the weights for all nodes at random.
- Execute a forward pass using the current weights for each training example, and measure the contribution of each node as it moves from left to right. The value of the last node is the final product.
- Using a loss function, compare the final result to the original goal in the training data and measure the error.
- Make a backwards pass from right to left and use backpropagation to spread the error to each particular node. Calculate each weight's contribution to the error and use gradient descent to adjust the weights. Reverse the error gradients, beginning with the last sheet.

Performance metrics are used to evaluate the performance of any model, to test the performance of the proposed approach the following metrics are used:

- Accuracy: Accuracy refers to the closeness of the measurements to a particular value. It was calculated using Eq. (1):

$$\text{Accuracy} = \frac{\text{TP} + \text{TN}}{\text{TP} + \text{FP} + \text{TN} + \text{FN}} \tag{1}$$

- Precision: is the number of valid instances among all the positive data used. It was calculated using Eq. (2):

$$\text{Precision} = \text{Precision} = \frac{TP}{TP + FP} \tag{2}$$

- Recall: these also measures the valid instances that were retrieved. It was calculated using Eq. (3):

$$\text{Recall} = \frac{TP}{TP + FN} \tag{3}$$

- F1-Score: This is the harmonic mean of the precision and recall. It was calculated using Eq. (4):

$$\text{F1 - Score} = \frac{Precision \cdot Recall}{Precision + Recall} \tag{4}$$

## 4 Results and Discussion

Majorly, this section narrates the results obtained from the data exploration stage, an overview of the ANN model generated with Phython's Keras and the performance evaluation of the proposed credit card fraud detection model.

### 4.1 Data Exploration

Understanding the features available in the dataset will determine what can be done with the dataset. Data exploration allows us to visualize the content of the dataset so as to know the relationship between the features. It was while exploring the dataset, that we observed that it has 284315 legit credit card transactions and 492 fraudulent credit card transactions. Furthermore, the presence or absence of missing values were also examined. Interestingly, the dataset has no missing values. After-wards, the distribution of the amount and the time each credit card transactions occur was also visualize. This is shown in Fig. 2. Afterwards, the histogram diagram of the fraudulent transactions and the non-fraudulent transactions were also visualized as shown in Fig. 3. This was done to access how the data were distributed. From the distributions, we can have an idea of how skewed the features are. Further distributions of the other features present in the dataset were also visualized. Furthermore, the correlation of the 28 features was visualized using a Heatmap as shown in Fig. 4. The correlation matrix also reveals that there is no correlation between either of the V1 to V28 PCA elements. Class, on the other hand, has both positive and negative associations with the V elements, but no association with Time or Number. It was thanks to this visualization that we were able to see the need to reduce the data's skewness.

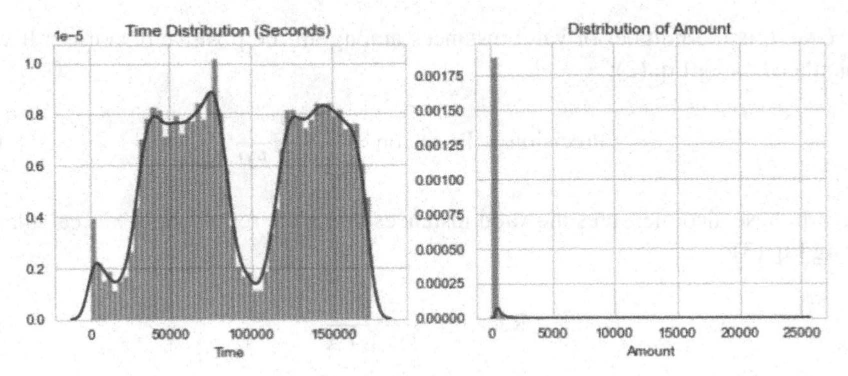


**Fig. 2.** Time and amount distribution

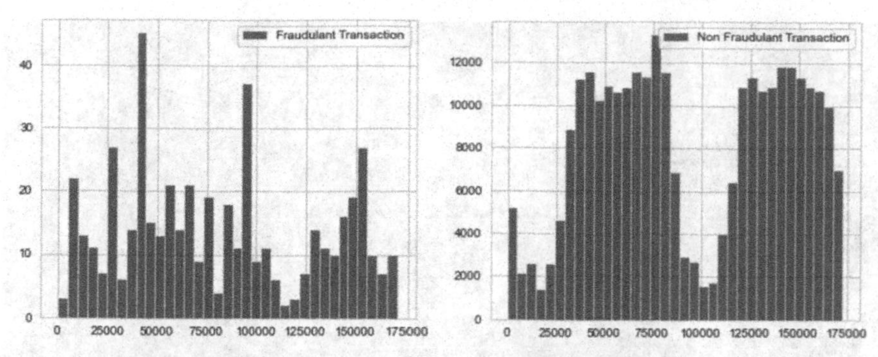


**Fig. 3.** Histogram of fraudulent and non-fraudulent transactions

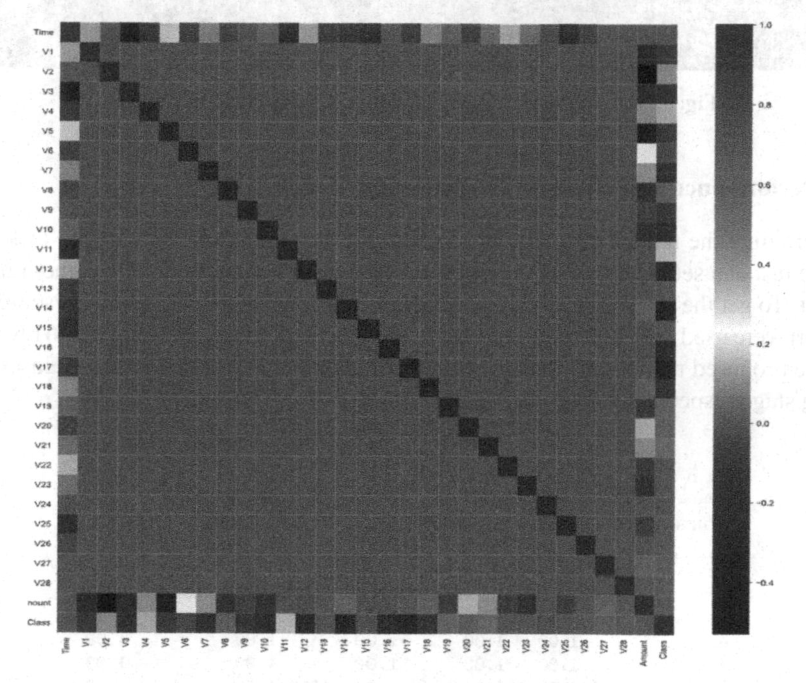


**Fig. 4.** Heatmap

## 4.2 ANN Model Built with Phython's Keras

Keras is a deep learning framework that allows quick prototyping and runs on both CPU and GPU. An overview of the ANN model generated with Keras is shown in Fig. 5. From this visualization, it was observed that the resulting model consists of four layers: the first three layers that uses Relu activation function have 256 nodes while the last layer which is the outer layer has one node and uses the sigmoid activation function. The number of trainable and untrainable parameters were also revealed.

```
Model: "sequential_2"
_________________________________________________________________
Layer (type)                 Output Shape              Param #
=================================================================
dense_8 (Dense)              (None, 256)               7936
_________________________________________________________________
batch_normalization_6 (Batch (None, 256)               1024
_________________________________________________________________
dropout_6 (Dropout)          (None, 256)               0
_________________________________________________________________
dense_9 (Dense)              (None, 256)               65792
_________________________________________________________________
batch_normalization_7 (Batch (None, 256)               1024
_________________________________________________________________
dropout_7 (Dropout)          (None, 256)               0
_________________________________________________________________
dense_10 (Dense)             (None, 256)               65792
_________________________________________________________________
batch_normalization_8 (Batch (None, 256)               1024
_________________________________________________________________
dropout_8 (Dropout)          (None, 256)               0
_________________________________________________________________
dense_11 (Dense)             (None, 1)                 257
=================================================================
Total params: 142,849
Trainable params: 141,313
Non-trainable params: 1,536
_________________________________________________________________
```

**Fig. 5.** An overview of ANN model built with Phython's Keras

## 4.3 Performance Evaluation of the Proposed Model

After training the model with the training set and the validation set, the next task is to use the test dataset set to get an unbiased evaluation of a final model fit on the training dataset. To get the performance of the model, Accuracy, Precision, Recall, F1score and Support were used as the evaluation metrics. Results obtained as shown in Fig. 6 revealed that the proposed model achieved an accuracy of 100% and 99.95% for the training and testing stage respectively (Fig. 7).

```
Train Result:
================================================
Accuracy Score: 100.00%
_______________________________________________
Classification Report:
                   0      1  accuracy  macro avg  weighted avg
precision       1.00   1.00      1.00       1.00          1.00
recall          1.00   1.00      1.00       1.00          1.00
f1-score        1.00   1.00      1.00       1.00          1.00
support    159204.00 287.00      1.00  159491.00     159491.00
_______________________________________________
Confusion Matrix:
 [[159204      0]
 [     1    286]]
```

**Fig. 6.** Performance evaluation result for the training stage

```
Test Result:
================================================
Accuracy Score: 99.95%
_______________________________________________
Classification Report:
                  0      1  accuracy  macro avg  weighted avg
precision      1.00   0.89      1.00       0.94          1.00
recall         1.00   0.81      1.00       0.90          1.00
f1-score       1.00   0.85      1.00       0.92          1.00
support    85307.00 136.00      1.00   85443.00      85443.00
_______________________________________________
Confusion Matrix:
 [[85293    14]
 [   26   110]]
```

**Fig. 7.** Performance evaluation result for the testing stage

# 5 Conclusion

The widespread use of cashless transactions has resulted in an influx of transaction data, necessitating the use of advanced machine learning models to detect fraud. Fraud identification is usually a supervised learning process that classifiers do. Classifier prediction accuracy is determined by the quality of the data used to train them. The massive transaction data generated by consumer purchases is used to train classifiers. This massive volume of data serves as a vast training base for the classifier, allowing it to perform well. The use of supervised techniques is based on a compilation of previous transactions for which the transaction mark is identified. The mark is either genuine or fake in credit card fraud identification issues. The sticker is normally discovered after the fact, either as a result of a consumer complaint or as a result of a credit card issuer audit. Supervised approaches use branded past transactions to learn a fraud prediction model, which returns the possibility of a new transaction becoming a fraud for every new transaction. This study explored the use of ANN for credit card fraud detection. The dataset used contains 284,807 where 284, 315 of these are legit credit card transactions while the remaining 492 are fraudulent credit card transactions. The model achieved a 100% and 99.95% classification accuracy during training and testing respectively. This showed that ANN model could be efficiently used to predict credit card fraudulent transactions.

**Acknowledgements.** Authors appreciate Covenant University Centre for Research, Innovation and Development for sponsoring the publication of this article.

# References

1. Zhang, X., Han, Y., Xu, W., Wang, Q.: HOBA: a novel feature engineering methodology for credit card fraud detection with a deep learning architecture. Inf. Sci. **557**, 302–316 (2021). https://doi.org/10.1016/j.ins.2019.05.023
2. Card Fraud Worldwide 2010–2027. https://nilsonreport.com/publication_chart_and_graphs_archive.php?1=1&year=2019. Accessed 20 Apr 2021
3. Unisys Security Index (2017). http://www.app5.unisys.com/library/cmsmail/USI/UnisysSecurityIndex_Global.pdf

4. Bagga, S., Goyal, A., Gupta, N., Goyal, A.: Credit card fraud detection using pipeling and ensemble learning. Procedia Comput. Sci. **173**(2019), 104–112 (2020). https://doi.org/10.1016/j.procs.2020.06.014
5. Makki, S., Assaghir, Z., Taher, Y., Haque, R., Hacid, M.S., Zeineddine, H.: An experimental study with imbalanced classification approaches for credit card fraud detection. IEEE Access **7**, 93010–93022 (2019). https://doi.org/10.1109/ACCESS.2019.2927266
6. Carneiro, N., Figueira, G., Costa, M.: A data mining-based system for credit-card fraud detection in e-tail. Decis. Support Syst. **95**, 91–101 (2017). https://doi.org/10.1016/j.dss.2017.01.002
7. Montague, D.: Essentials of Online Payment Security and Fraud Prevention. Wiley, Hoboken (2010). Essentials Series. https://books.google.pt/books?id=3IJCmhWztBIC
8. Phua, C., Gayler, R., Lee, V., Smith-Miles, K.: On the communal analysis suspicion scoring for identity crime in streaming credit applications. Eur. J. Oper. Res. **195**(2), 595–612 (2009)
9. Bolton, R.J., Hand, D.J.: Statistical fraud detection: a review. Stat. Sci. **17**(3), 235–249 (2002)
10. Rtayli, N., Enneya, N.: Enhanced credit card fraud detection based on SVM-recursive feature elimination and hyper-parameters optimization. J. Inf. Secur. Appl. **55**, 102596 (2020). https://doi.org/10.1016/j.jisa.2020.102596
11. Johnson, J.M., Khoshgoftaar, T.M.: Survey on deep learning with class imbalance. J. Big Data **6**(1), 1–54 (2019). https://doi.org/10.1186/s40537-019-0192-5
12. Walke, A.: Comparison of supervised and unsupervised fraud detection. In: Alfaries, A., Mengash, H., Yasar, A., Shakshuki, E. (eds.) Advances in Data Science, Cyber Security and IT Applications: First International Conference on Computing, ICC 2019, Riyadh, Saudi Arabia, December 10–12, 2019, Proceedings, Part I, pp. 8–14. Springer, Cham (2019). https://doi.org/10.1007/978-3-030-36365-9_2
13. Krawczyk, B.: Learning from imbalanced data: open challenges and future directions. Prog. Artif. Intell. **5**(4), 221–232 (2016). https://doi.org/10.1007/s13748-016-0094-0
14. de Sá, A.G.C., Pereira, A.C.M., Pappa, G.L.: A customized classification algorithm for credit card fraud detection. Eng. Appl. Artif. Intell. (2018). https://doi.org/10.1016/j.engappai.2018.03.011
15. Akila, S., Srinivasulu, R.U.: Cost-sensitive risk induced Bayesian inference bagging (RIBIB) for credit card fraud detection. J. Comput. Sci. **27**, 247–254 (2018). https://doi.org/10.1016/j.jocs.2018.06.009
16. Osho, O., Mohammed, U.L., Nimzing, N.N., Uduimoh, A.A., Misra, S.: Forensic analysis of mobile banking apps. In: Misra, S., et al. (eds.) ICCSA 2019. LNCS, vol. 11623, pp. 613–626. Springer, Cham (2019). https://doi.org/10.1007/978-3-030-24308-1_49
17. Osho, O., Musa, F.A., Misra, S., Uduimoh, A.A., Adewunmi, A., Ahuja, R.: AbsoluteSecure: a tri-layered data security system. In: Damaševičius, R., Vasiljevienė, G. (eds.) Information and Software Technologies: 25th International Conference, ICIST 2019, Vilnius, Lithuania, October 10–12, 2019, Proceedings, pp. 243–255. Springer, Cham (2019). https://doi.org/10.1007/978-3-030-30275-7_19
18. Jambhekar, N.D., Misra, S., Dhawale, C.A.: Mobile computing security threats and solution. Int. J. Pharm. Technol. **8**(4), 23075–23086 (2016)
19. Jambhekar, N.D., Misra, S., Dhawale, C.A.: Cloud computing security with collaborating encryption Indian. J. Sci. Technol. **9**(21), 95293 (2016)
20. Christiana, A.O., Dokoro, H.A., Oluwatobi, A.A.A.N., Oluwatobi, A.E.: Modified advanced encryption standard algorithm for information security. Symmetry **11**(12), 1–17 (2019). https://doi.org/10.3390/sym11121484
21. Akande, N.O., Abikoye, C.O., Adebiyi, M.O., Kayode, A.A., Adegun, A.A., Ogundokun, R.O.: Electronic medical information encryption using modified Blowfish algorithm. In: Misra, S., et al. (eds.) ICCSA 2019. LNCS, vol. 11623, pp. 166–179. Springer, Cham (2019). https://doi.org/10.1007/978-3-030-24308-1_14

22. Carcillo, F., Borgne, Y.L., Caelen, O., Kessaci, Y., Oblé, F.: Combining unsupervised and supervised learning in credit card fraud detection. Inf. Sci. **557**, 317–331 (2021). https://doi.org/10.1016/j.ins.2019.05.042
23. De Sá, A.G.C., Pereira, A.C.M., Pappa, G.L.: Engineering applications of artificial intelligence a customized classification algorithm for credit card fraud detection. Eng. Appl. Artif. Intell. **72**, 21–29 (2018). https://doi.org/10.1016/j.engappai.2018.03.011
24. Forough, J., Momtazi, S.: Ensemble of deep sequential models for credit card fraud detection. Appl. Soft Comput. J. **99**, 106883 (2021). https://doi.org/10.1016/j.asoc.2020.106883
25. Li, Z., Huang, M., Liu, G., Jiang, C.: A hybrid method with dynamic weighted entropy for handling the problem of class imbalance with overlap in credit card fraud detection. Exp. Syst. Appl. **175**, 114750 (2021). https://doi.org/10.1016/j.eswa.2021.114750
26. Jain, Y., Tiwari, N., Dubey, S., Jain, S.: A comparative analysis of various credit card fraud detection techniques. Int. J. Recent Technol. Eng. **7**, 402–407 (2019)

# Machine Learning Classification Based Techniques for Fraud Discovery in Credit Card Datasets

Roseline Oluwaseun Ogundokun[1], Sanjay Misra[2(✉)], Opeyemi Eyitayo Ogundokun[3], Jonathan Oluranti[4], and Rytis Maskeliunas[5]

[1] Department of Computer Science, Landmark University, Omu Aran, Nigeria
Ogundokun.roseline@lmu.edu.ng

[2] Department of Computer Science and Communication, Ostfold University College, Halden, Norway

[3] Directorate of Financial Services, Agricultural and Rural Management Training Institute, Ilorin, Nigeria

[4] Center of ICT/ICT, Covenant University, Ota, Ogun State, Nigeria
jonatha.oluranti@covenantuniversity.edu.ng

[5] Kaunas University of Technology, Kaunas, Lithuania
Rytis.Maskeliunas@ktu.lt

**Abstract.** The frequency of credit card-based online payment frauds has increased rapidly in recent years, forcing banks and e-commerce companies to create automated fraud detection systems that perform mining on massive transaction logs. Machine learning appears to be one of the most promising techniques for detecting illegal transactions since it uses supervised binary classification algorithms appropriately trained using pre-screened sample datasets to differentiate between fraudulent and non-fraudulent cases. This study aims to concentrate on machine learning (ML) methods thereby proposing a credit card fraud discovery scheme to detect fraud. The ML techniques employed are Decision Tree (DT) and K-Nearest Neighbor (KNN) ML classification techniques. The performance outcomes of the two ML classification techniques are evaluated depending on accuracy, precision, specificity, recall, f1-score, and false-positive rate (FPR). The area under the ROC curve (AUC) of the receiver operating characteristics (ROC) curve was similarly drawn built on the confusion matrix for both classifiers. The two classification techniques were evaluated and compared using the performance metrics mentioned earlier and it was demonstrated that the KNN technique outperformed that of the DT with a greater ROC curve value of 91% for KNN and 86% for DT. It was concluded that KNN is considered a better ML classification technique that can be employed to discover credit card fraudulent activities.

**Keywords:** Credit card fraud · Machine learning · Decision tree · K-Nearest Neighbor · Classification


H. Florez and M. F. Pollo-Cattaneo (Eds.): ICAI 2021, CCIS 1455, pp. 26–38, 2021.
https://doi.org/10.1007/978-3-030-89654-6_3

## 1 Introduction

Credit cards (CC) have presently been progressively prevalent, especially with the upsurge of e-commerce. Credit card fraud (CCF) is a big challenge, even though CC purchases make many sorts of commercial dealings easier. CCF, not only costs fiscal establishments and banks lots of money, but it similarly leads to a lot of worry and tension in the existence of the persons who are impacted. According to current figures, the worldwide fiscal damage instigated by CC theft in 2018 was 27.85 billion dollars, up 16.2% from 23.97 billion dollars in 2017. If current trends continue, credit card theft will cost the economy more than $35 billion by 2023 [1]. CCF lost to delivering and managing electronic transaction establishments that can be reduced or avoided with effective fraud monitoring and stoppage. Additionally, efficient fraud detection software may boost consumer confidence and minimize complaints. Machine learning is used in the majority of CCF discovery methods [2]. To tackle the problem of CCF, ML has several mature approaches [3, 4], which include supervised learning [5, 6], semi-supervised learning [2, 7], and unsupervised learning [7]. Despite considerable study [7–11], a flawless and competent resolution remains elusive [12].

Several obstacles and challenges facing fraud detection systems include [13–16]: (1) Due to imbrication of data, numerous transactions might appear to be deceitful when they are legitimate businesses. When a fraudulent business looks to be legitimate, the reverse occurs. (2) Data are unbalanced, such as credit card fraud detection data. This indicates that only a small proportion of all CC transactions are deceitful. (3) Adaptability: Structures must be capable of familiarizing to newfangled fraud types. To get the utmost results, an efficient fraud detection approach should be capable of dealing with these issues. A competent fraudster will always come up with new and innovative ways to carry out his task because successful fraud tactics lose their effectiveness over time as they become more widely recognized. Fraud examination (misappropriation discovery) and operator comportment examination are the two broad categories of CCF recognition approaches [15, 17] (Anomaly discovery). The first set of approaches deals with transaction-level supervised categorization. Based on past historical data, these techniques classify transactions as fraudulent or legitimate. This method has been shown to successfully detect the majority of previously identified fraud schemes (known fraud tricks). The second technique is based on account behavior and is based on unsupervised methods. A transaction is flagged as fraudulent in this manner if it deviates from the user's usual behavior (user profile). We don't anticipate fraudsters to act in similar means as the account holder or to be aware of the holder's comportment method. Different enough actions are identified as frauds when fresh behaviors are compared to this model. Even though operator comportment examination approaches are effective in detecting fresh scams, they are plagued by intensifying false alarm rates [17, 18].

Data mining (DM) procedures such as clustering examination; statistics like time series examination; ML technique such as neural network (NN); and artificial intelligence (AI) such as swarm optimization have all been used to implement existing fraud detection [13, 14, 16]. Conventional arithmetical approaches build classification models by approximating parameters to match the data, but ML approaches permit learning the model's specific organization from the data [19, 20].

Consequently, the framework of models acquired using arithmetical approaches are comparatively understandable, not difficult to commentate, and is likely to under-fit the data, whereas replicas created through ML techniques are often complex, difficult to elucidate, and likely to over-fit the data. Underfitting and overfitting data is a tradeoff between a model's descriptive power and frugality, whereas descriptive power leads to extreme forecasting accuracy and frugality typically ensuring the model's generalizability and interpretability. Recent research has demonstrated that data mining approaches based on artificial intelligence (AI) outperformed conventional arithmetical approaches for developing forecasting models [16, 21]. Clustering procedures are divided into hierarchical and partitioned procedures built on their abstraction structure [22]. In an endeavor to recuperate normal clusters that are accessible in the data, hierarchical clustering procedures build to order of divisions, whereas partitioned clustering procedures construct a solitary division of the data with a stated or predictable amount of non-overlapping clusters [22–27]. Among the different clustering algorithms, the K-means procedure is understandable and utmost extensively employed. The k-means technique is employed to reduce data grouping complications. This procedure is affected by the preliminary cluster centers, which are chosen at random. The sophisticated foraging behavior of honey bee swarms inspired the ABC algorithm [28]. ABC has several benefits over other optimization methods, including the use of fewer control parameters and the ability to handle both restricted and unrestrained situations [29]. ABC method was recently created to address clustering issues and has shown capable outcomes in terms of conversion speed and convergence to the optimum result [30, 31].

To improve the classification accuracy and detection rate as well as reduce the false positive rate of credit card discovery, this study hence intends to concentrate on machine learning (ML) methods thereby proposing a credit card fraud discovery scheme to detect fraud. The ML approaches employed are DT and K-Nearest Neighbor (KNN) ML classification techniques. The performance outcomes of the two ML classification techniques are evaluated depending on accuracy, precision, specificity, recall, f1-score, and false-positive rate (FPR). The ROC curve was drawn built on the confusion matrix for both classifiers.

The remaining segment of the article is pre-arranged as thus: Sect. 2 presented the review of related works. Section 3 discussed the materials and methods used for the execution of the study. Section 4 presented the results and implementation of the research and the article was concluded in Sect. 5 with the study conclusion presented.

## 2 Literature Review

There have been several prevailing investigations on CCF discovery approaches, for instance, a variety of research methodologies and fraud recognition strategies, with a focus on neural networks, DM, and distributed data mining. CCF is detected using a variety of methods. After conducting a literature review on several ways CCF recognition, it could be inferred that there are numerous additional approaches in Machine Learning that may be used to identify credit card fraud. SVM, DT, logistic regression (LR), gradient boosting (GB), KNN, and other Machine Learning procedures are employed to identify credit fraud and a few of the researches that have employed these ML techniques are discussed as follows:

SVM, artificial neural networks (ANN), Bayesian networks (BN), hidden Markov model (HMM), KNN, fuzzy logic (FL) system, and DT were among approaches [10] investigated by Jain, Tiwari, Dubey, & Jain [32]. They found that the techniques KNN, DT, and SVM offer an average degree of accuracy in their study. Among all the methods, FL and LR have the lowest accuracy. The detention rate of NNs, NB, fuzzy systems, and KNN had an extreme accuracy value. At the middle level, DT based on LR, SVM, gave a high detection rate (DR). ANN and Nave Bayesian Networks are two methods that outperformed each other across the board. Training costs for these techniques involved a lot of money. For all algorithms, there was a significant flaw. The disadvantage was that these algorithms may not produce consistent results in all situations. With one sort of dataset, they produced superior results, but with another, they produced bad results. Small datasets yield great results from algorithms like KNN and SVM, and raw and unsampled data yielded outstanding accuracy from methods like LR and FL systems.

Naik & Kanikar [33] researched in 2019 on a variety of algorithms such as NB, LR, J48, and Adaboost. Amid the classification procedures, NB was used. The Bayes theorem was used in this algorithm. The Bayes theorem determines the likelihood of an event occurring. The linear regression algorithm and the LR technique are quite alike. The linear regression method was employed to estimate or envisage values. For classification, LR was commonly employed. For the classification function, the J48 method was utilized to construct a DT. J48 is an ID3 extension (Iterative Dichotomieser). Machine Learning's J48 is one of the most commonly utilized and studied domains and the constant and category variables are the focus of this method. Adaboost is a binary classification method that is one of the utmost frequently employed ML techniques. The procedure's main purpose is to improve the decision tree's performance. This was also how the regression was classified. The Adaboost algorithm uses fraud scenarios to distinguish between fraudulent and non-fraudulent transactions. According to the authors' findings, both the Adaboost and Logistic Regression yielded the greatest accuracy. Because they are both accurate, the time factor was used to select the superior algorithm. They determined that the Adaboost algorithm was effective at detecting CCF when the time component was taken into account.

Sahayasakila, Aishwaryasikhakolli, & Yasaswi [34] introduced the Whale Optimization Approaches (WOA) and SMOTE, which are two key algorithmic techniques (Synthetic Minority Oversampling Techniques). They primarily sought to enhance merging swiftness and resolve the challenge of data unevenness. The SMOTE and WOA techniques are used to solve the challenge of class imbalance. The SMOTE methodology separates all synthetic transactions, which are then re-sampled to ensure data correctness and optimization by utilizing the WOA method. The method similarly boosts the system's concurrence speed, dependability, and competency.

Navanushu & Yunus Sait [35] presented a study that employed DT, RF, SVM, and LR. They used an extremely skewed dataset to work on this sort of dataset. Accuracy, sensitivity, specificity, and precision are used to evaluate the performance of the study. The accuracy of LR was 97.7%, DT was 95.5%, RF was 98.6%, and SVM classifier was 97.5%, according to the results. They determined that, among all the algorithms employed in the study, the RF method has the maximum accuracy and is the best algorithm for detecting fraud. They also concluded that the SVM procedure has a data unevenness challenge and does not perform better in detecting CCF.

The main purpose of fraud detection is to identify the fraudulent actions and if this is done, it aids in characterizing the behavior of the fraudster in the specific fraud act and the historical dataset. Therefore, to detect new fraudulent activities and continually adopt the new credit card fraud activities, we proposed an ML-based classification technique. The goal of the proposed study is to increase the detection rate and accuracy and at the same time reduce the false positive rate (FPR) on CCF activities.

# 3 Materials and Methods

## 3.1 Dataset

The credit card fraud dataset used for the implementation in this study was gathered from the Kaggle database repository. The dataset can be gotten from this link: https://www.kaggle.com/rahulmakwana/creditcard-fruad-detection. The dataset was uploaded to Kaggle by Rahul Makwana in 2020. The dataset comprises 31 numerical features. The dataset comprises 284807 transactions. The overall amount of the sample employed for testing was 85, 442 since the test set of data accounted for 30% of the whole dataset.

## 3.2 Proposed Method

This study employed two ML classification techniques which are DT and KNN. The core objective of this study is to classify the CCF dataset that has together with the fraud and non-fraud transactions in it by employing the two proposed ML classifiers. The classifiers' outcomes are thereby evaluated and likened with each other to establish the classifier that superlatively identifies CCF transactions. The proposed system block diagram for the CCF discovery is shown in Fig. 1. The CCF dataset was first gathered and was passed to the next phase which is the data preparation phase where the data was cleansed and normalized. The dataset is also structured and organized after which it was passed to the testing and training phases where the datasets were split into testing and training. The training dataset was later passed to the ML classifiers DT and KNN for classification after which the classified datasets are then evaluated to deduce the proposed system performance.

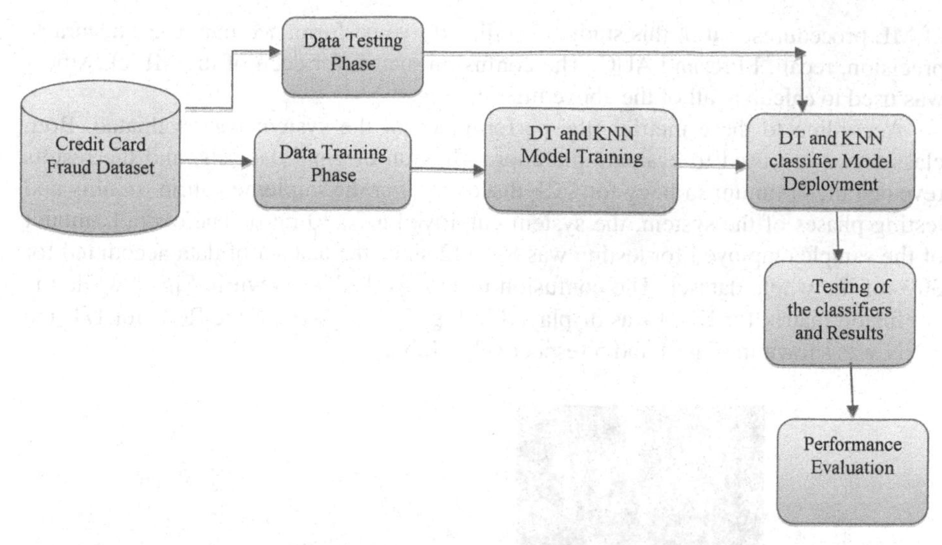


**Fig. 1.** Proposed system process flow

### 3.3 Performance Evaluation

The study employed numerous metrics of system classification performance commonly deduced in the literature. The accuracy of the positive (fraud) and negative (non-fraud) situations was measured using recall and specificity matrices. Naturally, there must be a balance between these true positives and true negatives. The various performance indicators are listed in Table 1 concerning the confusion matrix, where positive values correspond to fraud instances and negative values to non-fraud situations [36]. The study employed four performance metrics for the evaluation of the proposed system. The metrics are accuracy, precision, recall, false-positive rate (FPR), and AUC of the classifiers ROC.

**Table 1.** Confusion matrix

| | Predicted positive | Predicted negative |
|---|---|---|
| Actual positive | True positives | False negatives |
| Actual negative | False positives | True negatives |

## 4 Result and Discussions

Dissimilar measures for technique assessment were employed to evaluate which technique is superlatively appropriate for the challenge of identifying fraud transactions. Accuracy, recall, and precision are the most often used metrics for assessing the results of ML procedures, but in this study, we utilized five performance measures: accuracy, precision, recall, FPR, and AUC. The confusion matrix for each of the ML classifiers was used to calculate all of the above metrics.

According to these metrics, the performance of the system was evaluated. Both classifiers were used to evaluate the approaches on original datasets, and the results revealed the optimum strategy for CCF discovery. For the implementation training and testing phases of the system, the system employed a 70:30 ratio. The overall amount of the sample employed for testing was 85, 442 since the test set of data accounted for 30% of the whole dataset. The confusion matrix for DT is shown in Fig. 2 while the confusion matrix for KNN was displayed in Fig. 3. The AUC of the ROC for DT and KNN was shown in Figs. 4 and 5 respectively (Table 2).

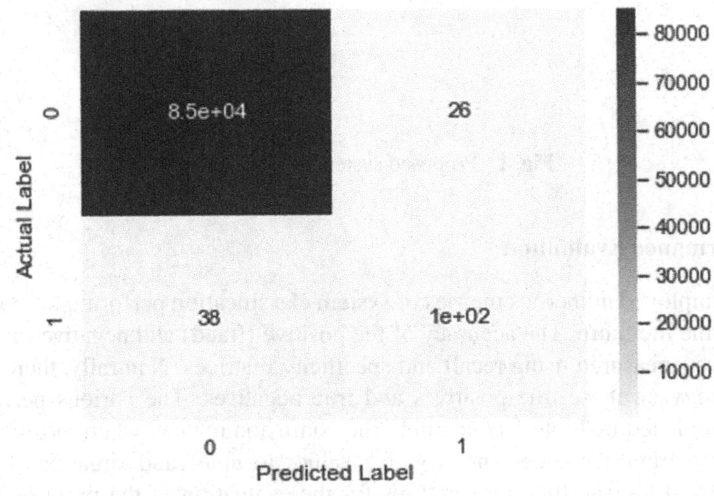


**Fig. 2.** Confusion matrix for DT

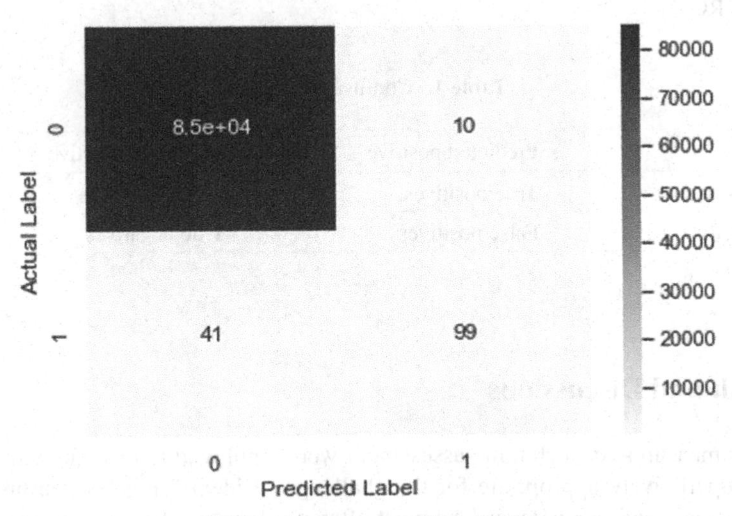


**Fig. 3.** Confusion matrix for KNN

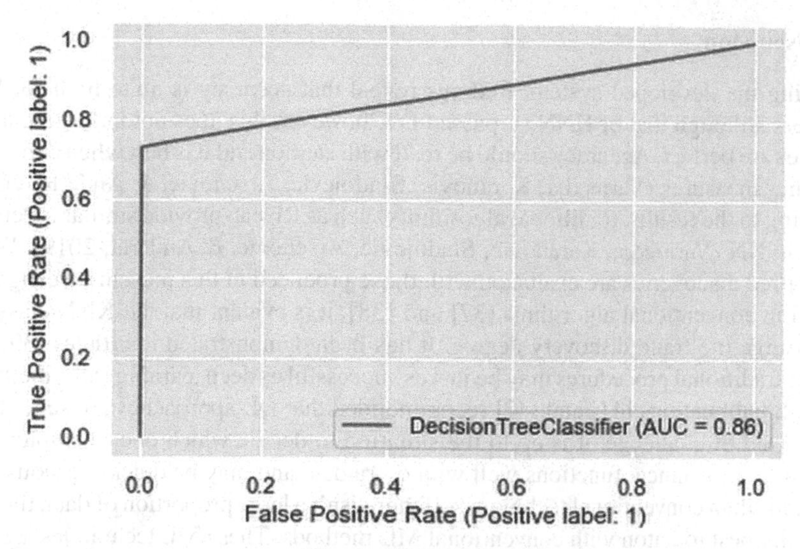


**Fig. 4.** AUC for DT

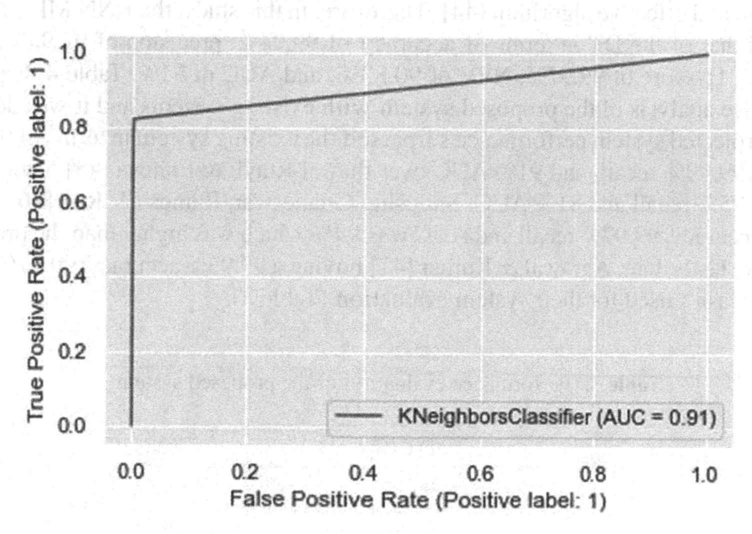


**Fig. 5.** AUC for KNN

**Table 2.** Proposed classifiers confusion matrix

| Classifiers | TP | TN | FP | FN |
|---|---|---|---|---|
| DT | 85000 | 100 | 38 | 26 |
| KNN | 85000 | 99 | 41 | 10 |

## 4.1 Discussion

Analyzing the developed system, findings reveal that accuracy is quite high for both classifiers although that of KNN surpassed DT, however, this does not indicate that the outcomes are perfect. Accuracy should be read with caution and it is best when combined with other measures (Varmedja, Karanovic, Sladojevic, Arsenovic, & Anderla, 2019). According to the results, traditional algorithms such as RF can provide similar outcomes to a basic NN (Varmedja, Karanovic, Sladojevic, Arsenovic, & Anderla, 2019). When the acquired discoveries are evaluated with those produced in this present investigation employing conventional algorithms [37] and [38], it is evident that the KNN classifier may upsurge the fraud discovery degree. It has been demonstrated in articles [39] and [40] that traditional procedures may be just as successful as deep learning (DL) methods. Although publications [41] and [42] recommended that DL approaches are superlative for this kind of challenge, it is up to the situation to decide which one to employ. DL technique, for instance, functions well with extra data and may be quickly espoused to other fields than conventional techniques. If there isn't a large proportion of data, though, it's usually best to baton with conventional ML methods. These ML techniques are also not difficult to comprehend and not expensive, both monetarily and computationally [43]. It is said that an algorithm with higher accuracy, precision, recall, f1-score, and AUC is an efficient and effective algorithm [44]. Therefore, in this study, the KNN ML classifier surpassed that of the DT in terms of accuracy of 99.94%, precision of 99.95%, recall of 99.99%, f1-score of 99.97%, NPV of 90.83%, and AUC of 91%. Table 4 displays a comparative analysis of the proposed system with existing systems and it was deduced that the projected system performance surpassed the existing system in terms of 99.94% accuracy, 99.99% recall, and 91% AUC over that of Rtayli & Enneya [45] having 99% accuracy, 95% recall and 81% AUC; Sailusha, Gnaneswar, Ramesh & Rao [46] having 99.91% accuracy, 99.97% recall and AUC was 94% which was higher than the proposed system and lastly Jain, Agrawal & Kumar [47] having a 99.93% accuracy, 99.97% recall but AUC wasn't used for their system evaluation (Table 3).

**Table 3.** Performance evaluation of the proposed system

| Measures | DT (%) | KNN (%) |
|---|---|---|
| Accuracy | 99.92 | 99.94 |
| Precision | 99.96 | 99.95 |
| Sensitivity | 99.97 | 99.99 |
| Specificity | 72.46 | 70.71 |
| F1-score | 99.96 | 99.97 |
| False positive rate (FPR) | 0.2754 | 0.2929 |
| AUC | 86 | 91 |

**Table 4.** Comparative analysis with state-of-the-art

| Authors | Year | Method | Accuracy | Sensitivity | Specificity |
|---|---|---|---|---|---|
| Rtayli & Enneya [45] | 2020 | RFE, HPO and SMOTE | 99% | 95% | 81% |
| Sailusha, Gnaneswar, Ramesh & Rao [46] | 2020 | Random Forest | 99.91% | 99.97% | 94% |
| Jain, Agrawal & Kumar [47] | 2020 | Decision Tree | 99.93 | 99.97 | N/A |
| **Proposed System** | **2021** | **KNN**<br>**DT** | **99.94**<br>**99.92** | **99.99**<br>**99.97** | **70.71**<br>**72.46** |

## 5 Conclusion and Future Research Direction

Fraudulent credit card transactions are a major corporate issue. These types of scams can result in significant financial and personal losses. As a result, businesses are investing an increasing amount of money in creating new concepts and methods for detecting and preventing fraud. This paper's main objective was to evaluate two machine learning methods for detecting fraudulent transactions. This was determined using a variety of measures, including recall, accuracy, and precision. It is critical to have high recall, accuracy, and precision values for this type of situation. As a consequence of the comparison, it was discovered that the KNN technique produces the best results to that of the DT in terms of accuracy of 99.94%, the precision of 99.95%, recall of 99.99%, f1-score of 99.97%, NPV of 90.83% and AUC of 91%, i.e., it better identifies whether transactions are fraudulent or not. It was also discovered that DT outperformed KNN in terms of specificity 72.46% and FPR of 0.2754. It is therefore concluded that DT outperformed that of KNN because it has a higher specificity of 72.46% and at the same time a lower FPR of 0.2754 and the KNN classifier outperformed that of DT classifier in terms of sensitivity of 99.99% and accuracy of 99.94%.

To improve outcomes, more study should be done on alternative ML methods, for instance, genetic algorithms and several kinds of stacked classifiers, as well as comprehensive feature selection techniques.

## References

1. Tingfei, H., Guangquan, C., Kuihua, H.: Using variational autoencoding in credit card fraud detection. IEEE Access **8**, 149841–149853 (2020)
2. Salazar, A., Safont, G., Vergara, L.: Semi-supervised learning for imbalanced classification of credit card transactions. In: 2018 International Joint Conference on Neural Networks (IJCNN), pp. 1–7. IEEE (July 2018)
3. Gao, J., Zhou, Z., Ai, J., Xia, B., Coggeshall, S.: Predicting credit card transaction fraud using machine learning algorithms. J. Intell. Learn. Syst. Appl. **11**(3), 33–63 (2019)
4. Yee, O.S., Sagadevan, S., Malim, N.: Credit card fraud detection using machine learning as a data mining technique. J. Telecommun. Electron. Comput. Eng. (JTEC) **10**(1–4), 23–27 (2018)

5. Roy, A., et al.: Deep learning detecting fraud in credit card transactions. In: 2018 Systems and Information Engineering Design Symposium (SIEDS), pp. 129–134. IEEE (April 2018)
6. Xuan, S., Liu, G., Li, Z., Zheng, L., Wang, S., Jiang, C.: Random forest for credit card fraud detection. In: 2018 IEEE 15th International Conference on Networking, Sensing and Control (ICNSC), pp. 1–6. IEEE (March 2018)
7. Carcillo, F., Le Borgne, Y.A., Caelen, O., Bontempi, G.: Streaming active learning strategies for real-life credit card fraud detection: assessment and visualization. Int. J. Data Sci. Anal. **5**(4), 285–300 (2018)
8. Carcillo, F., Le Borgne, Y.-A., Caelen, O., Kessaci, Y., Oblé, F., Bontempi, G.: Combining unsupervised and supervised learning in credit card fraud detection. Inf. Sci. **557**, 317–331 (2021). https://doi.org/10.1016/j.ins.2019.05.042
9. Eshghi, A., Kargari, M.: Introducing a method for combining supervised and semi-supervised methods in fraud detection. In 2019 15th Iran International Industrial Engineering Conference (IIIEC), pp. 23–30. IEEE (January 2019)
10. Karimi Zandian, Z., Keyvanpour, M.R.: MEFUASN: a helpful method to extract features using analyzing social networks for fraud detection. J. AI Data Min. **7**(2), 213–224 (2019)
11. Tran, L., Tran, T., Tran, L., Mai, A.: Solve fraud detection problems by using graph-based learning methods. arXiv preprint arXiv:1908.11708 (2019)
12. Morgan, J.P.: Payments Fraud and Control Survey. Kirchhain, Germany. https://www.afponline.org/publications-data-tools/reports/survey-research-economic-data/Index/. Accessed 2016
13. Fashoto, S.G., Owolabi, O., Adeleye, O., Wandera, J.: Hybrid methods for credit card fraud detection using K-means clustering with hidden Markov model and multilayer perceptron algorithm. Br. J. Appl. Sci. Technol. **13**(5), 1–11 (2016)
14. Philip, N., Sherly, K.K.: Credit card fraud detection based on behavior mining. TIST Int. J. Sci. Technol. Res. **1**, 7–12 (2012)
15. Ishu, T., Mrigya, M.: Credit card fraud detection. Int. J. Adv. Res. Comput. Commun. Eng. **5**(1), 39–42 (2016)
16. Al-Khatib, A.: Electronic payment fraud detection techniques. World Comput. Sci. Inf. Technol. J. (WCSIT) **2**(4), 137–141 (2012)
17. Tripathi, K.K., Pavaskar, M.A.: Survey on credit card fraud detection methods. Int. J. Emerg. Technol. Adv. Eng. **2**(11), 721–726 (2012)
18. Dheepa, V., Dhanapal, R.: Behavior-based credit card fraud detection using support vector machines. ICTACT J. Soft Comput. **2**(4), 391–397 (2012)
19. Hastie, T., Trevor, H., Robert, T., Friedman, J.H.: The Elements of Statistical Learning: Data Mining, Inference, and Prediction. Springer, New York (2001). https://doi.org/10.1007/978-0-387-21606-5
20. Azeez, N.A., Asuzu, O.J., Misra, S., Adewumi, A., Ahuja, R., Maskeliunas, R.: Comparative evaluation of machine learning algorithms for network intrusion detection using Weka. In: Chakraverty, S., Goel, A., Misra, S. (eds.) Towards Extensible and Adaptable Methods in Computing, pp. 195–208. Springer, Singapore (2018). https://doi.org/10.1007/978-981-13-2348-5_15
21. Oladele, T.O., Ogundokun, R.O., Kayode, A.A., Adegun, A.A., Adebiyi, M.O.: Application of data mining algorithms for feature selection and prediction of diabetic retinopathy. In: Misra, S., et al. (eds.) ICCSA 2019. LNCS, vol. 11623, pp. 716–730. Springer, Cham (2019). https://doi.org/10.1007/978-3-030-24308-1_56
22. Vaishali, V.: Fraud detection in credit card by clustering approach. Int. J. Comput. Appl. **98**(3), 29–32 (2014)
23. Siddhi, D., Vidhi, S., Jay, V.: Credit card fraud detection using a hybrid approach. Int. J. Adv. Res. Comput. Commun. Eng. **5**(5), 287–289 (2016)

24. Madhav, P., Anil, K., Varun, B.: Credit card fraud detection using an efficiently enhanced k-mean clustering algorithm. Int. J. Eng. Comput. Sci. **4**(2), 10367–10374 (2015)
25. Nadisha, A., Rakendu, R., Surekha, M.: A hybrid approach to detecting credit card fraud. Int. J. Sci. Res. Publ. **5**(11), 304–314 (2015)
26. Vadoodparast, M., Razak, H.: Fraudulent electronic transaction detection using a dynamic model. Int. J. Comput. Sci. Inf. Secur. **13**(2), 1–10 (2015)
27. Mortazavi, E., Ahmadzadeh, M.: A hybrid approach for automatic credit approval. Int. J. Sci. Eng. Res. **5**(8), 614–619 (2014)
28. Mohd, A., Yuk, Y., Wei, C., Noorhaniza, W., Ahmed, M.: ABC based data mining algorithms for classification tasks. Can. Cent. Sci. Educ. **5**(4), 217–231 (2011)
29. Rinkal, S., Samir, K., Hiteshkumar, N.: Artificial bee colony algorithm, a comparative approach for optimization algorithm and application: a survey. Int. J. Fut. Trends Eng. Technol. **4**(1), 17–21 (2014)
30. Faiza, A., Azuraliza, A.: A cluster-based deviation detection task using the artificial Bee colony algorithm. Int. J. Soft Comput. **2**(7), 71–78 (2012)
31. Deoshree, D., Snehlata, S.: Classification model using optimization technique a review. Int. J. Comput. Sci. Netw. **6**(1), 42–48 (2017)
32. Jain, Y., Tiwari, S.N., Jain, S.: A comparative analysis of various credit card fraud detection techniques. Int. J. Recent Technol. Eng. **7**(5S2), 402–407 (2019)
33. Naik, H., Kanikar, P.: Credit card fraud detection based on machine learning algorithms. Int. J. Comput. Appl. **182**(44), 8–12 (2019)
34. Sahayasakila, V., Aishwaryasikhakolli, D., Yasaswi, V.: Credit card fraud detection system using smote technique and whale optimization algorithm. Int. J. Eng. Adv. Technol. (IJEAT) **8**(5), 190–192 (2019)
35. Khare, N., Sait, Y.: Credit card fraud detection using machine learning models and collating machine learning models. Int. J. Pure Appl. Math. **118**(20), 825–838 (2018). ISSN 1314-3395
36. Abdulsalam, S.O., et al.: Performance evaluation of ANOVA and RFE algorithms for classifying microarray dataset using SVM. In: Themistocleous, M., Papadaki, M., Kamal, M.M. (eds.) EMCIS 2020. LNBIP, vol. 402, pp. 480–492. Springer, Cham (2020). https://doi.org/10.1007/978-3-030-63396-7_32
37. Mishra, A., Ghorpade, C.: Credit card fraud detection on skewed data using various classification and ensemble techniques. In: 2018 IEEE International Students' Conference on Electrical, Electronics and Computer Science (SCEECS), pp. 1–5. IEEE (February 2018)
38. Navamani, C., Krishnan, S.: Credit card nearest neighbor-based outlier detection techniques. Int. J. Comput. Tech **5**(2), 56–60 (2018)
39. Kazemi, Z., Zarrabi, H.: Using deep networks for fraud detection in credit card transactions. In: 2017 IEEE 4th International Conference on Knowledge-Based Engineering and Innovation (KBEI), pp. 0630–0633. IEEE (December 2017)
40. Dhankhad, S., Mohammed, E., Far, B.: Supervised machine learning algorithms for credit card fraudulent transaction detection: a comparative study. In: 2018 IEEE International Conference on Information Reuse and Integration (IRI), pp. 122–125. IEEE (July 2018)
41. Wang, C., Wang, Y., Ye, Z., Yan, L., Cai, W., Pan, S.: Credit card fraud detection based on whale algorithm optimized BP neural network. In: 2018 13th International Conference on Computer Science & Education (ICCSE), pp. 1–). IEEE (August 2018)
42. Pumsirirat, A., Yan, L.: Credit card fraud detection using deep learning based on auto-encoder and restricted Boltzmann machine. Int. J. Adv. Comput. Sci. Appl. **9**(1), 18–25 (2018)
43. Deeplearningbook.org: Deep Learning (2019). https://www.deeplearningbook.org/. Accessed 11 Jan 2019
44. Su, T., Sun, H., Zhu, J., Wang, S., Li, Y.: BAT: Deep learning methods on network intrusion detection using NSL-KDD dataset. IEEE Access **8**, 29575–29585 (2020)

45. Rtayli, N., Enneya, N.: Enhanced credit card fraud detection based on SVM-recursive feature elimination and hyper-parameters optimization. J. Inf. Secur. Appl. **55**, 102596 (2020)
46. Sailusha, R., Gnaneswar, V., Ramesh, R., Rao, G.R.: Credit card fraud detection using machine learning. In: 2020 4th International Conference on Intelligent Computing and Control Systems (ICICCS), pp. 1264–1270. IEEE (May 2020)
47. Jain, V., Agrawal, M., Kumar, A.: Performance analysis of machine learning algorithms in credit cards fraud detection. In: 2020 8th International Conference on Reliability, Infocom Technologies and Optimization (Trends and Future Directions) (ICRITO), pp. 86–88. IEEE (June 2020)

# Object Detection Based Software System for Automatic Evaluation of *Cursogramas* Images

Pablo Pytel(✉), Matías Almad, Rocío Leguizamón, Cinthia Vegega(✉), and Ma Florencia Pollo-Cattaneo(✉)

Grupo de Estudio en Metodologías de Ingeniería en Software (GEMIS), Facultad Regional Buenos Aires, Universidad Tecnológica Nacional, Buenos Aires, Argentina
{ppytel,cvegega,fpollo}@frba.utn.edu.ar

**Abstract.** The aim of this work is to describe the tasks performed to carry out the development of a software system capable of detecting and recognizing the symbols of *Cursogramas* in images by using a Deep Learning model that has been trained from scratch. In this way, we seek to assist teachers of an undergraduate subject to automatically evaluate diagrams made as part of the practical exercise of their students. For this purpose, in addition to having carried out a process of understanding the problem and identifying the available data, tasks of technology selection and construction of each of the components that are part of the system are also carried out. Therefore, although the problem domain belongs to the field of university education, this work is more related to the engineering and technological aspect of the application of Artificial Intelligence to solve complex problems.

**Keywords:** *Cursogramas* · Object detection · Deep learning · Artificial intelligence

## 1 Introduction

A *Cursograma* is a work tool, which allows to represent graphically the movement of documents that correspond to a certain administrative procedure [1]. Its main objective is to be able to represent a routine, without falling into the complexity of the graph, since this can lead to misinterpretation [2]. Given its usefulness, this method is taught within the subject 'Sistemas y Organizaciones' in the first level of the career 'Information Systems Engineering' [3] within the Facultad Regional Buenos Aires of Universidad Tecnológica Nacional (Argentina). This annual course is part of the integrating core and crosses the curriculum at different levels [4].

As it usually happens with any type of practical exercise, in order to achieve a good handling of these diagrams, students must perform a large number of exercises. But, they also need to have feedback on the correction of the exercises. Since there are no automatic methods that allow revision, the evaluation work is carried out manually by teachers and assistants. In many cases, the proposal provided by the students may


H. Florez and M. F. Pollo-Cattaneo (Eds.): ICAI 2021, CCIS 1455, pp. 39–54, 2021.
https://doi.org/10.1007/978-3-030-89654-6_4

present some minor differences. For this reason, the correction of each diagram requires a considerable amount of time, since the resolution must be analyzed in depth, in addition to paying special attention to the analysis process that the students carry out at the time of the resolution.

In this context, the objective of this work is to describe the tasks carried out for the implementation of a software system that allows the automatic evaluation of *Cursograma* diagrams made by students in the practical work of the subject. Such a tool would generate benefits for both teachers and students. For this purpose, first in Sect. 2 the description of the problem is presented with its main particularities that must be considered to carry out the implementation of the software system. Then, in Sect. 3, the proposed solution is presented describing each of the main components that have been necessary to develop in order to achieve the objective. The results obtained are shown in Sect. 4. Finally, Sect. 5 presents the conclusions and future lines of work.

## 2 Problem Description

The subject 'Sistemas y Organizaciones' belongs to the first year of the career and is therefore mandatory for students who have passed the entrance course to the career. In the year 2020 the number of new students has been approximately 1400, so there are courses with more than 80 students. This has motivated the subject chair to apply Artificial Intelligence technologies to assist the teaching-learning process [5]. Among the objectives of this Intelligent System is a software system in charge of reviewing, correcting and automatically evaluating the *Cursograma* diagrams made by the students as part of the practical exercise of the course.

As a result of several requirements elicitation sessions, applying an engineering process similar to the one proposed in [6], it has been possible to obtain the necessary information to determine the requirements and constraints to be considered in the development of this software system.

On the one hand, the symbols used in *Cursogramas* have been surveyed and can be found in [7]. Later, an "expert" of the chair has also been trained to know the main rules that define the valid ways in which the symbols should be connected to each other.

On the other hand, the main functionality is identified as the automatic evaluation of images that include the *Cursogramas* made by the students. For this purpose, not only the symbol template with the rules previously mentioned must be taken into account, but also the particularities of the statement of the exercise solved by the students. Likewise, the operation to be included in the system is defined, from the reception of the images from the students to the return with their revision. This review should provide the student with a quantitative evaluation of the exercise as well as a set of qualitative observations on corrections. This means that, for each image reviewed, the mistakes and problems encountered should be marked, indicating, in each case, the nature of the error.

Once the problem to be solved has been determined, a bibliographic search has been carried out to try to find similar developments that could be used or adapted to meet the elicited requirements. In this sense, several proposals have been found where Artificial Intelligence is used for the automatic evaluation of diagrams both in the field of education, as well as, for professional usage. However, the proposals found are oriented towards

UML-type diagrams [8–12] Engineering CAD Drawing [13], and Digital Logic Circuit diagrams [14] among others, which have different characteristics and symbols from *Cursogramas* and then are not useful for the considered problem. Consequently, it has been decided to implement a totally new software system oriented to the particularities of the subject in question.

# 3 Proposed Solution

Considering all the requirements and restrictions that the software system must fulfill, work has begun on its design and development. In order to fulfill the objective of the software system to be developed, it is essential to apply a technology that automatically allows recognizing the symbols presented in the *Cursograma* image, and also their relative location in the diagram in order to determine how they are connected to each other. The construction of this Detection Model is described in Sect. 3.1. Having achieved a Symbol Detection Model, then the development of the main component is stared. This component deals with the evaluation of the *Cursograma* exercises solved by the students for a specific exercise. The details about the functionalities of this Evaluation program are presented in Sect. 3.2.

## 3.1 Construction of the Symbol Detection Model

Since it is preferred to use a 'Machine Learning' algorithm [15] that can "learn" to recognize the symbols, the application of Deep Learning Neural Networks [16], which are better known as 'Deep Learning', is selected. From the great variety of existing models for Deep Learning networks [17], a variety of Convolutional Neural Networks or CNN [18, 19] called 'Object Detection Model' [20] has been selected for this work. Since there are a large number of possible architectures for implementing Object Detection Models, specifically, three types are considered, the Faster R-CNN Inception version 2 [21], the SSD MobileNet version 2 [22] and the R-FCN ResNet 101 [23]. Although models of these three architectures are available that already "know" how to detect objects in an image, none are useful for working with *Cursograma* symbols. Therefore, only zero-trained architectures can be used by applying the corresponding algorithm to determine the set of values for a set of parameters in a way that represents the behavior of a set of available data [15, 29, 30].

However, in order to learn to detect the symbols, this training algorithm requires that, a set of additional "annotations". In our case, for each image of a *Cursograma* we need an XML file [24] which indicates the coordinates of a region (or "box") where each symbol is located and a "label" indicating the type of symbol in question. Since this information is not available (and its manual generation requires a lot of effort), it has been decided to develop an ad-hoc program to automatically generate random cases of *Cursogramas*. For each case, the diagram will be recorded in a PNG type image, and all the complementary information in the XML file. In order to carry out this case generation, the standardized format of symbols and rules defined by an "expert teacher" will be used.

Once this program has been developed, it is used to generate the cases (each one consisting of a PNG image and an associated XML file) for the construction of the model for the detection of the *Cursograma* diagram symbols. In the first instance, 1,000 cases are generated, 90% of which are used for training and the rest for validation. With them, the three types of architectures mentioned above are trained and validated. As can be seen in Table 1, none of them generate acceptable results. The SSD architecture has the worst results, while Faster R-CNN has the best results although it tends to detect fewer symbols present in the image than R-FCN so its recall is lower.

Since the processing speed is not as important as the model accuracy, it is decided to discard the SSD architecture and continue the training with the other two architectures by adding more examples. After adding 4,000 new cases (thus generating a total of 5,000), the architectures are re-trained and validated, obtaining the results shown in Table 1. As can be seen, now R-FCN is the one with the best results, being acceptable. After analyzing in detail, it is discovered that most of the errors have to do with the detection of the symbol that corresponds to a 'Decision'. Although the 'Decision' symbol is quite a very simple symbol to recognize, two connections (generally one vertical and one horizontal) come out of this symbol, which seems to confuse the model. Therefore, it is decided to add another 500 new examples of diagrams where decisions appear in order to retrain the R-FCN architecture. As can be seen in the last row of Table 1, an almost perfect model is obtained.

**Table 1.** Results of the symbol detection model's first training and validation.

| Cases generated | Model architecture | Validation metrics | | |
|---|---|---|---|---|
| | | Accuracy | Precision | Recall |
| 1,000 | SSD | 51.6% | 74.1% | 63.1% |
| | Faster R-CNN | 63.7% | 88.5% | 69.4% |
| | R-FCN | 56.3% | 63.7% | 82.9% |
| 5,000 | Faster R-CNN | 59.7% | 90.2% | 63.9% |
| | R-FCN | 99.5% | 99.6% | 99.9% |
| 5,500 | R-FCN | 99.9% | 99.9% | 100% |

Nevertheless, after a meeting where the expert teachers of the subject review the first results, they complained that that all the symbols always have the same letters and numbers, for example the documents have the same type ("F0"), while the controls and operations appear with "1". Another shortcoming that they also consider important is the lack of examples of document copies, where each copy may have a different circuit. All these issues have more to do with the diagram generator program, and therefore a new version is developed that corrects them.

As the previously trained model is used to handling only symbols with "hardcoded" letters and numbers, a new one must be train to handle these changes using a new batch of generated cases (each one including a PNG and XML files). The new batch includes 11,934 cases for training and 1,326 cases for validation.

Despite of using the same architecture R-FCN, in order to obtain a reliable model, several training sessions have been performed. In each of these sessions, several training parameters has been tried to change to improve the model performance. Based on a later result analysis it is detected that the critical one has been the quantity of training steps. As it can be seen in Table 2, by increasing the quantity of cycles performed during the training, the model accuracy and recall change considerably. However, after the 50,000 steps milestone, the model performance starts again to decline because the model is over-trained and thus loses its generalization capabilities. On that account, the 50,000 training steps is selected to be used for the evaluation program. Although, its performance is affected by 'false positives', they are related to symbol connectors so this is not considered a problem to carry out the evaluation of the diagrams, and the construction of the Symbol Detection Model is considered as successfully completed.

**Table 2.** Results of the symbol detection model's second training and validation.

| Model architecture | Training steps | Validation metrics | | |
|---|---|---|---|---|
| | | Accuracy | Precision | Recall |
| R-FCN | 10,000 | 97.5% | 99.9% | 97.5% |
| | 20,000 | 98.5% | 99.9% | 98.5% |
| | 50,000 | 99.5% | 99.9% | 99.5% |
| | 100,000 | 99.2% | 99.9% | 99.2% |

### 3.2 Implementation of the Evaluation Program

The Evaluation program is the software component that deals with the evaluation of the *Cursograma* exercises solved by the students. This is performed using another diagram that is used as a reference at the time of the review. This other diagram must obviously comply with both the general rules mentioned above, as well as the particularities of the exercise statement.

These two diagrams, the one provided by the students to be evaluated and the one provided by the teachers for references, are provided in the form of PNG images. Therefore, to carry out the evaluation, first the symbol Detection Model is applied on each image obtaining a list of detected symbols. Also, when a detected symbol includes a letter and/or number (e.g. a document), they are recognized using the open source Tesseract Optical Character Recognition (OCR) Engine [25, 28]. As a result, two lists are obtained (one for the students' image and another for teacher's) that for each detected symbol includes: the symbol class, its relative position in the diagram and the OCR string with the corresponding letters and numbers.

When the lists are available, they are processed by comparing them using an ad-hoc algorithm. For each the students' symbol the following rules are analyzed:

- If there is a match in the teacher's list where a symbol has the *same class, relative position and OCR string*, then the students' symbol is marked as *correct*.

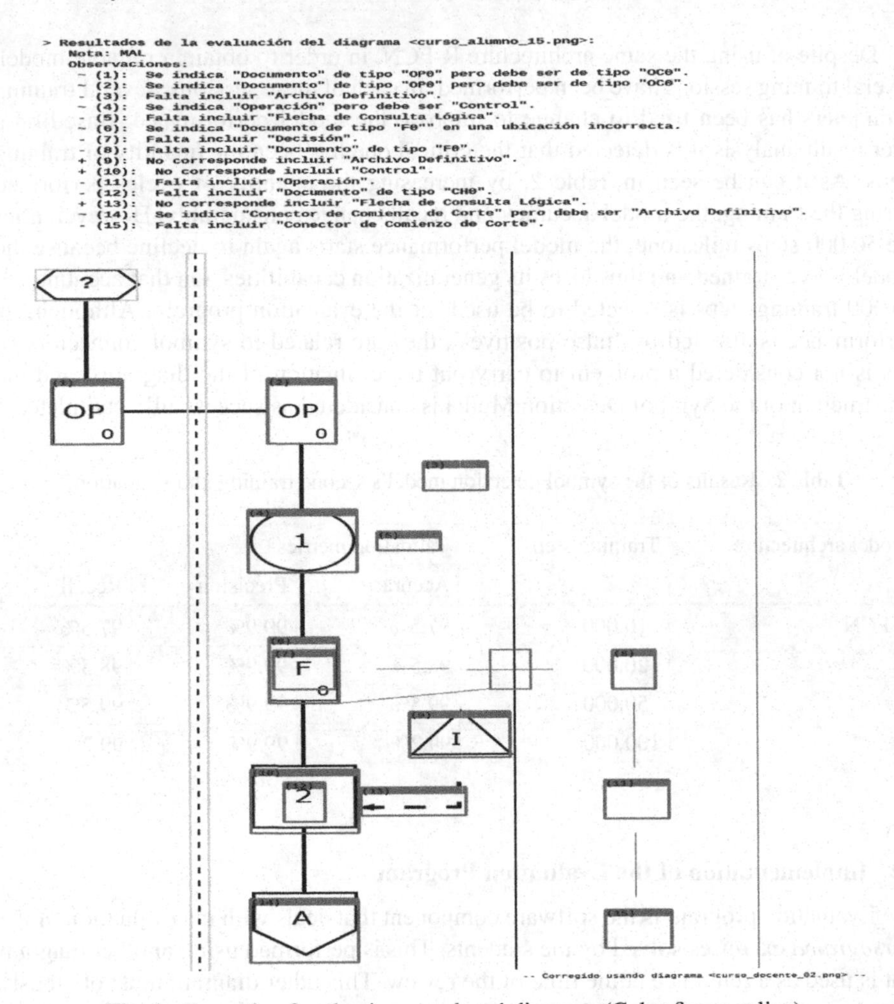
> Resultados de la evaluación del diagrama <curso_alumno_15.png>:
Nota: MAL
Observaciones:
~ (1): Se indica "Documento" de tipo "OP0" pero debe ser de tipo "OC0".
~ (2): Se indica "Documento" de tipo "OP0" pero debe ser de tipo "OC0".
- (3): Falta incluir "Archivo Definitivo".
~ (4): Se indica "Operación" pero debe ser "Control".
- (5): Falta incluir "Flecha de Consulta Lógica".
~ (6): Se indica "Documento de tipo "F0"" en una ubicación incorrecta.
- (7): Falta incluir "Decisión".
- (8): Falta incluir "Documento" de tipo "F0".
+ (9): No corresponde incluir "Archivo Definitivo".
+ (10): No corresponde incluir "Control".
- (11): Falta incluir "Operación".
- (12): Falta incluir "Documento" de tipo "CH0".
+ (13): No corresponde incluir "Flecha de Consulta Lógica".
~ (14): Se indica "Conector de Comienzo de Corte" pero debe ser "Archivo Definitivo".
- (15): Falta incluir "Conector de Comienzo de Corte".
?
OP
0
OP
0
1
F
0
I
2
A
-- Corregido usando diagrama <curso_docente_02.png> --

**Fig. 1.** Example of evaluating a students' diagram. (Color figure online)

- If there is a match in the teacher's list where a symbol has the *same class and relative position but different OCR string*, then the students' symbol is marked as *mistake in symbol text.*
- If there is a match in the teacher's list where a symbol has the *same class and OCR string but different relative position*, then the students' symbol is marked as *mistake in symbol location.*
- If there is a match in the teacher's list where a symbol has *different class but the same relative position and OCR string*, then the students' symbol is marked as *mistake in symbol class.*
- If there is not a match in the teacher's list, then the students' symbol is marked as *mistake of over indicated symbol.*

Finally, when the all the entries in the students' list are processed, if there is a teacher's symbol that *has not been matched*, then this teacher's symbol is marked as *mistake of missing symbol*.

The results of this evaluation algorithm are highlighted in the students' image marking them with colors (green color if it is correct, red color otherwise) as it can be seen in Fig. 1. In case of "missing" teacher's symbols, the location is established by a special function that uses position of other symbols matched in both diagrams as landmarks. Furthermore, a grade for the exercise is "calculated" and recorded in the header of the image with the corresponding observations where the detected mistakes are explained.

## 4 Results

This section presents the results of the tests performed to confirm that the developed Evaluation program works correctly. This is achieved by running the component on many different diagrams to review. In this case, it has been decided to use again completely randomly generated cases as explained in Sect. 4.1. This means that none of the test cases used here correspond to real exercises of the subject but, anyway, they are considered useful to check how the software behaves with different situations. Using these generated cases, in Sect. 4.2 a study case is utilize to explain the results of evaluating a students' diagram. Later, in Sect. 4.3 its robustness is verified by contrasting the results of normal and "noisy" images. Finally, in Sect. 4.4, the evaluation results are analyzed by expert professors of the subject to decide if the diagrams have been evaluated correctly.

### 4.1 Generation of Test Cases

In order to generate these test cases that are available in the repository [26], the following procedure used is as follows:

First, 5 different diagrams are generated, each with a different level of complexity, which are taken as reference images provided by the teacher. That is, these images (that stored in the "Teacher_diagrams" folder of [26]) are taken as if they were the correct resolution of an exercise and are used as a basis for the revision of the students' diagrams.

On the other hand, the students' diagrams are also generated automatically to obtain 10 new images. To these images a copy of the 5 reference images are added in order to be able to corroborate that the software works correctly in cases where there is no error. Moreover, a noise function is applied to these 15 images to generate images that will be used to verify the capabilities of the detection model. This function applies a kind of 'Salt-and-Pepper' noise [27] that changes the value of random pixels in order to generate unanticipated disturbances in the image. These changes are few but they are very noisy. As a result, the effect is similar to sprinkling white and black dots on the image. All the "noisy" images have the suffix "DAn" in their file name so they can be easily identified. In this way, a total of 30 students' exercises to evaluate are obtained and store in the "Students_diagrams" folder of [26].

Once all the test images have been generated, the automatic evaluation program is run for each combination, so that each of the 30 students' images is reviewed against each of the 5 reference teacher's images. Upon completing the evaluation of all combinations, it is observed that the evaluation program takes between 5 and 8 s to process each image. In total then 150 evaluations are carried out and recorded in the "Evaluation_results" folder of [26]. Note that to facilitate their organization and search, in the name of these images the identifier of the teacher's reference image is indicated first and then the image corresponding to the student's exercise. Thus, for example, in Fig. 1, the results of evaluating the "noisy" student's image '02' by using as reference the teacher's image '01'.

Finally, as the observations written in the evaluation image's header are in spanish, to assist non-spanish speakers understand the detected mistakes, a multiple color version of each evaluation is generated. To highlight the different evaluation results, the color **Green**is used to identify the symbols that are correct, **Orange**to show a mistake of class, location or OCR string, **Violet**for those that are over indicated and **Red**for those that are missing. This version of the example shown previously in Fig. 1 can be seen below in Fig. 2. Also, the multiple is available for all the images in the "Evaluation_results" folder of [26] and are identified with the suffix "-mc" in their file name.

## 4.2 Analysis of a Case Study

The objective of the case study is demonstrating the results of evaluating a students' diagram by analyzing the example shown in Fig. 1 and 2. This example is generated after evaluating the students' diagram '15' (shown in the left column of Table 3) based on the teacher's reference diagram '02' (shown in the right column of the same table).

As it can be noted, in spite of having a similar beginning, after the third symbol the students' and teacher's diagrams have a lot of differences. Therefore, it is clear why that the evaluation result image has a bad grade ("*MAL*" in spanish) and a lot of symbols highlighted in non-green color. But, in fact, this example is interesting because it includes all the types of evaluation mistakes, and that is why it is used as a case study.

In order to analyze the evaluation results, below each one of highlighted marks of Fig. 2 from top to bottom is explained. Also, for the purpose of facilitating this explanation, the Table 4 is presented where Fig. 2 is separated into parts and matched to the corresponding parts of the teacher's reference image.

1) At the beginning of both diagrams, the first symbol is a "not identified process", then it is a perfect match and the symbol is highlighted as *correct* in **green**color.
2) This "not identified process" is in an external sector of the organization, which is separated from the internal sector by a dotted line that is drawn in both diagrams, so it is also a perfect match and the symbol is highlighted as *correct* in **green**color.
3) As a result of this "not identified process", a document is generated that is transferred to an internal sector of the organization. Despite that this transfer is carried out correctly, the type of document indicated (which is recognized by the OCR engine) is incorrect. It can be noted that the teacher's documents are an "OC-0" but the students' are an "OP-0". As a result, the *mistake in symbol text* is described in the image's header (items 1 and 2) and highlighted in **orange**color.
4) Later, in the internal sector a "control" must be performed, but the students use an "operation" symbol. Then, there is a *mistake in symbol class* which is highlighted in **orange**color and described as item 4 of the image's header.

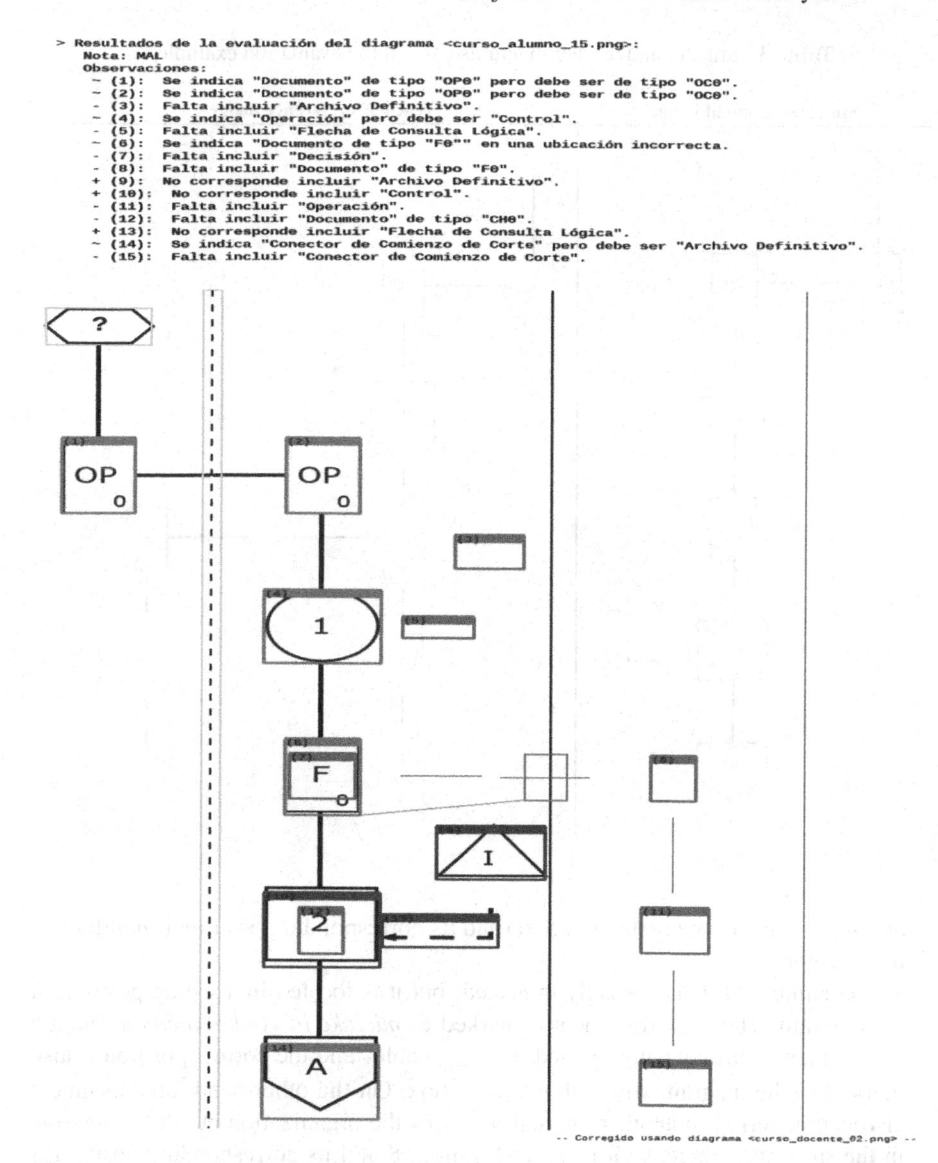

> Resultados de la evaluación del diagrama <curso_alumno_15.png>:
Nota: MAL
Observaciones:
~ (1): Se indica "Documento" de tipo "OP0" pero debe ser de tipo "OC0".
~ (2): Se indica "Documento" de tipo "OP0" pero debe ser de tipo "OC0".
- (3): Falta incluir "Archivo Definitivo".
~ (4): Se indica "Operación" pero debe ser "Control".
- (5): Falta incluir "Flecha de Consulta Lógica".
~ (6): Se indica "Documento de tipo "F0"" en una ubicación incorrecta.
- (7): Falta incluir "Decisión".
- (8): Falta incluir "Documento" de tipo "F0".
+ (9): No corresponde incluir "Archivo Definitivo".
+ (10): No corresponde incluir "Control".
- (11): Falta incluir "Operación".
- (12): Falta incluir "Documento" de tipo "CH0".
+ (13): No corresponde incluir "Flecha de Consulta Lógica".
~ (14): Se indica "Conector de Comienzo de Corte" pero debe ser "Archivo Definitivo".
- (15): Falta incluir "Conector de Comienzo de Corte".
?
OP
0
OP
0
1
F
0
I
2
A
-- Corregido usando diagrama <curso_docente_02.png> --


**Fig. 2.** Example of evaluating a students' diagram with multiple colors. (Color figure online)

5) Moreover, the previously "control" must apply a "logical query" to a "permanent file", that the students have omitted. Then, these two symbols are marked as *missing* (items 3 and 5 of the image's header) and their corresponding positions in the students' image are highlighted in **red** color.
6) After a "control" there is always a "decision", but the students have not included it. The position of this "decision" should be where the document "F0" is located, but this document will be used afterwards. Therefore, the "decision" is also marked as

**Table 3.** Student and teacher diagrams used in the evaluation example.

| Students' original image | Teacher's reference image |
|---|---|
| 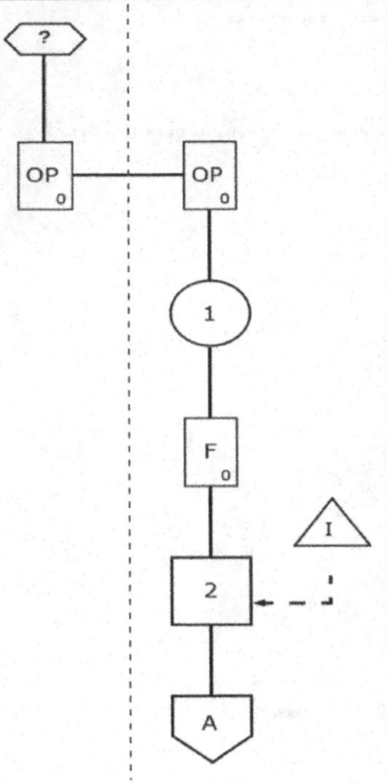 ? OP 0 OP 0 1 F 0 I 2 A | 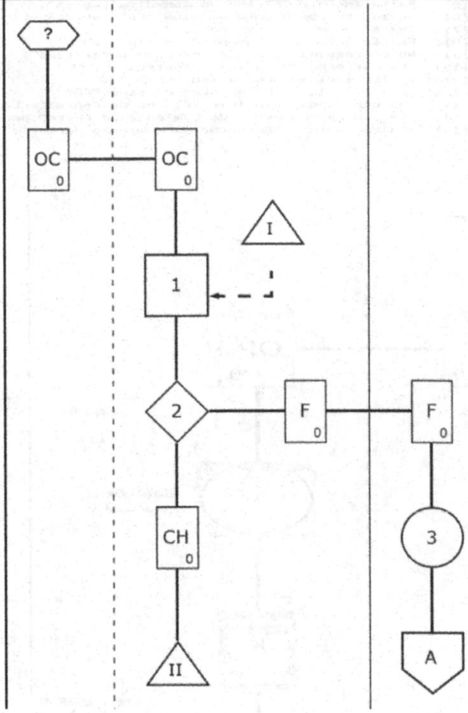 ? OC 0 OC 0 I 1 2 F 0 F 0 CH 0 3 II A |

*missing* (item 7 of the image's header) and its corresponding position is highlighted in **red**color.

7) The document "F0" is correctly indicated, but it is located in a wrong position of the diagram. Then, the document is marked as *mistake in symbol location* (item 6 of the image's header), highlighted in **orange**color and the correct position is also marked in the diagram with a thin **orange**box. On the other hand, this document also is transferred to another internal sector of the organization but this is *missing* in the students' image, so it is marked in item 8 and its corresponding position is highlighted in **red**color.
8) In the new internal sector of the organization, two more symbols are *missing* (an "operation" and a "cut start connector"). Therefore, both are marked as *missing* (items 11 and 15 of the image's header) and their corresponding positions are highlighted in **red**color.
9) Finally, in the old internal sector, there are three types of mistakes:

   – First, the document "CH0" is *missing*, so it is marked in item 12 of the image's header and its corresponding position is highlighted in **red**color.

**Table 4.** Analysis of the evaluation of the students' diagram example.

| # | Students' Evaluation Result part | Corresponding Teacher's reference part |
|---|---|---|
| 1 | ? | ? |
| 2 | | |
| 3 | OP 0 OP 0 | OC 0 OC 0 |
| 4 | 1 | 1 |
| 5 | | I |
| 6 | F 0 | 2 |
| 7 | F 0 | F 0 F 0 |
| 8 | | 3 A |
| 9 | I 2 A | CH 0 II |

- Secondly, the students have included a "cut start connector" instead of a "permanent file", so this is marked as a *mistake in symbol class* (item 14) and highlighted in orangecolor.
- And third, there is a "control" that uses a "logical query" to a "permanent file" that are marked as *mistake of over indicated symbol* (items 9, 10 and 13) and highlighted in violetcolor.

This last mistake is interesting to be discussed in a little more detail. Although one might come to think that these three symbols correspond to the mistakes previously indicated in **(3)** and **(4)**, it can be noted that their location are well below the diagram. If the Evaluation program will try to use them as a correct landmark, it would have to mark all the previous symbols as over indicated, and all the following symbols as missing. Since the idea is to try to reuse as much as possible the symbols available in the students' diagram, it is considered that this kind of review is correctly performed.

### 4.3 Verification of the Robustness with "Noisy" Images

A question that has arisen during development is how the Evaluation program will behave with images that are not 100% "perfect" (i.e. images that present some degree of pixel error or noise). The main concern was directed to confirm if the previously trained Symbol Detection Model would be able to detect the symbols in this type of images. In order to perform this verification (as has been described in Sect. 4.1) a "noisy" version of the students' images have been generated, and then evaluated by the software as a normal exercise diagram.

As a result, for each exercise there is a normal version and a "noisy" version that have been compared manually to determine if the Evaluation program generates different results. After reviewing all the pairs of images, differences have been detected in the evaluation of noisy images, but, surprisingly the differences were not due to the Symbol Detection Model.

Despite presenting different noise levels, the trained model is able to accurately detect the symbols present in the evaluated images. But, there is an issue with the Tesseract OCR Engine used to recognize the characters included in some symbols. For example, as it can be seen in the left column of Table 5, when the students' diagram '04' is evaluated using the teacher's diagram '04' as reference, obviously there are not mistakes (all the symbols are in greencolor) because they are the same image. However, when the "noisy" version of the student's diagram is processed (right column of Table 5), three orangecolor mistakes are indicated in the "document" symbols. These errors are related to the *mistake in symbol text* because the OCR Engine cannot recognize correctly the letters and numbers in each symbol. Despite of the black pixels, a human can read that the documents' strings are "RE-0", "NP-0" and "OP-0", but the OCR Engine recognizes them as "RE" (without any number), "NP.N8" and "OP.OO". As the Evaluation program trusts in the strings provided by the OCR process, then mistakes are marked incorrectly.

Given that the evaluation of images with this degree of noise should not be normal in real diagrams, and that the OCR engine is an external element used by the software, it is considered that this issue does not invalidate the development or the tests presented in this article. However, it has been decided to initiate a review of other available OCR engines that are more robust to be applied in a future version of the Evaluation program.

## 4.4 Discussions of the Evaluation Results

In this section, the main comments received by the expert teachers of the subject after reviewing the evaluations made by the software system are presented.

From their point of view, the software works in an acceptable manner, being able to correctly detect 100% of the symbols present in each image, which allows it to highlight different types of mistakes consistent with the reference image. Also, the observations indicated by the software system are considered as useful for the students. In fact, this will allow students to better understand their mistakes and try to avoid them in future exercises.

**Table 5.** Examples of comparing the evaluation of students' normal and "noisy" images.

| Evaluation of students' image | Evaluation of students' "noisy" image |
|---|---|
| 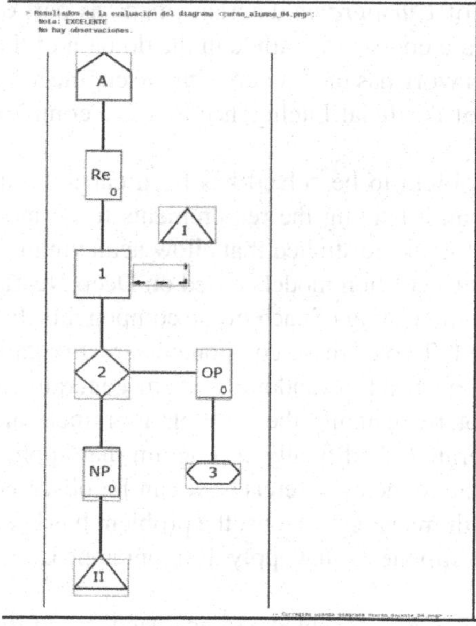 > Resultados de la evaluación del diagrama <curso_alumno_04.png>:<br>Nota: EXCELENTE<br>No hay observaciones. | 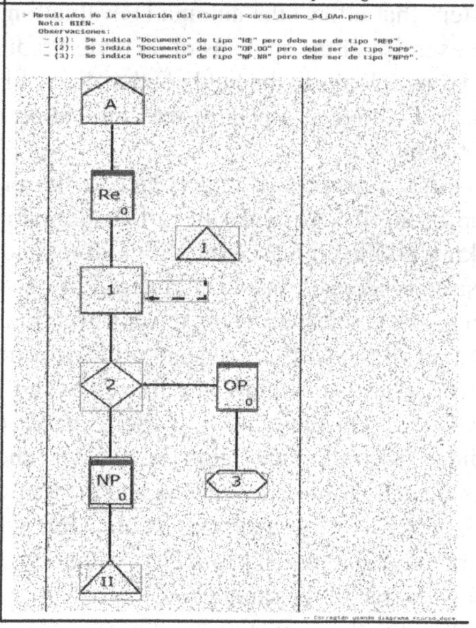 > Resultados de la evaluación del diagrama <curso_alumno_04_DAn.png>:<br>Nota: BIEN-<br>Observaciones:<br>- (1): Se indica "Documento" de tipo "RE" pero debe ser de tipo "RE0".<br>- (2): Se indica "Documento" de tipo "OP.OO" pero debe ser de tipo "OP0".<br>- (3): Se indica "Documento" de tipo "NP.N8" pero debe ser de tipo "NP0". |

On the other hand, it is believed that the grades assigned by the system in most of the cases are correct. From the expert teachers' point of view, it is considered that in around 80% of the cases, the assigned grade is consistent with the errors found. However, there are some errors that perhaps should be different, since there are mistakes that the software considers to be minor when in fact they should be serious mistakes. Such is the case of the "timing lines" that are important in a diagram since they indicate that the routine being plotted is performed every certain quantity if time or by certain specific condition (e.g. when a shortage of merchandise is detected). Given its importance, this should be considered a serious mistake that indicates that the student does not understand that this routine is not always performed. Likewise, a "decision" should always be placed under a "control". If this "decision" is not placed by the student, then it is a serious conceptual error and should be considered with more relevance than just considering that the symbol is missing. The same happens if the diagram is not finished with the corresponding end symbols that could be a "file", a "destruction" action, a "process" or a "connector". Therefore, it should be possible to differentiate them from other types of missing symbols.

Finally, it has been detected an issue associated to the location of some missing symbols. Although the location is correctly identified by the detection model, on some occasions, the location of the missing symbols is not 100% correct and might overlap with other symbol. This has been requested to be solved in future versions, since otherwise the students will not be able to understand the mistake they made.

## 5 Conclusions

In this paper we have described the tasks carried out for the implementation of a software system that allows the automatic evaluation of *Cursograma* diagrams made as part of the exercise of the students of an undergraduate course. So, although the domain of the problem belongs to the field of education, this work has more to do with the engineering and technological aspect of the application of Artificial Intelligence to solve complex problems.

First, a process of understanding the problem to be solved has been carried out, identifying the particularities of its domain and defining the requirements to be met. Taking all this into account, different technologies are studied that allow achieving the desired objective. As a result the use of object detection models based on Deep Neural Networks is selected. Then, the construction and testing of each of the components that are part of the software system are performed. Two of these components are auxiliary but necessary to achieve the third. On the one hand, a random diagram generator is developed. On the other, a model capable of recognizing the symbols is trained and validated with the diagrams previously generated. And finally, a program that applies this detection model to review and correct the students' exercises. It can be observed that, to try to achieve a complex objective with many restrictions, the problem has been worked on in a modular way, with separate components that apply different approaches but converge to the same goal.

Finally, the results of the diagram evaluator are presented and analyzed. From the results obtained, it can be concluded that the software system works in an acceptable manner by being able to recognize different types of symbols, identify differences and correct four types of mistakes. But, although much progress has been made in principle, there is still a lot of work to be done before this software can be made operational for use by students. Consequently, as future lines of work, the following stand out:

- review other OCR engines that may have greater robustness when performing character recognition in noisy images;
- correct the issue associated with the location of missing symbols improving the system's ability to avoid overlaps with other objects;

- identify and implement the different rules that used by teachers to assign the grades to exercises taking into account the seriousness of the mistake detected; and
- continue working on improving the robustness of the object Detection Model so that it supports a greater variety in the characteristics of the diagrams to be processed.

## References


1. Pollo-Cattaneo, M.F.: The Organization and its Information Systems, 1st edn. Editorial CEIT, Buenos Aires (2017). ISBN 978-987-1978-36-6
2. IRAM (1974). IRAM 34503: Administrative procedures. General guidelines for the design of forms for graphic representation, 1st edn. Argentine Institute for Standardization and Certification (in force since 03/05/1974)
3. UTN-FRBA: Analytical program of the chair 'Systems and Organizations' - Plan 2008 (2008). https://tinyurl.com/y7xx33y5. Accessed May 2021
4. UTN: Curricular design of the Information Systems Engineering career - Plan 2008. Ordinance UTN-1150 (2008). https://tinyurl.com/yb22zm2q. Accessed June 2021
5. Cohen, P.R., Feigenbaum, E.A.: The Handbook of Artificial Intelligence, vol. 3. Butterworth-Heinemann (2014)
6. Vegega, C., Pytel, P., Pollo-Cattaneo, M.F.: Elicitation of requirements for the construction of predictive models based on intelligent systems in education. In: Proceedings of XXV CACIC 2019, pp. 980–989 (2019). ISBN 978-987-688-377-1
7. Chair SyO: Template of Symbols for use in Cursogramas to be used in the subject 'Systems and Organizations' of UTN FRBA (2020). https://tinyurl.com/y6jdnfy3. Accessed June 2021
8. Outair, A., Lyhyaoui, A., Tanana, M.: Towards an automatic evaluation of UML class diagrams by graph transformation. Int. J. Comput. Appl. **95**(21), 36–41 (2014). https://doi.org/10.5120/16721-7063
9. Lino, A. D.P., Rocha, A.: Automatic evaluation of ERD in e-learning environments. In: 2018 13th Iberian Conference on Information Systems and Technologies (CISTI), pp. 1–5 (2018)
10. Schäfer, B., Keuper, M., Stuckenschmidt, H.: Arrow R-CNN for handwritten diagram recognition. Int. J. Doc. Anal. Recogn. (IJDAR) **24**(1–2), 3–17 (2021). https://doi.org/10.1007/s10032-020-00361-1
11. Anwer, S., El-Attar, M.: An evaluation of the statechart diagrams visual syntax. In: 2014 International Conference on Information Science & Applications (ICISA), pp. 1–4 (2014)
12. Vachharajani, V., Pareek, J., Gulabani, S.: Effective label matching for automatic evaluation of use-case diagrams. In: 2012 IEEE 4th International Conference on Technology for Education, pp. 172–175 (July 2012)
13. Elyan, E., Jamieson, L., Ali-Gombe, A.: Deep learning for symbols detection and classification in engineering drawings. Neural Netw. **129**, 91–102 (2020)
14. Altun, O., Nooruldeen, O.: SKETRACK: stroke-based recognition of online hand-drawn sketches of arrow-connected diagrams and digital logic circuit diagrams. Sci. Program. **2019**, 1–17 (2019). https://doi.org/10.1155/2019/6501264
15. Alpaydin, E.: Introduction to Machine Learning. MIT Press (2014)
16. Schmidhuber, J.: Deep learning in neural networks: an overview. Neural Netw. **61**, 85–117 (2015)
17. LeCun, Y., Bengio, Y., Hinton, G.: Deep learning. Nature **521**(7553), 436 (2015)
18. Albawi, S., Mohammed, T.A., Al-Zawi, S.: Understanding of a convolutional neural network. In: 2017 International Conference on Engineering and Technology (ICET), pp. 1–6. IEEE (2017)

19. Zhou, D.X.: Universality of deep convolutional neural networks. Appl. Comput. Harmon. Anal. **48**(2), 787–794 (2020)
20. Zeiler, M.D., Fergus, R.: Visualizing and understanding convolutional networks. In: Fleet, D., Pajdla, T., Schiele, B., Tuytelaars, T. (eds.) ECCV 2014. LNCS, vol. 8689, pp. 818–833. Springer, Cham (2014). https://doi.org/10.1007/978-3-319-10590-1_53
21. Szegedy, C., Vanhoucke, V., Ioffe, S., Shlens, J.: Rethinking the inception architecture for computer vision. In: IEEE Conference on Computer Vision and Pattern Recognition (CVPR), pp. 2818–2826 (2016)
22. Sandler, M., Howard, A., Zhu, M., Zhmoginov, A., Chen, L.C.: MobileNetV2: inverted residuals and linear bottlenecks. In: Proceedings of the IEEE Conference on Computer Vision and Pattern Recognition, pp. 4510–4520 (2018)
23. Huang, J., Rathod, V., et al.: Speed/accuracy trade-offs for modern convolutional object detectors. In: Proceedings of the IEEE Conference on Computer Vision and Pattern Recognition, pp. 7310–7311 (2017)
24. Khandelwal, R.: COCO and Pascal VOC data format for Object detection. Medium (2019). https://tinyurl.com/y5fgyb9n. Accessed June 2021
25. Smith, R.W.: History of the Tesseract OCR engine: what worked and what didn't. In: Document Recognition and Retrieval XX, vol. 8658, p. 865802. International Society for Optics and Photonics (February 2013)
26. GEMIS: Randomly Generated Cases with its Results (2021). https://tinyurl.com/7k8mhrtw. Accessed June 2021
27. Bovik, A.C. (ed.): The Essential Guide to Image Processing. Academic Press (2009)
28. Velosa, F., Florez, H.: Edge solution with machine learning and open data to interpret signs for people with visual disability. CEUR Workshop Proc. **2714**, 15–26 (2020)
29. Zuluaga, J.Y., Yepes-Calderon, F.: Tensor domain averaging in diffusion imaging of small animals to generate reliable tractography. ParadigmPlus **2**(1), 1–19 (2021)
30. Espinosa, C., Garcia, M., Fernando Yepes-Calderon, J., McComb, G., Florez, H.: Prostate cancer diagnosis automation using supervised artificial intelligence. a systematic literature review. In: Florez, H., Misra, S. (eds.) ICAI 2020. CCIS, vol. 1277, pp. 104–115. Springer, Cham (2020). https://doi.org/10.1007/978-3-030-61702-8_8

# Sign Language Recognition Using Leap Motion Based on Time-Frequency Characterization and Conventional Machine Learning Techniques

D. López-Albán[1(✉)], A. López-Barrera[1], D. Mayorca-Torres[1,3], and D. Peluffo-Ordóñez[2,3]

[1] Universidad Mariana, Pasto 520001, Colombia
{diegoanlopez,manuellopez,dmayorca}@umariana.edu.co
[2] Mohammed VI Polytechnic University, 47963 Ben Guerir, Morocco
peluffo.diego@um6p.ma
[3] SDAS Research Group, 47963 Ben Guerir, Morocco
{dagoberto-mayorca,diego.peluffo}@sdas-group.com

**Abstract.** The abstract should briefly summarize the contents of the paper in Sign language is the form of communication between the deaf and hearing population, which uses the gesture-spatial configuration of the hands as a communication channel with their social environment. This work proposes the development of a gesture recognition method associated with sign language from the processing of time series from the spatial position of hand reference points granted by a Leap Motion optical sensor. A methodology applied to a validated American Sign Language (ASL) Dataset which involves the following sections: (i) pre-processing for filtering null frames, (ii) segmentation of relevant information, (iii) time-frequency characterization from the Discrete Wavelet Transform (DWT). Subsequently, the classification is carried out with Machine Learning algorithms (iv). It is graded by a 97.96% rating yield using the proposed methodology with the Fast Tree algorithm.

**Keywords:** Sign language · Leap motion · Discrete wavelet transform · Machine learning

## 1 Introduction

Worldwide, about 5% of the population (430 million people) suffer from disabling hearing loss [1]. Sign language is the natural language of deaf people, it has linguistic rules that involve nonverbal language based on corporal gesticulation, and hands. The gestures involved in sign language are structured in a very complex way, as they convey important information and feelings of human communication with multiple parts of the body. However, the communication is based mainly on the manual expressions of the orientation of the hand and its


H. Florez and M. F. Pollo-Cattaneo (Eds.): ICAI 2021, CCIS 1455, pp. 55–67, 2021.
https://doi.org/10.1007/978-3-030-89654-6_5

position relative to the body and the local configuration of the fingers [2]. Most of the population does not understand sign language, it is a barrier that prevents expressing the communicative and non-communicative needs of the human being.

Technological advances have allowed the development of devices with the need to recognize the gesture-spatial configuration of the hands; being the instrumented gloves, and the optical sensors the most used today. Among the optical sensors are mainly: RGB cameras [3], Kinect [4], and Leap Motion (LMC) [5,6]. The LMC device is a novel optical tracking module that describes the physiology of the hand from 26 three-dimensional coordinates [8], and that also can work with interfaces with high degree of precision, and repeatability [6].

These devices require, on the other hand, the use of classifier algorithms that allow accurate recognition of gestures associated with sign language [7]. Some of these algorithms employ artificial intelligence (AI), divided into two large groups: conventional Machine Learning (ML) techniques and more sophisticated Deep Learning (DL) techniques.

Some of the ML techniques used for dynamic gesture recognition are multi-class support vector machine (SVM) [8,9], linear binary SVM [10], Multilayer Perceptron (MLP). Some DL based models are Convolutional Neural Network (ConvNet) [11], recurrent neural network Long Short-Term Memory (DeepConvLSTM) [12]. Conventional techniques are of interest to the scientific community because they require lower computational cost and perform well in classification. The use of these implies an additional characterization stage to form a set of input data, such as time series type.

Based on the various studies proposed on Sign Language such as [13,14] there are different methodologies that require a high computational load, which makes a high-capacity equipment necessary when using DL. Therefore, we propose a methodology that uses ML classifiers and feature extraction from the Dataset "American Sign Language - Leap Motion - 25 subjects - 60 signs" [15]. In this sense, the characterization DWT captures the information in the frequency domain in a localized way in time and follows the morphology of the time series. Despite the fact that several studies have been developed regarding the recognition of sign language, even these methods do not have an adequate compromise between computational cost and precision. That is why the development of a sign language recognition methodology is proposed using time-frequency characteristics and conventional classifiers, which allows the recognition of ASL.

This article is structured as follows: Sect. 2 presents the procedures used for the development of a computational model for sign language recognition, Sect. 3 shows the results obtained. Lastly, Sect. 4 explains the research conclusions. One of the contributions developed by this study is that it provides a methodology that allows the use of conventional ML algorithms for the classification of Datasets with a high number of gestures, allowing the low computational cost, generating a new alternative study.

## 2 Materials and Methods

For the development of the methodological proposal based on conventional ML techniques, it is necessary to develop some stages such as: Database selection, cleaning up data and segmentation, data transformation and sampling, feature extraction, and classification & experiments is shown in Fig. 1.

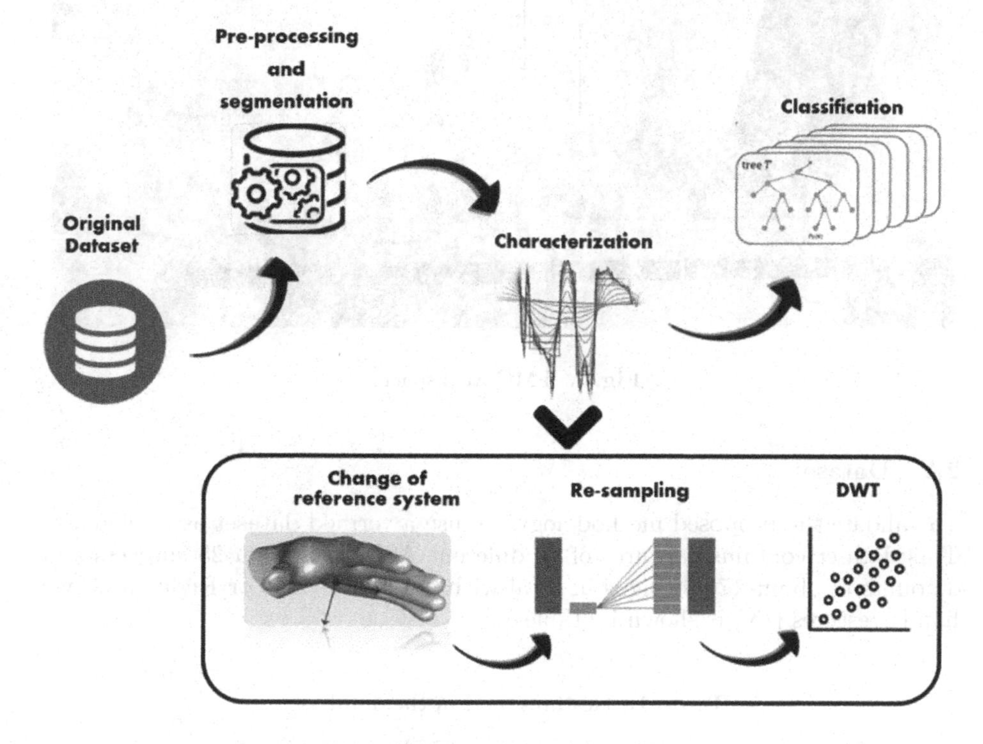


**Fig. 1.** Proposed methodology.

### 2.1 User Disposition for Data Acquisition

Regarding the workspace, it describes the space that the user has to perform gestures captured by the LMC device, [16] says that "The interaction box is a boundary area, where the expected precision sensor is the highest. The minimum distance along the Y axis is 82.5 mm and the maximum distance is 317.5 mm. The z-axis of the interaction box, which is perpendicular to the longest side of the sensor, is in the range of −73.5 mm to 73.5 mm and the range along the x-axis composes the depth with 147 mm.", said workspace is shown in Fig. 2.

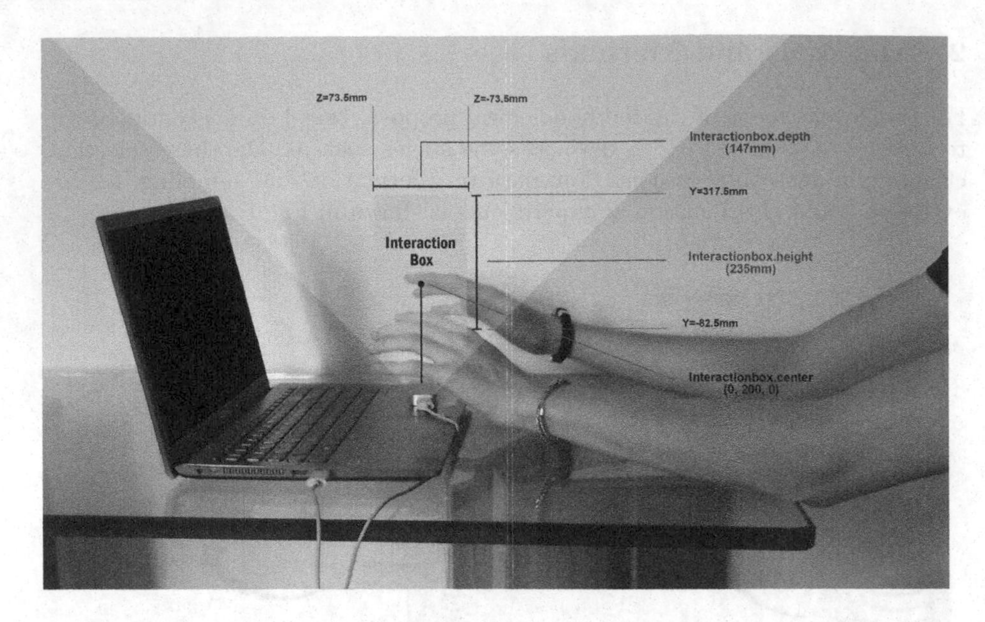


**Fig. 2.** LMC workspace.

### 2.2 Dataset

To validate the proposed methodology, we use a verified dataset as a reference. This dataset contains captures of 60 different ASL signs from 25 subjects and amounts to about 17,000 signs in total with one hand (Left or Right) and two hands gestures [15], is shown in Table 1.

**Table 1.** Gestures used in the database.

| Number of hands | Gesture | | | | | | | |
|---|---|---|---|---|---|---|---|---|
| 1 | 0 | 5 | 10 | Cat | Dog | Hot | Pig | Water |
| | 1 | 6 | Blue | Cereal | Drink | Hungry | Please | Where |
| | 2 | 7 | Brush | Come | Good | Milk | Red | Why |
| | 3 | 8 | Bug | Dad | Green | Mom | Thanks | Yellow |
| | 4 | 9 | Candy | Deaf | Happy | Orange | Warm | |
| 2 | Big | Coat | Cry | Go | Neutral | Socks | What | Work |
| | CarDrive | Cold | Egg | Hurt | Shoes | Stop | When | |
| | Clothes | Cost | Finish | More | Small | Store | With | |

The data are the position in the space provided by the LMC application programming interface (API) for either hand or both hands for each of its reference points [15], i.e. the distances from the device to each of the points shown in Fig. 3, with measurements at each instant of time and represented in more than 82 features.

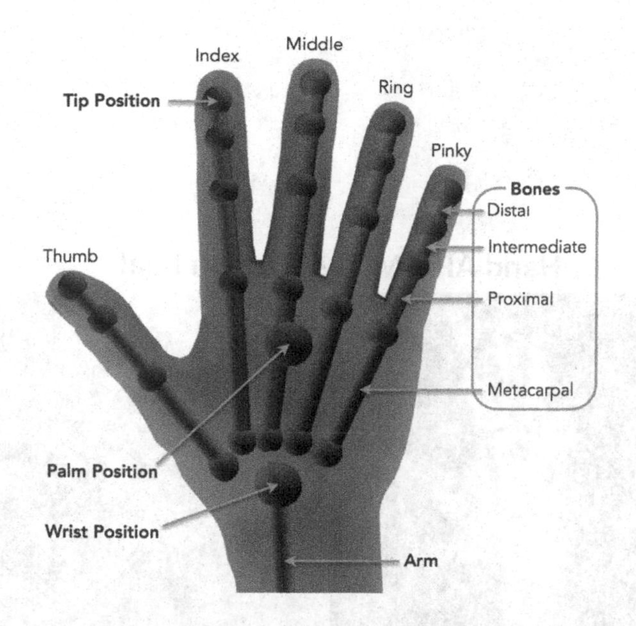


**Fig. 3.** Positional tracking data available as reference points

The characteristics are determined by the data of the 3 rectangular coordinates taken as independent data for each of the 26 reference points, as well as some additional characteristics of the time the gesture was captured. Participants performed each gesture multiple times during a time interval for each capture. We slightly modified the original Dataset records to adapt them to the proposed methodology which are described later. To learn more about the Dataset, a statistical summary is made that allows visibility of the ranges in the data to be used, which is shown in the Table 2.

**Table 2.** Statistical summary of the database.

| Repetitions | Statistical data | Total |
|---|---|---|
| For gesture | Average | 281.5 |
| | Min | 223 |
| | Max | 345 |
| For participants | Average | 11.26 |
| | Min | 8.92 |
| | Max | 13.8 |

**Hand Data from Dataset**

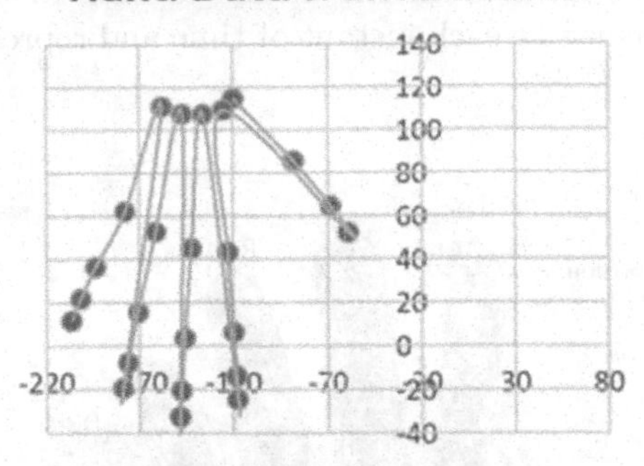


**Hand API LMC vs Hand in Real**

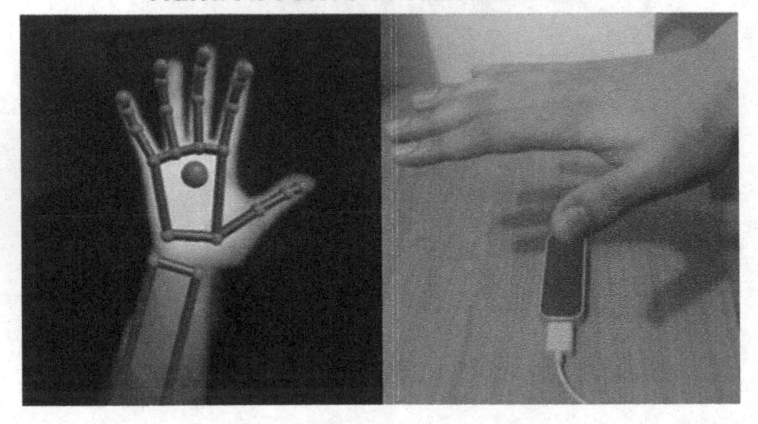

**Fig. 4.** Comparative of hand data from dataset, digital and real

Figure 4 depicts the visualization of the capture of the gesture "5" graphed in the YZ plane from the database. In order to visualize the information present in the database, are recreated the gesture made in a real environment together with the representation of the hand in digital form from the LMC API.

### 2.3 Data Cleansing and Segmentation

Once the database is selected, the ML cycle starts with the detection of noise, missing values, null data, and inconsistent data. In addition, the Dataset must have the size, structure, and format for the defined architecture.

The 25 folders correspond to each of the participants. There are two files for each gesture within each folder, one for the right-hand capture and one for the left-hand capture. Sixty gestures were captured, and therefore, each folder contains 120 files. Each file contains a certain number of repetitions of the gesture. The spreadsheet file represents the gesture repetitions using frames (rows), each with more than 80 features (columns). The features are the coordinates of the reference points and other characteristics, such as time (Fig. 5). We carried out a frame-by-frame search looking for empty frames and coordinates of reference points that exceeded the average range of the multiple repetitions of the participants. We conclude the data cleansing by pruning these null frames.

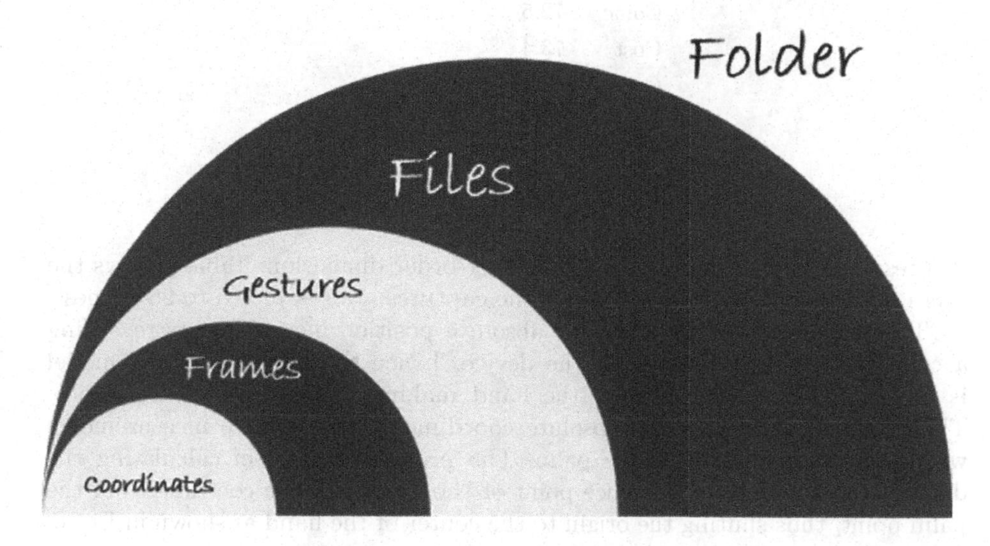


**Fig. 5.** Data hierarchy for comparison and filtering

As each capture contains multiple realizations of the gesture in each gesture file, we estimate the performed repetitions per capture from the total repetitions of each gesture in the 25 subjects, as shown in Table 3. We cut the information of each capture in equal proportions, according to each case.

**Table 3.** Partial view of gesture repetitions.

| Gesture | Total repetitions | Repetitions per capture |
|---|---|---|
| Cold | 345 | 13.8 |
| Eat | 296 | 11.84 |
| Cost | 290 | 11.6 |
| Cry | 280 | 11.2 |
| Hood | 278 | 11.12 |
| Deaf | 223 | 8.92 |

## 2.4 Data Transformation and Sampling

Since the time series of the gestures do not have the same dimension, either by duration or by sampling rate, we calculate the average frame rate of each gesture

**Table 4.** Partial view of the gestures average frame rate.

| Gesture | Average frame rate |
|---|---|
| Cold | 61.7 |
| Come | 72.5 |
| Cost | 73.4 |
| Cry | 78.6 |
| Dad | 77.4 |
| Deaf | 95.7 |

and resample all the captures to the higher-order dimension. Table 4 shows the average frames per gesture, by which the captures are resampled to 96 frames.

The data captured by LMC are absolute position measurements regarding a coordinate system located on the device, hence the behavior of the model is limited by the distance from the hand making the gesture to the device. Therefore, we transform the absolute coordinates into relative measurements with respect to a point in the palm. The process consists of calculating the difference between each reference point of the hand and the coordinates of the palm point, thus shifting the origin to the center of the hand as shown in Fig. 6.

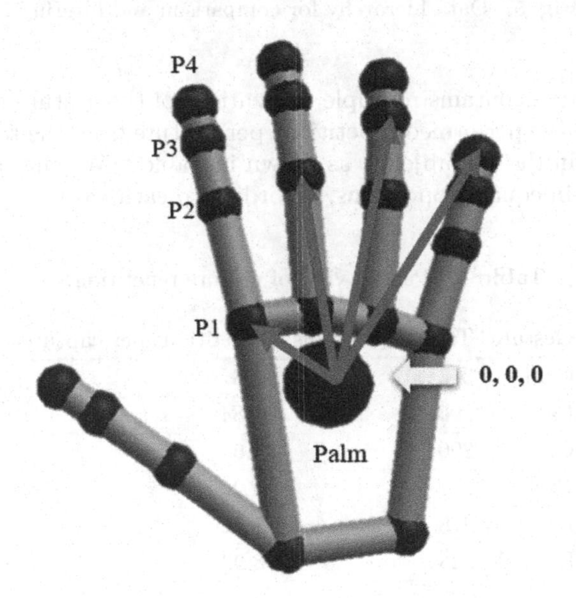


**Fig. 6.** Relative Measurements with respect to the palm.

In the original dataset structure, each column represents the X, Y, and Z coordinates independently for each of the 26 coordinate points of the hand, as shown in Table 5. Consequently, the data is restructured to the form of a time series, that means the X, Y, and Z positions of all the points are arranged contiguously at time t, as shown in Table 6.

**Table 5.** Original dataset structure.

| Frame 1 | | | | | | | |
|---|---|---|---|---|---|---|---|
| P1X | P1Y | P1Z | ... | Palm X | Palm Y | Palm Z | ... |
| X | X | X | ... | X | X | X | ... |
| X | X | X | ... | X | X | X | ... |
| X | X | X | ... | X | X | X | ... |

**Table 6.** Dataset proposed restructuring.

| Frame 1 | Frame 2 | ... | Frame 1 | Frame 2 | ... | Frame 1 | ... |
|---|---|---|---|---|---|---|---|
| P1X - Palm X | P1X - Palm X | ... | P1Y - Palm Y | P1Y - Palm Y | ... | P1Z - Palm Z | ... |
| X | X | X | ... | X | X | X | ... |
| X | X | X | ... | X | X | X | ... |
| X | X | X | ... | X | X | X | ... |

### 2.5 Feature Extraction

Due to the high number of data in the Original Dataset and the need for a data characterization with the most important features of the time-dependent signals, we perform the DWT. It represents the signal coming from the gesture as a nonlinear signal, generating invariance to the geometric transformations of time as a result. In doing so, we achieve a high degree of discriminability by decomposing the signal into frequency intervals, i.e., representing the same signals in coefficients with smaller amounts of data. Using DWT, we represent each temporal signal with a total of 118 features for all frames instead of having 82 data of characteristics for each frame used in the gesture.

### 2.6 Classification

A preliminary step in the classification process is to apply a Balancing Technique, in this case is Data Elimination Balancing Technique. Then, we train the model with the C# Visual Studio ML Model Builder Toolbox, an interface that automatically explores different algorithms and their ML configuration parameters.

**The classification algorithms selected are Stochastic Dual Coordinate Ascent (SDCA), Light Gradient Boosting Machine (LGBM), and FastTreeOva, due to Toolbox explore recommendation.** To evaluate the classification performance of the different algorithms, the experimental conditions of Data Restructuring, Data Balancing and frequency transformation proposed with DWT are tested. To give a reference to the influence of the proposed changes, the following experiments are carried out: 1) Training without Data Restructuring, without Data Balancing and without DWT 2) Training only with Data Restructuring 3) Training with Data Restructuring, without Data Balancing, with DWT 4) Training with Data Restructuring, with Data Balancing, and with DWT.

## 3 Results and Discussion

As a result of the data preprocessing and data characterization stages, we observe in Fig. 7 the process for 5 samples of the "0" gesture.

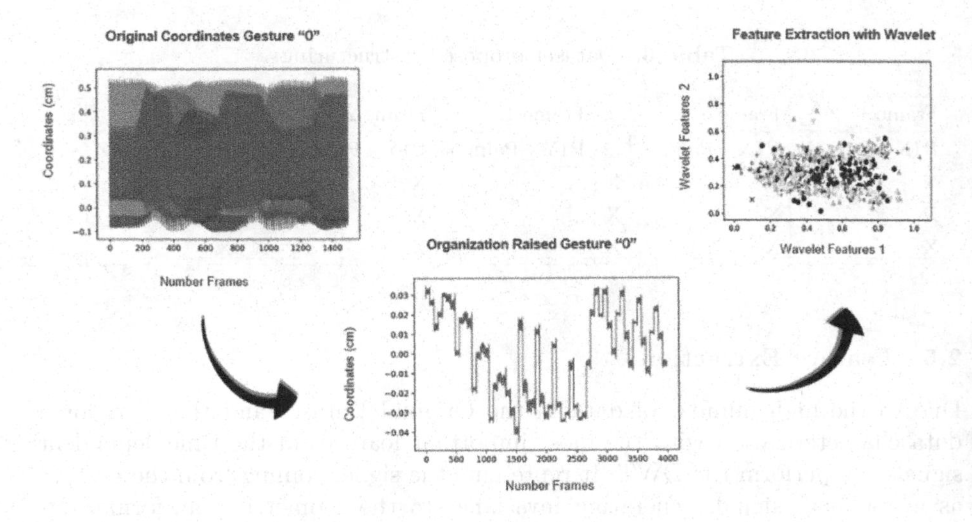


**Fig. 7.** Proposed dataset restructuring visualization.

All experiments realized in the classification stage are shown in Table 7.

**Table 7.** Comparison experiments.

| Classification performance | | | | Experiment conditions | | |
|---|---|---|---|---|---|---|
| Experiment number | SDCA | LGBM | Fast Tree | Data Restructuring | Data Balancing | DWT |
| 1 | **44.44%** | 22.22% | 22.22% | No | No | No |
| 2 | **63.64%** | 50% | – | Yes | No | No |
| 3 | 87.41% | **89.88%** | – | Yes | No | Yes |
| 4 | – | 97.71% | **97.96%** | Yes | Yes | Yes |

Initially, the database comprised three-dimensional coordinates of different hand reference points, with the coordinate system having the origin in the device. Data restructuring refers to changing the origin of the coordinate system to the center of the palm and reorganizing the data into a time series structure.

With the time series obtained, the experiments with different combinations are carried out, obtaining in Table 7 where the experiments with the most relevant results are indicated. Therefore, it is observed that in experiment 1 the classification of the database is performed without data restructuring, without data balancing, and without DWT, which allows to have as a reference the beginning of subsequent experiments. In the 2nd experiment the data restructuring condition is applied, showing better results with respect to experiment 1, given that the overall performance improved the classification is considered for subsequent experiments always apply data restructuring. In subsequent experiments(3 and 4) continue to apply different features obtaining better results to achieve 97.96% classification experiments are presented in the paper.

Finally, we compare the selected method with some methods from the reference study [13]. This paper employ the same Dataset with their propose models for database classification using DL and data balancing techniques that employ Data Augmentation (DA). These algorithms are DeepConvLSTM (+DA), ConvNet (+DA), ConvNet.

Once the method is selected witch is the EXPERIMENT 4 with FAST TREE Algorithm, we compare it with the models from the reference study. Table 8 shows the comparison between our proposed model based on conventional ML techniques and the reference models.

**Table 8.** Accuracy comparison between reference models and proposed model.

| Reference models | Accuracy |
|---|---|
| DeepConvLSTM(+DA) | 91.10% |
| ConvNet(+DA) | 89.30% |
| ConvNet | 87.40% |
| Proposed model and fast tree | 97.96% |

## 4 Conclusions

This study compares different ML models for classifying sign language. It builds on previous work to validate the proposed methodology and develops a low-resource system that implements time series and wavelet scattering delivering acceptable results. There are possibilities for future work to obtain new results, such as the support of sign language experts, automatic segmentation technique, and the possibility to realize the recognition of other sign languages.

Compared with DL network models [15], the use of conventional ML models and their easy-to-use algorithms avoids the excessive consumption of computational resources. Furthermore, ML models allows streamlined training based on good filtering and structuring of the data, as presented in the proposed methodology.

## References

1. World Health Organization. Deafness and hearing loss. https://www.who.int/news-room/fact-sheets/detail/deafness-and-hearing-loss. Accessed 27 May 2021
2. Al-Hammadi, M., et al.: Deep learning-based approach for sign language gesture recognition with efficient hand gesture representation. IEEE Access **8**, 192527–192542 (2020). https://doi.org/10.1109/ACCESS.2020.3032140
3. Cheok, M.J., Omar, Z., Jaward, M.H.: A review of hand gesture and sign language recognition techniques. Int. J. Mach. Learn. Cybern. **10**(1), 131–153 (2017). https://doi.org/10.1007/s13042-017-0705-5
4. Zafrulla, Z., Brashear, H., Starner, T., Hamilton, H., Presti, P.: American sign language recognition with the kinect. In: Proceedings of the 13th International Conference on Multimodal Interfaces, pp. 279–286, November 2011. https://doi.org/10.1145/2070481.2070532
5. Chong, T.W., Lee, B.G.: American sign language recognition using leap motion controller with a machine learning approach. Sensors **18**(10), 3554 (2018). https://doi.org/10.3390/s18103554
6. Weichert, F., Bachmann, D., Rudak, B., Fisseler, D.: Analysis of the accuracy and robustness of the leap motion controller. Sensors **13**(5), 6380–6393 (2013). https://doi.org/10.3390/s130506380
7. Lei, L., Dashun, Q.: Design of data-glove and Chinese sign language recognition system based on ARM9. In: 2015 12th IEEE International Conference on Electronic Measurement & Instruments (ICEMI), vol. 3, pp. 1130–1134. IEEE, July 2015. https://doi.org/10.1109/ICEMI.2015.7494440
8. Marin, G., Dominio, F., Zanuttigh, P.: Hand gesture recognition with leap motion and kinect devices. In: 2014 IEEE International Conference on Image Processing (ICIP), pp. 1565–1569. IEEE, October 2014. https://doi.org/10.1109/ICIP.2014.7025313
9. Shin, H., Kim, W.J., Jang, K.A.: Korean sign language recognition based on image and convolution neural network. In: Proceedings of the 2nd International Conference on Image and Graphics Processing, pp. 52–55, February 2019. https://doi.org/10.1145/3313950.3313967
10. Weerasekera, C.S., Jaward, M.H., Kamrani, N.: Robust asl fingerspelling recognition using local binary patterns and geometric features. In: 2013 International Conference on Digital Image Computing: Techniques and Applications (DICTA), pp. 1–8. IEEE, November 2013. https://doi.org/10.1109/DICTA.2013.6691521
11. Ravi, S., Suman, M., Kishore, P.V.V., Kumar, E.K., Kumar, M.T.K., et al.: Multi modal spatio temporal cotrained CNNs with single modal testing on RGB-D based sign language gesture recognition. J. Comput. Lang. **52**, 88–102 (2019). https://doi.org/10.1016/j.cola.2019.04.002
12. Su, Y., Qing, Z.: Continuous Chinese sign language recognition with CNN-LSTM. In: Proceedings of SPIE 10420, Ninth International Conference on Digital Image Processing (ICDIP 2017), 104200F, 21 July 2017. https://doi.org/10.1117/12.2281671
13. Hernandez, V., Suzuki, T., Venture, G.: Convolutional and recurrent neural network for human activity recognition: Application on American sign language. PLoS ONE **15**(2), e0228869 (2020). https://doi.org/10.1371/journal.pone.0228869

14. Shanmuganathan, V., Yesudhas, H.R., Khan, M.S., Khari, M., Gandomi, A.H.: R-CNN and wavelet feature extraction for hand gesture recognition with EMG signals. Neural Comput. Appl. **32**(21), 16723–16736 (2020). https://doi.org/10.1007/s00521-020-05349-w
15. Hernandez, V., Suzuki, T., Venture, G.: American Sign Language classification - LeapMotion - 25 subjects - 60 signs. Mendeley Data. V1 (2018). https://doi.org/10.17632/8yyp7gbg6z.1
16. Vysocký, A., Grushko, S., Oščádal, P., Kot, T.; Babjak, J., Jánoš, R., Sukop, M., Bobovský, Z.: Analysis of precision and stability of hand tracking with leap motion sensor. Sensors 2020, 20, 4088 (2020) https://doi.org/10.3390/s20154088

# Solar Radiation Prediction Using Machine Learning Techniques

Luis Alejandro Caycedo Villalobos(✉),
Richard Alexander Cortázar Forero(✉), Pedro Miguel Cano Perdomo(✉),
and José John Fredy González Veloza(✉)

Fundación Universitaria Los Libertadores, Bogotá, Colombia
{lacaicedov,racortazarf,pmcanop,jjgonzalezv02}@libertadores.edu.co

**Abstract.** The proposal of a solar radiation estimation model using Machine Learning is submit, the processing of meteorological data measured by satellite and data measured on the land is made. The model uses two solutions using an artificial neural network and robust linear regression the climatic variables used as input to the model are solar radiation, temperature and clarity index, all get from satellite data. The main aim of this work is to propose a model that allows using the satellite data to get an estimate of the behavior of the solar resource on the ground, reducing the error between the satellite data and the data measured on the ground. The results of the model got by training an artificial neural network with hidden layers are submit, here the normal distributions of the data reported by the satellite and the data got by the proposed model are submit. In addition, the results of the daily average got by the model and the daily average values measured on land are submit. I conclude it by proposing a second estimation model using robust linear regression. A proposed model adjusted to the assumptions made during the regression process and acceptable results to those got by the satellite and reported by other works are got.

**Keywords:** Estimate · Solar radiation · Machine learning

## 1 Introduction

Solar radiation is the energy emitted by the sun. It propagates as electromagnetic waves through space and reaches the earth's surface after passing through the atmosphere; consider the most abundant and important source of energy for life on the Earth causes natural phenomena and chemical reactions for the growth of plants and animals. The energy captured by the earth from the sun is the primary source of renewable energy that we have available, the amount of energy that is captured annually is approximately 1.6 million KWh, which is much more than what it consumed worldwide [1]. In order to reduce dependence on electrical energy produced by non-renewable sources, the use of solar energy has gained

Supported by organization Fundación Universitaria Los Libertadores.


H. Florez and M. F. Pollo-Cattaneo (Eds.): ICAI 2021, CCIS 1455, pp. 68–81, 2021.
https://doi.org/10.1007/978-3-030-89654-6_6

great importance in recent years, proof of this is the increase in photovoltaic - PV solar systems as part of the generation of electrical energy, according to Enerdata in 2019, the generation of electrical energy from solar energy had an increase of 24% compared to 2018 [2]. With the current high demand for electricity, there has been great interest worldwide to integrate solar energy systems into the electricity grid, in order to improve the quality supplied in some places and reduce the costs associated with dependence on the grid conventional electric. However, implementing these systems presents an obstacle. The dimensioning of a PV system is closely related to the climatic conditions [3], for which it required reliable data sources that allow making a forecast of the solar resource.

Thanks to the Institute of Hydrology, Meteorology and Environmental Studies (IDEAM), there is irradiance data measured in different meteorological stations in Colombia, but not just any user has access to this data, so it is necessary to have an additional source that allows get reliable data that provide information for estimating the values on land [5]. An alternative is the access that users have to the data from NASA's PowerViewer database, but in order for it to be used, the error in the measurement of solar radiation taken by satellite and recorded in the database must be reduce data [6]. The present work proposes two linear regression models of the daily average radiation reported by the satellite, getting a normal distribution model that predicts the values of daily radiation with its maximums and minimums, taking as a reference the values recorded by IDEAM in the different stations weather of Colombia. For this reason, it proposed two models that can be use as a planning tool for the design of PV systems, for which it made use of the Python and R programming language.?

This paper is organized [14] as follows: Sect. 2 related work; Sect. 3 Estimation model using Machine learning; Sect. 4 Estimation model results using Machine learning. Conslusions are presented in the last section

## 2 Related Work

Solar radiation prediction provides the ability to optimize the operation of solar-powered systems, improve quality, optimize production costs, and estimate the amount of energy that the system can deliver. Several studies similar to those conducted in this work have conducted previously.

In [4] the author used meteorological data stored in GRIB and netCDF files to generate a predictive model in Python using the sklearn library, specifically the linear regression method between different irradiance data used, in which a coefficient of determination of 0.82 and a mean absolute error of 4.16.

In [5] a statistical model generated for the estimation of solar radiation applying alternate meteorological data. IDEAM provided the data used for the study, from a station that has the measurement of solar radiation, temperature, relative humidity and hour of sunshine. Angstrom-Prescott and Gueymard statistical regressions used, in which it evidenced little relationship between the variables, the best fit achieved was through the linear regression between relative humidity and solar radiation with a coefficient of determination of 11.14%.

In [6] the author proposes a hybrid predictor model implemented in classification-grouping stages using Fuzzy Logic and for the estimation of Neural Networks and State Vector Machines, the main aim was to create a model that considered the conditions geographical areas of Colombia. The input variables to the model included temperature, wind speed, clarity index, total precipitation, relative humidity, and atmospheric pressure, all got from NASA's PowerViewer database. The correlation coefficients got by this model are close to 1 and the mean square error (RMSE) is between 0.04 and 0.09, showing that the model shows an acceptable performance in all the cities evaluated.

In [7] a model proposed to predict hourly solar radiation using linear regression and neural networks. Solar radiation got with mathematical models, temperature, atmospheric pressure and relative humidity used as input variables to the model. They compared the applied models to determine which one fitted the best with data from five meteorological stations in Tucumán, Argentina. The results got show that an average error of 11% achieved with linear regression and 7.84% with neural networks.

## 3 Estimation Model Using Machine Learning

The present work proposes a model that allows reducing the error in the measurement of solar radiation taken from the PowerViewer database, taking as a reference the data recorded by IDEAM in its meteorological stations near the geographic region of interest, getting from this proposal a normal distribution model for solar radiation and its standard deviation on a particular day and a linear regression model for estimating the daily mean value. The meteorological data used to generate the model proposed in this work were get from the database of the Institute of Hydrology, Meteorology and Environmental Studies (IDEAM) and from the NASA PowerViewer database, which are contained in CSV files and netCDF, respectively. The data from measurements of solar radiation, clarity index and temperature in the meteorological stations in the principal cities of Colombia is to have as a reference. The time window used in this job is between December 4, 2014 to August 31, 2020. It developed the proposed models from two different Machine Learning concepts and integrate their results to propose the estimation of solar radiation on the ground from the data recorded by the satellite and available in the PowerViewer database. The first model seeks to get through artificial neural networks - ANN the estimation of the normal distribution of solar radiation on the ground, which allows having average, maximum and minimum values during a day. The second model seeks to estimate through robust linear regression the daily average value of solar radiation on the ground from the data recorded by the satellite. The data of measurements on the ground provided by the meteorology station at the Jose Maria Cordova airport in the city of Medellin - Colombia by IDEAM, are contain in CSV files and through the PANDAS python library it is possible to visualize, analyze and process sets of data that are known as data frame. The data frame is reporte in hourly mode and when preprocessed it to have to daily average values to be compared with the satellite data file, the records in which the equipment did not capture information are eliminate, the rows with records equal to 0, values that correspond to the hours where there is no incident solar radiation on the equipment; It filtered data using a moving average filter to smooth fluctuations produced by the data recording device. The data available in the PowerViewer database, the official page of NASA's Powerviewer and consulted from the page http://power.larc.nasa.gov/data-Access-viewer/, are contain in a netCDF file that contains the metadata with all the meteorological variables registered, in this case the python netCDF4 library is used to open and read this dataset. The database built with data from the two sources, IDEAM and PowerViewer, comprises 1638 records reported between 2014 and 2020. It described the fields for each record in Table 1. They performed the analysis of the behavior of each variable using the R program. The graphical results are presented in Figs. 1, 2 and 3. Figure 1 shows the extreme, atypical data with left bias affecting the measures of central tendency. In Fig. 2 the atypical data and left bias observed for the variables Sat and RH2M, the variables of precipitation (PRECTOT) and percentage of error between the measurement on the ground and the satellite, extreme values observed that skews the data to the right.

**Table 1.** Database record fields

| Variable | Description | Units |
|---|---|---|
| Data | average daily radiation measured in surface | $W/m^2$ |
| MM | moving average applied to data | $w/m^2$ |
| Daily | daily average radiation measured on surface | $KW/m^2$ |
| Data_filtered_mm | moving average of Data | $KW/m^2$ |
| Sat | average radiation reported by the satellite | $KW/m^2$ |
| error_sat | error percentage between Sat and Daily | |
| Prectot | precipitación | mm |
| RH2M | Relative humidity at 2 m height | |
| T2M_range | temperature range at 2 m height | °C |
| Ws50M_range | wind speed range at 50 m | m/s |
| KT | lightness or insolation index | |

In Fig. 3 the data corresponding to the temperature range with the least amount of atypical data are to observe, the variable range of wind speeds presents the highest atypical data and for the variable of the clarity index the extreme, atypical data they are minimal. The correlations between the variables described in Table 1 are present in Fig. 4, there it can be observe very high correlations between several variables, which show a great tendency to linearity between them, without However, it must be clarified that the perfect correlations between variables are present because there are redundant data in the database that present the same variables but in different units. As of the variables

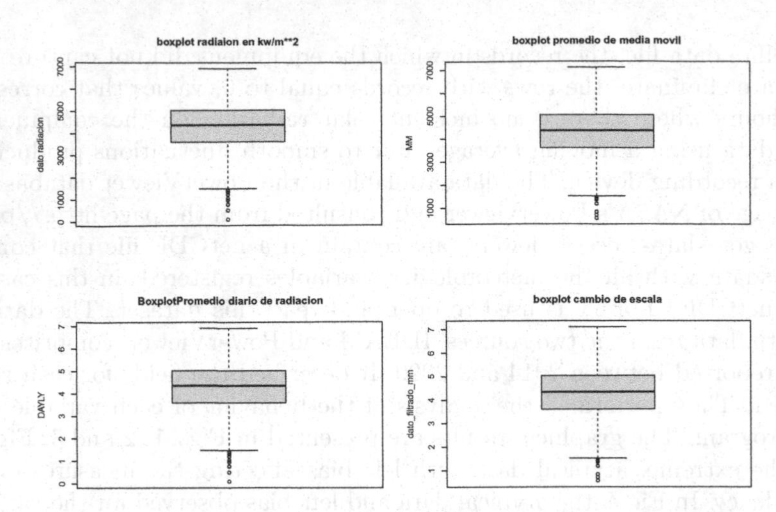


**Fig. 1.** Ground radiation analysis, moving average, daily ground radiation average, scale change

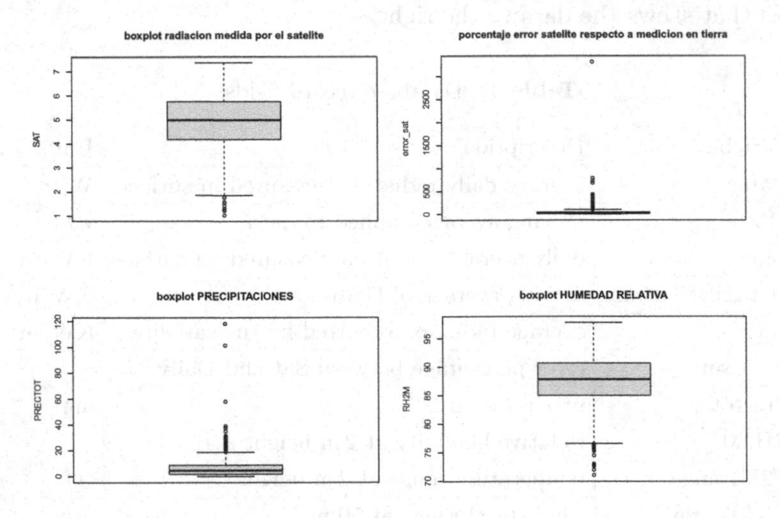


**Fig. 2.** Analysis of mean radiation by satellite, satellite error with respect to measurement on the ground, precipitation, relative humidity

$DATA-DAILY, MM-Filtereddata$, and there is a strong correlation between the variables $SAT-KT$, although in this case not It happens for the same reason as in the previous clarification. These variables are closely related and there is a strong tendency towards linearity between them. Also, there is a great correlation and therefore a linear trend with the variables $RH2M-T2MRANGE$, with the exception that it is an inverse linear trend. In [6] get evidenced that solar radiation presents a high correlation with other climatic variables, especially with temperature and clarity index, for which the input variables in the

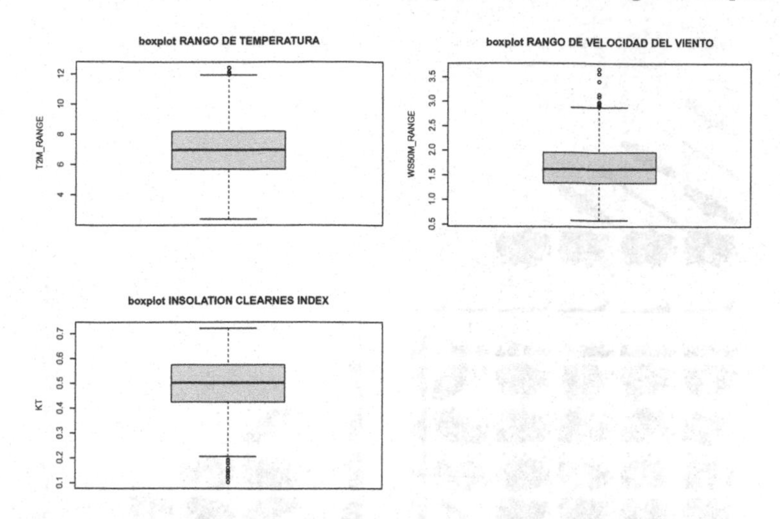


**Fig. 3.** Analysis of temperature range, wind speed range, clarity index

proposed models will be together with the solar radiation taken by the satellite. The prediction model presented in this work proposes the use of an artificial neural network - ANN, in problems where it is require making a prediction in time series using climatic variables as input to the model, it has been the ANN which provide a better response compared to other methods [12], in [13] it developed a model for forecasting the price of electrical energy by using ANN with a multilayer perceptron configuration. This architecture has proven to be one of the most useful to solve this type of problem. According to the information consulted in [8], the architecture most used in solar energy prediction applications is the multilayer perceptron, which is the most popular non-linear ANN architecture used to solve applied science problems. For this reason, the model selected for estimating solar radiation was an ANN with simple multilayer perceptron architecture [12,13]. The structure of the proposed ANN corresponds to: a. input layer: different climatic variables, b. hidden layer: contains the activation function, c. output layer: target variable. It represented the described structure in Fig. 5. According to the information consulted in [3], it shown that solar radiation presents a high correlation in the data recorded by the satellite for the temperature variables, the clarity index, for which these climatic variables used together with the solar radiation taken by the satellite as the predictor characteristics represented as [X] of the ANN. The target variable represented as [Y] is the average solar radiation measured on the ground. The next part of the process comprises creating data sets, which correspond to a training set, equivalent to 70% and a test set equivalent to 30% of the selected data.

To estimate the average of the solar radiation on the surface, a Machine Learning model proposed with the robust linear regression technique (rlm), because with the multiple regression technique (lm), when doing the analysis for the residuals, the model did not comply with the assumptions of normality,

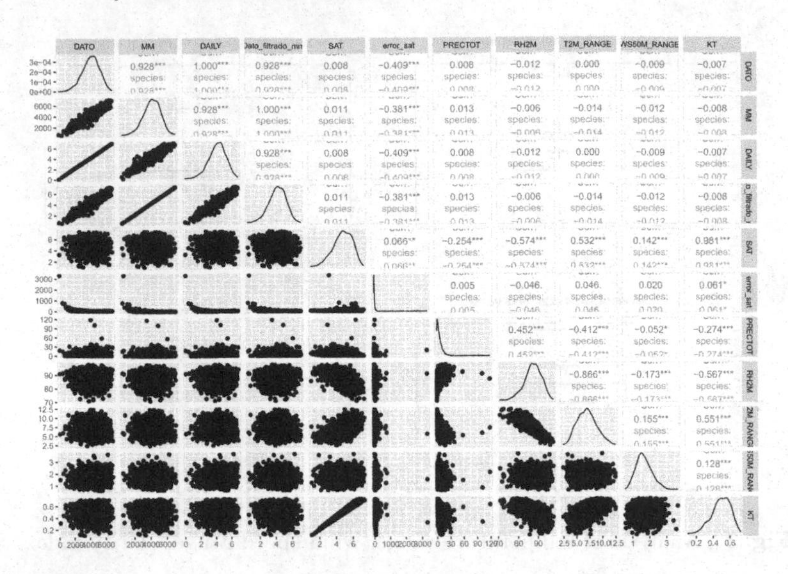


**Fig. 4.** Scatter diagram and correlation of the variables used

homoscedasticity, independence of residuals and multicollinearity. For the model (lm) it got that the significant variables for the regression were five: satellite radiation (SAT), percentage error of the solar radiation measurement between the station and satellite measurement (*errorsat*), relative humidity (RH2M), temperature range (*T2MRANGE*) and lightness index (KT). Although the (*lm*) towards a good forecast for the estimated radiation values with mean square errors (RSME) that oscillated between 1.1 $KW/m^2$ and 0.87 $KW/m^2$ as Outliers were being treated, with the particularity that each time the RSME improved,

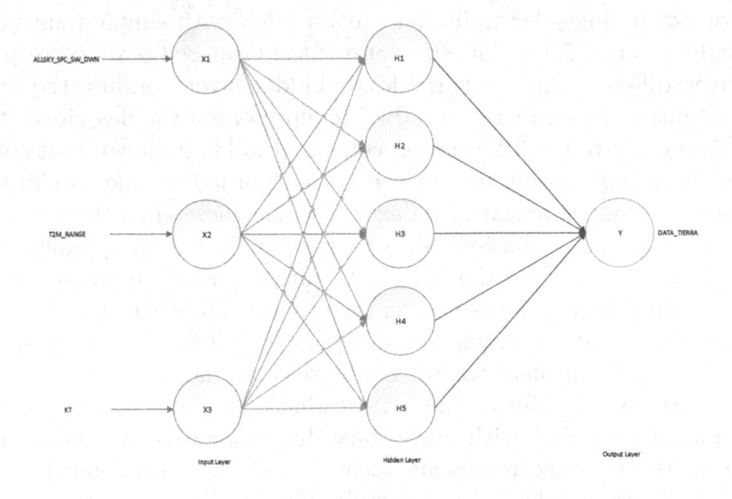


**Fig. 5.** Proposed multilayer perceptron neural network structure

the problem of non-compliance with assumptions became more acute, especially the assumption of normality. To correct these problems, it made a modification to the variable (*errorsat*), applying the logarithm function, since this variable is the one with the greatest bias because of the maximum extreme data it has. It also made a transformation to the response variable (DAILY), with the boxcox function, raising the variable to the exponent (1, 4). These transformations to the variables evidently improved the fulfillment of the assumptions, although the normality test continued to fail. As an additional procedure trying to make adjustments to the model, the variable (*errorsat*) eliminated, and the variable (DAILY) left with the exponent suggested by the boxcox function. This change guaranteed the fulfillment of the assumptions, but seriously affected the predictions made by the model, and increased my RSME from 0.87 $\mathrm{KW/m^2} to$ 1.7 $\mathrm{KW/m^2}$, this increase in RSME It attributed to the fact that for the multiple regression model, the discarded variable had a significance greater than 95%. After doing all these tests, we opted for the development of a robust regression model, because this type of regression is less affected by extreme values. For the creation of the robust regression model, the same variables of the multiple linear regression used and besides these variables included: average moving average of radiation (MM), precipitation (PRECTOT) and range of wind speeds ($WS50MRANGE$). considering the T tests, a change made to the intercept because it is not significant, I also eliminate the (rlm), some variables that can affect the model because of multicollinearity problems, because they present the same information but in different units, This is the case of the variable: solar radiation (DATA), which contains the same information as the response variable (DAILY), except that they have different units. A similar situation occurs with the variable moving average of radiation (MM) and filtered data of moving average of radiation ($Data\_filtered\_mm$). For this reason, the variables MM, SAT, *errorsat*, PRECTOT, RH2M, $T2MRANGE$, $WS50MRANGE$, KT taken. In the residuals histogram of (rlm), Fig. 6, a normal trend observed, which shows signs of normality of the residuals, Fig. 7, the same observed in Fig. 8 normal QQ with confidence intervals. According to the tests of: Kolmogorov-Smirnov, shapiro-Wilk, Ks.test the p-value = 0.2655 with a 95% confidence there is not enough statistical evidence to reject the null hypothesis in favor of the alternative hypothesis.

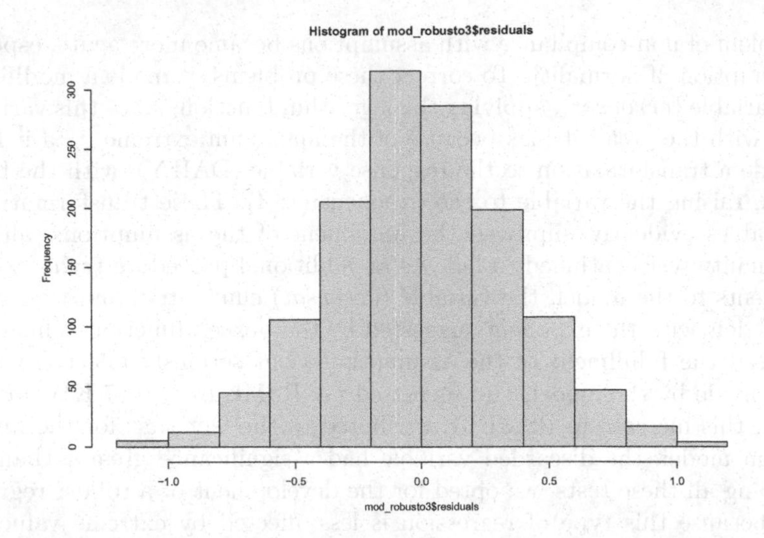


**Fig. 6.** Histogram of residuals for the robust linear regression model

The proposed model (rlm) complies with the assumption of normality, if the robust regression model (rlm) fulfills the assumption of normality, it shows that linear regression is a good choice as a model to predict solar radiation on the surface using environmental variables recorded by meteorological stations. To validate the assumption of homoscedasticity, the ncvTest (robust mod3) of the model performed, getting a p-value = 0.14579, with a 95% confidence, there is not enough statistical evidence to reject the null hypothesis in favor of the alter-

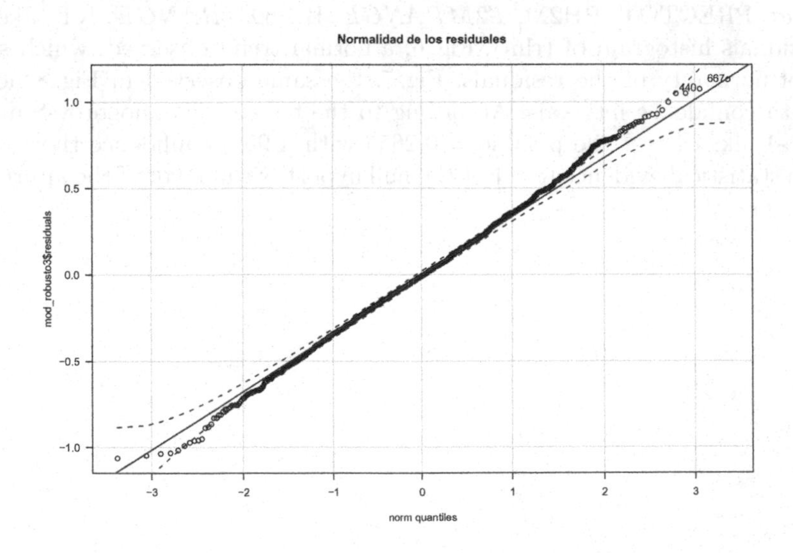


**Fig. 7.** Normality of the residuals of the proposed robust linear regression model

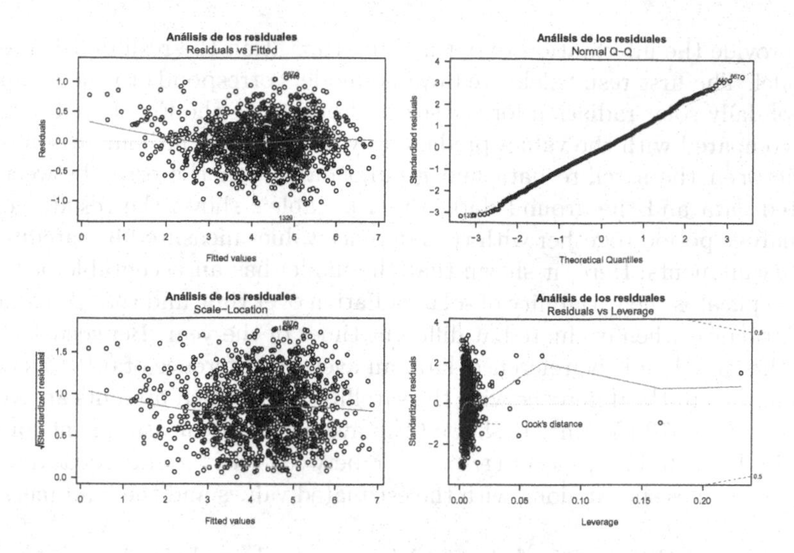


**Fig. 8.** Analysis of residuals for the proposed robust linear regression model. to. Residual Vs Fitted, b. normal Q-Q, c. Scale Vs Location, d. Residual Vs Leverage.

native hypothesis, the proposed model complies with the assumption of constant variance. The model is homoelastic, which guarantees that the regression is adequate for forecasting. It is important that someone distribute constantly the variance in the errors throughout all the observations. This characteristic in a regression increases the efficiency and reliability of the model. To validate the independence assumption, the test performed with the Durbin Watson statistic, the DW statistic is close to two (DW = 2.365) which is a good indicator and there is a p-value = 1, with a confidence of 95%, there is not enough statistical evidence to reject the null hypothesis in favor of the alternative hypothesis. The proposed model fulfills the assumption of independence. Failure to comply with the autocorrelation error test for the model can explained because of the phenomenon of solar radiation and the other associated variables, it has a space-time domain, which was not used in the analysis and subsequent estimation of the values of daily solar radiation. Using the temporal variable recommended for later work, in order to improve the estimation and reliability of the model. Because of the randomness of this natural phenomenon, there is a spatiotemporal correlation, which gives indications that from data analytics without considering physical linear models, it can become more efficient to forecast a model in time series.

## 4 Estimation Model Results Using Machine Learning

The results got from the estimation model using ANN with multilayer perceptron configuration between the target variable and the predictor variables carried out for one year, divided into sub-periods of four months, gave the following results, which provide the information to optimal interpretation the results delivered by the model. The first result delivered by the model corresponds to the estimated values of daily solar radiation for the selected period, in the Fig. 9 are the target values compared with the values predicted by the model. The comparison of the error between the satellite data and ground data with the error between the predicted data and the ground data, Fig. 10. Table 2 shows the results got for each analysis period together with the reference values measured by satellite and land measurements. Here, it shown that the model has an acceptable behavior, since it replicates the variability of solar radiation over time and that it maintains its performance when evaluated at different times of the year. Between the data got by the model and the reference data, an average difference of 0.1916 kW/m$^2$ got, compared to the difference with the satellite data, which presents an average difference of 0.5879 kW/m$^2$. It showed the average error measurement for each period in Table 3. The percentage of error between the satellite-measured and land-based values shown along with the estimated values and the land-measured values.

It presented the results of the model proposed by robust linear regression in Fig. 11. The RSME value for the estimates made by the model is 0.3927409 KW/m$^2$, for this case, this value shows that the predicted value is around more or less 0.4 KW/m$^2$ of the real value measured by surface stations. Based on the results presented, it was possible to show that the robust regression complies with the assumptions of normality, homoscedasticity and independence, and cannot comply with the assumption of autocorrelation errors. This explained because of the phenomenon that has a space-time domain, which was not used in the anal-

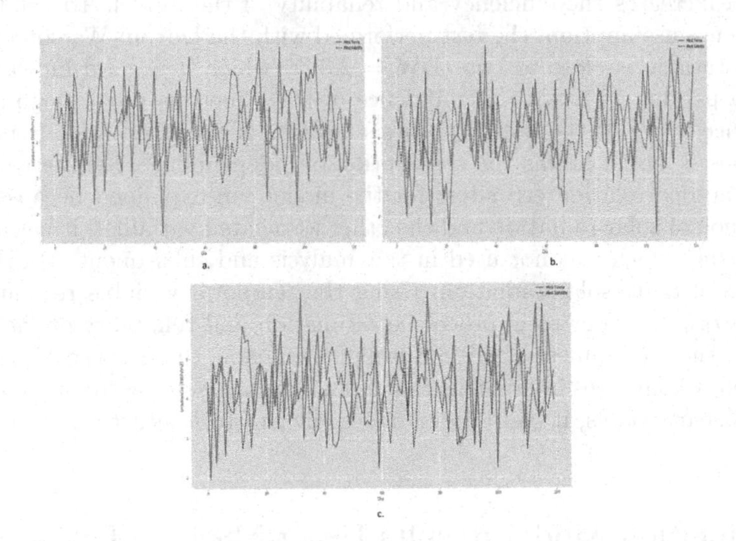


**Fig. 9.** ANN Model, target values compared to model predicted values separated in 4-month periods over a year. a) Radiation data for four-month period 1- Julian day 1–120. b) Radiation data four-month period 2- Julian day 120–240. c) Radiation data four-month period 3- Julian day 240–360

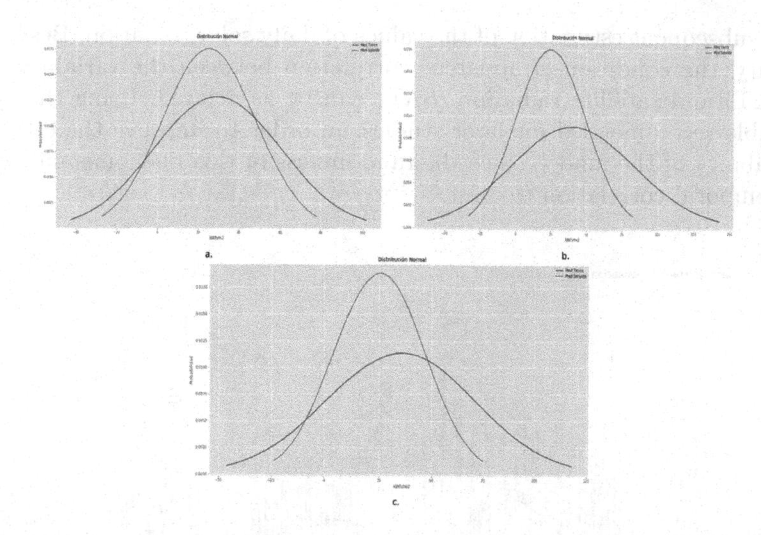

**Fig. 10.** ANN Model, comparison of the normal distribution of the estimated data, separated into 4-month periods over a year. a) Comparison of the normal distribution for four months 1- Julian day 1–120. b) Comparison of the normal distribution four-month period 2- Julian day 120–240. c) Comparison of the normal distribution four-month period 3- Julian day 240–360.

**Table 2.** Results of the estimation model - ANN in the evaluated time periods

| | First four months | Second four months | Third four months |
|---|---|---|---|
| Measurement | Average in kW/m$^2$ | Average in KW/m$^2$ | Average in KW/mm$^2$ |
| Land | 4.572559 | 4.247772 | 4.434122 |
| Satellite | 4.640138 | 5.305578 | 5.063395 |
| Estimation - RNA | 4.297978 | 4.035805 | 4.255731 |
| Difference (earth-satellite) | 0.067579 | 1.057806 | 0.629273 |
| Difference (earth-estimate RNA) | 0.274581 | 0.211967 | 0.178391 |

**Table 3.** Comparative average error of the ANN model results in the evaluated time periods

| | First four months | Second four months | Third four months |
|---|---|---|---|
| Comparative | Average error (%) | Average error (%) | Average error (%) |
| Satellite - land | 29.268876 | 40.742404 | 35.88887 |
| estimate - land | 25.389562 | 25.141444 | 25.619838 |
| Decrease | 3.879314 | 15.60096 | 10.269049 |

ysis and subsequent estimation of the values of daily solar radiation. Besides this peculiarity, the condition of apparent correlation between the variables clarity index (KT) and satellite radiation (SAT) can be associated. Using the temporal variable recommended for later studies, in order to improve the estimation and reliability of the model, since the randomness of this phenomenon causes a spatio-temporal correlation to exist.

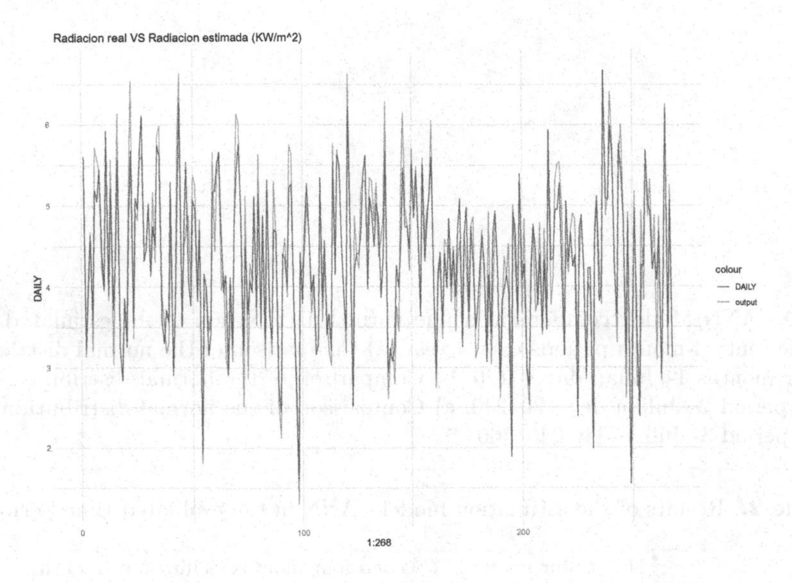


**Fig. 11.** Result of the robust linear regression model Vs real values of solar radiation for one year. In red the actual data of the average radiation measured by the meteorological station are represented. The data estimated by the proposed robust linear regression model are represented in Blue (Color figure online)

## 5 Conclusions

The work presented, proposed a model for estimating direct solar radiation using data consulted from the NASA PowerViewer database, getting an error rate between the proposed model and the values measured on the ground of 25.61%. Which is less than the 40.74% taken by the satellite. The data used for the development of the model selected according to the correlation they present with solar radiation, these data correspond to the clarity index and temperature, provided by the PowerViewer database, the best fit achieved allowed to have an Average error between the aim variable and the estimated data of 25.38%, regarding the error of the satellite data corresponding to 40.74%, which shows that an average decrease of 9.91% achieved. The model captures the variability of the solar resource and maintains a trend towards the average value of the target data, getting an average error in the estimate of +/− 216.6 $W/m^2$ that is lower than that got by the satellite corresponding to 581.3 $W/m^2$, which is considered an improvement in the value of solar radiation calculated using the proposed model. The RSME for the estimates made by the robust regression model is 0.3632, which is approximately equivalent to an error value in the estimated radiation of 0.6 KW/$m^2$, compared to other works that do the validation of the model with the same metric [11]. Therefore, this confirms that the choice of the robust regression model to estimate the solar radiation on the surface was a wise decision.

## References


1. Secretaria de Energía Argentina (2008) - Coordinación de Energías Renovables - Dirección Nacional de Promoción - Subsecretaría de Energía Eléctrica. Energías Renovables 2008 - Energía Solar
2. Enerdata - Estadísticas Energéticas Mundiales. (11 de Abril del 2021). Anuario estadístico mundial 2020. https://datos.enerdata.net/energias-renovables/eolica-solar-produccion.html
3. Obando-Paredes, E., Vargas-Cañas, R.: Desempeño de un sistema fotovoltaico autónomo frente a condiciones medioambientales de una región en particular, Rev. la Acad. Colombia. Ciencias Exactas, Físicas y Nat. **40**(154), 27–33 (2016)
4. Bella-Santos, J.: Herramientas python para la predicción de energías renovables (trabajo de pregrado). Universidad Autónoma de Madrid, Madrid, España (2018)
5. Vélez-Pereira, A., Vergara, E., Barraza, W., Agudelo, D.: Determinación de un modelo paramétrico para estimar la radiación solar. Ingenium **7**(18), 11–17 (2013)
6. Obando Paredes, E. Modelo de pronóstico de radiación solar basado en Machine Learning. 2018
7. Jimenez, V., Will, A., Rodriguez, S. Estimación de Radiación Solar Horaria Utilizando Modelos Empíricos y Redes Neuronales Artificiales. Ciencia y Tecnología. 1. 10.18682/cyt.v1i17.608. 2017
8. Tymvios. Filippos S, Michaelides. Silas, Chr. Skouteli, Chara S. Estimation of Surface Solar Radiation with Artificial Neural Nerworks. Modeling Solar Radiation at the Earth's Surface. Springer - Verlag Berlin Heidelberg. 2008
9. Boland, J.: Time Series Modelling of Solar Radiation. Springer - Verlag Berlin Heidelberg, Modeling Solar Radiation at the Earth's Surface (2008)
10. Mora-López, L.: A new Procedure to Generate Solar RAdiation Time Series from Machine Learning Theory. Modeling Solar Radiation at the Earth's Surface. Springer, Heidelberg (2008)
11. Ordoñez-Palacios, L.-E., León-Vargas, D.-A., Bucheli-Guerrero, V.-A., Ordoñez-Eraso, H.-A.: Solar Radiation Prediction on Photovoltaic Systems Using Machine Learning Techniques. Revista Facultad de Ingeniería, vol. 29 (54), e11751 (2020). https://doi.org/10.19053/01211129.v29.n54.2020.11751
12. Brentan, B., et al.: Hourly water demand forecasting using nonlinear autoregressive with exogenous artificial neural networks -narx (2015)
13. Villada, F., Cadavid, D.R., Molina, J.D.: Electricity price forecasting using artificial neural networks. Revista Facultad De Ingeniería Universidad De Antioquia **44**, 111–118 (2014)
14. Misra, S.: A step by step guide for choosing project topics and writing research papers in ICT Related Disciplines. In: Information and Communication Technology and Applications, Third International Conference, ICTA 2020, Minna, Nigeria, 24–27 November 2020

# Towards More Robust GNN Training with Graph Normalization for GraphSAINT

Yuying Wang(✉) and Qinfen Hao

Institute of Computing Technology, Chinese Academy of Sciences, Beijing, China
wangyuying@ict.ac.cn

**Abstract.** Graph Neural Networks (GNNs) field has a dramatically development nowadays due to the strong representation ability for data in non-Euclidean space, such as graphs. However, with the larger graph datasets and the trend of more complex algorithms, the stability problem appears during model training. For example, GraphSAINT algorithm will not converge in training with a probability range from 0.1 to 0.4. In order to solve this problem, this paper proposes an improved GraphSAINT method. Firstly, a proper graph normalization strategy is introduced into the model as a neural network layer. Secondly, the structure of the model is modified based on the normalization strategy to normalize the original input data and the input data of the middle layer. Thirdly, the training process and the inference process of the model are adjusted to fit this normalization strategy. The improved GraphSAINT method successfully eliminates the instability and improves the robustness during training. Besides, it accelerates the training procedure convergence of the GraphSAINT algorithm and reduces the training time by about a quarter. Furthermore, it also achieves an improvement in the prediction accuracy. The effectiveness of the improved method is verified by using the citation dataset of Open Graph Benchmark (OGB).

**Keywords:** Graph Neural Networks (GNNs) · Graph normalization · Training stability · Link prediction

## 1 Introduction

The Graph Neural Networks (GNNs), which was firstly proposed in 2009 [1], has been developed rapidly in recent years due to the powerful processing ability for data in non-Euclidean space, for example the graph data [2–6]. Nowadays, GNNs are widely used in many areas such as social networks [7, 8], drug discovery [9] and recommendation [10, 11]. As mentioned in [12], there have been mainly four GNNs categories so far including Recurrent Graph Neural Networks (RecGNNs) [13, 14], Convolutional Graph Neural Networks (ConvGNNs) [15–23], Graph Autoencoders (GAEs) [24–27] and Spatial-temporal Graph Neural Networks (STGNNs) [28–30]. Among them, ConvGNNs realize the generalization of the convolution from grid data to graph data, whose typical model is Graph Convolution Network (GCN) [31, 32].

In order to improve the generalization performance of GCN for new nodes, the Graph SAmpling based INductive learning meThod (GraphSAINT) algorithm [33] is proposed.


H. Florez and M. F. Pollo-Cattaneo (Eds.): ICAI 2021, CCIS 1455, pp. 82–93, 2021.
https://doi.org/10.1007/978-3-030-89654-6_7

GraphSAINT can realize the effective training of the deep GCNs by using a special mini-batch construction way. This algorithm obtains a set of subgraphs by sampling the original training graph and then builds a GCN based on the subgraphs. Therefore, the graph sampling strategy is the main contribution of GraphSAINT. Besides, this strategy also alleviates the problem of the neighbor explosion, so that the number of neighboring nodes no longer increases exponentially with the number of layers in GraphSAINT. Moreover, compared with GraphSAGE [34], GraphSAINT also enhances the processing capability for large graphs by applying the subgraph sampling method. Therefore, although GraphSAINT uses the same inductive framework as GraphSAGE, GraphSAINT and GraphSAGE are different on sampling method: GraphSAINT samples multiple subgraphs from original dataset to construct minibatch for training; while GraphSAGE adopts the neighbor-node sampling method to generate the node embeddings.

Although GraphSAINT solves the problem of neighbor explosion and has the stronger generalization ability than GCN, the use of the graph sampling strategy makes it difficult training the network. In general, nodes which have the higher influence on each other should be selected to form a subgraph with the higher probability. This ensures that the nodes can support each other within the subgraph. However, such sampling strategy leads to different node sampling probabilities and introduces bias in the mini-batch estimator. In [33], the normalization techniques are developed to deal with this issue so that the feature learning will not give priority to the more frequently sampled nodes. As a result, GraphSAINT effectively solves the problems of instability and non-convergence faced in the training process, and obtains a good performance improvement on the classification task.

However, experiments show that the stability problem of GraphSAINT in the training process appears when GraphSAINT is applied to solve the link prediction task [35–38] in different application areas [39–43] on the citation dataset of the standard Open Graph Benchmark (OGB) [44]. Link prediction is widely used in many aeras such as recommendation system [45–47], biological networks [48] and knowledge graph completion [49]. Different with node classification, the main task of link prediction is to judge whether two nodes in a network are likely to have a link. This stability problem in the training process means that the normalization techniques in [33] are insufficient to improve the training quality for the link prediction task. Thus, it is difficult to avoid falling into non-convergence during training, which appears that the value of the training loss suddenly rises and remains forever unchanged. This is a problem that has a big impact on the GNN model development.

From the analysis of the CNN training method, we had some new discovery in training stability. Stochastic gradient descent and its variants such as momentum [50] and Adagrad [51] have been widely used to train the neural networks. The training process is complicated. Besides, as the network gets deeper, small changes to the network parameters will amplify [52]. In the process of constantly adapting to the new distribution, the distributions of layers' inputs present a problem called covariate shift [53–58], which is harmful for the neural network convergence. Ioffe and Szegedy proposed the Batch normalization (BN) mechanism to reduce the internal covariate shift and accelerate the convergence of the deep neural nets [52]. This mechanism makes use of the mean and variance to normalize the data values over each mini-batch, which allows us to set a higher learning rate and drop the Dropout [59]. However, this effective BN mechanism has not been widely used in GNNs due to the fewer network layers [12]. Nowadays, as the graphs become larger and the tasks become more complex, the GNN models are more complicated and the difficulty of training the GNN models is also increasing rapidly, which leads to the stability problem during training. Thus, the application of the BN mechanism is of great significance to the robustness of training on large graphs.

Therefore, in order to solve the stability problem in the training process of the link prediction task, we propose an improved GraphSAINT method by adjusting the BN strategy to the special training and inference process of GraphSAINT based on the OGB in this paper. By applying the normalization strategy during training, we achieve the elimination of the instability during training successfully. Moreover, we also realize a reduction in the training time and gain an increase in accuracy under the premise of maintaining the original link prediction accuracy. The effectiveness of our method is validated by the citation dataset of the OGB.

Inspired by [60], the paper is organized as follows: Firstly, Sect. 2 is the related work about the GraphSAINT, especially its sampling strategy and the typical batch normalization techniques; Next, Sect. 3 describes our improved GraphSAINT and the training method; Then, Sect. 4 shows the comparative experiment results based on the citation dataset; At last, conclusions are given in Sect. 5.

## 2 Related Work

### 2.1 The Sampling Strategy of GraphSAINT

GCN achieves one-hop neighbors' information aggregation by using the adjacent matrix [31, 32]. However, because of the using of the adjacent matrix, when a new node is added into the graph, we must adapt the adjacent matrix and re-train the GCN model based on the new adjacent matrix of the adapted graph data to obtain all node's new embeddings. Therefore, as mentioned in Sect. 1, GCN causes a lack of generalization performance for unseen nodes. Besides, it has a high time cost. To overcome this shortcoming, GraphSAGE is proposed [34]. In this method, the neighbors of the target node are sampled and a new aggregation function is learned to aggregate neighbor nodes and generate the embedding vector of a target node, which avoids applying the adjacent matrix and reduces training costs by sampling the nodes of each layer of GNN.

Furthermore, compared with GraphSAGE, GraphSAINT makes it possible to solve the learning tasks on the large graphs by designing a sampler called SAMPLE to obtain subgraphs [33]. Besides, this subgraph sampling method can also improve the generalization performance like GraphSAGE. With the help of the sampling strategy, GraphSAINT can deal with the Neighbor Explosion problem better.

In order to preserve the connectivity features of the graph, bias in the mini-batch estimation will almost inevitably introduced by the sampler. Therefore, in [33], the self-designed normalization techniques are introduced to eliminate deviations. The key step is to estimate the sampling probability of each node, edge, and subgraph.

The sampling probability distribution $P(u)$ of the node $u$ is

$$P(u) \propto \left\| \tilde{\mathbf{A}}_{:,u} \right\|^2 \quad (1)$$

where $\mathbf{A}$ is the adjacency matrix and $\tilde{\mathbf{A}}$ is the normalized one, that is $\tilde{\mathbf{A}} = \mathbf{D}^{-1}\mathbf{A}$, $\mathbf{D}$ is the diagonal degree matrix.

The sampling probability distribution $P_{u,v}^{(l)}$ of the edge $(u, v)$ in the $l^{th}$ GCN layer is

$$P_{u,v}^{(l)} \propto \frac{1}{\deg(u)} + \frac{1}{\deg(v)} \tag{2}$$

The sampling probability distribution $P_{u,v}$ of the subgraph is

$$P_{u,v} \propto \mathbf{B}_{u,v} + \mathbf{B}_{v,u} \tag{3}$$

where $\mathbf{B}_{u,v}$ can be interpreted as the probability of a random walk to start at $u$ and end at $v$ in $L$ hops, $\mathbf{B}_{v,u}$ can be interpreted as the probability of a random walk to start at $v$ and end at $u$ in $L$ hops, $\mathbf{B} = \tilde{\mathbf{A}}^L$, $L$ means $L$ layers which can be represented as a single layer with edge weights. Thus, the sampling probabilities of each node, edge, and subgraph are all well-estimated. Then, the subgraphs obtained by sampling will be used for GraphSAINT training.

### 2.2 The Typical Normalization Technique

The normalization techniques are proposed to eliminate the Internal Covariate Shift, which is caused by the change in the distributions of internal nodes of a deep network in the process of training, and offer the faster training [53, 58].

The typical normalization technique for mini-batch presented in [52] basically follows the mathematical statistics: Firstly, the mini-batch mean is calculated based on the values of each data point over a mini-batch; Next, the mini-batch variance is calculated based on the mini-batch mean; Then, each data point over a mini-batch can be normalized by subtracting the mini-batch mean and then dividing by the mini-batch variance; Finally, the scale and shift parameters are introduced and learned for each data point over a mini-batch.

## 3 Methodology

As mentioned in Sect. 1, the original normalization techniques of GraphSAINT, which are effective for the node classification task, are insufficient to improve the training quality for the link prediction task. Therefore, an improved GraphSAINT training algorithm is proposed.

Since the sampled subgraphs in GraphSAINT are based on the connectivity rules of the nodes, it can get the edge sampling probability with the smallest variance. In contrast, for node selection, it uses the Random node sampler. Therefore, the node feature data of each sampled subgraph do not obey the standard normal distribution. Assume the graph dataset to be processed is a whole graph $\zeta = (V, \xi)$ with N nodes $v \in V$, edges $(v_i, v_j) \in \xi$. For the node $v_i$ in a sampled subgraph $\zeta_s$ of $\zeta$ according to SAMPLE, its feature $h_{i,s}$ has $d$ elements. In order to normalize the distributions of the inputs to reduce the internal covariate shift, the input node feature vector can be normalized by

$$\hat{h}_{i,s} = \frac{h_{i,s} - \mu_s}{\sqrt{\sigma_s^2}}$$

$$\mu_s = \frac{1}{d}\sum_{i=1}^{d} h_{i,s}\sigma_s^2 = \frac{1}{d}\sum_{i=1}^{d}\left(h_{i,s} - \mu_s\right)^2 \quad (4)$$

where $\mu_s$ and $\sigma_s^2$ are the mean and the variance for the node $v_i$ in the node feature dimension and are computed over the training data set.

Therefore, by means of the node-wise normalization technology in each subgraph, each node feature vector is normalized by making its mean zero and variance 1. Besides, based on the typical normalization principle, the training time can also be effectively shortened by applying the node-wise normalization technology in each subgraph.

**Algorithm 1** The improved GraphSAINT training algorithm on citation dataset in OGB

**Input**: Training graph $\zeta=(V,\xi)$; GraphSAINT sampler;

**Output**: Loss and accuracy; GCN model with trained weights;

1: Pre-processing: Directed graph to undirected graph; Sampled subgraph $\zeta_s$ of $\zeta$

.

2: **for** each subgraph $\zeta_s$ **do**

3: GCN construction on $\zeta_s$ and **before** each active layer RELU **do**

4: Apply the normalization technology on the output of the convolutional layer

5: Forward propagation to calculate loss according to MRR.

6: Backward propagation to update weights.

7: **end for.**

The whole training process of the improved GraphSAINT is illustrated in Algorithm 1. Before the training starts, we perform a pre-processing on $\zeta$ to convert the directed graph to the undirected graph and obtain the sampled subgraph $\zeta_s$ with the given SAMPLE [33]. Then an iterative training process is conducted via SGD to update model weights. Each iteration uses an independently subgraph $\zeta_s$. Next, the original GCN is modified through applying the normalization technology on the output of the convolutional layer, which is also the original input of the RELU layer. Finally, the modified GCN on $\zeta_s$ is built to generate embeddings and the loss can then be calculated according to Mean Reciprocal Rank (MRR). In MRR, the score of the first matched result is 1, the score of the second matched result is 1/2, and the score of the *nth* matched result is 1/n. If there is no matching sentence, the score is 0. The final score is the sum of all scores.

As mentioned in Sect. 2.1, GraphSAINT uses the subgraphs obtained by the subgraph sampling method for training, while it uses the whole graph data to calculate the output result during inference. Therefore, the normalization operation can be added independently during inference.

## 4 Experiments

In this section, in order to verify the effectiveness of the improved GraphSAINT algorithm, we choose the Link prediction task based on the citation dataset of OGB (ogbl-citation). The ogbl-citation dataset is a directed graph and can be viewed as a 'subgraph' of the citation network called MAG [61]. In this dataset, each node represents a paper, whose title and abstract are encoded into a 128-dimensional word2vec features, and each directed edge indicates the citation relationship between two papers.

The link prediction task means that we need to predict missing citations based on the exiting citations on the graph. Two of each source paper's references are randomly dropped and the model is required to achieve the ranking of the missing two references in front of other 1000 references that are also randomly sampled from all the papers and not referenced by the source papers. According to this, MRR is chosen as the evaluation metric [44]. Besides, we use the two dropped edges of all source papers respectively for validation and testing. Naturally, the training set contains the rest of the edges.

**Table 1.** Results for GraphSAINT on citation dataset.

| GraphSAINT (Citation) | | MRR | | |
|---|---|---|---|---|
| | | Training | Validation | Test |
| Traditional (official) | | $0.8626 \pm 0.0046$ | $0.7933 \pm 0.0046$ | $0.7943 \pm 0.0043$ |
| Traditional | Convergence | 0.8690 | 0.8031 | 0.8048 |
| | Non-convergence | 0.0010 | 0.0010 | 0.0010 |
| Improved | | $0.9001 \pm 0.0014$ | $0.8335 \pm 0.0020$ | $0.8344 \pm 0.0023$ |

The official results of the traditional GraphSAINT on the citation dataset are given in Table 1: the MRR value of the training set is $0.8626 \pm 0.0046$, the MRR value of the validation set is $0.7933 \pm 0.0046$ and the MRR value of the test set is $0.7943 \pm 0.0043$. However, in the training of our recurrence experiment of the traditional GraphSAINT, we found that GraphSAINT algorithm will not converge in training with a probability range from 0.1 to 0.4. For a RUN where loss converges as shown in Fig. 1(a), the MRR results are consistent with the official results as shown in Table 1, i.e., the training result can reach about 0.8690, the validation result can reach about 0.8031 and the test result can also reach about 0.8048. But for a RUN where loss does not converge, the MRR results are also shown in Table 1 and the loss curve trained under different epochs is basically as shown in Fig. 1(b).

As shown in Fig. 1, we can see that one RUN contains 200 epochs. Besides, in Fig. 1(b), after the $78^{th}$ epoch, the loss is suddenly and sharply increased to 34.5388 and remains unchanged. Therefore, some measures need to be taken to solve the problem of non-convergence in the training process.

After applying the improved GraphSAINT, our loss curve during training is shown by the solid line in Fig. 2. We can see that the solid loss curve is convergent. Besides, compared with the dotted line in Fig. 2, which is the loss curve during training of the traditional GraphSAINT as shown in Fig. 1(a), the improved GraphSAINT has a more stable convergence during training and a faster convergence rate. Moreover, all the three MRR values have an improvement as shown in Table 1: the training results can reach $0.9001 \pm 0.0014$, the validation results can reach $0.8335 \pm 0.0020$ and the test results can also reach $0.8344 \pm 0.0023$. Thus, the effectiveness of our improved GraphSAINT is verified.

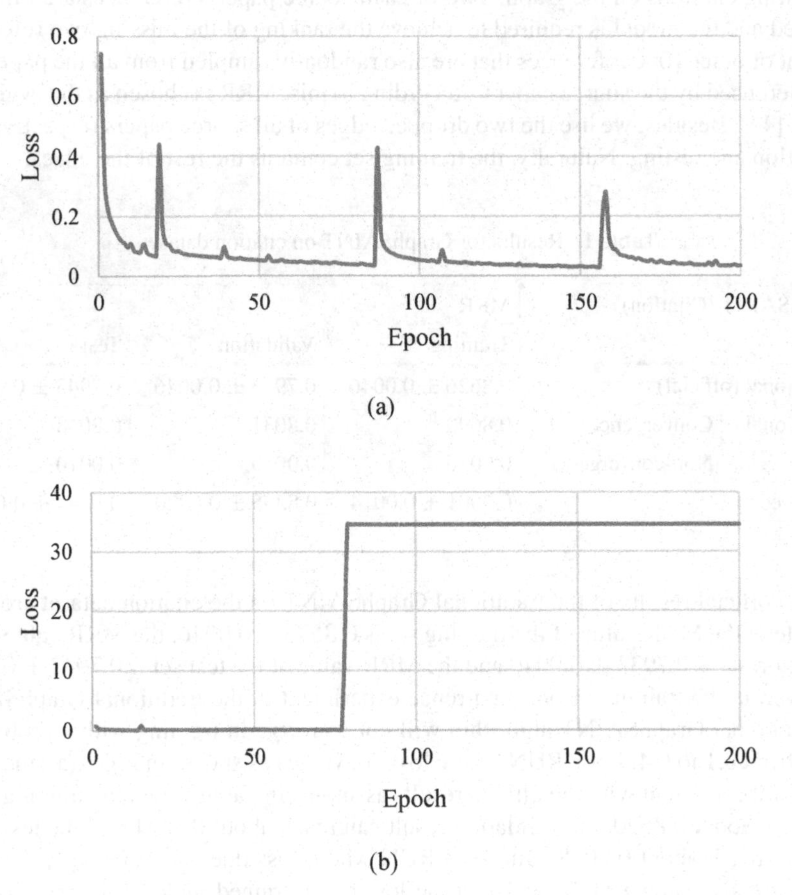


**Fig. 1.** The loss curve during training of the traditional GraphSAINT (a) with convergence (b) without convergence

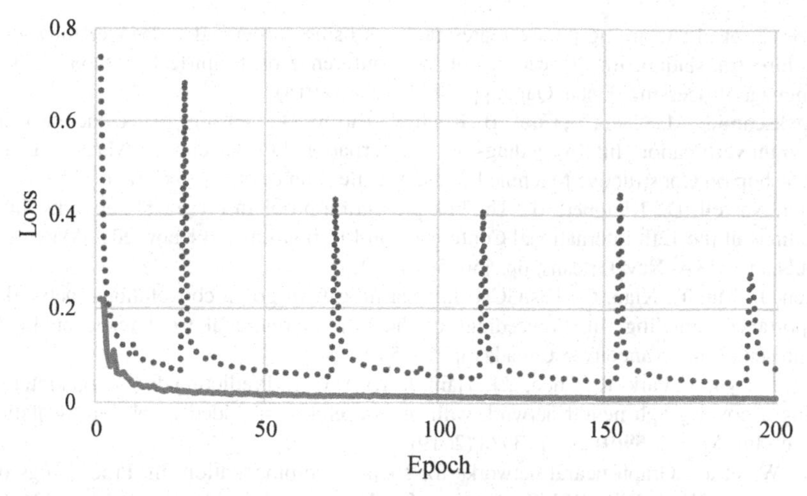


**Fig. 2.** The loss curve during training of the improved GraphSAINT (the solid line) and the traditional GraphSAINT with convergence in Fig. 1(a) (the dotted line).

## 5 Conclusions

The stability problem during training of graph neural network is crucial. In this paper, we focus on this stability problem and propose an improved GraphSAINT by applying the normalization strategy. The proposed method not only improves the robustness of the training process of the GraphSAINT, but also accelerates the convergence of the model. In the future, more attention will be paid to the distributed training methods for large graph datasets.

**Acknowledgement.** This research was funded by the Fifth Innovative Project of State Key Laboratory of Computer Architecture (ICT, CAS) under Grant No. CARCH 5403.

## References

1. Scarselli, F., Gori, M., Tsoi, A.C., Hagenbuchner, M., Monfardini, G.: The graph neural network model. IEEE Trans. Neural Netw. **20**(1), 61–80 (2009)
2. Shuman, D.I., Narang, S.K., Frossard, P., Ortega, A., Vandergheynst, P.: The emerging field of signal processing on graphs: extending high-dimensional data analysis to networks and other irregular domains. IEEE Signal Process. Mag. **30**(3), 83–98 (2013)
3. Zang, C., Cui, P., Faloutsos, C.: Beyond Sigmoids: the NetTide model for social network growth, and its applications. In: Proceedings of the International Conference on Association for Computing Machinery's Special Interest Group on Knowledge Discovery and Data Mining, San Francisco, CA, USA, pp. 2015–2024 (2016)
4. Yan, S., Xu, D., Zhang, B., Zhang, H.J., Yang, Q., Lin, S.: Graph embedding and extensions: a general framework for dimensionality reduction. IEEE Trans. Pattern Anal. Mach. Intell. **29**(1), 40–51 (2007)

5. Cho, K., et al.: Learning phrase representations using rnn encoder–decoder for statistical machine translation. In: Proceedings of the Conference on Empirical Methods in Natural Language Processing, Doha, Qatar, pp. 1724–1734 (2014)
6. Brockschmidt, M., Chen, Y., Cook, B., Kohli, P., Tarlow, D.: Learning to decipher the heap for program verification. In: Proceedings of the International Conference on Machine Learning Workshop on Constructive Machine Learning, Lille, France, pp. 1–5 (2015)
7. Liben-Nowell, D., Kleinberg, J.: The link-prediction problem for social networks. In: Proceedings of the 12th International Conference on Information and Knowledge Management, Louisiana, USA, New Orleans, pp. 556–559 (2003)
8. Chen, J., Ma, T., Xiao, C.: FastGCN: fast learning with graph convolutional networks via importance sampling. In: Proceedings of the 6th International Conference on Learning Representations, Vancouver, Canada, pp. 1–15 (2018)
9. Lim, J., Ryu, S., Park, K., Choe, Y.J., Ham, J., Kim, W.Y.: Predicting drug–target interaction using a novel graph neural network with 3D structure-embedded graph representation. J. Chem. Inf. Model. **59**(9), 3981–3988 (2019)
10. Fan, W., et al.: Graph neural networks for social recommendation. In: Proceedings of the International World Wide Web Conferences, San Francisco, United States, pp. 417–426 (2019)
11. Wu, S., Tang, Y., Zhu, Y., Wang, L., Xie, X., Tan, T.: Session-based recommendation with graph neural networks. In: Proceedings of the AAAI Conference on Artificial Intelligence, Hawaii, USA, Honolulu, pp. 346–353 (2019)
12. Wu, Z., Pan, S., Chen, F., Long, G., Zhang, C., Yu, P.S.: A comprehensive survey on graph neural networks. IEEE Trans. Neural Netw. Learn. Syst. **32**(1), 4–24 (2021)
13. Li, Y., Tarlow, D., Brockschmidt, M., Zemel, R.: Gated graph sequence neural networks. In: Proceedings of the International Conference on Learning Representations, Caribe Hilton, San Juan, Puerto Rico, pp. 1–20 (2016)
14. Dai, H., Kozareva, Z., Dai, B., Smola, A., Song, L.: Learning steady-states of iterative algorithms over graphs. In: Proceedings of the International Conference on Machine Learning, Vienna, Austria, pp. 1114–1122 (2018)
15. Bruna, J., Zaremba, W., Szlam, A., LeCun, Y.: Spectral networks and locally connected networks on graphs. In: Proceedings of the International Conference on Learning Representations, Banff, Canada, pp. 1–14 (2014)
16. Levie, R., Monti, F., Bresson, X., Bronstein, M.: CayleyNets: graph convolutional neural networks with complex rational spectral filters. IEEE Trans. Signal Process. **67**(1), 97–109 (2017)
17. Micheli, A.: Neural network for graphs: a contextual constructive approach. IEEE Trans. Neural Netw. **20**(3), 498–511 (2009)
18. Gilmer, J., Schoenholz, S., Riley, P.F., Vinyals, O., Dahl, G.E.: Neural message passing for quantum chemistry. In: Proceedings of the International Conference on Machine Learning, Sydney, Australia, pp. 1263–1272 (2017)
19. Monti, F., Boscaini, D., Masci, J., Rodola, E., Svoboda, J., Bronstein, M.: Geometric deep learning on graphs and manifolds using mixture model CNNs. In: Proceedings of the Internaltional Conference on Computer Vision and Pattern Recognition, Honolulu, Hawaii, pp. 5115–5124 (2017)
20. Zhang, J., Shi, X., Xie, J., Ma, H., King, I., Yeung, D.Y.: GaAN: gated attention networks for learning on large and spatiotemporal graphs. In: The Conference on Uncertainty in Artificial Intelligence, Monterey, California, USA, pp. 1–11 (2018)
21. Chen, J., Zhu, J., Song, L.: Stochastic training of graph convolutional networks with variance reduction. In: Proceedings of the International Conference on Machine Learning, Vienna, Austria, pp. 941–949 (2018)

22. Ying, Z., You, J., Morris, C., Ren, X., Hamilton, W., Leskovec, J.: Hierarchical graph representation learning with differentiable pooling. In: Proceedings of the Conference on Neural Information Processing Systems, Montréal, Canada, pp. 4801–4811 (2018)
23. Chiang, W.L., Liu, X., Si, S., Li, Y., Bengio, S., Hsieh, C.J.: Cluster-GCN: an efficient algorithm for training deep and large graph convolutional networks. In: Proceedings of the International Conference on Association for Computing Machinery's Knowledge Discovery and Data Mining, Anchorage, AK, USA, pp. 257–266 (2019)
24. Yu, W., et al.: Learning deep network representations with adversarially regularized autoencoders. In: Proceedings of the Association for Computing Machinery's Association for the National Conference on Artificial Intelligence, New Orleans, Louisiana, USA, pp. 2663–2671 (2018)
25. Cao, N., Kipf, T.: MolGAN: an implicit generative model for small molecular graphs. In: Proceedings of the International Conference on Machine Learning workshop on Theoretical Foundations and Applications of Deep Generative Models, Vienna, Austria, pp. 1114–1122 (2018)
26. Pan, S., Hu, R., Long, G., Jiang, J., Yao, L., Zhang, C.: Adversarially regularized graph autoencoder for graph embedding. In: Proceedings of the International Joint Conference on Artificial Intelligence, Stockholmsmässan, Stockholm, pp. 2609–2615 (2018)
27. Ma, T., Chen, J., Xiao, C.: Constrained generation of semantically valid graphs via regularizing variational autoencoders. In: Proceedings of the Conference on Neural Information Processing Systems, Montréal, Canada, pp. 7110–7121 (2018)
28. Seo, Y., Defferrard, M., Vandergheynst, P., Bresson, X.: Structured sequence modeling with graph convolutional recurrent networks. In: Cheng, L., Leung, A.C.S., Ozawa, S. (eds.) ICONIP 2018. LNCS, vol. 11301, pp. 362–373. Springer, Cham (2018). https://doi.org/10.1007/978-3-030-04167-0_33
29. Yu, B., Yin, H., Zhu, Z.: Spatio-temporal graph convolutional networks: A deep learning framework for traffic forecasting. In: Proceedings of the International Joint Conference on Artificial Intelligence, Stockholmsmässan, Stockholm, pp. 3634–3640 (2018)
30. Guo, S., Lin, Y., Feng, N., Song, C., Wan, H.: Attention based spatial-temporal graph convolutional networks for traffic flow forecasting. In: Proceedings of the Association for Computing Machinery's Association for the National Conference on Artificial Intelligence, Honolulu, Hawaii, USA, pp. 922–929 (2019)
31. Kipf, T.N., Welling, M.: Semi-supervised classification with graph convolutional networks. In: Proceedings of the 5th International Conference on Learning Representations, Toulon, France, pp. 1–14 (2017)
32. Defferrard, M., Bresson, X., Vandergheynst, P.: Convolutional neural networks on graphs with fast localized spectral filtering. In: Proceedings of the Conference on Neural Information Processing Systems, Barcelona, Spain, pp. 3844–3852 (2016)
33. Zeng, H., Zhou, H., Srivastava, A., Kannan, R., Prasanna, V.: GraphSAINT: graph sampling based inductive learning method. In: Proceedings of the 8th International Conference on Learning Representations, pp. 1–19. Formerly Addis Ababa Ethiopia: Virtual Conference (2020)
34. Hamilton, W., Ying, Z., Leskovec, J.: Inductive representation learning on large graphs. In: Proceedings of the Conference on Neural Information Processing Systems, California, USA, Long Beach, pp. 1024–1034 (2017)
35. Lü, L., Zhou, T.: Link prediction in complex networks: a survey. Physica A Stat. Mech. Appl. **390**(6), 1150–1170 (2011)
36. Zhang, M., Chen, Y.: Link prediction based on graph neural networks. In: Proceedings of the Conference on Neural Information Processing Systems, Montréal, Canada, pp. 5165–5175 (2018)

37. Lichtenwalter, R.N., Chawla, N.V.: Vertex collocation profiles: subgraph counting for link analysis and prediction. In: Proceedings of the 21st International Conference on World Wide Web, Lyon, France, pp. 1019–1028 (2012)
38. Li, R.H., Jeffrey, X.Y., Liu, J.: Link prediction: the power of maximal entropy random walk. In: Proceedings of the 20th International Conference on Association for Computing Machinery's Information and Knowledge Management, Glasgow, UK, pp. 1147–1156 (2011)
39. Ying, R., He, R., Chen, K., Eksombatchai, P., Hamilton, W.L., Leskovec, J.: Graph convolutional neural networks for web-scale recommender systems. In: Proceedings of the International Conference on Association for Computing Machinery's Knowledge Discovery and Data Mining, London, United Kingdom, pp. 974–983 (2018)
40. Chen, H., Liang, G., Zhang, X.L., Giles, C.L.: Discovering missing links in networks using vertex similarity measures. In: Proceedings of the Twenty-Seventh Conference on Annual Association for Computing Machinery's Applied Computing, Trento, Italy, pp. 138–143 (2012)
41. Monti, F., Bronstein, M., Bresson, X.: Geometric matrix completion with recurrent multi-graph neural networks. In Proceedings of the Conference on Neural Information Processing Systems, California, USA, Long Beach, pp. 3697–3707 (2017)
42. Ahn, M.W., Jung, W.S.: Accuracy test for link prediction in terms of similarity index: the case of WS and BA models. Physica A Stat. Mech. Appl. **429**(1), 3992–3997 (2015)
43. Hoffman, M., Steinley, D., Brusco, M.J.: A note on using the adjusted Rand index for link prediction in networks. Soc. Netw. **42**, 72–79 (2015)
44. Hu, W., et al.: Open graph benchmark: datasets for machine learning on graphs. In: Proceedings of the 34th Conference on Neural Information Processing Systems, Vancouver, Canada, pp. 1–34 (2020)
45. Aiello, L.M., Barrat, A., Schifanella, R., Cattuto, C., Markines, B., Menczer, F.: Friendship prediction and homophily in social media. Assoc. Comput. Machinery's Trans. Web **6**(2), 1–33 (2012)
46. Tang, J., Wu, S., Sun, J.M., Su, H.: Cross-domain collaboration recommendation. In: Proceedings of the International Conference on Knowledge Discovery and Data Mining, Beijing, China, pp. 1285–1293 (2012)
47. Akcora, C.G., Carminati, B., Ferrari, E.: Network and profile based measures for user similarities on social networks. In: Proceedings of the IEEE International Conference on Information Reuse & Integration, Las Vegas, NV, USA, pp. 292–298 (2011)
48. Turki, T., Wei, Z.: A link prediction approach to cancer drug sensitivity prediction. BMC Syst. Biol. **11**(5), 13–26 (2017)
49. Nickel, M., Murphy, K., Tresp, V., Gabrilovich, E.: A review of relational machine learning for knowledge graphs. Proc. IEEE **104**(1), 11–33 (2016)
50. Sutskever, I., Martens, J., Dahl, G., Hinton, G.: On the importance of initialization and momentum in deep learning. In: Proceedings of the 30th International Conference on Machine Learning, Georgia, USA, Atlanta, pp. 1139–1147 (2013)
51. Duchi, J., Hazan, E., Singer, Y.: Adaptive subgradient methods for online learning and stochastic optimization. J. Mach. Learn. Res. **12**(61), 2121–2159 (2011)
52. Ioffe, S., Szegedy, C.: Batch normalization: accelerating deep network training by reducing internal covariate shift. In: Proceedings of the 32nd International Conference on Machine Learning, Lille, France, pp. 448–456 (2015)
53. Shimodaira, H.: Improving predictive inference under covariate shift by weighting the log-likelihood function. J. Stat. Plan. Infer. **90**(2), 227–244 (2000)
54. Wiesler, S., Ney, H.: A convergence analysis of log-linear training. In: Proceedings of the Conference on Neural Information Processing Systems, Granada, Spain, pp. 657–665 (2011)

55. Wiesler, S., Richard, A., Schlüter, R., Ney, H.: Mean-normalized stochastic gradient for large-scale deep learning. In: IEEE International Conference on Acoustics, Speech, and Signal Processing, Florence, Italy, pp. 180–184 (2014)
56. Nair, V.H., Geoffrey, E.: Rectified linear units improve restricted Boltzmann machines. In: Proceedings of the 30th International Conference on Machine Learning, Haifa, Israel, pp. 807–814 (2010)
57. Bengio, Y., Glorot, X.: Understanding the difficulty of training deep feedforward neural networks. In: Proceedings of the 13th International Conference on Artificial Intelligence and Statistics, Sardinia, Italy, pp. 249–256 (2010)
58. Raiko, T., Valpola, H., LeCun, Y.: Deep learning made easier by linear transformations in perceptrons. In: Proceedings of the International Conference on Artificial Intelligence and Statistics, La Palma, Canary Islands, pp. 924–932 (2012)
59. Srivastava, N., Hinton, G., Krizhevsky, A., Sutskever, I., Salakhutdinov, R.: Dropout: a simple way to prevent neural networks from overfitting. J. Mach. Learn. Res. **15**(1), 1929–1958 (2014)
60. Misra, S.: A Step by Step Guide for Choosing Project Topics and Writing Research Papers in ICT Related Disciplines. In: Misra, S., Muhammad-Bello, B. (eds.) ICTA 2020. CCIS, vol. 1350. Springer, Cham (2021). https://doi.org/10.1007/978-3-030-69143-1_55
61. Wang, K., Shen, Z., Huang, C., Wu, C., Dong, Y., Kanakia, A.: Microsoft academic graph: when experts are not enough. Quant. Sci. Stud. **1**(1), 396–413 (2020)

# Data Analysis

# A Complementary Approach in the Analysis of the Human Gut Microbiome Applying Self-organizing Maps and Random Forest

Valeria Burgos[1(✉)], Tamara Piñero[1], María Laura Fernández[2], and Marcelo Risk[1]

[1] Instituto de Medicina Traslacional e Ingeniería Biomédica (IMTIB) - CONICET, Hospital Italiano de Buenos Aires, Instituto Universitario del Hospital Italiano, Buenos Aires, Argentina
valeria.burgos@hospitalitaliano.org.ar

[2] CONICET - Universidad de Buenos Aires, Instituto de Física del Plasma (INFIP), Buenos Aires, Argentina

**Abstract.** The human gastrointestinal tract is colonized by millions of microorganisms that make up the so-called gut microbiota, with a vital role in the well-being, health maintenance as well as the appearance of several diseases in the human host. A data mining analysis approach was applied on a set of gut microbiota data from healthy individuals. We used two machine learning methods to identify biomedically relevant relationships between demographic and biomedical variables of the subjects and patterns of abundance of bacteria. The study was carried out focusing on the two most abundant human gut microbiota groups, *Bacteroidetes* and *Firmicutes*. Both subsets of bacterial abundances together with the metadata variables were subjected to an exploratory analysis, using self-organizing maps that integrate multivariate information through different component planes. Finally, to evaluate the relevance of the variables on the biological diversity of the microbial communities, an ensemble-based method such as random forest was used. Results showed that age and body mass index were among the most important features at explaining bacteria diversity. Interestingly, several bacteria species known to be associated to diet and obesity were identified as relevant features as well. In the topological analysis of self-organizing maps, we identified certain groups of nodes with similarities in subject metadata and gut bacteria. We conclude that our results represent a preliminary approach that could be considered, in future studies, as a potential complement in health reports so as to help health professionals personalize patient treatment or support decision making.

**Keywords:** Gut microbiome · Self-organizing maps · Random forest · Precision medicine · Bioinformatics


H. Florez and M. F. Pollo-Cattaneo (Eds.): ICAI 2021, CCIS 1455, pp. 97–110, 2021.
https://doi.org/10.1007/978-3-030-89654-6_8

## 1 Introduction

Data science comprises different scientific fields of knowledge to target the analysis of complex and massive data. In particular, the increased interest on the application of machine learning algorithms to extract hidden associations or patterns in electronic health records, processing of medical images, prediction of a health situation or classification of patients has demonstrated the need for machine learning tools for reliable decision-making in healthcare and handling of biological data. The human gastrointestinal tract harbors millions of microorganisms which includes bacteria, archaea, fungi and viruses, interacting in symbiotic relationships between the host and each microbial community. This is known as the gut microbiota, while the collective genome of all symbiotic and pathogenic microorgnisms represents the gut microbiome. The establishment of a large part of the component communities that will remain in the adult life occurs at birth and during the first years of life [1,2] and its composition is shaped not only by the host genetics but also by environmental factors, nutritional status, age and lifestyle. Importantly, the gut microbiome plays an essential role in a number of health-beneficial functions (digestion, synthesis of essential vitamins and amino acids, absorption of calcium, magnesium and iron, fermentation of indigestible components, protection against pathogens, etc.) [3].

The rates of growth and survival of its component populations may fluctuate in response to temporary stressors, such as changes in diet or the consumption of antibiotics [4]. This potential for dynamic restructuring involves two important characteristics of the gut microbiome: plasticity and resilience [5,6]. Ongoing research in human and animal models highlights the importance of a healthy gut microbiome since persistent disbalances in composition and stability, known as dysbiosis, are associated to the onset and progression of chronic diseases that include obesity, irritable bowel syndrome, diabetes, cancer, and neurological diseases such as Parkinson's, among others [7].

The generation of biological knowledge from the large flow of data generated by new technologies in biomedical sciences has accelerated their transformation into data-centered fields. Thus, the study of the human gut microbiome represents a major challenge since it requires an interdisciplinary work between computer science and medicine. The interaction between these two fields will help obtain knowledge about gut bacteria interactions in human health and disease.

In the present work, we analyzed microbiome abundance data and the associated metadata using a machine learning approach: we used the visualization capabilities offered by self-organizing maps to identify patterns of multivariate data stored in multiple layers and additionally, we applied random forest to model the prediction of microbial diversity. For each analysis, we focused on the abundance levels of two major groups of the human gut bacteria, such as *Bacteroidetes* and *Firmicutes.*

## 2 Methods

### 2.1 Dataset

Microarray profiling data of human gut microbiota and anonymized metadata were obtained from the Dryad Digital Repository, as described by [8]. Briefly, the data matrix contained 1172 intestinal samples of western adults. In each sample, bacterial abundances were quantified using the HITChip phylogenetic microarray. This technology allows the assessment of relative abundances of gut bacteria through signal intensities of the targeted 16S rRNA gene, frequently used for the identification of poorly described or non cultured bacteria. Data contained hybridization signals for 130 genus-like phylogenetic groups. Subject metadata included age, sex, nationality, probe-level Shannon diversity, BMI group and subjectID. Geographical origin of the study subjects were: Central Europe (Belgium, Denmark, Germany, the Netherlands), Eastern Europe (Poland), Scandinavia (Finland, Norway, Sweden), Southern Europe (France, Italy, Serbia, Spain), United Kingdom/Ireland (UK, Ireland) and the United States (US). We used VIM and tidyverse R packages [9,10] to check for the presence of missing values (NAs). Records containing NAs were carefully removed without causing bias in the dataset. During the cleaning process, the category 'Eastern Europe' was turned out since it was represented by only one complete case. The final dataset to be used was represented by 1056 complete patient records containing 130 bacterial abundance data and subject metadata.

### 2.2 Self-organizing Maps (SOMs)

Self-organizing maps (also known as Kohonen maps) represent an optimal option to organize multidimensional data in a two-dimensional space by using a neural network. SOM uses the vector space as a model to represent data in a two-dimensional lattice: each value through N samples could be referred to as a data point in an N-dimensional space. Thousands of data points would therefore form data clouds in space, with a intrinsic topology due to geometric relationships. From this it follows that the greater the similarity in the data value level, the closer is the geometric space they occupy. To visualize the trained SOM, we used heatmaps for each variable to plot the degree of connectivity between adjacent output neurons through the use of a color intensity panel. In the case of multivariate datasets, the visualization of different heatmaps allows an overall analysis of the relations between the variables since maps are linked to each other by position: in each map, a node in a given location corresponds to the same unit in another map. The SOM map can be implemented in different topologies. Data were divided into two subsets by major groups of gut bacteria (*Bacteroidetes* and *Firmicutes*). We used a regular hexagonal 2D grid consisting of 750 neurons, in $30 \times 25$ grids. Data was logarithmically-scaled before training. We used the kohonen R package, which provides a standardized framework for SOMs.

### 2.3 Random Forest

Random forest (RF) is an ensemble learning method that can solve regression and classification problems. The algorithm uses a random subset of the training samples for each tree and a random subset of predictors in each step during the training process. These two sources of randomization make the algorithm robust to correlated predictors and more reliable at obtaining average outputs into a model. Data were divided into two subsets by major groups of gut bacteria (*Bacteroidetes* and *Firmicutes*). Bacterial abundance data and metadata were used as RF regressors to generate a diversity prediction model, which was performed using the RF regression algorithm provided by the R interface for h2o [11]. We used 10-fold cross validation for training the regression models and their performance were evaluated using Mean Absolute Error (MAE) as the error metric. After parameter tuning (mainly focused on the number of trees) through cross validation, the best RF regression model was selected.

## 3 Results

### 3.1 Characteristics of the Study Population

The degree of obesity is a relevant aspect in gut microbiome studies in terms of its influence on the microbiota composition [12,13]. This parameter, that can be obtained through the body mass index (BMI), indicates the nutritional status of an individual. Descriptive analysis of the study population showed that lean individuals were homogeneously distributed in all age groups, while overweight, obese and severe obese categories were more abundant in 45–60 year-old individuals. A large proportion of the underweight population was represented between 20–30 years old (Fig. 1).

The distribution of the different BMI categories in each geographic region showed that lean individuals represented approximately half of the proportions for all locations. For Scandinavia, Southern Europe, UK/Ireland (UKIE) and the US, the following proportion was represented by overweight subjects. In contrast, in Central Europe, the second proportion after lean individuals was represented by obese individuals. Morbid obese subjects were present only in Central Europe (Fig. 2).

### 3.2 SOM Analysis

Each component plane or map in a SOM represents one type of data: a two-dimensional lattice for each metadata variable (BMI, nationality, age, sex and diversity) as well as for each bacteria (whose relative abundance is represented in expression levels of the 16S rRNA gene). Since each map preserves shape and density, exploration of the geometric relationships between nodes allows a direct identification of similarities and differences between the layers.

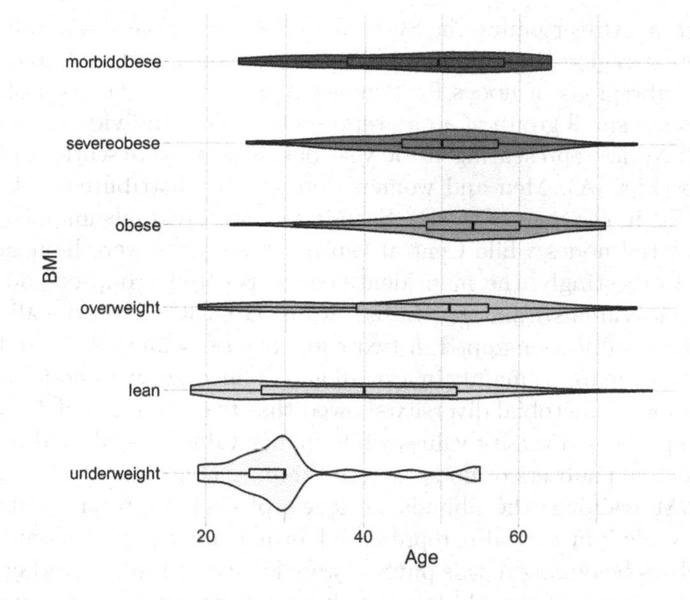


**Fig. 1.** Distribution of the BMI categories across age intervals in the study population.

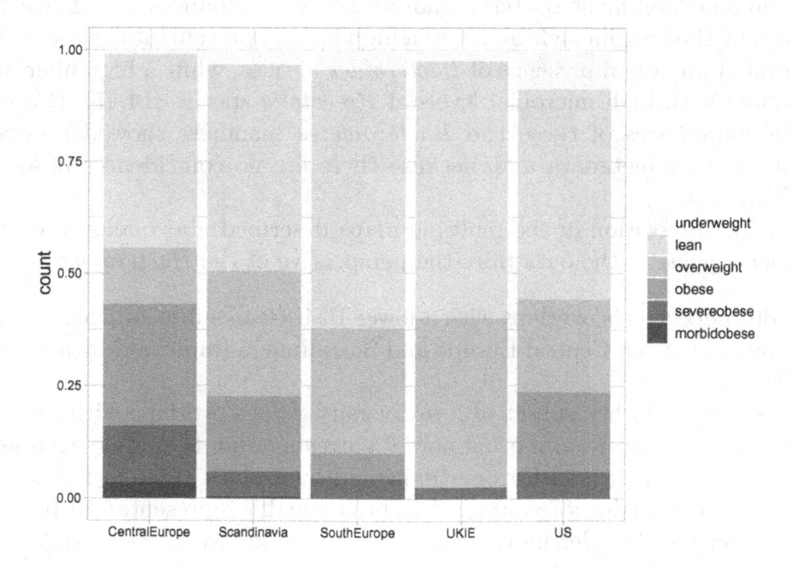


**Fig. 2.** Proportion of each BMI category of the study subjects across different geographic locations.

***Bacteroidetes*.** After running the SOM algorithm using the *Bacteroidetes* subset, the different regions in each map indicated that the age distributed into two well-defined subregions of nodes for younger ages (around 20 years old), quite separated from a small group of nodes representing older individuals (older than 65 years old). Nodes representing 40–50 year old subjects were scattered throughout the map (Fig. 3A). Men and women were clearly distributed in two large, separated areas in the map (Fig. 3B). Scandinavian individuals mapped in a few groups of isolated nodes while Central European subjects were homogeneously distributed. Interestingly, the map identified an isolated group of nodes corresponding to individuals from the United States (Fig. 3C). Additionally, underweight and lean subjects mapped in two wide groups, while severe and morbid obese individuals mapped mainly in a small and defined group of nodes (Fig. 3E). The distribution of microbial diversity showed that two small and defined groups of nodes mapped low diversity values while higher values distributed into larger and clearly defined subsets of nodes across the map (Fig. 3D).

After SOM training, the abundance levels of each bacteria species of the *Bacteroidetes* phylum was also represented in a map. In the present dataset, several members belonging to this phylum were identified but no overlay between any bacterial map was observed (data not shown). However, since many members of the *Bacteroidetes* community have a relevant role in the host health, we chose to analyze two prominent bacteria whose relative abundances are known to be influenced by the host lifestyle and diet: a high fat and protein intake is associated with elevated microbial presence of *Bacteroides* species, while a high fiber intake is associated with high microbial levels of *Prevotella* species [14,15]. It appears that the abundances of these two *Bacteroidetes* members showed no overlay between any host metadata map because there are no coincidences in location (Fig. 3F and G).

The superimposition of the multiple maps described above allows to obtain some clear aspects of the data from the perspective of the *Bacteroidetes* subset:

- Low diversity values overlaps with a lower BMI (lean subjects) corresponding to young men from Central Europe and Scandinavia (indicated by a red circle in Fig. 3A–E).
- Interestingly, another subset of low diversity values overlaps with high BMI values (that is, severe to morbid obese) corresponding to middle-aged female individuals from Central Europe (indicated by a black circle in Fig. 3A–E).
- The small proportion of young subjects is equally represented in both men and women, while older individuals correspond exclusively to men.
- There are no women older than 65 years.
- US nationality corresponds to lean women in the range of medium values of microbial diversity.

***Firmicutes*.** When subject metadata was trained using the *Firmicutes* subset, the different layers showed a more disperse variable distribution in the maps: middle-age individuals were homogeneously represented throughout the lattice while younger (around 20 years old) and older ages (older than 65 years) mapped

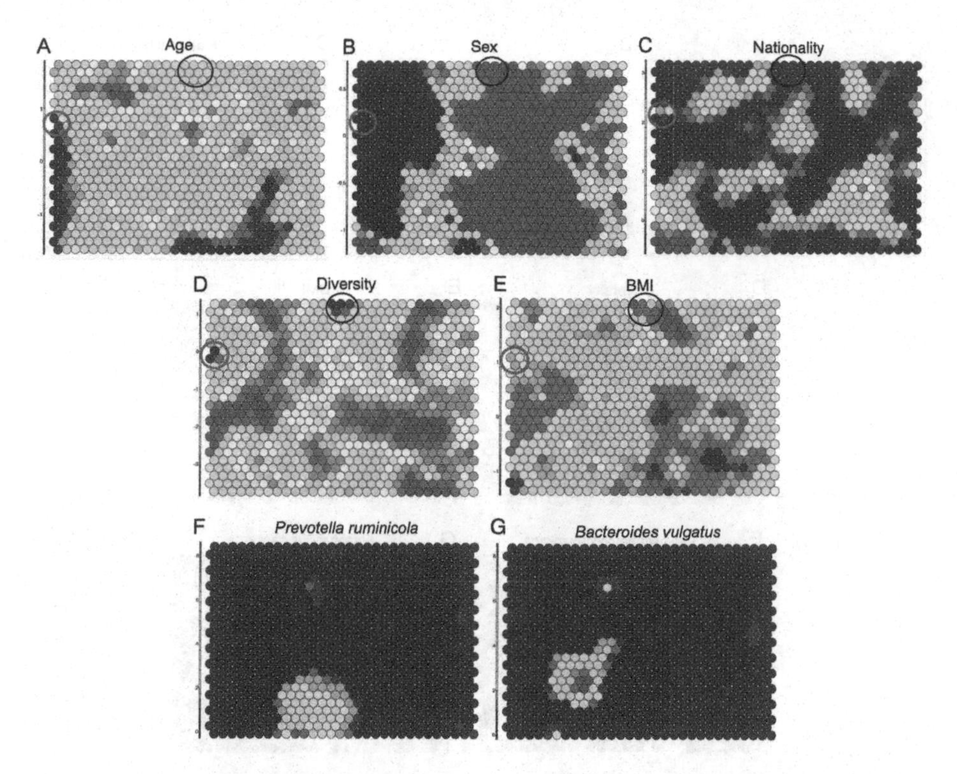


**Fig. 3.** SOM results for metadata variables associated with *Bacteroidetes*. The color index of each map is established based on the values for each variable. A) Age, scale adjusted for a range from 18 to 77 years old, blue = younger, red = older adults. B) Sex, blue = male, red = female. C) Nationality, color gradient beginning at Central Europe (level 0, color blue), Scandinavia (level 1, light blue), Southern Europe (level 2, green), UK/Ireland (level 3, orange) and the US (level 4, red). D) Diversity, scale adjusted for a range of 4.7 to 6.3 diversity index values beginning at blue (lowest diversity) to red (highest diversity). E) BMI, color gradient beginning at underweight (color blue), lean (light blue), overweight (green), obese (yellow), severe obese (orange) and morbid obese (red). In F) *P. ruminicola* and G) *B. vulgatus* the color gradient represents the level of abundance, blue = low, red = high. (Color figure online)

in small, discrete groups of nodes (Fig. 4A). Men and women were not as clearly separated in their node distribution as in the *Bacteroidetes* subset (Fig. 4B). Regarding nationality, a node pattern similar to *Bacteroidetes* was observed (Fig. 4C). High microbial diversity values predominated throughout the map while only three nodes mapped for low diversity values (Fig. 4D). Additionally, lean and underweight subjects predominated in most of the nodes and only a very small group of nodes grouped the highest BMI values, corresponding to the severe obese category (Fig. 4E).

The phylum *Firmicutes* is made up of around 250 different genera of bacteria, such as *Lactobacillus*, *Bacillus*, *Clostridium*, *Enterococcus*, and *Ruminicoccus*,

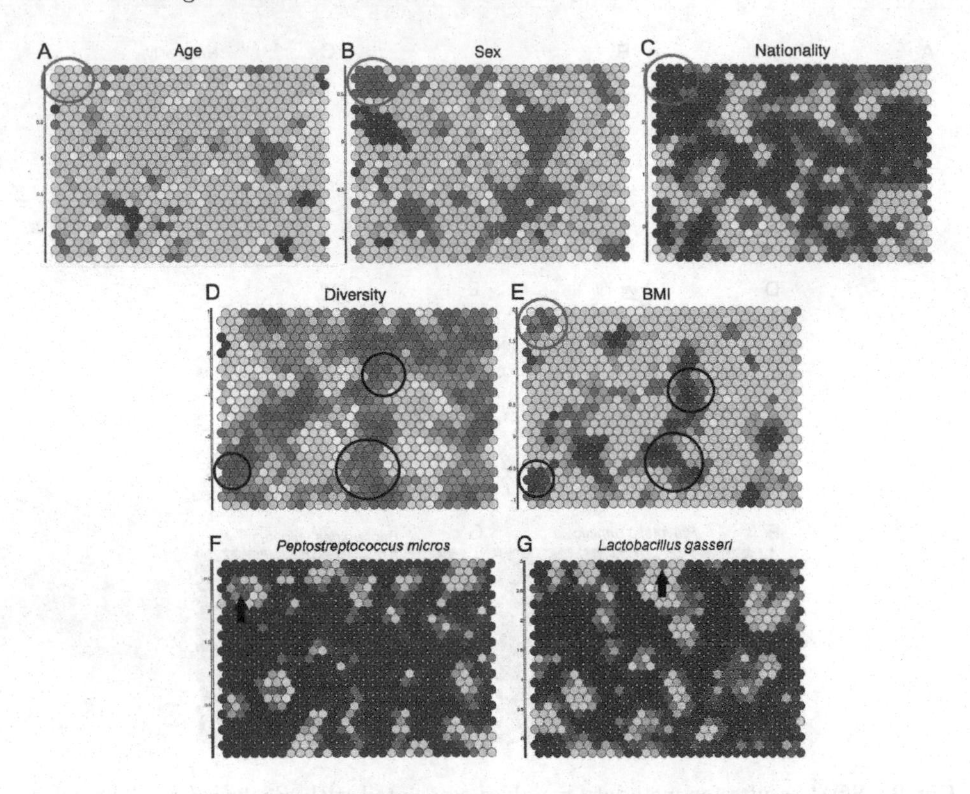


**Fig. 4.** SOM results for metadata variables associated with *Firmicutes*. The color index of each map is established based on the values for each variable. A) Age, scale adjusted for a range from 18 to 77 years old, blue = younger, red = older adults. B) Sex, blue = male, red = female. C) Nationality, color gradient beginning at Central Europe (level 0, color blue), Scandinavia (level 1, light blue), Southern Europe (level 2, green), UK/Ireland (level 3, orange) and the US (level 4, red). D) Diversity, scale adjusted for a range of 4.7 to 6.3 diversity index values beginning at blue (lowest diversity) to red (highest diversity). E) BMI, color gradient beginning at underweight (color blue), lean (light blue), overweight (green), obese (yellow), severe obese (orange) and morbid obese (red). In F) *P. micros* and G) *L. gasseri* the color gradient represents the level of abundance, blue = low, red = high. (Color figure online)

among other important members. In the present dataset, 74 members belonging to this phylym were identified and consequently, each generated a heatmap after SOM training (data not shown). However, superimposing multiple maps revealed that only one node corresponding to overweight individuals slightly coincided with higher values of abundance of a single species, *Peptostreptococcus micros* (indicated by an arrow in Fig. 4F). This represents an interesting result since several other members of *Firmicutes* have previously been associated to obesity [16,17]. Additionally, an overlay between high microbial diversity and higher values of abundance for *Lactobacillus gasseri* was observed (indicated by an arrow in Fig. 4G).

The map overlay between diversity and BMI categories allowed to observe that:

- Severe to morbid obese individuals were middle-age women from Central Europe (indicated by a red circle in Fig. 4A, B, C and E).
- Many of the maximum diversity values superimpose with the lowest BMI categories (indicated by a black circle in Fig. 4D and E).

### 3.3 Random Forest

Diversity is an important variable in microbiome research because it describes the richness (the number of classes) and distribution of the component microorganisms among classes. Understanding diversity in the intestinal microbial community also allows us to understand the impact of factors on bacteria distribution, such as the use of antibiotics, type of diet, degree of obesity, medical interventions and environmental factors, among others (reviewed in [18]). We developed a RF regression model of microbiome diversity. A very useful visual way to interpret RF results of the prediction model is through a ranking list of feature importance, which refers to the relative influence of each feature on the target variable. It considers whether a variable was selected to split and how much the squared error (over all trees) improved as a result.

***Bacteroidetes***. We observed that the age of the individuals was the most important factor in the regression model to predict the diversity of the gut microbiota. Regarding the bacterial abundances, it was observed that two species had the greatest relative importance in the identification of diversity: *Prevotella ruminicola* and *Bacteroides ovatus*. On the other hand, of the remaining metadata variables, the BMI of individuals had a remarkable position in the ranking of importance, followed by the geographic location. Gender was not important in the prediction of diversity. While the list of importance for each variable is informative, a better interpretation is obtained after scaling between 0 and 1 in a descending order of importance (Fig. 5A).

***Firmicutes***. When data was RF-trained considering this subset, the first three positions of the list were occupied by the BMI of the individuals, followed by Nationality and Age. Regarding the abundances of the 74 types of bacteria, it was observed that two members had the greatest relative importance in predicting diversity: *Lachnobacillus bovis* and *Granulicatella* (Fig. 5B).

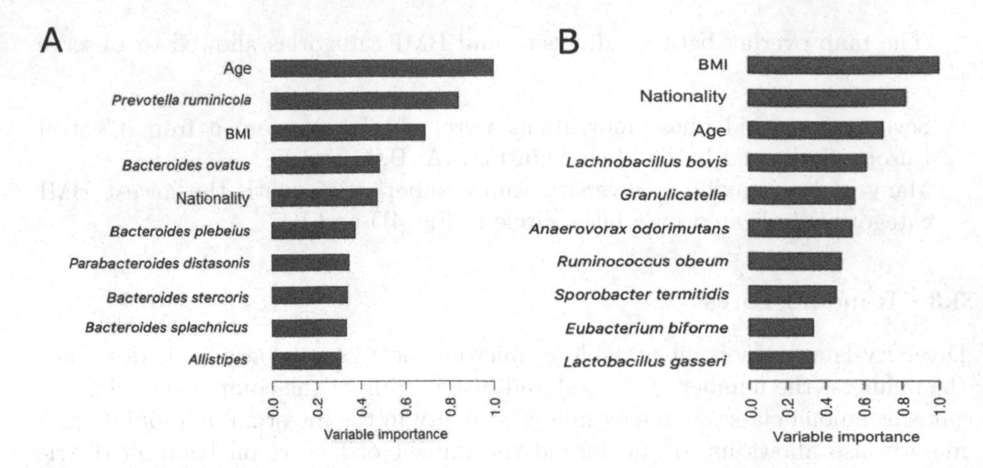


**Fig. 5.** Scale of the relative importance of the first ten variables on the diversity of *Bacteroidetes* (A) and *Firmicutes* (B) after implementing random forest.

# 4 Discussion

We sought to characterize the relations between subject metadata and specific bacterial members of the gut microbiome in a thousand western adults through the use of two machine learning methods that provide robust analytical visualizations, such as self-organizing maps and random forest. Bacteria of the human intestinal microbiome are taxonomically classified into six large groups or phyla, which in turn are subclassified into classes, orders, families, genera and species. The present work addressed the analysis on two of the most abundant phyla, *Bacteroidetes* and *Firmicutes*, which represent 90% of the gut microbiota [19].

The configuration of the SOM outcome maintains the topological structure of the input multidimensional data, in which similar values are mapped in the same or near node in a two dimensional map. Such topological preservation is of particular significance in the exploratory phase of omics data mining since there is generally no *a priori* knowledge of data structure. We presented the visualization of different superimposed heatmaps that allowed the exploration of relationships between input variables. This way of presenting SOM outcomes is similar to previous studies [20].

Our results showed that only one species of *Firmicutes*, *Peptostreptococcus micros*, was slightly associated to nodes that grouped overweight individuals, mostly middle-aged women. Notably, several previous studies have shown that *Peptostreptococcus micros* (later classified as *Parvimonas micra*) is one of various colorectal cancer (CRC) microbial markers [13,21]. This bacteria has also been reported to have a pathogenic role in periodontal diseases [22]. Considering the behavior of the oral microbiome during periodontal infection and its influence in health complications, such as diabetes, cardiovascular disease, and obesity [23], the significance of the results obtained here provides a basis for future studies on the possible role of gut bacteria as biological markers of a developing overweight condition. Among bacteria with known beneficial roles, we observed that high abundance of *Lactobacillus gasseri* was related to nodes that grouped high diversity values. This result supports the previously reported role of *Lactobacillus gasseri* in the management of obesity and probiotic properties [24,25]. The topological structures of the metadata variables were slightly different depending whether *Bacteroidetes* or *Firmicutes* subsets were used for SOM training. For both age and sex, mapping distribution of the study subjects was more effective using the *Bacteroidetes* data. In the case of the different BMI categories, subject distribution was more effective using *Firmicutes* data.

Although the SOM results presented here allowed us to gain insight into the different regions of matching information underlying host metadata and the relative abundance of bacteria, we consider that a deeper approach of the use of SOM is needed, in terms of parameter configuration, such as size, dimensionality, shape, learning rate, among others. Considering that the relationships between gut microbiome and host BMI are dynamic and complex, self-organizing maps provide an excellent tool of visualization and dimensionality reduction that could serve as a complementary tool in a biomedical report.

Analysis of the microbiome diversity in the human body is essential to understand the structure, biology and ecology of its component communities. This analysis represents a critical first step in microbiome studies. When supervised learning through a regression random forest algorithm was used to determine which variables were important in the prediction of microbial diversity, we observed that both age and BMI category of the individuals appeared as the most relevant in the regression models generated for *Bacteroidetes* and *Firmicutes* subsets. The contribution of these physiological factors in shaping the gut microbiome has been reported previously: with age, the beneficial functions provided by a healthy gut microbiome begin to decrease in association to an increasing frequency of inflammatory processes and disease, especially in the elderly. Regarding the influence of BMI, various studies that compare the intestinal microbiota between obese and lean individuals indicate that the variation in the degree of diversity is associated with body weight (obese individuals present a low diversity, which means a higher BMI) [26–29].

Random forest regression also indicated that two *Bacteroidetes* species were the most relevant on diversity: *Prevotella ruminicola* is involved in the response of individuals to dietary supplements [30] and *Bacteroides ovatus* is a dominant species in the human intestine, previously identified as a next generation probiotic due to its preventive effects on intestinal inflammation (reviewed in [31]). On the other hand, two members of the *Firmicutes* group were identified as important at predicting diversity in this subset: *Lachnobacillus bovis* and *Granulicatella* whose relative abundances are reported to be influenced by the type of diet and the use of antibiotics in obese individuals [32,33]. Hence, some of the results obtained in both RF regression models are consistent with published microbiome research, indicating the robustness of the RF regression algorithm. A further analysis is needed that involves a complete parameter tuning so as to characterize the most accurate RF setting for a microbiome project.

In the last decade, the impulse provided by innovative developments in technology and the generation of large volumes of microbiome data has caused an increase in the use of machine learning methods in this field, such as microbial ecology, identification of certain bacteria to cancer and forensics, among others [34–36]. We consider that our study represents the start of a contribution to the vast field of microbiome research, although we need further refinements of the methodology used in order to validate the obtained models and improve performances.

The accumulating research of the gut microbiome and its influence on health and disease has accelerated the need for integration of multidisciplinary fields in its analysis. In general, health professionals (medical doctors, nurses, biochemists) are not prepared to work in all the steps along the data analysis process (cleaning, filtering, choice of algorithms, interpretation, etc.). Therefore, data science and machine learning can contribute to the translation of innovative results into valuable knowledge that provide decision support in microbiome-based precision medicine.

## 5 Conclusions

We used two robust computer science-based methods, such as self-organizing maps and random forest, to study the relationships between gut microbiome data and host information. Our results represent a preliminary approach that could be considered, in future studies, as a potential complement in health reports so as to help health professionals to individualize patient treatment or support decision making. Additionally, this work contributes to the increasingly growing area of gut microbiome interactions on human health and disease. However, further studies using other machine learning algorithms to validate the results obtained here are required.

## References

1. Dominguez-Bello, M.G., et al.: Delivery mode shapes the acquisition and structure of the initial microbiota across multiple body habitats in newborns. Proc. Natl. Acad. Sci. **107**(26), 11971–11975 (2010)
2. Koenig, J.E., et al.: Succession of microbial consortia in the developing infant gut microbiome. Proc. Natl. Acad. Sci. **108**(Supplement 1), 4578–4585 (2011)
3. Rowland, I., et al.: Gut microbiota functions: metabolism of nutrients and other food components. Eur. J. Nutr. **57**(1), 1–24 (2017). https://doi.org/10.1007/s00394-017-1445-8
4. Kiewiet, M., et al.: Flexibility of gut microbiota in ageing individuals during dietary fiber long-chain inulin intake. Molecular Nutrition & Food Research, p. 2000390 (2020)
5. Ruggles, K.V., et al.: Changes in the gut microbiota of urban subjects during an immersion in the traditional diet and lifestyle of a rainforest village. Msphere, vol. 3(4) (2018)
6. Liu, H., et al.: Resilience of human gut microbial communities for the long stay with multiple dietary shifts. Gut **68**(12), 2254–2255 (2019)

7. Floch, M.H., Ringel, Y., Walker, W.A.: The microbiota in gastrointestinal pathophysiology: implications for human health, prebiotics, probiotics, and dysbiosis. Academic Press (2016)
8. Lahti, L., Salojärvi, J., Salonen, A., Scheffer, M., De Vos, W.M.: Tipping elements in the human intestinal ecosystem. Nat. Commun. **5**(1), 1–10 (2014)
9. Kowarik, A., Templ, M.: Imputation with the R package VIM. J. Stat. Softw. **74**(7), 1–16 (2016). https://doi.org/10.18637/jss.v074.i07
10. Wickham, H., et al.: Welcome to the tidyverse. J. Open Source Softw. **4**(43), 1686 (2019). https://doi.org/10.21105/joss.01686, http://dx.doi.org/10.21105/joss.01686
11. LeDell, E., et al.: h2o: R interface for 'h2o'. R package version 3(0.2) (2018)
12. Haro, C., et al.: Intestinal microbiota is influenced by gender and body mass index. PLoS ONE **11**(5), e0154090 (2016)
13. Yu, J., et al.: Metagenomic analysis of faecal microbiome as a tool towards targeted non-invasive biomarkers for colorectal cancer. Gut **66**(1), 70–78 (2017)
14. De Filippo, C., Cavalieri, D., Di Paola, M., Ramazzotti, M., Poullet, J.B., Massart, S., Collini, S., Pieraccini, G., Lionetti, P.: Impact of diet in shaping gut microbiota revealed by a comparative study in children from Europe and rural Africa. Proc. Natl. Acad. Sci. **107**(33), 14691–14696 (2010)
15. Wu, G.D., et al.: Linking long-term dietary patterns with gut microbial enterotypes. Science **334**(6052), 105–108 (2011)
16. Koliada, A., et al.: Association between body mass index and firmicutes/bacteroidetes ratio in an adult ukrainian population. BMC Microbiol. **17**(1), 120 (2017)
17. Crovesy, L., Masterson, D., Rosado, E.L.: Profile of the gut microbiota of adults with obesity: a systematic review. Eur. J. Clin. Nutr. **74**(9), 1251–1262 (2020)
18. Reese, A.T., Dunn, R.R.: Drivers of microbiome biodiversity: a review of general rules, feces, and ignorance. MBio **9**(4), e01294-18 (2018)
19. Arumugam, M., et al.: Enterotypes of the human gut microbiome. Nature **473**(7346), 174–180 (2011)
20. Qian, J., et al.: Introducing self-organized maps (som) as a visualization tool for materials research and education. Results Mater. **4**, 100020 (2019)
21. Xu, J., et al.: Alteration of the abundance of parvimonas micra in the gut along the adenoma-carcinoma sequence. Oncol. Lett. **20**(4), 1 (2020)
22. Nagarajan, M., Prabhu, V.R., Kamalakkannan, R.: Metagenomics: implications in oral health and disease. In: Metagenomics, pp. 179–195. Elsevier (2018)
23. Goodson, J., Groppo, D., Halem, S., Carpino, E.: Is obesity an oral bacterial disease? J. Dent. Res. **88**(6), 519–523 (2009)
24. Selle, K., Klaenhammer, T.R.: Genomic and phenotypic evidence for probiotic influences of lactobacillus gasseri on human health. FEMS Microbiol. Rev. **37**(6), 915–935 (2013)
25. Mahboubi, M.: Lactobacillus gasseri as a functional food and its role in obesity. Int. J. Med. Rev. **6**(2), 59–64 (2019)
26. Turnbaugh, P.J., et al.: A core gut microbiome in obese and lean twins. Nature **457**(7228), 480–484 (2009)
27. Yatsunenko, T., et al.: Human gut microbiome viewed across age and geography. Nature **486**(7402), 222–227 (2012)
28. Dominianni, C., et al.: Sex, body mass index, and dietary fiber intake influence the human gut microbiome. PLoS ONE **10**(4), e0124599 (2015)
29. Bosco, N., Noti, M.: The aging gut microbiome and its impact on host immunity. Genes & Immunity, pp. 1–15 (2021)

30. Chung, W.S.F., et al.: Relative abundance of the prevotella genus within the human gut microbiota of elderly volunteers determines the inter-individual responses to dietary supplementation with wheat bran arabinoxylan-oligosaccharides. BMC Microbiol. **20**(1), 1–14 (2020)
31. Tan, H., et al.: Pilot safety evaluation of a novel strain of bacteroides ovatus. Front. Genet. **9**, 539 (2018)
32. Salonen, A., et al.: Impact of diet and individual variation on intestinal microbiota composition and fermentation products in obese men. ISME J. **8**(11), 2218–2230 (2014)
33. Reijnders, D., et al.: Effects of gut microbiota manipulation by antibiotics on host metabolism in obese humans: a randomized double-blind placebo-controlled trial. Cell Metab. **24**(1), 63–74 (2016)
34. Ai, D., Pan, H., Han, R., Li, X., Liu, G., Xia, L.C.: Using decision tree aggregation with random forest model to identify gut microbes associated with colorectal cancer. Genes **10**(2), 112 (2019)
35. Thompson, J., Johansen, R., Dunbar, J., Munsky, B.: Machine learning to predict microbial community functions: an analysis of dissolved organic carbon from litter decomposition. PLoS ONE **14**(7), e0215502 (2019)
36. Topçuoğlu, B.D., Lesniak, N.A., Ruffin, M.T., IV., Wiens, J., Schloss, P.D.: A framework for effective application of machine learning to microbiome-based classification problems. MBio **11**(3), e00434-20 (2020)

# An Evaluation of Intrinsic Mode Function Characteristic of Non-Gaussian Autorregresive Processes

Fernando Pose[1], Javier Zelechower[2], Marcelo Risk[1], and Francisco Redelico[1,3(✉)]

[1] Instituto de Medicina Traslacional e Ingeniería Biomédica, Hospital Italiano de Buenos Aires, Instituto Universitario del Hospital Italiano de Buenos Aires, CONICET, Perón 4190 - (C1199ABB), Ciudad Autónoma de Buenos Aires, Argentina
francisco.redelico@hospitalitaliano.org.ar

[2] Department of Information Systems Engineering, National Technological University, Buenos Aires, Argentina

[3] Departamento de Ciencia y Tecnología, Universidad Nacional de Quilmes, Roque Sáenz Peña 352, B1876BXD Bernal, Buenos Aires, Argentina

**Abstract.** Empirical mode decomposition (EMD) is a suitable transformation to analyse non-linear time series. This work presents a empirical study of intrinsic mode functions (IMFs) provided by the empirical mode decomposition. We simulate several non-gaussian autoregressive processes to characterize this decomposition. Firstly, we studied the probability density distribution, Fourier spectra and the cumulative relative energy to each IMF as part of the study of empirical mode decomposition. Then, we analyze the capacity of EMD to characterize, both the autocorrelation dynamics and the marginal distribution of each simulated stochastic process. Results show that EMD seems not to only discriminate autocorrelation but also the marginal distribution of simulated processes. Results also show that entropy based EMD is a promising estimator as it is capable to distinguish between correlation and probability distribution. However, the EMD entropy does not reach its maximum value in stochastic processes with uniform probability distribution.

**Keywords:** Empirical mode decomposition (EMD) · Entropy-complexity informational plane · Intrinsic mode function (IMF) · Stochastic Processes

## 1 Introduction

Empirical mode decomposition was introduced for the first time by, Huang and collaborators in [9]. It is a fully adaptive technique useful to analyze non-linear and non-stationary time series. The basic idea of EMD method is decompose the actual time series into a number of intrinsic mode functions (IMFs) as the sum of these IMFs will recover the original time series. Even this technique has been have a great impact in analysing time series, it has the drawback to be a totally


H. Florez and M. F. Pollo-Cattaneo (Eds.): ICAI 2021, CCIS 1455, pp. 111–124, 2021.
https://doi.org/10.1007/978-3-030-89654-6_9

empirical technique and, from our knowledge, it does not have a completely mathematical formulation that allow us to explore its properties, so their should be evaluated empirically [16].

Despite of non-gaussian process are ubiquitous within science and technology in such diverse areas as random number generators [11], modeling irregularly spaced transaction financial data [5], foreign exchange rate volatility modeling [8], studying nervous systems mechanism (Spike sorting) [6] among others, the properties of IMFs generated from them seems to be limited evaluated [16].

For the reasons above, we propose analyse the statistical features of the IMFs generated from autoregressive process of order 1 endowed with Gaussian, Exponential or Uniform marginal probability distribution as prototype of non-gaussian autoregressive processes [21]. We divide our analysis in two parts, in the first part, we analyse the time-frequency domain computing the probability distribution, mean, peaks, periods and the Fourier spectrum of each intrinsic characteristic function of the respective stochastic process and within the second part we extract energetic and entropy like features of the IMFs, the density energy of each IMF, the accumulated energy, EMD entropy and the relative position in the entropy-complexity plane shows the dynamical nature of each IMF.

The paper reads as follows: Sect. 2 briefly explain the EMD technique and the autoregressive process generation algorithm, Sect. 3 shows the statistical features of the IMF from the series explained in the previous section and finally Sect. 4 is devoted to conclusions.

## 2 Empirical Mode Decomposition Features and Stochastic Processes

We describe the stochastic processes simulated and the features from the respective IMF along with the EMD algorithm in order to make the article selfcontained and accessible for a wider audience.

### 2.1 Stochastic Processes

We present all the necessary results to simulate the stochastic processes analysed in the next section, readers are referred to the original paper of each random process [3,12] or to [22] for a deeper explanation. There should be better algorithm implementations, in terms of memory use and algorithmical complexity, however these algorithms are good enough for clarity and for reproducibility purposes.

A general linear stochastic process is modeled as generated by a linear aggregation of random shocks [2] as the autoregressive model of order $p$,

$$z_t = \phi_1 z_{t-1} + \phi_2 z_{t-2} + .... + \phi_p z_{t-p} + a_t \tag{1}$$

where the current value $z_t$ is expressed as a finite, linear aggregate of previous values of the process $\{z_{t-1}, z_{t-2}, \cdots, z_{t-p}\}$ and a random shock $a_t$, distributed with mean 0 and finite variance $\sigma^2$.

Three kind of autoregressive process are considered within this paper: Gaussian, Exponential and Uniform. The difference between them is random shock $a_t$ form. Within this paper, for sake of simplicity we set the order of the autoregressive process $p = 1$ leading to the first-order autoregressive process AR(1).

In the Gaussian process, Eq. 2 takes the form:

$$z_t = \phi_1 z_{t-1} + a_t \tag{2}$$

where the current value $z_t$ is expressed as a finite, linear aggregate of the previous values of the process $z_{t-1}$ and an independent and identically distributed (i.i.d.) random shocks $a_t$ that has marginal Gaussian distribution with mean 0 and variance $\sigma^2$. We use the standard R command [15] *arima.sim()* to simulate this process.

For the Exponential marginally distributed AR(1) process, we use the NEARA(1) model [12] what is schematic present in Algorithm 3 and Eq. 2 takes the form:

$$z_t = a_t + \begin{cases} \beta . z_{t-1} & \text{w.p } \alpha \\ 0 & \text{w.p } 1-\alpha \end{cases} \tag{3}$$

with

$$a_t = \begin{cases} e_t & \text{w.p } \frac{1-\beta}{1-(1-\alpha)\beta} \\ (1-\alpha).\beta.e_t & \text{w.p } \frac{\alpha\beta}{1-(1-\alpha)\beta} \end{cases} \tag{4}$$

where *w.p.* stands by *with probability*, $\alpha > 0$ and $\beta > 0$ are free correlation parameters such as $\rho = \alpha\beta$, providing that $\alpha$ and $\beta$ are not both equal to one, $a_t$ has a p Exponential distribution. In Algorithm 3 we sketch the main features of the simulation. We simulate a time series of length $N$ (line 1) staring with a realization of a random variable exponentially distributed with $\lambda = 1$ (line 2), we set the desire correlation coefficient, in our simulations (line 3) and compute $\alpha$ y $\beta$ accordingly (lines 4 y 5), within the loop (lines 7–14) we resolve Eqs. 3 and 4. In order to generate the random variable $a_t$, we generate an exponential random variable with $\lambda = 1$ (line 8) and the selected it according a Binomial distribution with parameter $\frac{1-\beta}{(1-\beta+\alpha*\beta)}$ (lines 10–12). Lines 13 and 14 shows Ec. 3 when finally the stochastic process is fully generated.

**Algorithm 1.** Exponential Autoregressive Process generation

1: $N \leftarrow$ *Set the desired time series spam)*
2: $x[1] \sim \text{Exp}(\lambda = 1)$
3: $\rho \leftarrow$ *Set the desired autocorrelation coefficient)*
4: $\alpha \leftarrow \frac{2*\rho}{(1+\rho)}$
5: $\beta \leftarrow \frac{1}{(2-\alpha)}$
6: $t \leftarrow 2$
7: **for** t := 2 to N **do**
8:
$$En \sim \text{Exp}(\lambda = 1)$$
9: $pr \leftarrow \frac{1-\beta}{(1-\beta+\alpha*\beta)}$ )
10: $bn \sim \text{Bin}(n = 1, p = pr)$
11: $a_t[bn == 1] \leftarrow En[bn == 1]$ )
12: $a_t[bn == 0] \leftarrow En[bn == 0]$ )
13: $bn \sim \text{Bin}(n = 1, p = \alpha)$
14: $x[t] < -a_t + bn * beta * x[t-1]$

**Algorithm 2.** Uniform Autorregresive Process generation

1: $N \leftarrow$ *Set the desired time series spam)*
2: $k \leftarrow$ *Set k, which determine the desired autocorrelation coefficient)*
3: $\rho \leftarrow \frac{1}{k}$)
4: $x[1] \sim$ *sample(seq(0,(k-1)/k,1/k) ,1,replace=TRUE)*
5: **for** t := 2 to N **do**
6:
$$x[t] < -rho * x[t-1] + sample(seq(0, (k-1)/k, 1/k), 1, replace = TRUE)$$

In the Uniform autoregressive model of order 1 UAR(1) [3], depicted in Algorithm 2. Eq. 2 follows,

$$z_t = \frac{1}{k} z_{t-1} + a_t \tag{5}$$

with $k \geq 2$. It has been shown in [3] that $z_t$ would shield continuous U(0,1) marginal distribution if the i.i.d. random shocks $a_t$ is sampled as a $1/k$ uniform distribution over the set $\{0, 1/k, 2/k, \ldots, \frac{k-1}{k}\}$. Algorithm 3 shows the pseudocode for the computer implementation of Eq. 5. The desired time spam and correlation coefficient are sets in line 1 and 2–3, respectively. In line 4 the Eq. 5 is realized for $t = 1$ and them in the loop for all time spam.

## 2.2 Empirical Mode Decomposition Features

In 1998, Huat et al. proposed empirical mode decomposition [10]. It is a fully adaptive technique that can be applied to nonlinear and non-stationary process. The basic idea of EMD method is decompose the complicated time series $x(t)$ into a number of IMFs as,

$$x(t) = \Sigma_{i=1}^{M} h_i(t) + r(t) \tag{6}$$

where $M$ is the cardinality of the IMF set and $r(t)$ is the residual after transformation.

EMD method through which EMD decomposes the original time series into a series of IMF components with different timescales is shown in Algorithm 3. Each IMF must satisfy the following two conditions [7]: (1) the number of extreme and the number of zero crossings must be equal or differ at most by one (in the entire signal length); (2) the average value of the two envelope defined by the local maximum and the minimum must be zero at any moment. Finally, the last level is the residue of the time series which is related with the trend from time series.

**Algorithm 3.** EMD Algorithm

1: $D(t) \leftarrow x(t)$
2: $i \leftarrow 1$
3: **while** D(t) is not monotonic **do**
4:
$E_{max}(t) \leftarrow interpolation(max\{x(t)\})$
5: $E_{min}(t) \leftarrow interpolation(min\{x(t)\})$
6: $m(t) \leftarrow (E_{max}(t) + E_{min}(t)/2$
7: $D(t) \leftarrow x(t) - m(t)$
8: **if** D(t) satisfy the conditions of IMF **then**
9:
**else**
$IMF_i(t) \leftarrow D(t)$
10: $r(t) \leftarrow x(t) - IMF_i(t)$
11: $x(t) = r(t)$
12: $i \leftarrow i + 1$
13: $x(t) \leftarrow D(t)$

## 2.3 Time-Frequency Related Features

In order to characterize the IMF set for several non-gaussian autoregressive processes we follow [20] and estimate:

1. Density probability distribution for each IMF using their respective histogram, using the Scott's rule [20].
   We use Scott's rule for determining the bin number because this rule is optimal in the sense that asymptotically minimizes the integrated mean squared error.
2. The number of peaks of each IMF attended in the lines 4 and 5 inside the while loop in Algorithm 3 applied to each process studied and which allows determines the mean period of the function by counting the number of peaks of the function [23].
3. Mean period in terms of the mean number of samples for each IMF and each process (Gaussian, Uniform and Exponential).

4. The effect produced by EMD method for each level of decomposition from the time series was calculated with the spectrum of frequencies $X(\Omega)$ using the Discrete Fourier Transform defined as

$$X(\Omega) = \sum_{n=-\infty}^{\infty} x[t]e^{-j\Omega n} \tag{7}$$

### 2.4 Energy-Entropy Related Features

In [7] the EMD-entropy and EMD-statistical complexity are proposed as follows, The total energy of each IMF is proposed as $E_i = \Sigma_{t=1}^{n}[h_i(t)]^2$ where $n$ is the length of the $i-th$ IMF and $h_i(t)$ denotes the value n of the i IMF. Finally, the total energy of signal is the sum over all total energies $E_i$, $E = \Sigma_{i=1}^{m} E_i$. Hence the relative energy of each IMF is defined as

$$p_i = \frac{E_i}{E} \tag{8}$$

and following the usual discrete form of Shannon entropy, the EMD-entropy is defined using $p_i$ in Eq. 8 $S_{EMD}(l) = -\Sigma_{i=1}^{l} p_i \times log(p_i)$ where $l$ should stand for the order of EMD-entropy. As the reconstruction could be done using the first $l, l < m$ IMF, and regarding all other IMF rest as the rest in Eq. 6 this new entropy has this free parameter. This entropy $H_{EMD}(l)$ is an unnormalized quantity, so to restrict its value to the $[0, 1]$ interval, it is redefined as,

$$H_{EMD}(l) = \frac{S_{EMD}(l)}{log(l)} \tag{9}$$

as $H_{EMD}(l)$ gets it maximum value when $P = (p_i, \dots, p_l)$ $= (1/l, \dots, 1/l)$. Another defined quantity in [13,14] is the statistical complexity, using the Jensen-Shannon divergence, i.e.

$$C_{EMD}(l) = Q_{EMD}(l)H_{EMD}(l) \tag{10}$$

where $Q_{EMD}(l) = Q0 \times JSD[P, P_e]$, $P_e$ is a reference probability distribution $P = (p_i, \dots, p_l) = (1/l, \dots, 1/l)$ and $Q0$ a normalization constant. Both quantities form the entropy-complexity plane
$H_{EMD}(l) \times C_{EMD}(l)$ plane, There is a vast literature regarding this plane [13,14,17–19]

## 3 Numerical Results and Discussion

The processes presented in the previous section were simulated, varying their autocorrelation coefficients value to evaluate their IMFs using the features described in the previous section as an evaluation methodology [4]. For the Gaussian processes five positively autocorrelated time series were simulated, with the corresponding color code $rho$ = ((yellow), 0, (magenta), 0.2, (cyan) 0.4, (red) 0.6, (green) 0.8, (blue) 0.9), all these processes were simulated with normally distributed shocks $a_t$ with $\sigma = 1$ and $\mu = 0$. The Exponential processes where simulated with suitable values for $\alpha$ and $\beta$ (see lines 4 and 5 in Algorithm 3) such for exponential process are $rho$ = (yellow) 0, (magenta) 0.125, (cyan) 0.25, (red) 0.5, (green) 0.75) and the autoregressive Uniform process was simulated for $rho$ = (yellow) 0, (magenta) 0.1, (cyan) 0.2, (red) 0.25, (green) 0.33, (blue) 0.50), plus the uncorrelated data with Uniform marginal distribution. All series have length $N = 10^{15}$ points. IEEE 754 double precision floating point numbers was used for all computations.

Figure 1 shows the probability density function for IMFs 1, 3 and 5 of each stochastic process. Surprisingly, Uniform distribution, no matter the autocorrelation coefficient (see Fig. 1a, 1d and 1g.,) does not lead an equiprobable distribution of $p_i$ (in Eq. 8) since the density probability distribution for the first IMF is bi-modal. This fact has a profound impact on the entropy estimation based on EMD and especially in the entropy-complexity plane as it can be seen later in this Section. Gaussian process behaves as reported previously [7] and exponential processes seem to be more likely as equiprobable distribution of $p_i$. By equiprobable distribution we mean that all the IMF has the same probability distribution

In Tables 1 and 2 show the number of peaks and the mean period respectively for each simulated stochastic processes. The results show that, for all processes, the mean period of any IMF component is almost exactly double that of the previous one. This result is consistent with the result obtained by Wu et al. (2004) and Flandrin et al. (2003) for Gaussian distributions. Also, the results indicate that when the autocrrelation increases the mean difference between any two IMF component increases, no matter the probability distribution. Standard deviation also increases as the autocorrelation coefficient increase, too. Table 1 shows that for times series where the value of autocorrelation is higher there are fewer peaks in each IMF component so that, the number of peaks can be related to how correlated the samples of time series are and no matter about the time series marginal distribution.

Figure 3 shows Fourier spectra (Eq. 7) from simulated stochastic processes for IMFs 1, 3 and 6. It can be observed that when the IMF component increases the content of information from the process is found in low frequency which corresponds with [23]. Furthermore, we can see in IMF 1 component that Exponential distribution has more frequency content in low frequency than Gaussian distribution and Uniform distribution. It is interesting to note that when the value of the autocorrelation coefficient increases, the frequency content increases in low frequency so that the information about autocorrelation from the processes is mainly in the lower part of the spectra. The study of the stochastic processes with Gaussian distribution indicates that when the IMF component increases the spectra is concentrated on low frequency and its amplitude increases as if the energy decreaces slowly in each IMF component. For a stochastic process with uniform or exponential distribution, it does not happen.

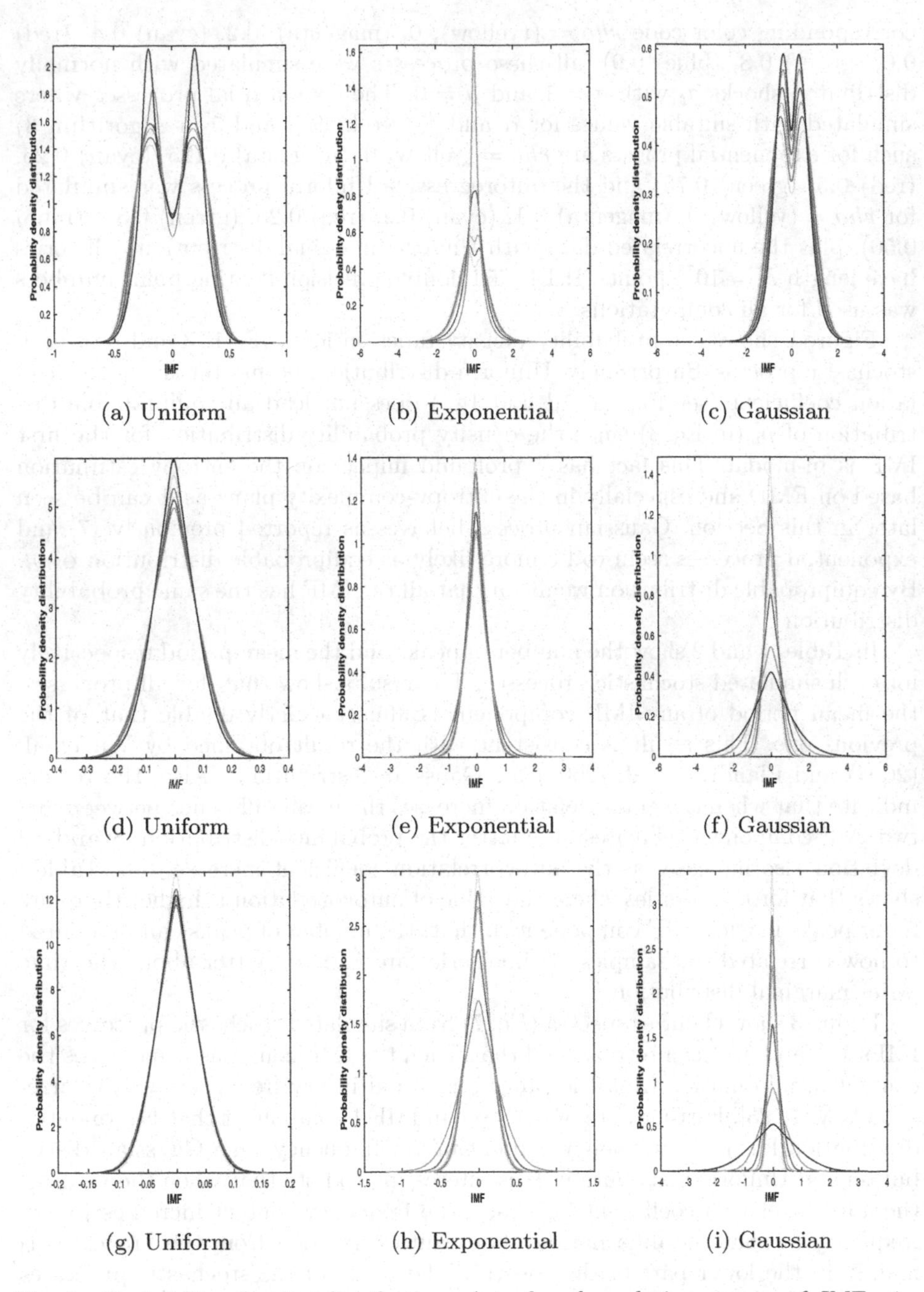


(a) Uniform (b) Exponential (c) Gaussian

(d) Uniform (e) Exponential (f) Gaussian

(g) Uniform (h) Exponential (i) Gaussian

**Fig. 1.** Probability density distribution plots for the relative energy of IMFs 1-3-6 (Eq. 8) for the simulated stochastic processes. Color codes are for $rho$ = $0, 0.10, 0.2, 0.25, 0.33, 0.5, 0.75$ are yellow, magenta, cyan, red, green, blue, respectivelly. (Color figure online)

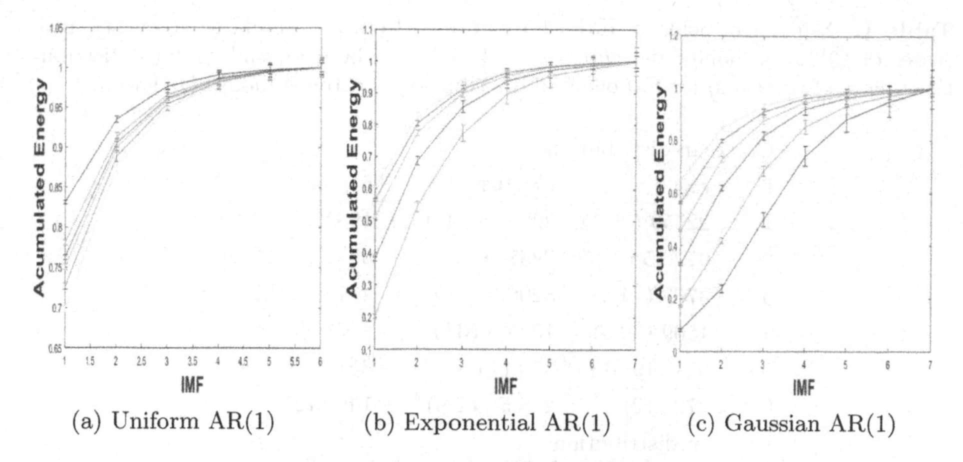


(a) Uniform AR(1) (b) Exponential AR(1) (c) Gaussian AR(1)

**Fig. 2.** Cumulative plots for the relative Energy (Eq. 8) for the simulated stochastic processes. Color codes are for $rho = 0, 0.10, 0.2, 0.25, 0.33, 0.5, 0.75$ are yellow, magenta, cyan, red, green, blue, respectivelly. (Color figure online)

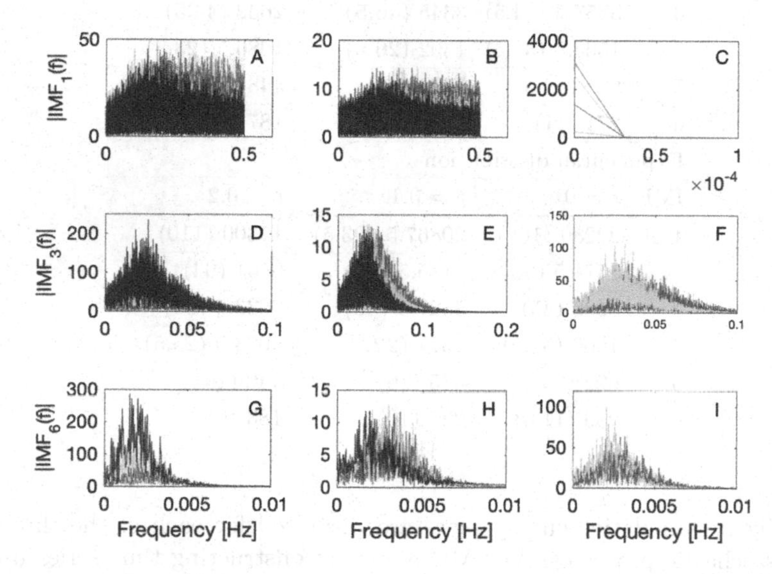


**Fig. 3.** Fourier spectra (Eq. 7) of IMFs 1,3 and 6 for simulated stochastic processes. A, D, G: Gaussian process. B, E, H: Uniform process. C, F, I: Exponential process. Color codes are for $rho = 0, 0.10, 0.2, 0.25, 0.33, 0.5, 0.75$ are yellow, magenta, cyan, red, green, blue, respectivelly. (Color figure online)

**Table 1.** Number of peaks of IMFs 1–6 obtained from the simulated stochastic processes of $10^{15}$ data points and $rho =$ (0, 0.4, 0.9) for Gaussian and Uniform distributions, $rho =$ (0,0.1,0.2) for Exponential distribution and the standard deviation.

| Gaussian distribution | | | |
|---|---|---|---|
| IMF | r = 0 | r = 0.4 | r = 0.9 |
| 1 | 22559 (90.5) | 20897.5 (136.5) | 19847.5 (252.5) |
| 2 | 9126.5 (73.5) | 7931 (65) | 6542.5 (473.5) |
| 3 | 3767 (44.5) | 3206.5 (52) | 2313.5 (152) |
| 4 | 1569 (30.5) | 1323.5 (31) | 892 (58) |
| 5 | 651 (19.5) | 550 (14.5) | 361 (25) |
| 6 | 272 (12) | 228.5 (12.5) | 149 (12) |
| Uniform distribution | | | |
| IMF | r = 0 | r = 0.4 | r = 0.9 |
| 1 | 22435 (88) | 20965 (115.5) | 16831 (127) |
| 2 | 9075.5 (53.5) | 8121 (72) | 6492.5 (63) |
| 3 | 3757.5 (44.5) | 3345 (46.5) | 2632 (44.5) |
| 4 | 1562.5 (30.5) | 1392 (26.5) | 1086.5 (23.5) |
| 5 | 651 (19.5) | 576.5 (18) | 448 (20) |
| 6 | 271 (11) | 238.5 (12) | 187 (10) |
| Exponential distribution | | | |
| IMF | r = 0 | r = 0.1 | r = 0.2 |
| 1 | 22289 (108) | 20867.5 (93.5) | 17300 (110) |
| 2 | 8771.5 (88.5) | 8135.5 (81) | 6561 (94) |
| 3 | 3606 (45) | 3327.5 (50) | 2633.5 (50) |
| 4 | 1500 (31.5) | 1377 (22) | 1073.5 (25.5) |
| 5 | 622 (20) | 575 (16) | 440 (16) |
| 6 | 259 (11.5) | 237 (10) | 183 (9) |

In Fig. 2, cumulative energy plots are displayed for each of the three simulated stochastic processes. As EMD allows reconstructing time series up to a certain $m$ in Eq. 6, this plot is just a summation of the relative energy values. It is shown that with only the first five IMFs we recover most than 80% of energy for all simulated processes. It is interesting to note that the Uniform stochastic processes retrieve the information with less IMFs than the other processes and the process which needs more IMF's to retrieving the information is the Gaussian. In turn, as the correlation increases, the number of IMFs necessary to retrieve the information also increases. This characteristic could be an advantage of this method since it would allow knowing the size of the entropy by means of an explicit criterion. In other words, if the researcher were to agree to recover 90% of the time series information, would search for the value of $m$ that makes

**Table 2.** Mean periods of IMFs 1–6 obtained from the simulated stochastic processes of $10^{15}$ data points and $rho$ = (0, 0.4, 0.9) for Gaussian and Uniform distributions, $rho$ = (0, 0.1, 0.2) for Exponential distribution in terms of the number of samples and the standard deviation.

| Gaussian distribution | | | |
|---|---|---|---|
| IMF | r = 0 | r = 0.4 | r = 0.9 |
| 1 | 1.4525 (0.0058) | 1.568 (0.0102) | 1.651 (0.021) |
| 2 | 3.5904 (0.0289) | 4.1316 (0.0339) | 5.0085 (0.3446) |
| 3 | 8.6987 (0.1026) | 10.2192 (0.1655) | 14.1639 (0.9013) |
| 4 | 20.8846 (0.4064) | 24.7586 (0.5791) | 36.7354 (2.354) |
| 5 | 50.3349 (1.5092) | 59.5782 (1.5695) | 90.7701 (6.2761) |
| 6 | 120.4706 (5.3175) | 143..4055 (7.7988) | 219.9195 (17.6223) |
| Uniform distribution | | | |
| IMF | r = 0 | r = 0.4 | r = 0.9 |
| 1 | 1.4606 (0.0057) | 1.5630 (0.0086) | 1.9449 (0.0146) |
| 2 | 3.6106 (0.0213) | 4.0350 (0.0357) | 5.0471 (0.049) |
| 3 | 8.7207 (0.1035) | 9.7961 (0.1362) | 12.4498 (0.2107) |
| 4 | 20.9715 (0.4103) | 23.5402 (0.4467) | 30.1592 (0.6552) |
| 5 | 50.3349 (1.5092) | 56.8396 (1.7537) | 73.1429 (3.2308) |
| 6 | 120.9151 (4.8920) | 137.3926 (6.8883) | 175.2299 (9.3773) |
| Exponential distribution | | | |
| IMF | r = 0 | r = 0.1 | r = 0.2 |
| 1 | 1.4701 (0.0071) | 1.5703 (0.0070) | 1.8941 (0.0120) |
| 2 | 3.7357 (0.0378) | 4.0278 (0.0400) | 4.9944 (0.0715) |
| 3 | 9.0871 (0.1134) | 9.8476 (0.1480) | 12.4428 (0.2348) |
| 4 | 21.8453 (0.4593) | 23.7967 (0.3799) | 30.5245 (0.7255) |
| 5 | 52.6817 (1.6971) | 56.9878 (1.5943) | 74.4727 (2.6967) |
| 6 | 126.5174 (5.6308) | 138.2616 (5.7875) | 179.0601 (8.8600) |

that recovery and then compute Shannon entropy using the $p_i$ set. We have arbitrarily selected the value of $m = 5$ and we let $m = 6$ as the residual energy in the plot.

Figure 4 shows three entropy-complexity planes. It is interesting that the Exponential processes are located in the right part and below the plane (see Fig. 3b) instead the uniform processes are located in the middle part of the plane (entropy = 0.6). This differs significantly from previous results referring to the same processes but located in the plane using the permutation entropy [22]. On the one hand, using the symbolization of Band and Pompe [1] to calculate the probabilities, the location of the three processes in the plane is practically indistinguishable, suggesting that the symbolization proposed by Band and

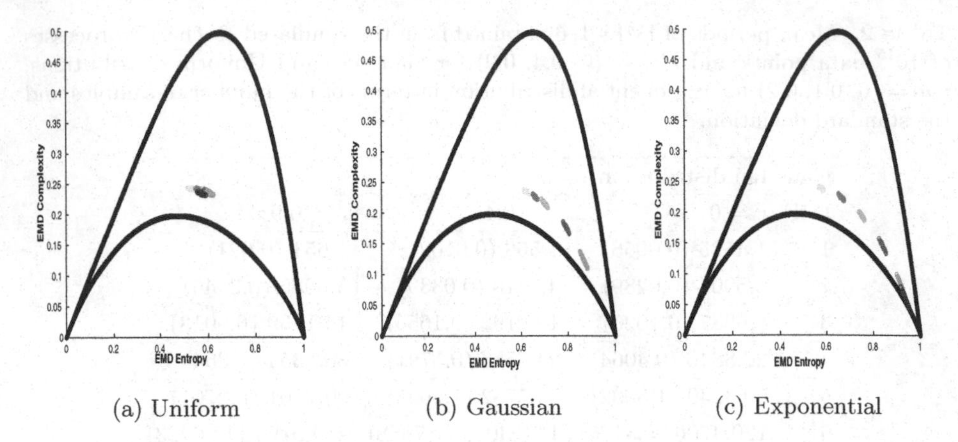


(a) Uniform (b) Gaussian (c) Exponential

**Fig. 4.** Several EMD Entropy-Complexity planes for the simulated stochastic processes. Color codes are for Uniform are $rho = (0, 0.1, 0.2, 0.25, 0.33, 0.50)$, for Exponential process are $rho = (0, 0.125, 0.25, 0.5, 0.75)$ and for Gaussian process are $rho = (0, 0.2, 0.4, 0.6, 0.8, 0.9)$ (Color figure online)

Pompe does not distinguish between probability distributions if not only between different correlations; on the other hand, in the methodology proposed in [7] entropy recognizes both characteristics, the marginal distribution of the stochastic process and autocorrelation. On the other hand, regarding the densities shown in Fig. 3, the order of the distributions is not as expected, one would have expected the maximum entropy to manifest in uniform stochastic processes, however, it is shown that these processes have the minimum entropy among the three analyzed and that the maximum entropy occurs in exponential stochastic processes.

## 4 Conclusions

Exploring features for IMF descriptions is an interesting task, useful in order to characterize actual time series. If a real time series is tough as a realization of an stochastic process endowed with a correlation structure and a marginal probability distribution, the IMF features could reveal some features of both.

Regarding the distribution if the energy for each IMF component and each stochastic process with different distribution is calculated, for a process with Gaussian distribution the energy loss in each IMF is lower compared with the previous IMF than the process with uniform or Exponential distribution. Thus, the EMD method in a Non-Gaussian process produces a large decrease in the energy when IMF component increment. We also observe Exponential distribution has more frequency content in low frequency than Gaussian and Uniform distribution.

According to the correlation value from distribution we observe for high values of autocorrelation coefficient that the frequency content increases in low frequency while for low values the frequency content decreases.

EMD also seems not to only discriminate autocorrelation but also the marginal distribution of simulated processes, as can be seen in Fig. 4. This would give it good performance as a classifier shown in [7] and should be explored. However, this strategy to calculate the probabilities shows a challenge and that is that the maximum entropy is not found in the scenario of greater uncertainty, that is, in the uniform distribution, as it can be seen in Fig. 4. This characteristic must be taken into account if the intention of the investigation is the characterization between chaos and noise. In particular, statistical complexity includes a measure of distance from a reference probability, and this reference is generally established as the equilibrium probability and this is not the case if we follow the proposal in [7].

This is a step ahead in using EMD to characterize several stochastic processes presents in science and technology. Even more research is needed in validating this approach with several probability distribution and correlation structures we think the results present in this contribution are general enough to a first glimpse into the subject.

## References

1. Bandt, C., Pompe, B.: Permutation entropy: a natural complexity measure for time series. Phys. Rev. Lett. **88**(17), 174102 (2002)
2. Box, G.E., Jenkins, G.M., Reinsel, G.C., Ljung, G.M.: Time Series Analysis: Forecasting and Control. Wiley, Hoboken (2015)
3. Chernick, M.R.: A limit theorem for the maximum of autoregressive processes with uniform marginal distributions. The Annals of Probability, pp. 145–149 (1981)
4. Davidson, E.J.: Evaluation Methodology Basics: The Nuts and Bolts of Sound Evaluation. Sage, Thousand Oaks (2005)
5. Engle, R.F., Russell, J.R.: Autoregressive conditional duration: a new model for irregularly spaced transaction data. Econometrica, pp. 1127–1162 (1998)
6. Farashi, S.: Spike sorting method using exponential autoregressive modeling of action potentials. World Acad. Sci. Eng. Technol. Int. J. Med. Health Biomed. Bioeng. Pharmaceutical Eng. **8**(12), 864–870 (2015)
7. Gao, J., Shang, P.: Analysis of complex time series based on EMD energy entropy plane. Nonlinear Dyn. **96**(1), 465–482 (2019)
8. Hafner, C.: Nonlinear time series analysis with applications to foreign exchange rate volatility. Springer Science & Business Media (2013)
9. Huang, N.E., Shen, Z., Long, S., Wu, M., Shih, H., Zheng, Q., Tung, C., Liu, H.: he empirical mode decomposition and hilbert spectrum for nonlinear and nonstationary time series analysis. Proc. Roy. Soc. A **545**(1971), 903–995 (1998)
10. Huang, N.E., et al.: The empirical mode decomposition and the hilbert spectrum for nonlinear and non-stationary time series analysis. Proc. Roy. Soc. London. Series A: Math. Phys. Eng. Sci. **454**(1971), 903–995 (1998)
11. Lawrance, A.: Uniformly distributed first-order autoregressive time series models and multiplicative congruential random number generators. J. Appl. Probability **29**, 896–903 (1992)
12. Lawrance, A., Lewis, P.: A new autoregressive time series model in exponential variables (near (1)). Advances in Applied Probability, pp. 826–845 (1981)

13. Lopez-Ruiz, R.: Complexity in some physical systems. Int. J. Bifurcation Chaos **11**(10), 2669–2673 (2001)
14. Lopez-Ruiz, R., Mancini, H., Calbet, X.: A statistical measure of complexity. arXiv preprint nlin/0205033 (2002)
15. R Core Team: R: A Language and Environment for Statistical Computing. R Foundation for Statistical Computing, Vienna, Austria (2020). https://www.R-project.org/
16. Rilling, G., Flandrin, P., Goncalves, P., et al.: On empirical mode decomposition and its algorithms. In: IEEE-EURASIP Workshop on Nonlinear Signal and Image Processing, vol. 3, pp. 8–11. NSIP-03, Grado (I) (2003)
17. Rosso, O., Larrondo, H., Martin, M., Plastino, A., Fuentes, M.: Distinguishing noise from chaos. Phys. Rev. Lett. **99**(15), 154102 (2007)
18. Rosso, O.A., Carpi, L.C., Saco, P.M., Ravetti, M.G., Plastino, A., Larrondo, H.A.: Causality and the entropy-complexity plane: Robustness and missing ordinal patterns. Physica A **391**(1), 42–55 (2012)
19. Rosso, O.A., Olivares, F., Zunino, L., De Micco, L., Aquino, A.L., Plastino, A., Larrondo, H.A.: Characterization of chaotic maps using the permutation bandt-pompe probability distribution. Eur. Phys. J. B **86**(4), 1–13 (2013)
20. Schlotthauer, G., Torres, M.E., Rufiner, H.L., Flandrin, P.: Emd of gaussian white noise: effects of signal length and sifting number on the statistical properties of intrinsic mode functions. Adv. Adapt. Data Anal. **1**(04), 517–527 (2009)
21. Sengupta, D., Kay, S.: Efficient estimation of parameters for non-gaussian autoregressive processes. IEEE Trans. Acoust. Speech Signal Process. **37**(6), 785–794 (1989)
22. Traversaro, F., Redelico, F.O.: Characterization of autoregressive processes using entropic quantifiers. Physica A **490**, 13–23 (2018)
23. Wu, Z., Huang, N.E.: A study of the characteristics of white noise using the empirical mode decomposition method. Proc. Roy. Soc. London. Series A: Math. Phys. Eng. Sci. **460**(2046), 1597–1611 (2004)

# Analysing Documents About Colombian Indigenous Peoples Through Text Mining

Jonathan Stivel Piñeros-Enciso(✉) and Ixent Galpin

Facultad de Ciencias Naturales e Ingeniera, Universidad de Bogotá Jorge Tadeo Lozano, Bogotá, Colombia
{jonathan.pinerose,ixent}@utadeo.edu.co

**Abstract.** The indigenous peoples of Colombia have a considerable social, political and cultural wealth. However, issues such as the decades-long armed conflict and drug trafficking have posed a significant threat to their survival. In this work, publicly available documents on the Internet with information about two indigenous communities, the Awá and Inga people from the Cauca region in southern Colombia, are analysed using text analytics approaches. A corpus is constructed comprising general characterisation documents, media articles and rulings from the Constitutional Court. Topic analysis is carried out to identify the relevant themes in the corpus to characterise each community. Sentiment analysis carried out on the media articles indicates that the articles about the Inga tend to be more positive and objective than the Awá. This may be attributed to the significant impact that the armed conflict has had on the Awá in recent years, and the productive projects of the Inga. Furthermore, an approach for summarising long, complex documents by means of timelines is proposed, and illustrated using a ruling issued by the Constitutional Court. It is concluded that such an approach has significant potential to facilitate understanding of documents of this nature.

**Keywords:** Text analytics · Sentiment analysis · Indigenous communities

## 1 Introduction

The indigenous populations of southern Colombia have historically suffered from drug trafficking and wars by illegal armed groups. These peoples live in a constant struggle to protect their land and rights [14,21]. Despite the violence and armed conflict pervasive throughout their recent history, they are characterised by their resilience [15]. Furthermore, they are known for their rich cultural and productive projects, and live in areas with high biodiversity. In this work, we analyse publicly available information on the Internet about these communities. We apply various text analytics techniques that characterise or summarise documents about these communities in a succinct and automated manner. We use Python for this work given the broad range of data wrangling and text analytics libraries available for this language.


H. Florez and M. F. Pollo-Cattaneo (Eds.): ICAI 2021, CCIS 1455, pp. 125–140, 2021.
https://doi.org/10.1007/978-3-030-89654-6_10

Documents on the Internet can be so diverse that the number of categories may result intractable. For this study, the scope of documents in this study is reduced to three categories. The categories defined are (1) documents which characterise the communities, limited to official documents and/or those issued by territorial entities (2) media articles, limited to articles within a specific time interval and by prominent media outlets, and (3) legal documents, limited to rulings from the Colombian Constitutional Court because these are of vital importance in the protection of human rights. Further details about the delimitation and extraction of the corpus are provided in Sect. 3.

Once the corpus has been determined, it is necessary to establish how to approach the analysis of the information. In this work, we go from the general to the specific. In Sect. 4, an analysis of the topics present in the entire corpus is carried out. This allows us to confirm the broad range of topics present in the extracted documents, from tragic violence to rich cultural aspects.

In Sect. 5, sentiment analysis is carried out on the texts from some Colombian media emblematic outlets. The documents are evaluated and classified according to whether they evoke positive or negative connotations, as the case may be. The purpose of this process is to check whether the characteristics of violence are evident, and the degree to which they are reflected in the collected material. Furthermore, an evaluation of the objectivity and subjectivity of these documents is carried out. This type of analysis can be useful when examining the media. This is important given that indigenous peoples in Colombia and other countries around the world have repeatedly denounced false accusations by the media [7], lack of access and participation in the media [12], and even attacks against fundamental rights by the media [10].

Section 6 presents an algorithm that identifies date patterns (both exact dates and time intervals) and uses entity recognition techniques to determine events, actors, organisations, and actions. The result of this algorithm is a visualisation that contains a timeline with a small summary of each event. This aims to enable the understanding of a sequence of events in a summarised and automated way. Given that legal texts are often long and difficult to understand, this algorithm is demonstrated using a Constitutional Court ruling as an example.

Section 7 concludes.

## 2 Related Work

There are few previous works involving the automated analysis of documents about indigenous communities. Colquhoun *et al.* [11] study the *Aboriginal and Torres Strait Islander* communities in Australia. This study aims to verify the correlation between maintaining cultural traditions and the happiness of the population. For this, they analyse qualitative data with the Leximancer[1] text analytics software. This allows the grouping of topics and the generation of graph visualisations. Shah *et al.* [25] explore sentiment analysis in the indigenous

[1] https://info.leximancer.com/.

languages of India. This is challenging due to the differences in the structures with languages such as English. It is worthy of mention that no works related to sentiment analysis or text analytics were found for indigenous communities in Colombia. It is also observed that globally, there is very little research that analyses documents about indigenous communities.

Most previous work involving automated sentiment analysis use a corpus obtained from social networks. Twitter seems to be popular due to its microblog structure and the generation of opinions in real-time [1,22,24]. There are several works whose focus is to establish a methodology for the analysis of sentiments over a corpus comprising media or news articles. Godbole *et al.* [13] explore positive or negative connotations in sports, weather, and stock market news. Raina *et al.* [23] emphasise the computational technique called Sentic Computing[2]. Balahur *et al.* [3] analyse different points of view that can be given when analysing a news item. On the other hand, there are practical cases that focus on the search for documents related to news. Bautin *et al.* [5] address the problem of language, translations and their effectiveness in this process; Li *et al.* [18] perform financial sentiment analysis for the Hong Kong Stock Exchange based on economic news and indicators; Kaya *et al.* [16] focuses on political journalism in Turkey.

Finally, previous work involving the summarisation of documents based on timelines is considered. Initial research in this area focused on statistical approaches for the generation of time entities and summaries. Swan *et al.* [26] propose a statistical model to find temporal characteristics in a text. Mani *et al.* [19] propose an experimental model to determine the chronology of events and [27] generate a visualisation of events in a timeline. Allan *et al.* [2] generate an experimental methodology to manage a corpus and propose abstract structures. Chieu *et al.* [9] propose a framework that assumes important events within a text collection and orders them chronologically. More recent work tends to use *machine learning* techniques. Yan *et al.* [29] focus on finding the most relevant news under the *web mining* technique. Kessler *et al.* [17] do not focus on events, rather on the normalisation of temporal entities and [28] relies on news headlines as a fundamental element of the search. Chen *et al.* [8] focuses on time series and uses neural networks to generate more effective summaries.

## 3 Definition of the Corpus

Documents that mention two well-known indigenous communities are chosen for the present study: the *Inga* and *Awá*. Both communities are affiliated to the Regional Indigenous Council of Cauca[3]. From each population, the following categories of information are selected to comprise the corpus, *viz.*,

[2] https://sentic.net/computing/.

[3] The CRIC (Consejo Regional Indígena del Cauca) it is an association of indigenous councils in Cauca, Colombia.

- **Characterisation documents**. Theses are official documents and/or those issued by territorial entities that characterise the population. These documents include information including demographic, social, and political aspects of the populations.
- **News articles** published over a range of five years (2015–2020) by three emblematic media outlets in Colombia, namely *Semana* magazine, and *El Tiempo* and *El Espectador* newspapers;
- **Constitutional Court rulings** concerning these communities over the last 30 years (1990–2020).

### 3.1 Characterisation Documents

A Python script is used to download the PDF documents that meet the search parameters. This process consists of searching for keywords, such as the name of the population and the word "characterisation". For this process, the wget[4] Python library is employed. Subsequently, a manual pruning of redundant or irrelevant documents is performed. In total, ten documents are obtained for each population in PDF format that contains the characterisations information.

### 3.2 News Articles

News articles are obtained through a *webscraping* process. An automated Google search is performed using the Python google-search[5] library. Filtering is carried out over three parameters: the date range (2015–2020), keywords (the name of the community and the word "indigenous" are used), and the information sources that correspond to the URLs of the aforementioned media. In this way, only the text corresponding to each news is obtained and saved in text files. As a result, 133 documents related to the Awá community and 139 to the Inga community are obtained, distributed as shown in Table 1.

**Table 1.** News Corpus

| Indigenous community | Source | Number of documents | Number of words | Number of characters |
|---|---|---|---|---|
| Awá | El Espectador | 100 | 78150 | 496754 |
| Awá | El Tiempo | 3 | 2140 | 13362 |
| Awá | Revista Semana | 30 | 38725 | 239875 |
| Inga | El Espectador | 79 | 80587 | 511989 |
| Inga | El Tiempo | 41 | 35111 | 217731 |
| Inga | Revista Semana | 19 | 19559 | 122689 |

[4] https://pypi.org/project/wget/.
[5] https://pypi.org/project/google-search/.

### 3.3 Constitutional Court Rulings

To obtain this corpus, it is necessary to perform a manual search on the Constitutional Court of Colombia website[6]. The page allows users to search by year and keyword. Ten documents are obtained for each community.

## 4 Overall Analysis of the Corpus

In order to perform an overall analysis of the corpus, the three categories of documents are loaded into a single Python `DataFrame`[7]. One script is run for the PDF documents and another for the text files. The text is encoded in UTF-8 format and special characters that could interfere with the information analysis are removed.

Subsequently, *tokenisation* is carried out, whereby words are extracted from the text as different types of entities (referred to as *tokens*). These tokens exclude punctuation marks and *stopwords*, the most common words in the language that, due to their high repetition, do not convey important meaning in isolation [20]. As a next step, *bigram* and *trigrams*, *i.e.*, pairs and groups of three words, respectively, that appear consecutively, are identified. Such words are deemed to pertain to the same context [20].

The final preprocessing step is *lemmatisation*, which involves obtaining the "root" of each word so that it appears in a normal form [4]. With this, words can be grouped according to their meaning even if they have different forms or conjugations. Subsequently, a document-term matrix is generated. The matrix indicates the number of times the words are repeated. This provides an initial measure of the relative importance of words [20].

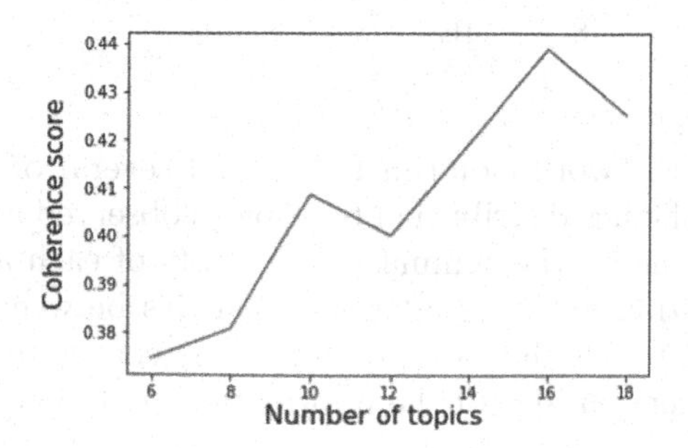


**Fig. 1.** Coherence of the model vs Number of topics

To identify topics in the corpus a *Latent Dirichlet Allocation* (LDA) model is created. LDA identifies latent factors that each topic has in common [6]. The

[6] http://www.corteconstitucional.gov.co.

[7] https://pandas.pydata.org/pandas-docs/stable/reference/api/pandas.DataFrame.html.

optimal number of topics for an LDA model is determined using the LdaMallet[8] library. The model is executed with different parameters, and the one that results in the highest coherence range is deemed to be optimal. This is a commonly used approach to determine the optimal number of topics to characterise a corpus. Figure 1 shows the coherence of the model against the number of topics evaluated. The highest coherence score (0.439) is found with 16, which would result in a rather unwieldy number of topics. Furthermore, when LDA is performed with 16 topics, it is possible to observe the redundancy of information, and several topics without representation within the documents. Trial and error suggested that the ideal number of topics is closer to 10, which is the value used. Figure 2 shows the distribution of the topics concerning the information extracted and gives an initial idea of the information present in each topic. Table 2 shows the number of documents per topic.

**Table 2.** Documents by topic

| Topic | Number of documents |
|---|---|
| 9 | 79 |
| 2 | 55 |
| 7 | 33 |
| 5 | 25 |
| 6 | 21 |
| 1 | 20 |
| 4 | 18 |
| 0 | 18 |
| 3 | 17 |
| 8 | 16 |

When analysing the word cloud in Fig. 2 and several of the documents by topic, the following themes distribution by topic is observed as shown in Table 3. This selection of themes is the summary of a study of each article classified by topic. The identification of topics allows similarities between documents to be found, the variety of topics that are present in the corpus to be understood, as well as providing a general idea of the information contained in the corpus.

## 5 News Article Sentiment Analysis

In this section, sentiment analysis is applied over the news article corpus, to understand whether positive or negative aspects are being reflected in the media when these indigenous communities are mentioned. As an initial step, word clouds are generated for each community to understand the recurring themes for each community, shown in Fig. 3.

[8] http://mallet.cs.umass.edu/.

Topic 0
desarrollar indigenas hacer territorio tradicional poblar formar resguardo cultural autoridad
Topic 1
salud contar ciudad medicinar inga cultura realizar encontrar gran sistema
Topic 2
solo ano arte llegar hacer tambien decir hablar vivir mas
Topic 3
nacional auto informar especial octubre riesgo poblar narino medir comunidad
Topic 4
comunidad indigena sentenciar area proyectar consultar ambiental cultural derecho desarrollar
Topic 5
reconocer cultural comunidad pueblo indigenas miembro derecho autoridad indigena sentenciar
Topic 6
procesar realizar territorio ver comunidad territorial lugar partir tierra actividad
Topic 7
hectareas personar presentar indigenas ano zona total municipio grupo registrar
Topic 8
inga hacer pueblo manera territorio dar partir poblar dentro cultural
Topic 9
mas campesino encontrar explicar paz hacer millón decir aguar indigenas

**Fig. 2.** Word cloud for the entire corpus

**Table 3.** Documents by topic

| Topic | Generality of topics |
|---|---|
| 0 | Territory, demographic aspects, economic activities |
| 1 | Science, university professionals and different academics fields |
| 2 | Productive projects and new ways of producing in the face of adversity |
| 3 | Violence, armed groups, recruitment of minors, threats |
| 4 | Territorial disputes, claim for environmental rights, indiscriminate mining |
| 5 | Rulings that protect indigenous peoples, their subsistence and ethnic and/or cultural identity |
| 6 | Land as a territory, its geographical characteristics, or agricultural, fauna and flora |
| 7 | Characterisation of the peoples and geographical aspects |
| 8 | Social aspects, religions, festivals and rituals |
| 9 | Oral Tradition, Music, festivals, peace initiatives and cultural activities |

(a) Awá community

(b) Inga community

**Fig. 3.** World Clouds showing main themes in new articles

Similarities are seen between the word clouds in Fig. 3. For example, the terms *indígena* (indigenous), *comunidades* (communities), *territorio* (territory), y *pueblos* (peoples) appear in both clouds, as each one mentions the name of its community. However, with respect to differences, it can be noted that the Inga word cloud contains a number of words with a positive connotation, for example, *vida* (life), *universidad* (university), *paz* (peace). On the other hand, the Awá word cloud has words like *armados* (armed), *victimas* (victims), *FARC* (Revolutionary Armed Forces of Colombia). This seems to suggest that news articles referring to the Awá have more negative aspects than those of the Inga.

To determine whether a news article has a positive or negative sentiment associated with it, the `nltk.sentiment`[9] library is used. This library allows numerical value to be assigned that reflects the general sentiment of each text, on a scale of $-1$ (the most negative sentiment possible) to 1 (the most positive sentiment possible). Since this package currently only works for the English language, the translation of the text is previously carried out with `Google Translate API`[10]. Table 4 confirms what was shown by the word clouds in Fig. 3: For the Awá, most of the documents are negative, a trend that is reversed in the case of the Inga.

**Table 4.** News article sentiment analysis

| | Number of articles with Positive sentiment | Number of articles with Negative sentiment |
|---|---|---|
| Awá | 54 | 79 |
| Inga | 100 | 39 |

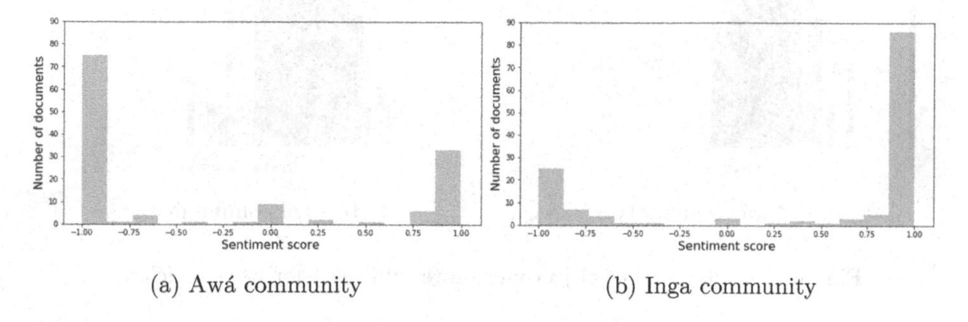


(a) Awá community (b) Inga community

**Fig. 4.** News article sentiment distribution

Figure 4 shows the distributions of sentiments across the new articles, for each community. Once again, it is observed that the documents where the Inga community is mentioned tend to have more positive sentiment scores compared to those of the Awá community, that have a greater tendency to a have a negative sentiment score.

Subsequently, the objectivity of the news corpus is analysed, to determine whether the narrative is closer to an opinion or fact. For this purpose, an objectivity score is calculated using the `Textblob`[11] library. Values closer to 0 indicate high subjectivity in the text, and values closer to 1 high objectivity. Figure 5

[9] https://www.nltk.org/api/nltk.sentiment.html#nltk-sentiment-package.
[10] https://cloud.google.com/translate/docs/basic/translating-text#translate_translate_text-python.
[11] https://textblob.readthedocs.io/en/dev/.

shows the objectivity values obtained for articles that mention each community. It is observed that most of the news articles for both populations are between 0 and 0.5, which indicates that there is a tendency towards subjectivity.

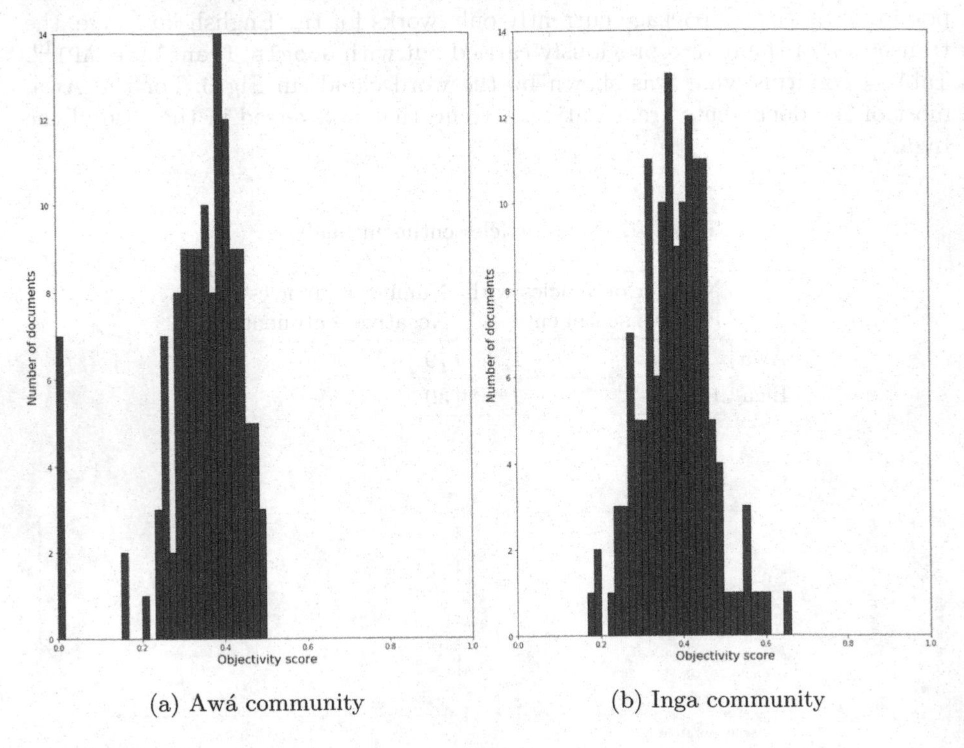


(a) Awá community (b) Inga community

**Fig. 5.** Distributions of objectivity scores obtained for news articles

Finally, we explore whether a correlation exists between sentiment score and objectivity score. Figure 6 shows the relationship between the two variables for each of the populations under study. According to the charts, for the Awá people, the Pearson correlation coefficient is 0.88, and for the Inga people, it is 0.87. This indicates that both variables are correlated to a high degree in the corpus analysed. These results suggest that the more positive a news article is, the more factual in nature it tends to be. Conversely, more negative articles tend to reflect opinions.

## 6 Timeline Generation

This section presents an approach to summarise a document by generating a timeline with the main events mentioned in the document. This is illustrated using the corpus of rulings of the Constitutional Court, whereby the events mentioned within a sentence are extracted. For this purpose, the text of the PDF

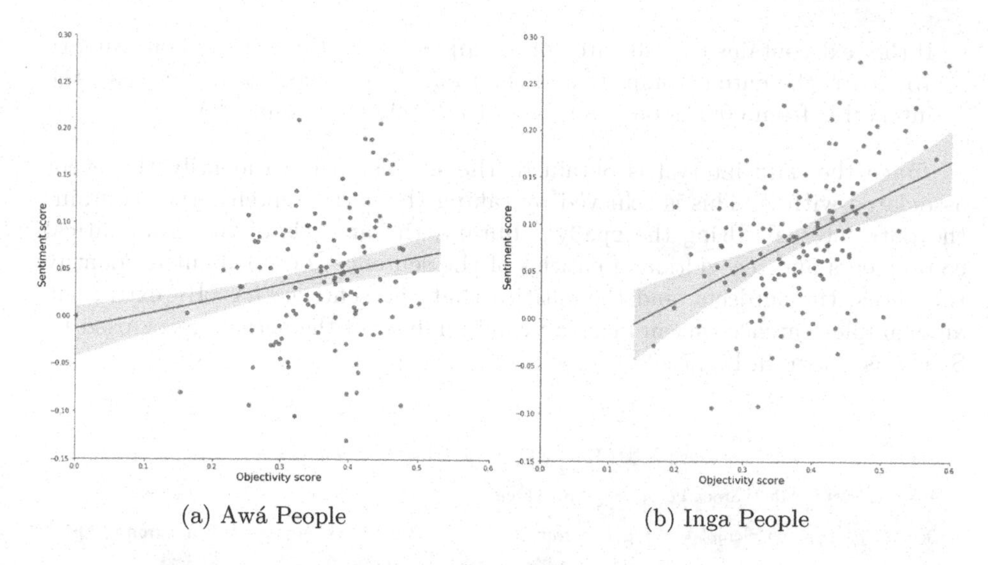


(a) Awá People (b) Inga People

**Fig. 6.** Objectivity score vs. Sentiment Score for news articles

`Sentencia T-630/16`[12], is extracted and cleaned. This ruling concerns with a legal dispute between indigenous communities and the Gran Tierra Energy Colombia Ltd petroleum company, and seeks to protect the right to prior consultation. It is chosen due to the complexity of the case and the length of the document. Important events are determined by the date in which they happen. However, since the nomenclature of a date can be presented in diverse ways in a text, it is necessary to identify various date patterns and time intervals, which is done with the algorithm described ahead.

To find a date written in a non-traditional format, such as: "the early morning of January 30, 2020" or "by mid-2015" the first thing that is searched in the text are years, between 1900 and 2099 that have the peculiarity that they only present four digits. Once the year has been found, in a range of next and previous characters the name of a month or a possible abbreviation is searched, for instance, "January" or "Jan". Finally, a digit with a maximum of two digits is sought that indicates a particular day, its respective written form or its ordinal representation, for example, "1", "one" or "first". However, the text may refer to a time range (that is, a range of dates) rather than a single date. The procedure to determine an interval is as follows:

- If the text mentions only one year, for example: "a tragedy happened in 2018", the date range is the entire year in mention from 1 January to 13 December 2018.
- If the text contains a year accompanied by the month, for example: "The letter is signed in January 2017", the range of dates includes all that comprise the month in question, in this case, from 1 January to 31 January from 2017

[12] https://www.corteconstitucional.gov.co/relatoria/2016/t-630-16.htm.

- If the text contains the full date, for example: "Everything started on January 15, 2019" the date is complete and the range is a whole day. In this case, the interval is from 00:00 h on January 15 to 23:59 on the same day.

Once the date interval is obtained, the next step is to identify the event associated with it. This is achieved by taking the entire sentence that contains the date interval. Using the `spaCy`[13] library, the analysis of the grammatical components and recognition of entities of the sentence is carried out to identify the verbs, the subjects, and the entities that the sentence has. By extracting an example sentence we can make a visual analysis of the entities recognised by Spacy as shown in Fig. 7.

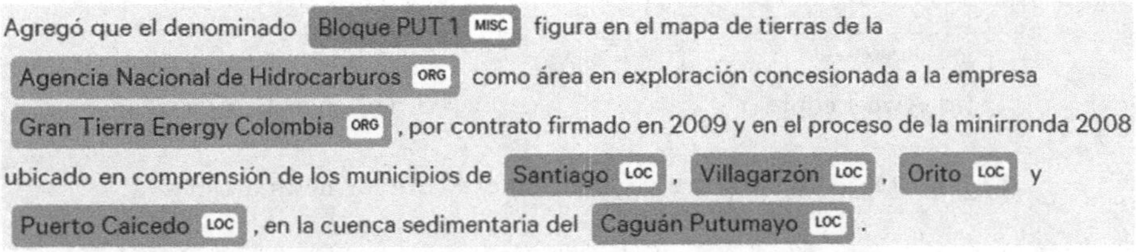

Agregó que el denominado Bloque PUT 1 MISC figura en el mapa de tierras de la Agencia Nacional de Hidrocarburos ORG como área en exploración concesionada a la empresa Gran Tierra Energy Colombia ORG , por contrato firmado en 2009 y en el proceso de la minirronda 2008 ubicado en comprensión de los municipios de Santiago LOC , Villagarzón LOC , Orito LOC y Puerto Caicedo LOC , en la cuenca sedimentaria del Caguán Putumayo LOC .

**Fig. 7.** Entity recognition example

The entities detected are the following:[14]

- LOC: Places, mountain ranges, bodies of water
- ORG: Designated as a corporate, governmental, or other organisation entity.
- MISC: Various entities, for example, events, nationalities, products or works of art.

The characteristics detected are the following:[15]

- NOUN: Subjects.
- VERB: Actions.

Finally, the sentence that is obtained is generally very long, so each sentence is summarised to allow for a succinct visualisation of the data. A function is created in Python that allows to make a summary of a paragraph of the text, starting from the original text. This function generates a frequency matrix of terms, excluding *stopwords*. Subsequently, a percentage weight is assigned to each word and thus each sentence of the text receives a value depending on the words they have. Phrases that exceed a higher weight than 0.5 are included in the summary. Table 5 shows the information collected to construct the timeline.

[13] https://spacy.io/.
[14] https://spacy.io/api/annotation.
[15] https://spacy.io/usage/linguistic-features.

**Table 5.** Data collected Time line

| Data | Datatype |
|---|---|
| Event | Int |
| Initial date | Datetime |
| Final date | Datetime |
| Text | String |
| Entities | String |
| Actions | String |
| Nouns | String |
| Summary | String |

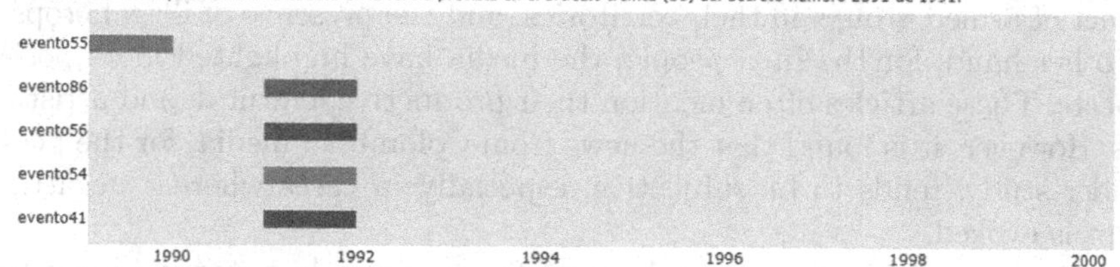
social y económica de las comunidades indígenas.
el decreto numero 1320 de 1998 introdujo un primer desarrollo normativo en esta materia al reglamentar la consulta previa de las comunidades indígenas
el artículo 2º del acuerdo numero 55 de 2003 de la Sala Plena del Consejo de Estado y el artículo 2.2.3.1.2.1 del decreto numero 1069 de 2015.
TERCERO: ORDENAR al Ministerio del Interior Dirección de Consulta Previa en el término de 4 meses contados a partir de la notificación de este sentencia
para regular diferentes aspectos relacionados con los pueblos indígenas y tribales en los países independientes.
6.1 La regulación normativa de la consulta previa El derecho a la consulta previa.
4 Una interesante descripción ilustrativa sobre las diferentes decisiones adoptadas por la Corte en el ámbito de la consulta previa.
TERCERO: Notifíquese a los interesados en la forma prevista en el artículo treinta (30) del decreto numero 2591 de 1991.
evento55
evento86
evento56
evento54
evento41
1990
1992
1994
1996
1998
2000

**Fig. 8.** Timeline 1

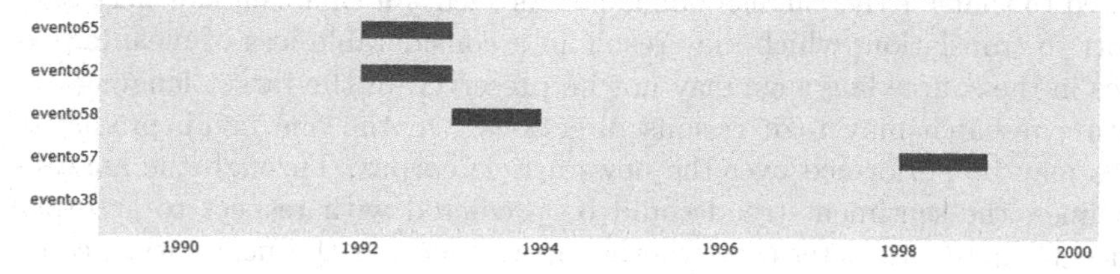
social y económica de las comunidades indígenas.
el decreto numero 1320 de 1998 introdujo un primer desarrollo normativo en esta materia al reglamentar la consulta previa de las comunidades indígenas
el artículo 2º del acuerdo numero 55 de 2003 de la Sala Plena del Consejo de Estado y el artículo 2.2.3.1.2.1 del decreto numero 1069 de 2015.
TERCERO: ORDENAR al Ministerio del Interior Dirección de Consulta Previa en el término de 4 meses contados a partir de la notificación de este sentencia
para regular diferentes aspectos relacionados con los pueblos indígenas y tribales en los países independientes.
6.1 La regulación normativa de la consulta previa El derecho a la consulta previa.
4 Una interesante descripción ilustrativa sobre las diferentes decisiones adoptadas por la Corte en el ámbito de la consulta previa.
TERCERO: Notifíquese a los interesados en la forma prevista en el artículo treinta (30) del decreto numero 2591 de 1991.
evento65
evento62
evento58
evento57
evento38
1990
1992
1994
1996
1998
2000

**Fig. 9.** Timeline 2

Figures 8 and 9 show part of the timeline resulting from events over time. Events automatically extracted from the ruling are reflected in the timeline. These are ordered by time and the summary assigned to that instant of time is used as a label. About 100 events are captured, but due to space reasons, only the first 10 events are shown here in chronological order.

## 7 Conclusions

This article has explored the potential of text mining to characterise various aspects of texts related to two indigenous peoples in southern Colombia. Through the analysis of topics carried out on the corpus, it is observed that these have a significant diversity of themes due to their cultural wealth and their processes of struggle and resistance. Although it is evident that violence caused by factors such as the armed conflict is an issue that affects indigenous people significantly, their productive projects and culture are also topics that stand out. It is noted that the text analytics techniques used here are generic, in the sense that they may be applied to other indigenous communities elsewhere, or other (non-indigenous) populations in general.

Our analysis indicates that for the period of study, news articles that mention the Awá tend to have a more negative sentiment. This may be because of the impact of armed groups in their territories, and the presence of illegal crops. On the other hand, for the Inga people, the media have highlighted more positive aspects. These articles often mention their productive, cultural, and artistic projects. However, it is found that the news from Colombian media, for the peoples' under study, tends to be subjective, especially in cases where a negative sentiment is evoked.

The generation of timelines from documents may provide a starting point to achieve effective summaries of legal documents. Time entities would enabled a chronological sequence of events to be established from a text unlike conventional analytical summaries. This is framed within a temporal process, which would allow us to change the order of the narrative and generate new data framed within a timeline.

An important lesson learnt from this work is that text mining tool support for the Spanish language has a long way to go. Existing libraries have a high degree of maturity for processing text in the English language. Some libraries offer multi-language support such as Spacy. However, functionality is more limited compared to that for the English language. As such, for Spanish, it is necessary to resort to translation, which may result in a considerable loss of meaning, as nuances in the source language may not be preserved in the target language.

Future research may take various directions. On the one hand, predictive analysis may be performed over the news article corpus. Through the analysis of timelines, the sentiment trend could be predicted with respect to a corpus classified by dates, to determine whether news tends in the near future could be positive or negative. This could assist human rights early warnings to be proactively issued for the protection of vulnerable communities. On the other hand, the incorporation of machine learning techniques to improve the temporal event identification algorithm may be explored. Furthermore, since the current approach is focused on rulings of the Constitutional Court, it may be generalised for other types of lengthy documents that require summarisation by means of a timeline. Exploring interactive visualisations for timelines also has the potential to be a promising research direction.

## References


1. Agarwal, A., Xie, B., Vovsha, I., Rambow, O., Passonneau, R.J.: Sentiment analysis of Twitter data. In: Proceedings of the Workshop on Language in Social Media (LSM 2011), pp. 30–38 (2011)
2. Allan, J., Gupta, R., Khandelwal, V.: Temporal summaries of new topics. In: Proceedings of the 24th Annual International ACM SIGIR Conference on Research and Development in Information Retrieval, pp. 10–18 (2001)
3. Balahur, A., et al.: Sentiment analysis in the news. arXiv preprint arXiv:1309.6202 (2013)
4. Bassi A., A.: Lematización basada en análisis no supervisado de corpus, August 2020. https://users.dcc.uchile.cl/~abassi/ecos/lema.html
5. Bautin, M., Vijayarenu, L., Skiena, S.: International sentiment analysis for news and blogs. In: ICWSM (2008)
6. Blei, D.M., Ng, A.Y., Jordan, M.I.: Latent Dirichlet allocation. J. Mach. Learn. Res. **3**, 993–1022 (2003)
7. del Cauca, C.R.I.: Denuncia pública sobre las falsas acusaciones hacia la minga por parte de los medios de comunicación locales, regionales y nacionales, March 2019. https://www.cric-colombia.org/portal/denuncia-publica-sobre-las-falsas-acusaciones-hacia-la-minga-por-parte-de-los-medios-de-comunicacion-locales-regionales-y-nacionales/
8. Chen, X., Chan, Z., Gao, S., Yu, M.H., Zhao, D., Yan, R.: Learning towards abstractive timeline summarization. In: IJCAI, pp. 4939–4945 (2019)
9. Chieu, H.L., Lee, Y.K.: Query based event extraction along a timeline. In: Proceedings of the 27th Annual International ACM SIGIR Conference on Research and Development in Information Retrieval, pp. 425–432 (2004)
10. de Colombia, C.C.: Sentencia t-500/16 acción de tutela contra medios de comunicación (2016), https://www.corteconstitucional.gov.co/relatoria/2016/t-500-16.htm
11. Colquhoun, S., Dockery, A.M.: The link between indigenous culture and wellbeing: Qualitative evidence for Australian aboriginal peoples (2012)
12. Doyle, M.M.: Acceso y participación de los pueblos indígenas en el sistema de medios de argentina. Disertaciones: Anuario electrónico de estudios en Comunicación Social 11(2), 30–49 (2018)
13. Godbole, N., Srinivasaiah, M., Skiena, S.: Large-scale sentiment analysis for news and blogs. ICWSM **7**(21), 219–222 (2007)
14. Hernández Delgado, E.: La resistencia civil de los indígenas del cauca. Papel político **11**(1), 177–220 (2006)
15. Hristov, J.: Indigenous Struggles for Land and Culture in Cauca. Colombia. J. Peasant Stud. **32**(1), 88–117 (2005)
16. Kaya, M., Fidan, G., Toroslu, I.H.: Sentiment analysis of Turkish political news. In: 2012 IEEE/WIC/ACM International Conferences on Web Intelligence and Intelligent Agent Technology, vol. 1, pp. 174–180. IEEE (2012)

17. Kessler, R., Tannier, X., Hagege, C., Moriceau, V., Bittar, A.: Finding salient dates for building thematic timelines. In: Proceedings of 50th Annual Meeting of the Association for Computational Linguistics (Vol 1: Long Papers), pp. 730–739 (2012)
18. Li, X., Xie, H., Chen, L., Wang, J., Deng, X.: News impact on stock price return via sentiment analysis. Knowl.-Based Syst. **69**, 14–23 (2014)
19. Mani, I., Wilson, G.: Robust temporal processing of news. In: Proceedings of the 38th Annual Meeting of the Association for Computational Linguistics, pp. 69–76 (2000)
20. Manning, C.D., Schütze, H., Raghavan, P.: Introduction to Information Retrieval. Cambridge University Press, Cambridge (2008)
21. Ortega, M.C.R.: El cauca lleva más de medio siglo azotado por la violencia: así es el territorio colombiano de las dos masacres esta semana, November 2019. https://cnnespanol.cnn.com/2019/11/01/cauca-masacres-indigenas-colombia-violencia-conflicto-armado/
22. Pak, A., Paroubek, P.: Twitter as a corpus for sentiment analysis and opinion mining. In: LREc, vol. 10, pp. 1320–1326 (2010)
23. Raina, P.: Sentiment analysis in news articles using sentic computing. In: 2013 IEEE 13th International Conference on Data Mining Workshops, pp. 959–962. IEEE (2013)
24. Saif, H., He, Y., Alani, H.: Semantic sentiment analysis of twitter. In: International Semantic Web Conference, pp. 508–524. Springer (2012)
25. Shah, S.R., Kaushik, A.: Sentiment analysis on Indian indigenous languages: a review on multilingual opinion mining. arXiv preprint arXiv:1911.12848 (2019)
26. Swan, R., Allan, J.: Automatic generation of overview timelines. In: Proceedings of the 23rd Annual International ACM SIGIR Conference on Research and Development in Information Retrieval, pp. 49–56 (2000)
27. Swan, R., Allan, J.: Timemine (demonstration session) visualizing automatically constructed timelines. In: Proceedings of the 23rd Annual International ACM SIGIRC, p. 393 (2000)
28. Tran, G., Alrifai, M., Herder, E.: Timeline summarization from relevant headlines. In: European Conference on Information Retrieval, pp. 245–256. Springer (2015)
29. Yan, R., Wan, X., Otterbacher, J., Kong, L., Li, X., Zhang, Y.: Evolutionary timeline summarization: a balanced optimization framework via iterative substitution. In: Proceedings of the 34th international ACM SIGIR Conference on Research and Development in Information Retrieval, pp. 745–754 (2011)

# Application of a Continuous Improvement Method for Graphic User Interfaces Through Intelligent Systems for a Language Services Company

Osvaldo Germán Fernandez[1,2](✉), Pablo Pytel[1,2](✉), and Ma Florencia Pollo-Cattaneo[1,2](✉)

[1] Programa de Maestría en Ingeniería en Sistemas de Información, Facultad Regional Buenos Aires, Universidad Tecnológica Nacional, Buenos Aires, Argentina
ppytel@frba.utn.edu.ar

[2] Grupo de Estudio en Metodologías de Ingeniería en Software (GEMIS), Facultad Regional Buenos Aires, Universidad Tecnológica Nacional, Buenos Aires, Argentina

**Abstract.** The Graphic User Interface represents the most common mechanism in human-computer interaction. Thus, its correct development is an utterly important task, which is reflected to the point where most software projects fail due to low user acceptance. The techniques, methodologies, tools, and principles to improve Graphical User Interfaces have been the object of interest of researchers and organizations over time. Due to the low success rate, a recent trend consists in experimenting with a greater number of interface alternatives, lowering their development costs through automation and Artificial Intelligence. Therefore, this paper presents a method that, by Interactive Genetic Algorithms, a semi-automatic continuous improvement process for landing pages is established. The proposed method is applied in a study case, consisting in a landing page that belongs to a translation services company. The method application provides positive results, enabling the landing page to reach the goals for which it was designed.

**Keywords:** Graphic User Interface · Landing page · Continuous improvement · Intelligent systems · Interactive Genetic Algorithms

## 1 Introduction

Even before the invention of writing, humanity has created interfaces that have served as support resources to satisfy their interaction needs [1], and over time, those have evolved alongside the cognitive human capacity [2]. With the advent of digital screens, a new notion of user interface was born, defined as the part of a computer that the user can see, hear or touch, or, in simpler terms, the one that the user can comprehend and control [3, 4]. This interface makes the user see what the product is capable of, which is why it is considered an essential part of any software application.

On the other hand, the Graphical User Interface (GUI) represents the most common mechanism in human-computer interaction, and given its prominence, it can influence


H. Florez and M. F. Pollo-Cattaneo (Eds.): ICAI 2021, CCIS 1455, pp. 141–160, 2021.
https://doi.org/10.1007/978-3-030-89654-6_11

the software application success. This is why the correct construction of GUIs is a vitally important task, crossed by a wide diversity of disciplines and activities [5–7]. Ramírez et al. [5] emphasizes the link between GUIs and the software projects success, mentioning that about 70% fail due to low acceptance by users. Thus, it is correct to mention that a properly designed interface is vital for the success of the application that contains it [6]. Several studies support the relationship between the GUI quality and the credibility of the application [7], the perception of integrity [8], customer loyalty [9], the intention to revisit [10], and lower costs in customer acquisition and increases in retention [11].

In addition, the GUI acquires greater relevance due to its impact on usability and ergonomics, the latter being understood as usability oriented to physiological aspects in such as touch and mobile devices [12, 38]. For this reason, it has been described that about 67% of users are more likely to use interfaces adapted for mobile devices [13].

Lastly, the benefits of an appropriate GUI are recognized by several organizations that provide software worldwide, basing their decisions in interface design disciplines [14].

Due to these aspects, GUIs have been the main interest object of researchers and organizations over time. However, the improvement proposals that emerged from this interest have not ensured their success, which is reflected in the fact that 80% of the time designers are wrong about what the user really wants [15] and that, for example, only a third of the ideas implemented by companies like Microsoft have improved the metrics for which they were implemented [16]. To overcome these obstacles, experimenting with a great amount of GUI variations has been lately considered, which results in higher construction costs.

In this context, the present work proposes a method that, if followed by an IT Professional, facilitates the implementation of a continuous improvement process for GUIs through Artificial Intelligence and automation, and thus lowering their construction costs. For that purpose, Sect. 2 describes the proposed method and Sect. 3 presents the results of the implementation in a study case. These results are then analyzed in Sect. 4. Finally, Sect. 5 presents the conclusions and future lines of work.

## 2 Proposed Method

Based on the problems identified in the previous section, the proposed solution consists on a method that seeks to guide IT Professionals and UX designers in the implementation of a continuous improvement process for web platforms through Artificial Intelligence based on Interactive Genetic Algorithms (described in Sect. 2.1).

The detailed method and analysis, in this case, focus on landing pages. A landing page is a visual platform where organizations can attract visitors in a more comfortable way to make sales [17]. These pages are very suitable for continuous improvement processes, mainly because, in many cases, they represent the first audience interaction with the application. For example, the landing page presented for the study case (Sect. 3) belongs to a translation services company.

In order to evaluate the result of the interaction, it is necessary to use well-established metrics within the industry [18, 35, 36]. Therefore, it is then appropriate to analyze the perception phenomenon in such platforms. The concept has been developed in various studies, analyzing aspects such as complexity [19, 20], symmetry [21], color [22], or several of these characteristics combined [6]. Likewise, user inherent characteristics such as sociodemographic variables have been described [23, 24]. However, for the present method, the approach proposed by Kohavi & Longbotham [15] is used, since this approach on perception can be measured through clearly defined indicators, which reflect the behavior of users on the web platform: conversion rate, sessions per user, session duration, bounce rate and pages visited per session.

The general structure of the proposed method is later described in Sect. 2.2.

## 2.1 Interactive Genetic Algorithms

A Genetic Algorithm builds on a population of individuals that represent possible solutions to a problem. Each of these individuals is maintained in the form of a "chromosome", which is merely a string of characters that encodes a solution to a problem [27]. Mirjalili [28] describes the following stages for a Genetic Algorithm:

1. *Initial Population:* The genetic algorithm begins with a randomly generated population. This population includes multiple solutions.
2. *Selection:* Natural selection is the main inspiration for this algorithm, which is why a fitness function is used to assign a score to each individual, based on its performance in the environment. The best individuals are selected for the next generation.
3. *Crossover:* After selecting the individuals in the previous operation, two solutions (parents) combine their characteristics in two new solutions (children).
4. *Mutation:* It is the last evolutionary operator, in which one or multiple genes are altered in order to maintain the population diversity by introducing randomness.

On this basis, an Interactive Genetic Algorithm (also called IGA) includes human evaluation in the optimization process [29]. During the selection operation, a fitness function is used, in which the users interact with the generated solutions to assign a value. Although there are some examples of successful IGA implementations in the GUI improvement domain [30], most of them consist on proprietary tools with limitations considered by their authors. Many of these limitations are not from the tool itself, but on how the tool is defined and applied to meet the stated goal. For this reason, embedding the tool in an engineering process is imperative. In fact, the effective integration of activities and media (such as, in this case, an AGI software tool) is inherent to Information Systems Engineering [31]. To achieve this, it is possible to apply knowledge from different Engineering disciplines such as Information Systems Engineering [31], Software Engineering [32] and Requirements Engineering [33].

For this proposed method, one landing page variation is one possible solution, which is encoded in a chromosome whose structure is composed of three elements:

- a Permutation gene (GP), that defines the relative order of a web element, and therefore its position,
- a Color gene (GC), where three genes for each color, corresponding to tone -H-, saturation -S- and color luminosity -L-, and

– a Style gene (GE), that reflects various modifications, such as typographic hierarchies.

In addition, other elements are stored, such as the chromosome generation and the chromosome performance result.

## 2.2 General Structure of the Proposed Method

The proposed method describes phases and activities necessary to identify the landing page goals, determining GUI elements whose attributes are modified. Subsequently, the method lists the phases and activities in order to build tools that automate elements intervention, as well as build tools that automate the evaluation and deploy of these intervened GUIs, under two architectural proposals, which depends on the project needs.

All these activities are performed by two main roles:

– *IT Professional* (also referred to as IT): This role is the main responsible for the construction of all the software tools mentioned previously (and even responsible for tool acquisition if those exist and such is considered pertinent). It is for this reason that this role skills and experience are strongly oriented to web development. In addition, it is convenient that the IT Professional has experience in software methodologies application as well as requirements gathering and interpretation.
– *UX designer* (also referred to as UX, "User Experience designer" or "Interaction Expert"): it is understood as someone whose work domain is the ways in which users interact with and through computers [25]. This broad definition considers that the role includes graphic visual interaction disciplines (such as graphic design) but exceeds it, including also holistic ergonomics knowledge and even product and business ideation [26]. In fact, under the presented method, the UX designer is responsible for identifying organizational goals and understanding in which ways the landing page collaborates with them. This information is essential to determine how the platform can be intervened by the tools that will be later built by the IT Professional.

It is also important to consider that, depending on the landing page and organization size and complexity, many people can perform these two roles. As such, this method also considers possible that both roles are fulfilled by the same person.

On the other hand, the proposed method aims to establish continuous improvement in GUIs, and is composed by ten phases (shown in Fig. 1):

(1) **Induction**: The goal of this phase is to identify the people involved in the continuous improvement project (among them, the main roles of "IT" and "UX"). In addition, landing page scope, its audience, its mission, its goals and its indicators should be identified.
(2) **Intervention Elements Selection:** This phase aims to identify the elements that will be intervened by the method, following established criteria and business restrictions.
(3) **Configuration:** This phase is utterly important, since it combines UX disciplines with those of Information Systems Engineering. With the information previously collected, the elements of the phase 2 are linked with their representation in the

source code. The method architecture and the parameters involved in the next phases are also defined in this phase.

(4) **Initial Population Generation:** In this phase, the chromosome structure is defined, according to the project characteristics. In addition, the first generation chromosomes are created.

(5) **GUIs modification:** This is the phase where the actual modifications take place. The phase describes how to create and intervene different landing page versions, according to the previously chosen architectural alternative.

(6) **Deployment:** The goal of this phase is to make the intervened GUIs be available to the audience, as well as their related measurement tools.

(7) **Measurement:** This phase remains for a defined amount of time with the objective of measuring user interactions, considering the audience type and indicators identified in phase 1. Pre-defined alerts can also occur to indicate anomalies in the method performance.

(8) **Evaluation:** Once the previous phase has concluded, performance scores are assigned to the intervened interfaces, using a Fitness Function.

(9) **Crossover and Mutation:** The goal of this phase is to obtain a new generation (that is, a new group of GUI variations) that own shared visual characteristics from previous GUIs with good performance. Furthermore, this phase adds new modifications to the new group of GUIs, in order to explore the effect of new perception phenomena on the audience.

(10) **Closing:** When the termination criteria are met, the resulting GUI changes should be documented and analyzed. The resulting GUI is then permanently set as the definitive landing page.

As it can be seen in Fig. 1, and as this is a Continuous Improvement Method, Crossover and Mutation leads to the GUIs Modification in a cyclic manner. Alternatively, the method provides termination criteria that could lead to the Closing phase.

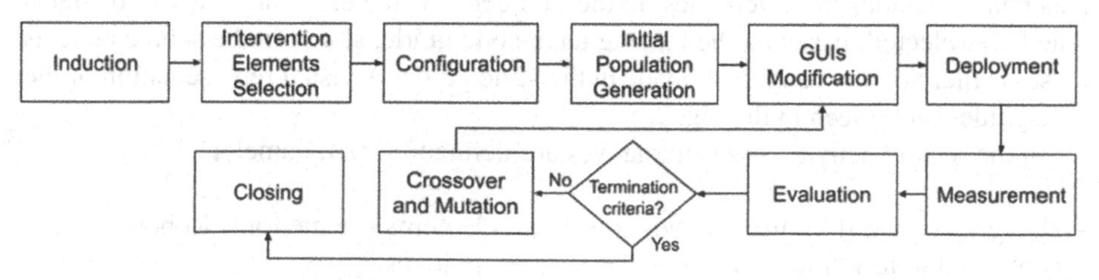


**Fig. 1.** Continuous improvement method for GUIs

## 3 Results

For the validation of the proposed method, a study case of a language services company is used [38]. In this case, the landing page plays a fundamental role in the client acquisition. The company is 7 years old and provides translation services, language conversational sessions and language online classes.

Although the company relies on online campaigns that are considered effective, with an established landing page that provides them with an acceptable Conversion Rate (measured with a value of 5.15% on January 10, 2021), the company is expecting to acquire a large amount of language service facilitators. For this reason, it is desired to increase the landing page Conversion Rate (leads generated divided by total sessions), as well as reduce the Bounce Rate (visitors without interaction, divided by total sessions). To achieve these goals, the company technology team will apply the proposed method to their landing page.

In the following sections, the application of each phase of the method is presented. Please note that during the results description, the company data is deliberately omitted in order to preserve the organization identity.

### 3.1 Induction

This phase is performed by IT and UX. The landing page is navigated entirely and, apart from the main page, five secondary pages are identified. It is decided that the intervention will occur on the main page only, documenting this decision in the induction record. In this case, the goals for the landing page are:

- Provide a better online experience, capturing the audience attention and potential customers. Its associated indicator is a Bounce Rate, expecting a value of less than 65% in the next 2 months.
- Attract new customers through the landing page. Its associated indicator is the action buttons Conversion Rate, expecting to reach a sustained value of 10% in the next 2 months.

### 3.2 Intervention Elements Selection

This phase includes three activities. In the first activity, the elements that will be intervened are selected, based on the landing page style guide, stored in the source code as a ".scss" file. So, no changes are made in the style guide. A visual representation of the style guide can be seen in the Fig. 2.

In the second activity, style alternatives are defined by UX, namely:

- *Typography*: On the current Landing Page all elements use the font "Roboto", and it is planned to test "Lato" and "Exo" as font alternatives.
- *Color palette*: Currently the color palette is inspired by the company brand manual, and consists of the main color "Ocean Green" (# 44986E), the secondary color "Indian Red" (# D25A5A) and the tertiary color "Desert Storm" (# F2EBE2) of the landing page (see Fig. 2). UX proposes 54 color palette alternatives, and HSL distances are calculated for each color palette with respect to the original palette. The thirty-four most suitable palettes (closest distance to original palette) will be used in the next phases.

**Fig. 2.** Original landing page (left) and visual representation of the style guide (right)

In the last activity, the permutation groups are established, where UX proceeds to identify the landing page structural composition, as shown in Fig. 3. The most prominent visual restrictions identified by UX are:

- The page main sections (with the most visual impact and influence on the conversion rate) are Sect. 1 and Sect. 2. These sections can only be swapped with each other.
- Subsequent Sects. (3, 4, and 5) can be swapped with each other, with the only restriction of keeping the background color interleaved, to facilitate the distinction between sections.
- Sect. 6 (footer) should always be kept as the last section.

Considering these restrictions and the elements visual hierarchy, UX decides to create four permutation groups (Fig. 3):

- Permutation group composed with the main sections: “Sect. 1” and “Sect. 2”.
- Permutation group with the secondary landing page sections: “Sect. 3”, “Sect. 4” and “Sect. 5”.
- Permutation group composed with the Carousel elements: “Carousel 3 images”.

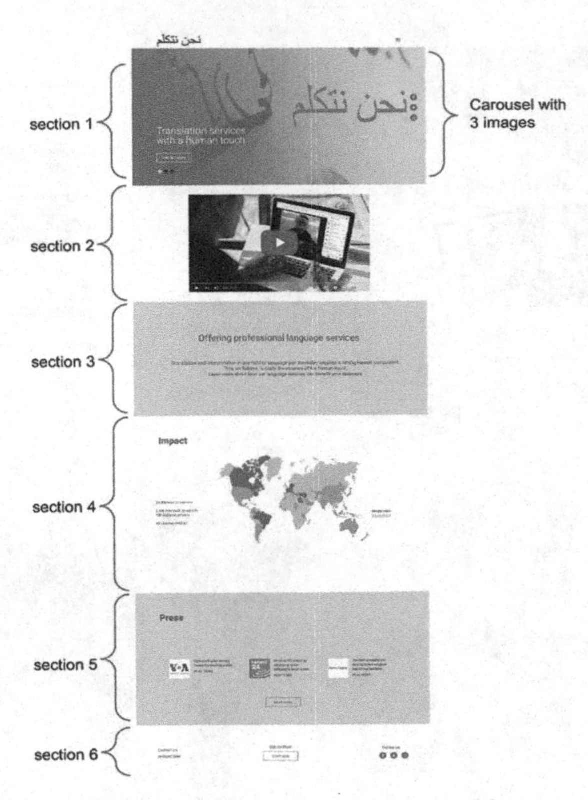


**Fig. 3.** Landing page structural composition

### 3.3 Configuration

The four activities corresponding to this phase are: (I) Architecture evaluation, (II) Obtaining and integrating the analytics tool, (III) Linking the Intervention Elements with their representation in source code and, finally, (IV) Parameterization.

In activity I, IT reviews the criteria established by the method and decides to use the offline architecture (this means, automatically modifying source code by parsing instead of editing the DOM).

Later in Activity II is expected to get a tool that fetches the landing page Conversion Rate and Bounce Rate. Then, Google Analytics is integrated onto the existing landing page, and each chromosome will have a unique tracking ID. After the analytics integration, a test is carried out in a period of three days, obtaining an initial bounce rate of 73%, conversion rate of 5.15%, sessions per user of 1.19, and session duration of 17 s with 145 evaluated sessions in total.

In Activity III, the Permutation Elements (Fig. 3) are spotted in the source code and labeled through HTML class names.

Finally, in Activity IV, IT and UX determine the parameters that will be used as input information for the method tools:

- *Individuals per generation*: Based on the available tracking IDs and the sessions evaluated in activity II and, the method will create 10 individuals per generation. As indicated in [34], this population size is considered small enough to gather many sessions per chromosome, while big enough to achieve good results.
- *Fitness criteria*: Based on the information stated in phase 1, Conversion Rate represents the 70% of the fitness function, and Bounce Rate represents the remaining 30%.
- *Generation duration*: Based on the number of sessions evaluated, it is determined that seven days is a suitable generation duration.
- *Audience reach for new and old individuals*: 20% of a previous generation (the 2 best individuals) will remain unchanged in the next generation. Then 8 new individuals will be created.
- *Termination criteria*: The method will reach its closing phase when the best individual of a generation reaches a Conversion Rate of 10% and a Bounce Rate of 65%.

### 3.4 Initial Population Generation

The three activities in this phase are: (I) Original chromosome description, (II) Chromosomes creation, and (III) Non-interactive fitness function: discard invalid candidates.

Firstly, in activity I, IT describes the chromosome structure (Table 1). Next, IT develops and implements micro services to create and modify these structures.

Secondly, in Activity II, the chromosomes that would be part of the first generation are created. This process is based on the registered permutation groups, and style elements that are chosen randomly. One tracking ID is obtained for each chromosome.

Finally, in Activity III, a function is developed to discard chromosomes whose fitness is so low that it compromises the landing page visual experience. This is based on the difference in luminosity units between colors in the palette (comparing primary, secondary and tertiary colors). Thus, low-contrast, visually conflicting chromosomes are replaced with new randomly (and valid) generated individuals.

### 3.5 GUIs Modification

The two activities in this phase are: (I) Architecture Replication, and (II) Source code intervention.

First, the services used to host the landing page are replicated ten times (one for each landing page) and an extra service is created to randomly redirect a new user to any of these individuals.

Finally, a toolkit that automatizes workflow is used, since it was already applied in the project. Automated tasks are created to modify the source code. CSS code is added to reflect the chromosomes style elements and color palette, and HTML elements are swapped to reflect the permutation groups.

**Table 1.** Chromosome structure

| Field name | Data type |
|---|---|
| ID | INTEGER |
| perm_group_1_carousel | STRING[] |
| perm_group_2_main_sections | STRING[] |
| perm_group_2_secondary_sections | STRING[] |
| main_color_H | INTEGER |
| main_color_S | INTEGER |
| main_color_L | INTEGER |
| secondary_color_H | INTEGER |
| secondary_color_S | INTEGER |
| secondary_color_L | INTEGER |
| tertiary_color_H | INTEGER |
| tertiary_color_S | INTEGER |
| tertiary_color_L | INTEGER |
| font_family | STRING |
| generation | INTEGER |
| times_requested | INTEGER |
| analytics_tracking_id | INTEGER |
| bounce_rate | FLOAT |
| sessions_per_user | FLOAT |
| session_duration | FLOAT |
| bounce_rate | FLOAT |
| total_score | FLOAT |

### 3.6 Deployment

IT works with the tool built in the previous phase to add the capacity to automatically deploy the individuals in each of the replicated hostings. A cookie is added and read in the user device to make sure they are redirected to the same individual, and thus assuring visual consistency among sessions.

### 3.7 Measurement

In this phase, two activities are performed: (I) Obtaining results, (II) Results recording.

First, in Activity I, the resources previously configured are used to create a tool that obtains the performance metrics for each chromosome. Then, in Activity II, after the seven days of the first generation, the performance results are obtained.

### 3.8 Evaluation

The three activities belonging to this phase are: (I) Determine and calculate the fitness function, (II) Evaluation of the termination criteria, and (III) Individuals selection.

In Activity I, the fitness function is defined as in Eq. 1:

$$F(X) = \frac{(A - R0a)}{(R1a - R0a)} * \alpha + \frac{(B - R0b)}{(R1b - R0b)} * \beta \tag{1}$$

X is an individual to be evaluated by the fitness function. For the performance dimension “a” (A), which refers to the Conversion Rate, R0a and R1a are their lower and upper performance limits respectively, and $\alpha$ is the fitness criteria defined in Configuration. This shows the measured performance of dimension “a” (A), the lower and upper performance limit of dimension “a” (R0a and R1a, respectively), and the coefficient defined by the aptitude criterion in phase configuration, activity IV. The same rationale applies to dimension “b” (Bounce Rate), and respective limits (R0b and R1b) and weight ($\beta$).

To set lower performance limits, the first-generation lowest measurements are considered, while the upper performance limits are set by the values stated in the landing page goals in phase 1 (Table 2).

**Table 2.** Lower and upper performance limits.

| | Conversion rate | Bounce rate |
|---|---|---|
| Lower performance limit | 0.03 | 0.879 |
| Upper performance limit | 0.10 | 0.65 |

Therefore, the generic fitness function defined in Eq. 1 becomes:

$$F(X) = \frac{(A - 0.03)}{(0.10 - 0.03)} * 0.7 + \frac{(B - 0.879)}{(0.65 - 0.879)} * 0.3 \tag{2}$$

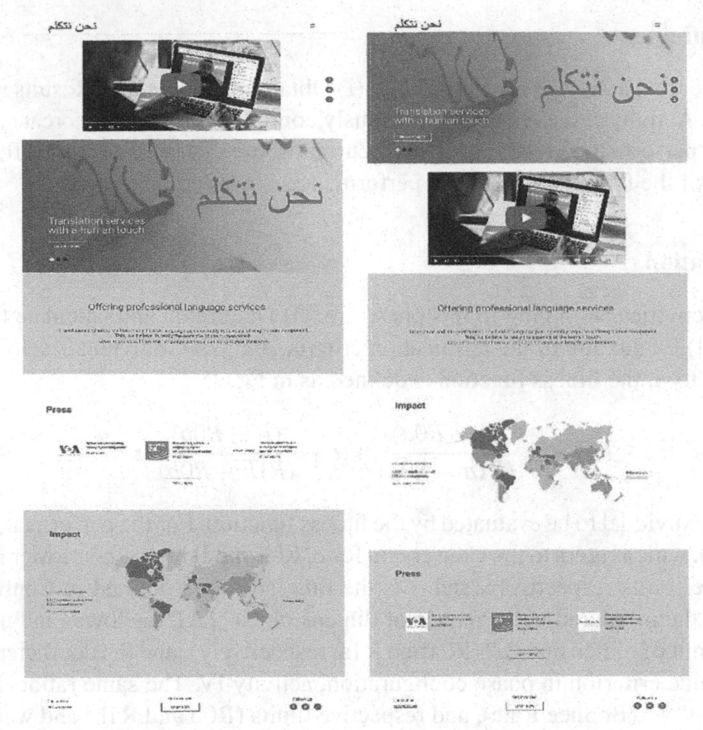


**Fig. 4.** Best performing landing page (left) and worst performing landing page (right) from the first generation

The Activity II confirms that the best individual performance does not meet the termination criteria established in the Configuration, presenting a Conversion Rate of 0.061 and a Bounce Rate of 0.686, and more generations are needed. The best and worst performing landing pages from the first generation can be seen in Fig. 4.

In Activity III, first generation individuals are chosen by the "ranking" selection method. The Chromosome Creator tool developed in the phase 4 assign copies to the next generation: The three worst performing chromosomes are assigned with 0 copies, whilst the three best performing chromosomes are assigned with 2 copies.

### 3.9 Crossover and Mutation

In this phase, three activities are carried out: (I) Crossover, (II) Mutation and (III) Inclusion of the null hypothesis (if needed).

First, in activity I, IT works on the Chromosome Creator tool to generate new chromosomes. As previously defined, two copies belonging to the best performing individuals remaining as they are. Then, random binomial crossover is performed in the rest of the copies, creating chromosomes with the shared characteristics of their parents combined.

Then, in Activity II, the chosen chromosomes for the mutation experience an alteration in one randomly chosen gene. The gene value is replaced in the same way genes were created in phase 4.

Finally, activity III is not performed until the creation of the third generation.

### 3.10 Subsequent Generations

In this section, the subsequent generations (created in each iteration of the method) are analyzed. At this point, the method is running completely automated.

The second-generation starts with the GUIs modification phase and, after the deployment, the individuals' performance is measured trough 7 days. As a result, in this second generation, the best performing chromosome does not meet the termination criteria, counting with a Conversion Rate of 0.0882 and a Bounce Rate of 0.588. This means that more generations are needed. The best and worst performing landing pages from the second generation are shown in Fig. 5.

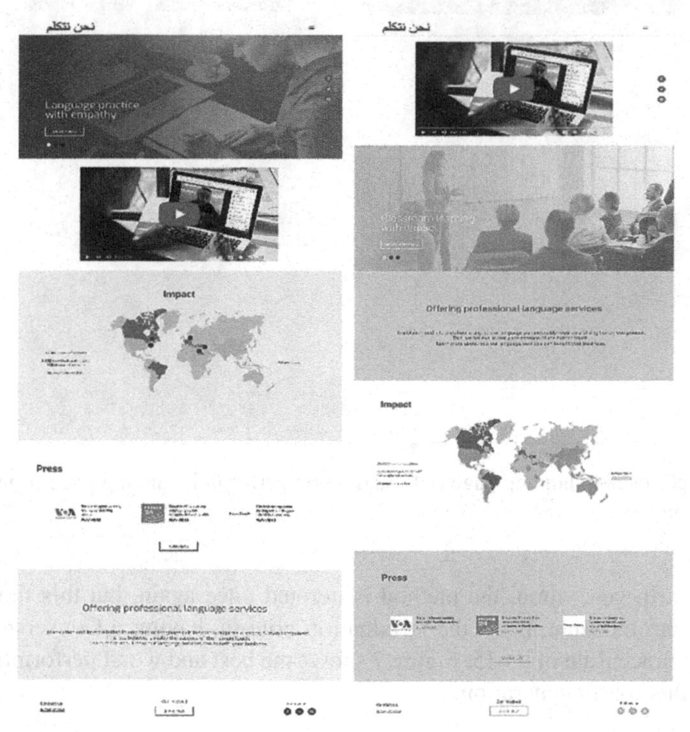


**Fig. 5.** Best performing landing page (left) and worst performing landing page (right) from the second generation

Therefore, the third generation is created with ten new chromosomes, and the method iterates again. Once again, the best performing chromosome does not meet the termination criteria, showing a Conversion Rate of 0.0968 and a Bounce Rate of 0.6129. In addition, in this generation, null hypothesis should be included. After replacing a chromosome for another already existing one, similar metrics are found in both cases, with the average of the sessions duration varying only slightly (11.28 vs 13.57). The best and worst pages from this generation are shown in Fig. 6.

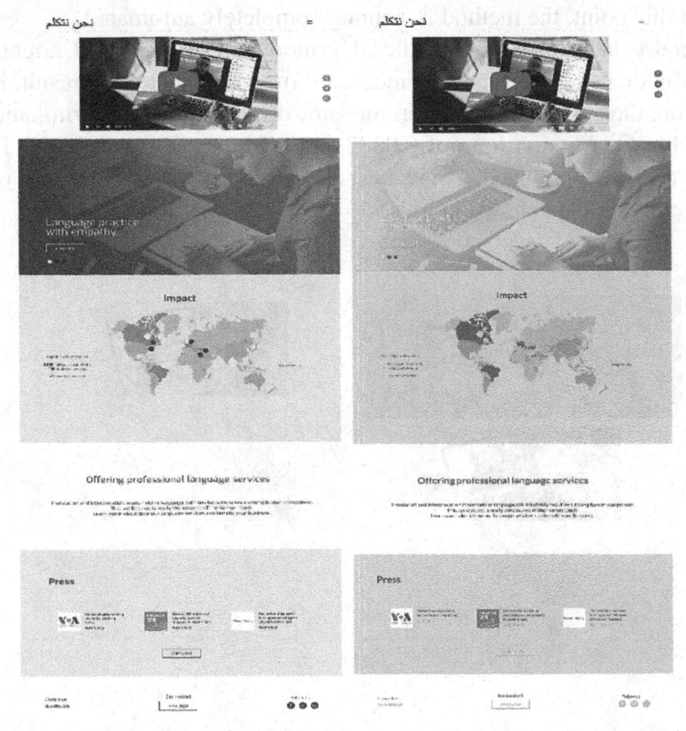


**Fig. 6.** Best performing landing page (left) and worst performing landing page (right) from the third generation

In the fourth generation, the method is iterated once again, but this time the best performing chromosome meets the termination criteria, having a Conversion Rate of 0.11 and a Bounce Rate of 0.615. Figure 7 shows the best and worst performing landing pages from this fourth generation.

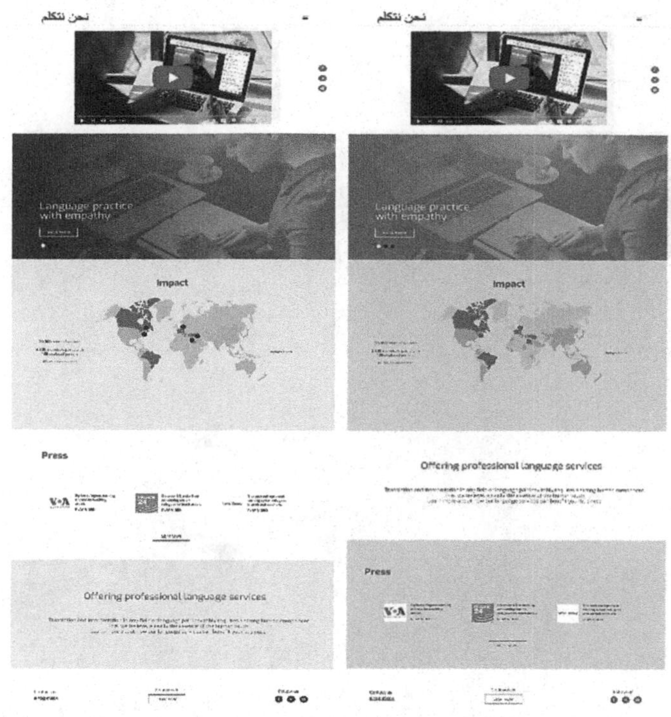


**Fig. 7.** Best performing landing page (left) and worst performing landing page (right) from the fourth generation

Although the method should continue with the Closing phase, two more generations are performed to obtain a more profound method analysis. In the fifth generation, ten new chromosomes are exposed. The corresponding performing landing pages from the fifth generation are shown in Fig. 8.

In addition, the sixth generation is performed to evaluate the performance of the most successful chromosome in the most successful generation. For this, chromosome 40 (in fourth generation) is selected, and ten chromosomes with the same characteristics are created. The results of these generations will be addressed later in Sect. 4.

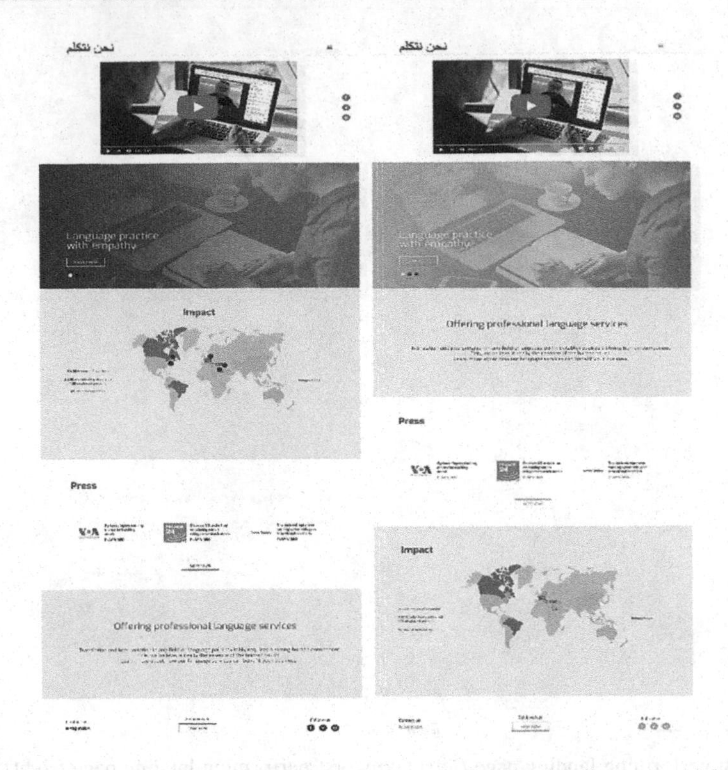


**Fig. 8.** Best performing landing page (left) and worst performing landing page (right) from the fifth generation

## 3.11 Closing

After six generations, the closing activities are performed by IT and UX. The landing page is presented to the project stakeholders, showing the changes introduced by the method tools and their impact on the landing page performance. The page final version is shown in Fig. 9 where it can be seen important changes compared to the initial version (available in Fig. 2). The changes proposed by chromosome 40 are reviewed, and the landing page's source code is updated to match these changes.

The landing page is then deployed and all services provided by the method tools are paused. The source code of these tools is stored in a repository, in case the method is implemented in the future, and all the chromosome records are stored in a document. Location and credentials to access this information is stated in the "Closing document".

**Fig. 9.** Final landing page (Chromosome 40)

# 4 Analysis of the Results

For two and a half months, the Continuous Improvement Method for GUIs was deployed. By doing this, after the fourth iteration, the landing page of a language services company has reached the goals for which it has been designed. The performances of the fifty generated landing pages along the six generations are shown in Fig. 10.

Reviewing these results by comparing the initial and final version, some improvements can be noticed:

- Average Conversion Rate has increased from 0.0515 to a sustained value between 0.1007 and 0.102.
- Average Bounce Rate, a secondary performance dimension, has decreased from 0.78 to a sustained value between 0.60 and 0.63.
- Other performance dimensions not considered in the landing page goals are shown in Fig. 10: Average Sessions per User have increased from 1.19 to 1.367 and Average Session Duration has increased from 17 s to 27.5 s. This emphasizes the holistic nature of the perception phenomena, where Conversion Rate or Bounce Rate improvements are also reflected in the overall user experience.

The null hypothesis included in the third generation reflects that the method experiments are consistent: two chromosomes sharing the same characteristics have similar performance. This is also reflected in the sixth generation.

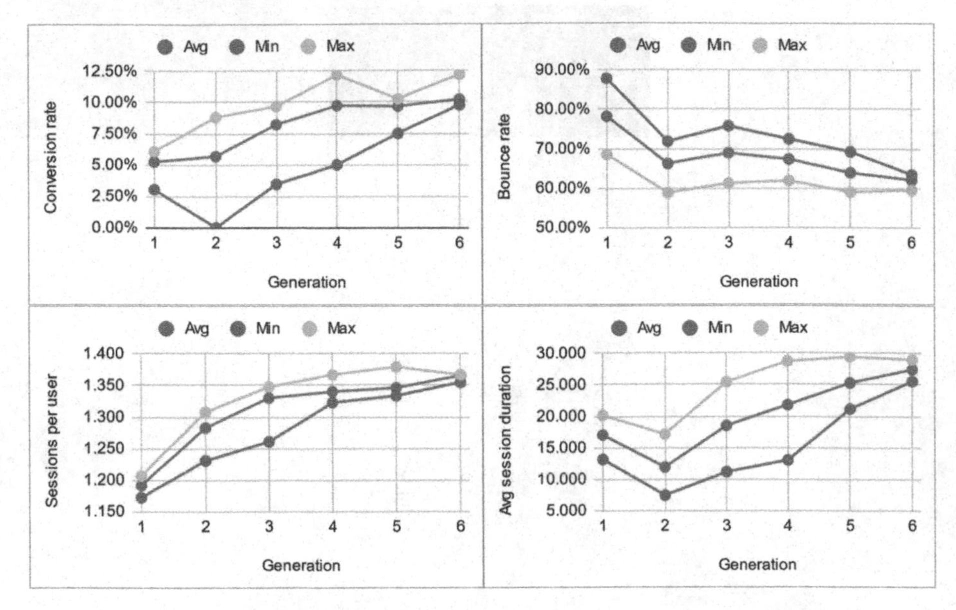


**Fig. 10.** GUIs performance along six generations

An interesting analysis can be done by observing the fifty generated GUIs. Some visual characteristics have gradually populated the generations, enabling their convergence (i.e. squared font families). Another visual example is the order of images in the carousel, where "conversational sessions" have been preferred over "translate services".

In addition, it has been determined that landing pages that included the promotional video at the top presented better performance, even when this meant that the action buttons were not easily available. This reinforces the previously stated concept that GUI design is a complex (and even counter intuitive) domain that could be improved through experimentation, automation and intelligent systems.

## 5 Conclusions

In this paper it is determined that the Continuous Improvement Method for GUIs can be applied with positive results, helping GUIs to achieve the goals for which they were designed. Therefore, this method assists individuals among the computer science and user experience disciplines who are involved in the design, development and implementation of web based GUIs. As a result of applying a semi-automatic method, using AGIs and feedback from end users, it establishes continuous improvement, in order to ensure low costs in development and implementation, while maximizing GUI performance.

Based on the above, further analysis and validations of the method could be carried out in other contexts. This could lead to the identification of common rules or patterns that govern the method evolution through its generations (such as the impact of different levels of exploration in performance or the impact of different selection, mutation or crossover techniques).

On the other hand, and based on the performance results of the method application in this and other cases, Artificial Intelligence algorithms could be implemented (such as ID3) to obtain hypotheses that relate the presence or absence of visual characteristics with improvements in the GUIs performance dimensions.

Finally, it is recommended to experiment with more complex chromosomes. This means, adding new Intervention Elements (besides permutation and style elements) or even adding new possible values to the aforementioned elements, to widen the possibilities of visual modification.

## References


1. Miller, S., Huber, S.: A Bíblia e Sua História: O Surgimento e o Impacto da Bíblia. Sociedade Bíblica do Brasil, Rio de Janeiro (2017)
2. Passos, J.E., Silva, T.L.: La Evolución Tecnológica y su Impacto en el Diseño de la Interfaz: Diplomado en Diseño de la Universidad Federal de Rio Grande do Sul (2013). http://www.bocc.ubi.pt/pag/passos-silva-2013-la-evolucion-tecnologica-impacto.pdf
3. Santaella, L.: Navegar no ciberespaço: O perfil cognitivo do leitor imersivo. Paulus, São Paulo (2004)
4. Albornoz, M.C., Berón, M., Montejano, G.: Interfaz gráfica de usuario: el usuario como protagonista del diseño. In: XIX Workshop de Investigadores en Ciencias de la Computación (WICC 2017, ITBA, Buenos Aires), pp. 570–574 (2017)
5. Ramírez, D., Cervantes, J., Molina, J., López, S.: Requirements engineering to design a brain-computer interface for cognitive training. In: 2018 7th International Conference On Software Process Improvement (CIMPS), pp. 79–84 (2018)
6. Moshagen, M., Tielsch, M.T.: Facets of visual aesthetics. Int. J. Hum.-Comput. Stud. **68**, 689–709 (2010)
7. Sbaffi, L., Rowley, J.: Trust and credibility in web-based health information: a review and agenda for future research. J. Med. Internet Res. **19**(6), e218 (2017)
8. Ponte, E.B., Carvajal-Trujillo, E., Escobar-Rodriguez, T.: Influence of trust and perceived value on the intention to purchase travel online: integrating the effects of assurance on trust antecedents. Tour. Manag. **47**, 286–302 (2015)
9. Hasan, B.: Perceived irritation in online shopping: the impact of website design characteristics. Comput. Hum. Behav. **54**, 224–230 (2016)
10. Tielsch, M.T., Blotenberg, I., Jaron, R.: User evaluation of websites: from first impression to recommendation. Interact. Comput. **26**(1), 89–102 (2014)
11. Watermark Consulting: The Watermark Consulting 2013 Customer Experience ROI Study (2013). https://www.watermarkconsult.net/blog/2019/01/14/customer-experience-roi-study/
12. Sánchez, W.: La usabilidad en Ingeniería de Software: definición y características. Ing-novación **1**(2), 7–21 (2011)
13. Pina, P.: What's perfect for me? How CPG shoppers are making the best choice (2018). https://www.thinkwithgoogle.com/consumer-insights/consumer-packaged-goods-industry/
14. Lee, A.: The Future of Design in Start-Ups Survey: 2016 Results (2016). https://www.nea.com/blog/the-future-of-design-in-start-ups-survey-2016-results

15. Kohavi, R., Longbotham, R.: Online controlled experiments and A/B testing. Encyclopedia Mach. Learn. Data Min. **7**(8), 922–929 (2017)
16. Kohavi, R., Crook, T., Longbotham, R.: Online experimentation at Microsoft (2009). https://exp-platform.com/experiments-at-microsoft/
17. Agung Zulkarnaen, A.: Pengembangan Web Landing Page CMS Wordpress Dengan Plugin Thrive Architect Pada Acara Entreprenuer Gathering 5. Technical report, studentstaff5, Perpustakaan Institut Teknologi Telkom Purwokerto (Unpublished) (2018)
18. Gafni, R., Dvir, N.: How content volume on landing pages influences consumer behavior: empirical evidence. In: Proceedings of the Informing Science and Information Technology Education Conference, La Verne, California, pp. 035–053 (2018)
19. Forsythe, A., Nadal, M., Sheehy, N., Cela-Conde, C.J., Sawey, M.: Predicting beauty: fractal dimension and visual complexity in art. Br. J. Psychol. **102**(1), 49–70 (2011)
20. Michailidou, E., Harper, S., Bechhofer, S.: Visual complexity and aesthetic perception of web pages. In: Proceedings of the 26th Annual ACM International Conference on Design of Communication, pp. 215–224 (2008)
21. Bringas Hidalgo, A.: Piscología. Esfinge, Mexico (2010)
22. Seckler, M., Opwis, K., Tuch, A.N.: Linking objective design factors with subjective aesthetics: an experimental study on how structure and color of websites affect the facets of users' visual aesthetic perception. Comput. Hum. Behav. **49**, 375–389 (2015)
23. Cyr, D., Head, M., Larios, H.: Colour appeal in website design within and across cultures: a multi-method evaluation. Int. J. Hum. Comput. Stud. **68**(1–2), 1–21 (2010)
24. Cyr, D., Head, M.: Website design in an international context: the role of gender in masculine versus feminine oriented countries. Comput. Hum. Behav. **29**(4), 1358–1367 (2013)
25. Sharples, M.: Human-computer interaction. In: Boden, M. (ed.) Artificial Intelligence, pp. 293–323. Oxford University Press, Oxford (1996)
26. Kolko, J.: Design thinking comes of age (2015). https://cdn.fedweb.org/fed-42/2892/design_thinking_comes_of_age.pdf
27. Srinivas, M., Patnaik, L.M.: Genetic algorithms: a survey. Computer **27**(6), 17–26 (1994)
28. Mirjalili, S.: Evolutionary Algorithms and Neural Networks. SCI, vol. 780. Springer, Cham (2019). https://doi.org/10.1007/978-3-319-93025-1
29. Cho, S.B.: Towards creative evolutionary systems with interactive genetic algorithm. Appl. Intell. **16**(2), 129–138 (2002)
30. Miikkulainen, R., et al.: Conversion rate optimization through evolutionary computation. In: Proceedings of the Genetic and Evolutionary Computation Conference, pp. 1193–1199 (2017)
31. Blanchard, B.S.: Ingeniería de sistemas. Isdefe, Madrid (1995)
32. Sommerville, I.: Ingeniería del Software. Pearson educación, Madrid (2005)
33. Macaulay, L.A.: Requirements Engineering. Springer, London (2012)
34. Quiroz, J.C., Louis, S.J., Shankar, A., Dascalu, S.M.: Interactive genetic algorithms for user interface design. In: 2007 IEEE Congress on Evolutionary Computation, pp. 1366–1373 (2007)
35. Florez, H., Garcia, E., Muñoz, D.: Automatic code generation system for transactional web applications. In: Misra, S., et al. (eds.) ICCSA 2019. LNCS, vol. 11623, pp. 436–451. Springer, Cham (2019). https://doi.org/10.1007/978-3-030-24308-1_36
36. Ndadji, M.M.Z., Tchendji, M.T., Djamegni, C.T., Parigot, D.: A language and methodology based on scenarios, grammars and views, for administrative business processes modelling. ParadigmPlus **1**(3), 1–22 (2020)
37. Mendez, O., Florez, H.: Applying the flipped classroom model using a VLE for foreign languages learning. In: Florez, H., Diaz, C., Chavarriaga, J. (eds.) ICAI 2018. CCIS, vol. 942, pp. 215–227. Springer, Cham (2018). https://doi.org/10.1007/978-3-030-01535-0_16
38. Gómez, P., Sánchez, M.E., Florez, H., Villalobos, J.: An approach to the co-creation of models and metamodels in enterprise architecture projects. J. Object Technol. **13**(3), 1–29 (2014)

# Descriptive Analysis Model to Improve the Pension Contributions Collection Process for Colpensiones

William M. Bustos C.[1] and Cesar O. Diaz[2](✉)

[1] Fundación Universidad Jorge Tadeo Lozano, Bogotá, Colombia
williamm.bustosc@utadeo.edu.co
[2] Universidad ECCI, Bogotá, Colombia
cdiazb@ecci.edu.co
http://www.utadeo.edu.co, http://www.ecci.edu.co

**Abstract.** The objective of this work is to show how, with the use of Big data techniques, the results of a descriptive analysis of the collection for pension contributions of the COLPENSIONES income management process; because the sustainability of the average premium regime must be guaranteed by the collection by the employers; in order for the pension system to be maintained. Payments are made through the Integrated Contribution Settlement Form–PILA by spanish Planilla Integrada de Liquidación de Aportes–, after registering with an information operator, which can be completed on the information operators' website or through an advisor to them. The information operator is the company in charge of facilitating the creation, modification, validation, correction and sending of PILA. Additionally, it directs information, records and payments to the different social security and parafiscal administrators, in a safe and timely manner. Therefore, we want to make known the importance of this process within the company with this article, from the information found in the payment tables for contributors during the years 2013 to 2017, thanks to COLPENSIONES. A descriptive analysis methodology of the collection process is implemented current situation and the construction of a model based on the different variables that can influence collection management, also validating the existence of methods for predicting collection or non-payment within the company. Through the CRISP DM methodology, apply each of the different stages that it contains within the development of the process.

**Keywords:** Pensions · Descriptive · Big data · Collection

## 1 Introduction

At the pension level, Colombia has wanted to propose a pension reform; The analysis presented by entities such as ANIF (National Associations of Financial Institutions) and ASOFONDOS (Colombian Association of Pension and

Special thanks to organization COLPENSIONES.


H. Florez and M. F. Pollo-Cattaneo (Eds.): ICAI 2021, CCIS 1455, pp. 161–172, 2021.
https://doi.org/10.1007/978-3-030-89654-6_12

Unemployment Fund Administrators) is indisputable to carry out, as part of the problems the Evasion of payments is presented, which consists of not making the payment of contributions to the Social Protection System. Therefore, taking the technological development that is taking place with the use of large information databases, and being able to give it the import value that these represent for both public and private entities. Big data techniques are being used as a tool for managing large volumes of information and the added value they give to information for decision making. The way to identify variables, trends and groupings on specific bases, becomes an interactive process in a transformation and use of information. All focused on implementing automated processes to improve response times and respective collection of the company [21]. Data analysis is the important underpinning of many projects worldwide in different sectors. Applying the above to a system as sensitive as the pension system in Colombia, generates greater importance in the strategies that are linked to risk analysis within the sector [19,20]. Big data techniques are not frequently used for the creation of planning in the collection processes in the pension sector, and it is intended to be useful in management to be able to create actions that reduce the non-payment of contributions in collection and power reduce delinquency in employers. Taking into account the importance of the pension sector, its behavior in the market and the influence it exerts on the Colombian economy, this work develops a methodology through a descriptive model allowing to adequately estimate the number of contributors per month who make payments, classified by city, Quote Value and in this way ensure that Colpensiones obtain tools to reduce evasion or non-payment for contributions from the emperors. This is possible, through the application of descriptive statistical methodologies and different Big Data techniques that have made it possible to identify and quantify levels of collection directed at all companies that make pension payments; For this, the collection made during the years 2013 to 2017 (Information provided by the Bank Payments Log) of the companies in question was taken as a basis, through them a collection analysis was carried out, based on the georeferencing and amounts of payments made. The main objective is to determine quantitative variables that group and determine higher collection areas, where it is determined where there is no coverage for collection and if collection policies and commercial management for the emperors should be changed. The article consists of five sections, where Sect. 1 is the introduction and the importance of the pension systems in Colombia is explained, in Sect. 2 the theoretical and normative framework is proposed and developed, on which the existing obligation to make the contribution payments, as well as the methodology proposed for the development of the model in the third section, some articles will be reviewed where descriptive models were made on the collection process, and finally section four presents the results of this work.

## 2 Methodology and Materials

### 2.1 Background

The Colombian health system created with Law 100 of 1993 [1] incorporates in its design several mechanisms aimed at making it an efficient and equitable system and, in the Latin American context, The universal character of affiliation to the General System of Social Security in Health (SGSSS) established by article 153 of the law, entails two types of obligations: on the one hand, the obligation of every employer to affiliate their workers to this System, and on the other, the obligation of the State to facilitate the affiliation to those who lack a link with an employer or the ability to pay.

Payments must be made through PILA, after registering with an information operator, which can be completed on the information operators' website or through an advisor to them.

The information operator is the company in charge of facilitating the creation, modification, validation, correction and sending to the PILA. Additionally, it directs the information, records and payments to the different social security and parafiscal administrators, in a safe and timely manner.

Law 100 of 1993 [1], establishes in its articles 13, 17 and 22 the following:

*ARTICLE 13.* Characteristics of the General Pension System. The General Pension System will have the following characteristics:

a Membership is mandatory except as provided for independent workers;
b The selection of any one of the regimes provided by the previous article is free and voluntary on the part of the affiliate, who for this purpose shall state his choice in writing at the time of joining or transferring. The employer or any natural or legal person who is unaware of this right in any way, will be entitled to the sanctions referred to in paragraph 1. Article 271 of this Law;
c The affiliates will have the right to the recognition and payment of disability, old-age and survivor benefits and pensions, in accordance with the provisions of this Law;
d Membership implies the obligation to make the contributions established in this Law;
e Members of the General Pension System may choose the pension scheme they prefer. Once the initial selection has been made, they may only be transferred from the regime once every 3 years, counted from the initial selection, in the manner indicated by the national government;
f For the recognition of the pensions and benefits contemplated in the two regimes. It will be taken into account: the sum of the weeks quoted prior to the effective date of this Law, the Social Security Institute or any box, fund or entity of the sector public or private, or the time of service as public servants, whatever the number of weeks quoted or the time of service.
g For the recognition of the pensions and benefits contemplated in the two regimes, the sum of the weeks contributed to any of them will be taken into account;

h In development of the principle of solidarity, the two regimes provided by article 12 of this Law guarantee to their affiliates the recognition and payment of a minimum pension in the terms of this Law;
i There will be a Pension Solidarity Fund designed to expand coverage by subsidizing population groups that, due to their characteristics and socioeconomic conditions, do not have access to social security systems, such as peasants, indigenous people, independent workers, artists, athletes and community mothers;
j No member may simultaneously receive disability and old-age pensions;
k The administrative entities of each one of the regimes of the General Pension System will be subject to the control and surveillance of the Banking Superintendency.

*ARTICLE 17.* Obligation of Contributions. During the validity of the labor relationship, mandatory contributions to the General Pension System Regimes must be made by the affiliates and employers, based on the salary that they earn.

Except for the provisions of article 64 of this Law, the obligation to contribute ceases at the moment the member meets the requirements to access the minimum old-age pension, or when the member retires due to disability or early.

*ARTICLE 22.* Obligations of the Employer. The employer will be responsible for the payment of his contribution and the contribution of the workers in his service. For this purpose, it will deduct from the salary of each affiliate, at the time of payment, the amount of the mandatory contributions and that of the voluntary contributions that the affiliate has expressly authorized in writing, and will transfer these sums to the entity chosen by the worker, together with those corresponding to their contribution, within the terms determined by the Government for this purpose. As we determine the importance and obligation that employers have within the process, we will evaluate in more detail each of the study variables that the entire collection process has, where we will define which are the most important. Topics and technologies are identified and ranked, and key use cases are highlighted [2]. Data mining presents many methodologies for carrying out projects, in this project we will focus on developing and publicizing the CRISP DM. Giving its full scope and identifying the different phases it contains [3].

1. Understanding the business: Understanding the objectives and requirements of the project. Definition of the Data Mining problem.
2. Understanding the data: Obtaining the initial set of data. Exploration of the data set. Identify the quality characteristics of the data. Identify the obvious initial results.
3. Data Preparation: Data Selection. Data cleansing
4. Modeling: Implementation in Data Mining tools
5. Evaluation: Determine if the results coincide with the business objectives. Identify the business issues that should have been addressed.

6. Deployment: Install the resulting models in practice. Settings for data mining repeatedly or continues.

Descriptive analysis as a research source will allow us to summarize and interpret some of the characteristics and properties of a data set; generating information such as the average, the median, the geometric mean, the variance, the standard deviation among others. These measures offer us key properties for the project. A critical issue in ensuring the long-term sustainability of pension funds is making reliable decisions. To design a more comprehensive performance measure+ [4], all of this starts from collection. This also impacts money returns, the valuations of financially solid companies, and those that are strengthening, respond better to the financing of pension plans [5].

## 3 Related Work

In Canada, the types of pension plans offered to Canadian employees are changing. As membership in traditional defined benefit pension plans declines [6], in Colombia we are taking the same path in search of improving our pension system.

Within the information that we have in Colombia, we can generate great changes in improving people's rights in pension issues, as in Europe, which It did an analysis that focuses on 20 European countries using a large number of data available, including a set of 20 possible explanatory variables for the period 2001–2015 [7].

Over the past decade, many countries have reformed their retirement systems by reducing the generosity of benefits, tightening provisions for obtaining a pension, introducing undefined benefit plan options, and even replacing plans. Many of these reforms have affected the post-employment benefits public workers will receive when they retire [8].

In response to the aging population, the United Kingdom (UK) government, like many others, has increased the State Pension Age. This has implied equalizing the state pension age of women and men, raising it from 60 to 65 years [9].

Mexico introduced a Defined Contribution Pension System in 1997. They analyzed the behavior of affiliated workers under the institutional design of the reformed system. Before the reform, 75% of affiliated workers could receive a lifetime annuity upon retirement; They projected that under the new rules only 30% of participants will be able to transform savings into pension income [10].

Due to the fails in the Colombian pension system, it can be reflected in fiscal changes, which leads to a pension reform, as stated by Olivia S. Mitchell and Robert S. Smith [11], where they explore the determinants of pension financing in the public sector. They evaluated the financial characteristics even related to the 80s were analyzed, and shows the great changes and variations in financing practices where it suggests that past financing has to be perpetuated, and that financing it is sensitive to tax pressure [12].

For Colombia, the level of coverage is highly variable determined by access to information, displacement of people, what most influence is the companies omitting payments; Garcia M.T.M. [13] validates the expansion of the coverage of private pension plans is considered an alternative to address the growing pension gap. In fact, due to recent reforms, lower public pension system replacement rates are expected for future generations of retirees. However, in most countries, private provision remains voluntary and observed coverage rates remain very low. Several factors could explain this evidence and various policy options have been suggested to increase coverage in private pension plans. However, the questions regarding both the decision of the type of occupational pension plan, the defined benefit or the defined contribution, as well as the characteristics of the personal pension plans, are more important and complex, since the implications to achieve the adequate retirement income adequacy are diverse. This document reviews these issues and the coverage of these pension plans. Another interesting article where you can see where the pension system could go in Colombia where it did not act in a balanced way in the changes or reforms on issues such as increased contributions, increased weeks or increased pension age, Diamond, Peter A. and Orszag, Peter R [14] review the financial position of Social Security, present a plan to save it, and discuss why Social Security income should not be diverted to individual accounts. Our approach preserves the value of Social Security by providing a basic level of benefits for workers and their families that cannot be decimated by stock market crashes or inflation, and that lasts for the life of the beneficiary; increases benefits for some particularly needy groups, such as those who have worked at low wages for long careers and widows and widowers with low benefits; It eliminates the long-term Social Security deficit without resorting to accounting gimmicks, putting the program and the federal budget on a stronger financial footing. Also as part of research in Colombia, the Universidad Libre, in an article [15], analyzes the Spanish public pension system, describing its components and the incidence of its changes since the 1978 Constitution.

# 4 Methodology

Colpensiones has 6,527,193 affiliates at the end of December 2017, continuing with the achievement of having a third of the affiliates to the general pension system in Colombia. Of the aforementioned affiliates, 2,349,505 correspond to contributing Affiliates and 4,177,688 are Non-contributing Affiliates, that is, they are citizens who did not make their mandatory contribution during the last month. During the 2017 period, 100,370 were affiliated [16]. Colpensiones managed to exceed the goal established for income for the year 2017 by 119% in relation to the programmed income for the different collection concepts, the total amount collected amounted to 15 trillion with respect to the projected income of 12.7 trillions. The main sources of income are Collection of Pension Contributions (contributions of 16%) for a value of 8.7 billion (distributed in 8,105,717 billion of the funds and 592,084 million corresponding to the 1.09% commission of the Administrator) [16]. As of December 2017, the number of companies and commercial establishments active in the commercial register of the Bogotá Chamber of Commerce is 728,784 [4], which would be obliged to make payments to the Social Protection System. The growth that uncertainty has had within Colombians in which they can achieve their pension and due to the levels of non-payment of contributions has generated great concern within the pension sector, the proposal to create a descriptive model to rethink strategies and decision making within the collection process. Statistical tools should be used in the creation of descriptive models for monitoring, coverage [17] of the collection. The database that will be used for the work corresponds to the one generated by the bank payment logs that this structure in the contributor payment table. For the ordering of the variables and the necessary calculations, the Statistical Analysis and Statistics software SAS Enterprise Guide will be used in its version number 9.4.

The database contains the historical contribution payments of 53,803,746 employers made for the years 2013 to 2017 with a monthly average of 896,729 contributors who make contributor payments, with a cut-off date in December 2017, where 32 variables are identified in relation to each employer registry of the entity.

*The dependent variable* will be the Quotation_Cycle that corresponds to the period or month for which the payment of contributions is in default or in default, this is used to calculate and evaluate the portfolio, within the debt settlement process it is taken as the starting point for the respective liquidation of arrears. The remaining variables will explain and determine the dependent variable explained in the proposed model of the 20 variables that are available, 6 that are better related to the structure of the proposed model will be taken.

1. Department: Place where the payment was made; 32departments, a capital district Bogotá and Exterior correspond to payments by affiliates outside of Colombia.
2. Type of Contributors: They define some characteristics of the contributors and the types of contributors that can be linked according to Resolution 634, which parameterizes the content of the Single Form or Integrated Form for the Settlement of Contributions.
3. Document Number: It is the identification of the contributor for the Unique Form or Integrated Form for the Settlement of Contributions.
4. Contribution Value: Corresponds to the contribution to the pension fund made by the employer, which is still 16% of the base contribution income.
5. Number of Workers: Determines the number of affiliates for whom the contribution payment is made.
6. Payment Date: Determines the date on which the contributor made the payment, which can validate whether it was made within the established dates or outside of them.

## 5 Results

The results generated from the project after discarding the variables that are involved in the collection, results in a descriptive model where useful strategies are obtained, which allow the increase collection reduction by contributions. An employer is considered to be any natural or legal person who has the obligation to make contributions to the General Pension System (SGP) on his behalf or on behalf of third parties, that is, they can be employers and/or independent persons. In Fig. 1 is possible to see the number of contributors per month who make payment. The graph illustrates the monthly variation of the population of employers that pay contributions, at the arithmetic mean of the sample is 1,005,922 contributors, clearly seeing the increase for the period 2017-05 due to changes in the regulations by entry Resolution 2082 of 2016 of the UGPP (Pension and Parafiscal Management Unit) in force.

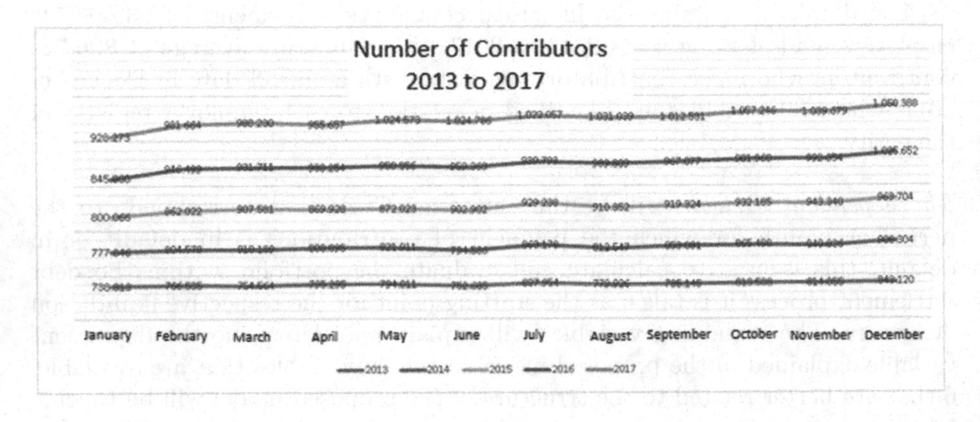


**Fig. 1.** Number of Contributors from 2013 to 2017 in thousands

In Fig. 2 shows the quote value per month of payment. The graph illustrates the monthly variation of the listed value made by payment of contributions, the arithmetic mean of the sample is $689, 142 Million pesos, clearly seeing the decrease for the period 2017-03 due to changes in the regulation by entry Resolution 2082 of 2016 of the UGPP (Pension and Parafiscal Management Unit) in force.

In Fig. 3 shows the distribution of payments by city per year. We can see that the largest amount of payments made during 2017 is in the city of Bogotá DE 3,553,769 Billions which corresponds to 42.2%, followed by Antioquia with 17% and Valle with 10.2%, these three departments correspond to the 69.8% of payments made in Colombia.

## 5.1 Geographical Zones

In the descriptive model used, the result was a scenario applicable to the general process of collection of contributions, which allows improving collection and reducing the evasion of payment of pension contributions. 74% of companies or employers have problems with data quality. This is demonstrated by the study made as we will see it below and it is possible to see in Fig. 4. Employers' contact

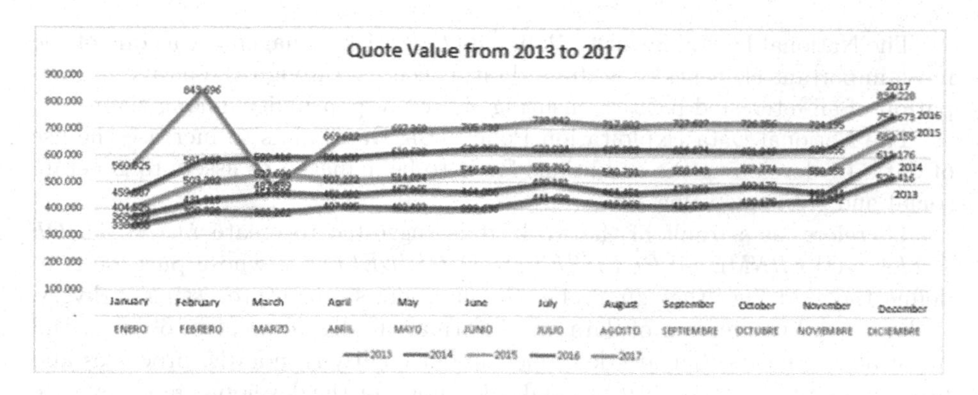


**Fig. 2.** Quote Value per Month from 2013 to 2017 in Millions

| DEPARTMENT | 2.013 | 2.014 | 2.015 | 2.016 | 2.017 | AVERAGE | INCREASE | PARTICIPATION 2017 |
|---|---|---|---|---|---|---|---|---|
| BOGOTA | 2.148.088 | 2.409.715 | 2.726.418 | 3.133.058 | 3.553.769 | 2.794.209 | 65,44% | 42,2% |
| ANTIOQUIA | 827.432 | 924.825 | 1.080.526 | 1.262.854 | 1.459.632 | 1.111.054 | 76,41% | 17,4% |
| VALLE | 487.844 | 556.007 | 638.021 | 745.582 | 859.861 | 657.463 | 76,26% | 10,2% |
| OTHERS DEPS | 1.524.516 | 1.751.746 | 1.984.939 | 2.191.023 | 2.538.163 | 1.998.077 | 66,49% | 30,2% |
| TOTAL | 4.987.880 | 5.642.292 | 6.429.903 | 7.332.516 | 8.411.426 | 6.560.803 | | |

**Fig. 3.** Distribution of Payments by City per Year from 2013 to 2017 in Millions

details, although we presented that 69.8% are distributed in 3 cities, validating in more detail the issue of Addresses only for the year 2017 corresponds to 12 million payments, it is found that there are 4,180,556 almost 35% of these addresses they do not match, the employers are not traceable. According to the Data Governance Technical Guide of the Ministry of Information Technologies and Communications -MINTIC [18], entities must choose to follow an adequate management of data governance, taking into account areas such as governance, quality, migration, life cycle and master data management. The National Government through Decree 1008 of 2018 established the general guidelines of the Digital Government Policy, which aims to promote the use and exploitation of information and communication technologies to consolidate a State and citizens that are competitive, proactive, and innovative, that generate public value in an environment of digital trust. One of the fundamental purposes of the Digital Government Policy is aimed at ensuring that public entities make decisions based on data. Decision-making based on data requires the implementation of an effective governance of data, from the definition of a data governance policy based on the adequate articulation of its fundamental components, such as the people who are part of a digital society, processes and technology. As a complement to the Digital Government Policy, there are documents from the National Council for Economic and Social Policy - CONPES, No. 3854 of April 2016 related to the National Policy of Digital Security and No. 3920 of April 2018 corresponding to the National Data Exploitation Policy (Big Data).

The National Digital Security Policy includes risk management as one of the most important elements to address digital security through the active participation of all interested parties, ensuring shared responsibility among them.

The National Data Exploitation Policy (Big Data) aims to increase the use of data, by developing the conditions for it to be managed as assets to generate social and economic value.

Therefore, as a result of the work, it is suggested to create *MANUAL OF DATA GOVERNMENT POLICIES AND GUIDELINES*, whose purpose is to define the guidelines that guarantee the creation, storage, processing, delivery, exchange and elimination of data and information, digital archival objects. and physical documents that are developed under quality standards, processes and procedures allowing the Entity to make decisions for the development of policies, regulations, plans, programs, projects, application development, among others, with the participation of citizens, users and stakeholders.

Where it has a scope that involves all areas, processes, and establishes the mechanisms to meet the information needs of all interested parties in the organization in terms of availability, security and quality of information, continuously improving the quality of the information. Data, information, digital archival objects and physical documents. The content of this document must include:

**Fig. 4.** Geographical zones of payments for 2017

- Background of the Data Governance Framework in COLPENSIONES
- Mission and Vision of Data Governance
- General principles
- General policies and guidelines
- Roles and responsibilities
- Data Governance Domains in COLPENSIONES
- Data governance activities

## 6 Conclusions and Future Work

It was identified that between the years 2013 to 2017 employers have increased by more than 200,000, making their contributions, maintaining an average of 1,002,922 contributors. It is also reflected that in all years the increase has occurred during the months of December 7%.

69.8% of the contributors belonged to the city of Bogotá with the highest percentage of payment 42%, the Valle del Cauca and Antioquia departments that complete the remaining percentage 27.8%, in these areas are the vast majority of companies in Colombia due to their geographical location and job creation.

Also as a complementary conclusion we see that the remaining 32 departments of Colombia only have 30% of the total collection for contributions; We can see that there is a lack of coverage within these cities that can be generated by a lack of collection points or labor informality.

As future work, an attempt will be made to explain if the Periodic Economic Benefits Program (BEPS) can replace the coverage or not; The BEPS program was created by the Government of Colombia, in 2015, through legislative act 01, with its voluntary savings plan for those people who will not meet the requirements to retire, but who have the capacity to generate some surplus over their expenses can improve coverage.

Additionally, include in the model the issue of informality as a social and economic problem where Colombia has a large gap; High informality exacerbates inequalities, because the informal sector has limited access to financing and public benefits, and intensifies the difficulties faced by the tax and pension system by reducing the contribution and tax base. Depending on the definition, informal employment represents between 50% and 70% of total employment, according to 2018 figures from DANE.

## References


1. Congreso de la República, T.: Ley 100 De 1993. Editorial MOMO Ediciones, Bogotá D.C. (1993)
2. Emerson, S., Kennedy, R., O'Shea, L., O'Brien, J.: Tendencias y aplicaciones del aprendizaje automático en finanzas cuantitativas. 8a Conferencia Internacional sobre Investigación en Economía y Finanzas (ICEFR 2019)
3. Chapman, J.C., Kerber, R., Khabaza, T., Reinartz, T., Shearer, C., Wirth, R.: CRISP-DM 1.0, Step-by-step Data Mining Guide (2000)

4. Lin, S.-W., Lu, W.-M., Lin, F.: Confiar decisiones al fondo de pensiones de servicio público: un modelo predictivo integrado con enfoque de red aditiva DEA. J. Oper. Res. Soc. (2020)
5. Cai, J., Luo, M. y Marcus, A. Salud financiera y valoración de planes de pensiones corporativos. J. Pension Econ. Financ., 1–32 (2019)
6. Gros, B., Sanders, B.: La búsqueda de la sostenibilidad en los planes de pensiones contingentes. CD Comentario del Instituto Howe 553 (2019)
7. Ntotsis, K., Papamichail, M., Hatzopoulos, P., Karagrigoriou, A.: Sobre la modelización de los gastos de pensiones en Europa, Comunicaciones en estadísticas: estudios de casos, análisis de datos y aplicaciones **6**(1), 50–68 (2020)
8. Gorina, E., Hoang, T.: Reformas de pensiones y rotación del sector público. J. Public Administration Res. Theory **30**(1), páginas 96–112 (2020)
9. Holman, D., Foster, L., Hess, M.: Desigualdades en la conciencia de las mujeres sobre los cambios en la Edad de Pensiones del Estado en Inglaterra y el papel de la capacidad cognitiva. Envejecimiento y sociedad **40**(1), 144–161 (2020)
10. Herrerías, R., Zamarripa, G.: Diseño institucional de sistemas de pensiones y comportamiento individual: ¿cómo responden los hogares? (2019)
11. Mitchell, O., Smith, R.: Pension funding in the public sector. Rev. Econ. Stat. **76**(2), 278–290 (1994)
12. Beetsma, R.M.W.J., Komada, O., Makarski, K., Tyrowicz, J.: La estabilidad política (in) de los sistemas de pensiones financiados. Documento de investigación del Centro de Amsterdam para Estudios Europeos (2020)
13. Garcia, M.T.M.: The coverage of occupational and personal pension plans. In: da Costa Cabral, N., Cunha Rodrigues, N. (eds.) The Future of Pension Plans in the EU Internal Market. Financial and Monetary Policy Studies, vol. 48. Springer, Cham (2019)
14. Diamond, P.A., Orszag, P.R.: Resumen de ahorro de la seguridad social: un enfoque equilibrado. Documento de trabajo del Departamento de Economía del MIT No. 04–21 (2004)
15. Nader Orfale, L., Pérez De La Rosa, S.: La crisis financiera del sistema público de pensiones en España. Advocatus **15**(30), 157–165 (2018)
16. Informe de Gestión 2017 Colpensiones, Pág. 6; 49. https://www.colpensiones.gov.co/Documentos/informes_de_gestion
17. Rutledge, M.S.: Pueden los trabajadores no tradicionales mayores encontrar cobertura de salud y jubilación? (2020)
18. Guía Técnica de Gobierno de Datos. Ministerio de las Tecnologías de la Información y Comunicación. Versión 1.0 (2014)
19. Florez, H., Sánchez, M., Villalobos, J.: A catalog of automated analysis methods for enterprise models. Springerplus **5**(1), 1–24 (2016). https://doi.org/10.1186/s40064-016-2032-9
20. Florez, H., Sánchez, M., Villalobos, J.: Modeling and Analyzing Imperfection in Enterprise Models. Engineering Letters (2020)
21. Ndadji, M.M.Z., Tchendji, M.T., Djamegni, C.T., Parigot, D.: A Language and Methodology based on Scenarios, Grammars and Views, for Administrative Business Processes Modelling. ParadigmPlus, pp. 1–22 (2020)

# User Response-Based Fake News Detection on Social Media

Hailay Kidu, Haile Misgna, Tong Li(✉), and Zhen Yang

Faculty of Information Technology, Beijing University of Technology, Beijing, China
{hailay.kidu,haile.misgna}@mu.edu.et, {litong,yangzhen}@bjut.edu.cn

**Abstract.** Social media has been a major information sharing and communication platform for individuals and organizations on a mass scale. Its ability to engage users to react to information posted on this media in the form of like, share, and comment made it a preferable information sharing platform by many. But the contents posted on social media are not filtered, fact checked or judged by an editorial body like any traditional news platform. Therefore, individuals, institutions and communities who consume news from social media are vulnerable to misinformation by malicious authors. In this work, we are proposing an approach that detects fake news by investigating the reaction of users to a post composed by malicious authors. Using features extracted by bag-of-words model and TF-IDF from text based replies (comments), and visual emotion responses in the form of categorical data, we built models that predicted news as fake or real. We have designed and conducted a series of experiments to evaluate the performance of our approach. The results show the proposed approach outperforms the baseline in all the six models. In particular, our models from random forest, logistic regression, and XGBoost algorithms produce a precision of 0.97, a recall of 0.99 and an F1 of 0.98.

**Keywords:** Fake news detection · Users' responses · Machine learning · Deep learning

## 1 Introduction

The capability of social media technologies to communicate and disseminate information on a massive scale, high speed, low cost, easy access and freedom to publish anything contributed in drawing mainstream news consumers and publishers into social media, such as Twitter, Facebook, blogs and others. The freedom to publish anything on those platforms is a double edged sword. The contents posted on those platforms are not filtered, fact checked or judged by an editorial body. Most people do not verify the source of the news before they read and share it, which can lead to the propagation of fake news quickly and may lead to it going viral. Individuals, institutions and communities who consume news from those platforms are vulnerable to malicious authors whose intention is misinforming readers.


H. Florez and M. F. Pollo-Cattaneo (Eds.): ICAI 2021, CCIS 1455, pp. 173–187, 2021.
https://doi.org/10.1007/978-3-030-89654-6_13

The expression "fake news" is popularized during the 2016 US elections [20]. Since then researchers in journalism and artificial intelligence are working on defining fake news and how to identify it automatically using the help of computer algorithms as early as possible. Allcott and Gentzkow [6] define fake news as intentionally written to mislead news consumers with information which are verifiably false.

The seriousness of the threats and fears caused because of fake news has a high impact in the integrity of journalism and politics. To mitigate those threats and fears Nakamura et al. [18] and Wang [30] proposed a benchmark datasets for fact checking. Abbasi and Liu [2], and Sitaula et al. [27] claim assessing the association of users with fake news is an important feature in detecting fake news. Choudhary and Arora [9] proposed a linguistic feature approach that employs syntax, sentiment, grammatical and readability evidence to characterize fake news from textual contents to design a set of features. Shu et al. [25] explores auxiliary information from tri-relationship, a relationship between publishers, users, and news content, to improve the process of fake news detection.

In all those works, features extracted from news content are used to detect fake news. Even though users' responses are honest reactions to a news content with less intention to deceive others, we found few works that considered users' reactions in the task of fake news detection. The works which considered user responses treated features extracted from users' reactions as auxiliary source of information. On social media platforms, like Twitter and Facebook, there is a high degree of user engagement on posted news. Comments are written and emotions are expressed against or in support of the post.

In this research paper, we investigated fake news detection on Twitter from the users' responses presented in the form of text and visual emotional reaction. To our knowledge there is no comparative study done on fake news detection approaches; news content-based approach and social context approach. In this work, we are proposing the use of users' responses which is a type of social context as a main source of information to detect fake news rather than using them as an auxiliary source of information. We proposed a new way of looking at the users' responses information in the study of fake news detection, and we did comparative analysis of fake news detection on users' responses and news content using machine learning algorithms.

The remaining sections of this article are structured as follows. In Sect. 2 related works are covered. Section 3 discusses the methodologies. Experimental results and analysis are presented in Sect. 4. In Sect. 5 discussion on the overall results is discussed. The last Sect. 6 makes concluding remarks and discusses future works.

## 2 Related Work

Fake news identification using computer algorithms has been not trivial. Shu and Liu [22] presented four unique challenges of fake news on social media: (1) it is not simple to detect fake news simply based on content, (2) the volume, variety and veracity characteristics of the social media data, (3) the background of social media users, and (4) the easiness to create malicious accounts. These challenges motivate researchers to study the mitigation of fake news propagation by detecting them as early as possible in order to prevent tremendous negative political and social impacts.

Shu and Liu [22] classify existing fake news detection approaches into two broad categories: fake news detection from news contents and fake news detection from social contexts.

## 2.1 News Content-Based Approach

News contents contain a great deal of information that can tell whether a given news is fake or real. The features that characterize the news content can be the source of the news, the headline, the main text, and visual information embedded in the news. From those features more discriminative characteristics of fake news are built. This approach is the most studied approach. A number of works employ techniques from natural language processing and machine learning, and apply it on the news content [3,7,11,12,28,29].

Wang [30] and, Nakamura et al. [18] proposed benchmark datasets for fake news detection. Nakamura et al. [18] present a large-scale multimodal fake news dataset which contains over 1 million samples containing text, image, metadata, and comments from a highly diverse set of resources. The dataset is a multiple labeled dataset with 2-way, 3-way, and 6-way classification. In addition, the dataset utilizes image data as evidence for text truthfulness or text data for image truthfulness.

Abbasi and Liu [2], and Sitaula et al. [27] claimed assessing credibility of users has a significant role in detecting fake news. Abbasi and Liu [2] proposed CredRank algorithm to measure users' credibility on social media. The credibility score is built from the behaviour of users on social media that is posted as a news content. The authors argue that a user with high credibility score is less likely to propagate fake news detection. Studying the role of user profile helps in detecting who will likely share a fake news [26].

Choudhary and Arora [9] proposed a linguistic model to find out the properties of news contents that will generate language-driven features. This model extracts syntactic, grammatical, sentimental, and readability features of particular news. The linguistic feature-driven model achieved the average accuracy of 86% for fake news detection and classification. Granik and Mesyura [11] proposed a simple approach for fake news detection using naive Bayes classifier on news contents, and achieved a classification accuracy of approximately 74%.

Khan et al. [15] argues that most fake news detection algorithms are trained in politics dataset and this will result in producing biased models. The authors combined three different datasets with diverse topics and investigated the performance of different machine learning models. Ahmed et al. [4] proposed n gram models for fake news detection and they applied TF-IDF for feature extraction. They conducted a comparative analysis on six machine learning models and they got 92% accuracy with linear SVM classifiers. Aslam et al. [7] proposed an ensemble-based deep learning model to classify the news into fake or real on LIAR dataset. NLP techniques were applied to extract text features from news content. A deep learning model, Bi-LSTM-GRU-dense model, on news content attributes achieved a better performance result.

Horne and Adali [14] discussed systematic, stylistic and other content differences between fake and real news. In conducting their fake news study, they investigated three separate data sets. They also include satire as a type of fake news that relies on absurdity other than sound arguments to make claims, but explicitly identifies itself as satire. Shu et al. [23] proposed a general data mining framework for fake news detection which includes two phases: (i) feature extraction and (ii) model construction. This work presented the narrow and broad definitions of fake news and clear direction from characterization to detection.

Yang et al. [31] proposed an unsupervised learning framework, which utilizes a probabilistic graphical model to model the truths of news and the users' credibility. An efficient collapsed Gibbs sampling approach is proposed to solve the inference problem. They conducted experiments on two real-world social media datasets, LIAR and BuzzFeed. Their experimental results demonstrated the effectiveness of the proposed framework for fake news detection on social media.

### 2.2 Social Context Approach

In this approach, the engagement of users to a news content via social media generates supplementary information that likely enhances the content-based models. A number of research articles address the challenge of fake news detection from different perspectives.

Shu et al. [24] argued that the performance of models built on news content only is not satisfactory, and it is suggested to incorporate user social engagements as auxiliary information to improve the fake news detection task. They constructed real-world datasets measuring users' trust level on fake news and select representative groups of both "experienced" and "naïve" users. And they performed a comparative analysis over explicit and implicit profile features between those user groups.

News content-based approach is the well researched approach. But targeting news contents for detection of fake news raises too many challenges. One of the main challenges for this approach is that news contents are carefully crafted to deceive readers. So, we argue that it is not trivial to target the mitigation of fake news detection from a news content angle. Shu et al. [25] discussed the helpfulness of social context approach in providing auxiliary information that may result in having better performance results. But we presented a different approach of using social context information in the process of fake news detection, i.e., using the users' responses information as the main data input in detecting fake news. The responses of users' contain rich information and the input data is presented in the form of text and visual emotion. To our knowledge, we have not found an article that addresses fake news detection from the perspective of users' responses. In this work, we proposed the user response-based approach for automatically classifying news as fake or real on Twitter.

## 3 Methodologies

The major question this paper addresses is whether we can automatically detect fake news by the textual and visual response of users that outperforms the news content approach. In this work, we proposed a new approach that targets users' reactions in detecting fake news. The proposed approach involves different tasks to shape the input before being used for model production. These tasks include data preprocessing, feature extraction and model production. Figure 1 and the following sections show the methods employed in this work.

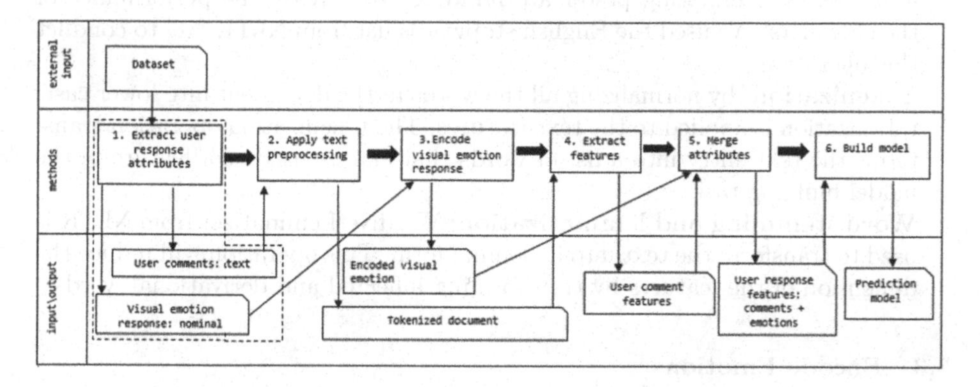


**Fig. 1.** A users' response-based fake news detection methodology. The first row shows the input data. The second row presents all the set of tasks employed in the proposed approach. And the last row presents the intermediate outputs and inputs of the tasks in row two.

### 3.1 Extract User Response Attributes

The dataset has attributes like idx, context_idx, text, reply, categories and mp4. In this stage, attributes that define users' response are selected. The attributes reply, mp4 and categories are attributes that show users' engagement to a tweet in the dataset we are experimenting with. Categories and mp4 represent the same type of information; categories attribute is a description of the GIF files in the mp4 attribute. Therefore, we considered categories attribute which represents the visual information and reply attributes that represents users comment as user response attributes for this work.

## 3.2 Apply Text Preprocessing

Before extracting features from the textual data, the data needs to be cleaned to a certain level to increase the performance of the models. Therefore, we applied data cleaning and transformation operations on the textual data, such as numbers and special characters removal, stopwords removal, tokenization, and stemming and lemmatization.

- **Removal of special and single characters:** symbols like ?, <, >, +, _, -, /, *, !, @, #, $, %, ;, &, (, ), {, }, [, ], ‘, “, ;, :, ˜, and more are removed. In addition to that, numbers and single characters are also removed because it is assumed that they don't have much significance in the fake news identification process.
- **Stopwords removal:** words that are frequently found in documents that have less discriminating power are removed to increase the performance of the classifiers. We used the English stopwords list from NLTK [16] to conduct this operation.
- **Tokenization:** by normalizing all the words in the document into lower case, tokenization is applied to the text features. The tokenization operation transforms the text entry into a list of words that can be used as a feature in the model building task.
- **Word stemming and lemmatization:** WordNetLemmatizer from NLTK is used to transform the text into its lemma form. This operation will reduce the dimension of the features by transforming inflected and derivational words.

## 3.3 Encode Emotions

The visual emotional information is described as categories attribute. This attribute is a nominal data presented in the form textual data format. Therefore, we encode this categorical data using one-hot encoding to transform the nominal textual input into a binary incidence data input. The one-hot encoding determines the presence or absence of emotions at every entry of the user. The encoded data shows an emotional reaction of user to a post. This encoding creates a categories matrix, $C$, that represents emotion $e$ expressed by user $u$.

$$C_{u,e} = \begin{cases} 1, \text{if the emotion } e \text{ is expressed by user } u \\ 0, \text{otherwise} \end{cases}$$

The entries are encoded 1 if the emotion is expressed by a given user, otherwise the entries are 0. In this matrix the rows represent the users and the columns represent the emotions.

## 3.4 Extract Features

The model building process and performance is highly dependent on the set of features used for the identification of fake news. This work mainly focuses on two attributes of the dataset, namely reply and categories. The categories attribute is nominal data whereas the reply is textual data. For this work, we used state-of-the-art feature extraction models for text classification, n-gram models. This model is used for text classification in short texts [10,11] and large contents [4,5].

**BoW Model with TF-IDF Weighting Scheme for Feature Extraction:** Bag-of-words (BoW) model is used to extract features from the reply attribute which is a text entry. The process of feature extraction starts from tokenization of the reply entries into a word vector. And then, we applied a TF-IDF weighting scheme to measure the relevance of each term.

$$tf(t,d) = \frac{f_{t,d}}{\sum_{t' \epsilon d} f_{t',d}} \tag{1}$$

The above formula is read as, term frequency of t in document d, tf(t,d), is the frequency of term t in document d, $f_{t,d}$, divided by the sum of the frequencies of all the words in the document, $\sum_{t' \epsilon d} f_{t',d}$.

$$idf(t,D) = log \frac{N}{|\{d \epsilon D : t \epsilon d\}|} \tag{2}$$

Inverse document frequency of a term t in document d is the log of total number of collections in the corpus N divided by the number of documents d that contain the term t. So then, TF-IDF is the product of the two equations.

$$TF - IDF = tf(t,d) * idf(t,D) \tag{3}$$

### 3.5 Merge Attributes

The user's reaction is defined by features extracted from the comments and visual emotions. We extracted 100 features from the comments and 44 features from the visual emotions. Both data are presented in matrix form. The first matrix represents features extracted from comments. The entries of this matrix are TF-IDF of each feature expressed by a user. The second matrix represents the emotion of users expressed towards a news post. The entries of this matrix represent the incidence of emotions. Each entry shows whether a given user expressed an emotion to a news post. By merging both matrices using NumPy [13] hstack function, we created a matrix that has 100 features from the text entry and 44 features from the nominal data. Therefore, a new form of data with 144 features is generated and used as an input to the algorithms to create fake news prediction models.

### 3.6 Build Model

To test our approach, we build models from six machine learning algorithms, namely random forest, logistic regression, support vector machine (SVM) and multinomial naive Bayes, XGBoost [8] and RNN, and investigate how the proposed approach performed. We used scikit learn API for building models from random forest, logistic regression, support vector machine, and multinomial naive Bayes [19]. For RNN, we used tensorflow to build as well as to test our model [1]. We trained our model in two types of input data; text and nominal data. For the RNN model, we used word embedding [17] as input and features from categories attribute where the output of the word embedding is concatenated before going to the activation layer. For the other algorithms, TF-IDF vectors of the reply attribute and feature vectors from categories attribute are concatenated and given as an input.

## 4 Experimental Results and Analysis

We conducted experiments on the proposed approach for fake news detection and compared it with a baseline, a method that detects fake news from news contents using a bag-of-words model. We did comparisons on 6 machine learning algorithms and the proposed approach outperformed in all the algorithms. In this work, we will answer the following research questions (RQ):

- RQ 1: Will users' responses provide more helpful clues than news content in fake news classification?
- RQ 2: Which one of the machine learning models performs best in detecting fake news from users' responses?

### 4.1 Dataset

This work uses a publicly available dataset prepared for the challenge of Fake-EmoReact 2021 Challenge[1]. The dataset is collected from Twitter and the records with at least one GIF response are considered in the preparation of the dataset. The challenge organizers prepared the dataset in 3 different JSON files;train.json, eval.json and dev.json. The train.json file is the only labeled dataset but both eval.json and dev.json are not labeled, they are the holdout data. train.json contains 168,521 entries and 6 features. The 31,799 (19%) entries are labeled as real and 136,722 (81%) of them are labeled as fake. There are only 227 unique tweets. The dataset has idx, context_idx, mp4, text, reply and categories attributes.

- **idx** is an identifier of a tweet
- **text** is the original tweet
- **context_idx** is an identifier of a reply for a given tweet
- **reply** is user response in the form of text
- **mp4** is a GIF user response that represents visual emotional reactions
- **categories** is a textual description of mp4 attribute

[1] https://sites.google.com/view/covidfake-emoreact-2021/.

### 4.2 Experimental Settings

To show detecting fake news from users' reactions is better than news content post, we conducted a comparison of fake news detection from news content (user tweet) and user emotion reaction in the following experimental settings and evaluation protocols.

**Parameter Settings:** The parameters are chosen after conducting many iterations to fine-tune the trade-off between speed and performance. The settings used in conducting the experiments are stated as follows:

- TF-IDF: One hundred unigram features are generated from the reply (text) attributes. The minimum document size is five and the maximum document size is 70%.
- Word embeddings: In this feature engineering task, we consider only 100 features. In other words, the vocabulary size of the embedding layer is 100.
- Random forest: The number of trees used to estimate the label is 100.
- Logistic regression, linear support vector machine, and naive Bayes we used the default parameters of scikit learn.
- RNN: The number of batches for training are 128, the number of iteration (epochs) are 25 and ADAM optimizer is used.

**Evaluation Protocol:** To perform the experiments, we first trained six algorithms in the training.json data. We did not split the dataset into training and testing data. We used train.json (training data from the challenge organizers) for training and dev.json (validation data from the challenge organizers) for testing. Every experimental results included in this work are results collected after they are tested on the dev.json file. This file is deliberately prepared by the organizers to test the performance of the models on held out data. All our performance scores are collected from CodaLab[2] after submitting the results of our models. And the performance of the models is presented using precision, recall and f-measure.

**Baseline:** We compare our proposed approach against approaches that detect fake news from news content using linguistic features.

- Detecting fake news from news content (**News content**): Ahmed et al. [4] proposed a methodology that detects fake news from news content using BoW model. We used the same feature extraction methodology for the classical machine learning algorithms.

[2] https://competitions.codalab.org/competitions/30741.

**Table 1.** The comparison of the proposed approach (users' responses) against the baseline (news content) using logistic regression, random forest, NB, linear SVC, XGBoost, and RNN.

| Algorithms | Metrics | News content [4] | User response (proposed) |
|---|---|---|---|
| Logistic regression | Precision | 0.57 | **0.97** |
| | Recall | 0.6 | **0.99** |
| | F1-score | 0.49 | **0.98** |
| Random forest | Precision | 0.56 | **0.97** |
| | Recall | 0.52 | **0.99** |
| | F1-score | 0.24 | **0.98** |
| NB | Precision | 0.61 | **0.97** |
| | Recall | 0.52 | **0.95** |
| | F-score | 0.5 | **0.96** |
| Linear SVC | Precision | 0.55 | **0.97** |
| | Recall | 0.56 | **0.98** |
| | F-score | 0.43 | **0.97** |
| XGBoost | Precision | 0.57 | **0.97** |
| | Recall | 0.53 | **0.99** |
| | F-score | 0.26 | **0.98** |
| RNN | Precision | 0.79 | **0.89** |
| | Recall | 0.91 | **0.96** |
| | F-score | 0.82 | **0.92** |

### 4.3 Performance Comparison (RQ1)

Table 1 shows the performance report of the six models on news content and users' responses approaches. This summary report shows the score of the two approaches using precision, recall and f1-score. According to our experiment, the proposed approach shows better performance. The users' response data is built from two different inputs, reply and GIF responses. Since the reply attribute is a text in nature, we conducted an experiment to see if it can produce a different result when it is compared with the data with the same data type. As it is shown on Fig. 2, the users' response from reply, and merged data of both reply and emotional information outperformed the baseline. RNN outperforms all the other classical machine learning models on news comment and user response(reply) data. This showed that word embedding is better in extracting discriminative features from text input data than the simple bag-of-words model.

We demonstrated the betterment of detecting fake news from users' comments (replies) and the whole users' responses (reply and visual emotional) information. The comparisons between news comment and users' responses on the textual input showed that more interesting textual features are extracted from users' responses.

However, on the proposed approach column the other models improve significantly when the categories attribute is merged with the replies attribute. The experimental results show that logistic regression, random forest and XGBoost models performed better with mixed data. The use of GIF emotions helped the

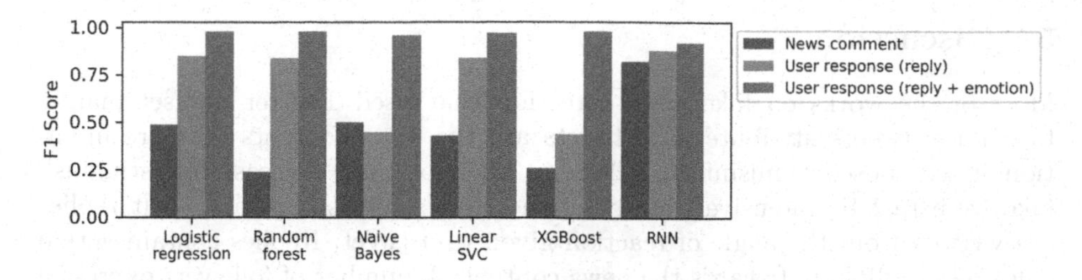


**Fig. 2.** Comparison of users' response-based vs. news content fake news detection (F1 score).

naive Bayes algorithm to jump from 0.50 F-1 score to 0.96. Similar increment patterns are shown for all the models with mixed data. In general, the use of emotions increased the performance of all the six algorithms. But the increase in performance based on categories is not that much on RNN. We have observed a similar result as [21] where the adding of LIWC features to features extracted from news content using word embeddings does not contribute any significant increase in performance. Therefore, we draw the same conclusion as [21] that the RNN model already learned most of the features from the text data and most of the features from categories attribute are redundant.

### 4.4 Which Algorithm Fits Best? (RQ2)

As it is seen on Fig. 3, our proposed approach is tested using five classical machine learning algorithms and one deep learning algorithm. Based on the experimental results, it has been shown that the ability of learning more patterns from the visual emotional information is demonstrated by the classical machine learning models. The merging of those features from users' textual and visual emotion reactions highly boosted the performance of the classical machine learning algorithms whereas the performance of RNN stayed with no significant increase. Logistic regression, random forest and XGBoost algorithms build better models using the basic feature extraction techniques.

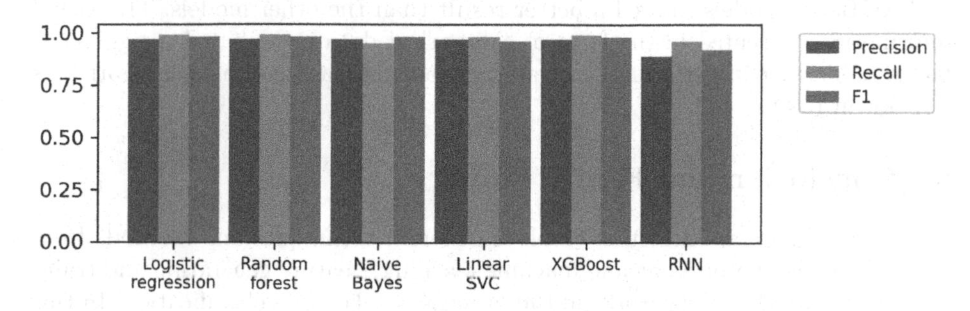


**Fig. 3.** An experimental result of users' response-based fake news detection.

## 5 Discussion

Most of the works on fake news detection that used Twitter dataset mainly target the tweets attribute. The tweets are the news contents that are intentionally composed to misinform followers. Therefore, it is not easy to distinguish fake from real by intensively studying the tweets (news contents). But replies are written from the angle of reaction given to a tweet. Replies contain actual emotion of followers towards the news content. A number of followers express a range of emotions in support and against the original news content.

The contents of replies are rich in content. And the contents are presented in the form of text and visual information. This information has been used as auxiliary information in the process of fake news detection. In this work, we want to highlight how much it is significant to consider users' responses in the study of fake news detection. To shed a light on the significance of this type of information, we compared it against news content where in major works used to detect fake news. We used this as a baseline for this work.

We used simple feature engineering techniques for feature extraction from the text, n-gram model using TF-IDF weighting scheme. By applying this method on both tweet and replies (text), we have demonstrated that the replies (text) outperformed the tweets in the process of fake news detection. Users' responses that contain both textual and emotional replies showed the best performance result in comparison with replies (text) and tweets (news content).

To demonstrate the users' responses contributes more interesting features than tweets in deep learning as well, we tested it using word embeddings and RNN. The deep learning model showed a good result in replies with textual and visual information. But the performance on textual replies is nearly the same with the replies that contain both textual and visual information. Therefore, the model did not learn not much new pattern from the visual information.

In summary, our investigation showed an important approach for the task of fake news detection from the perspective of users' responses. Specifically, we have observed classifiers that depend on visual emotion information perform consistently good on all the algorithms we tested. Figure 2 demonstrated the incorporation of visual information led to a better performance in the task of fake news detection. As it is shown on Fig. 3, random forest, logistic regression and XGBoost models showed a better result than the other models. The visual information presented in the form of categorical data contributed a significant factor of increase in performance on all the classical machine learning algorithms but not in RNN.

## 6 Conclusion and Future Work

Automatic fake news detection employs different techniques and methods from natural language processing and machine learning. Feature generation and training models are the major works in the process of fake news classification. In this paper, we addressed the task of automatic fake news identification from the perspective of who provides more clues. We proposed an approach of addressing fake news from the perspective of followers where their responses are manifested in the form of textual and visual emotion data. We built different classification models that showed users' responses contain more discriminative information in the task of fake news detection.

In this work, we only investigated the contents of the users' responses. We addressed the problem from a text classification perspective where simple syntactic approaches are used. For future work, we will explore more text classification algorithms. And it is useful to investigate domain-specific features, linguistic features, post-based detection, and network-based detection feature extraction methods. In algorithmic-wise, investigating transfer learning feature generation models, such as BERT, GloVe, etc. will be useful to extend our approach.

**Acknowledgments.** This work is partially supported by the National Natural Science of Foundation of China (No. 61902010, 61671030), the International Research Cooperation Talent Introduction and Cultivation Project of Beijing University of Technology (No. 2021C01), and the Project of Beijing Municipal Education Commission (No. KM202110005025).

## References

1. Abadi, M., et al.: TensorFlow: a system for large-scale machine learning. In: 12th {USENIX} Symposium on Operating Systems Design and Implementation ({OSDI} 2016), pp. 265–283 (2016)
2. Abbasi, M.-A., Liu, H.: Measuring user credibility in social media. In: Greenberg, A.M., Kennedy, W.G., Bos, N.D. (eds.) SBP 2013. LNCS, vol. 7812, pp. 441–448. Springer, Heidelberg (2013). https://doi.org/10.1007/978-3-642-37210-0_48
3. Ahmad, I., Yousaf, M., Yousaf, S., Ahmad, M.O.: Fake news detection using machine learning ensemble methods. Complexity **2020** (2020). https://doi.org/10.1155/2020/8885861
4. Ahmed, H., Traore, I., Saad, S.: Detection of online fake news using N-gram analysis and machine learning techniques. In: Traore, I., Woungang, I., Awad, A. (eds.) ISDDC 2017. LNCS, vol. 10618, pp. 127–138. Springer, Cham (2017). https://doi.org/10.1007/978-3-319-69155-8_9
5. Ahmed, H., Traore, I., Saad, S.: Detecting opinion spams and fake news using text classification. Secur. Priv. **1**, 1–15 (2018)
6. Allcott, H., Gentzkow, M.: Social media and fake news in the 2016 election. J. Econ. Perspect. **31**, 211–236 (2017)
7. Aslam, N., Ullah Khan, I., Alotaibi, F.S., Aldaej, L.A., Aldubaikil, A.K.: Fake detect: a deep learning ensemble model for fake news detection. Complexity **2021** (2021). https://doi.org/10.1155/2021/5557784
8. Chen, T., He, T., Benesty, M., Khotilovich, V., Tang, Y., Cho, H., et al.: XGBoost: extreme gradient boosting. R Package Version 0.4-2 **1**(4), 1–4 (2015)
9. Choudhary, A., Arora, A.: Linguistic feature based learning model for fake news detection and classification. Expert Syst. Appl. **169**, 114171 (2021)
10. Dilrukshi, I., De Zoysa, K., Caldera, A.: Twitter news classification using SVM. In: 2013 8th International Conference on Computer Science & Education, pp. 287–291. IEEE (2013)

11. Granik, M., Mesyura, V.: Fake news detection using Naive Bayes classifier. In: 2017 IEEE First Ukraine Conference on Electrical and Computer Engineering (UKRCON), pp. 900–903. IEEE (2017)
12. Hajek, P., Barushka, A., Munk, M.: Fake consumer review detection using deep neural networks integrating word embeddings and emotion mining. Neural Comput. Appl. **32**(23), 17259–17274 (2020). https://doi.org/10.1007/s00521-020-04757-2
13. Harris, C.R., et al.: Array programming with NumPy. Nature **585**, 357–362 (2020)
14. Horne, B., Adali, S.: This just. In: Fake News Packs a Lot in Title, Uses Simpler, Repetitive Content in Text Body, More Similar To Satire Than Real News. In: Proceedings of the International AAAI Conference on Web and Social Media, vol. 11 (2017)
15. Khan, J.Y., Khondaker, M., Islam, T., Iqbal, A., Afroz, S.: A benchmark study on machine learning methods for fake news detection. arXiv (2019)
16. Loper, E., Bird, S.: NLTK: the natural language toolkit. arXiv preprint cs/0205028 (2002)
17. Mikolov, T., Sutskever, I., Chen, K., Corrado, G., Dean, J.: Distributed representations of words and phrases and their compositionality. arXiv preprint arXiv:1310.4546 (2013)
18. Nakamura, K., Levy, S., Wang, W.Y.: r/Fakeddit: a new multimodal benchmark dataset for fine-grained fake news detection. arXiv preprint arXiv:1911.03854 (2019)
19. Pedregosa, F., et al.: Scikit-learn: machine learning in Python. J. Mach. Learn. Res. **12**, 2825–2830 (2011)
20. Quandt, T., Frischlich, L., Boberg, S., Schatto-Eckrodt, T.: Fake news. Int. Encyclopedia Journal. Stud. 1–6 (2019)
21. Rashkin, H., Choi, E., Jang, J.Y., Volkova, S., Choi, Y.: Truth of varying shades: analyzing language in fake news and political fact-checking. In: Proceedings of the 2017 Conference on Empirical Methods in Natural Language Processing, pp. 2931–2937 (2017)
22. Shu, K., Liu, H.: Detecting fake news on social media, vol. 11, pp. 1–129. Morgan & Claypool Publishers (2019)
23. Shu, K., Sliva, A., Wang, S., Tang, J., Liu, H.: Fake news detection on social media: a data mining perspective. ACM SIGKDD Explor. Newslett. **19**, 22–36 (2017)
24. Shu, K., Wang, S., Liu, H.: Understanding user profiles on social media for fake news detection. In: 2018 IEEE Conference on Multimedia Information Processing and Retrieval (MIPR), pp. 430–435. IEEE (2018)
25. Shu, K., Wang, S., Liu, H.: Beyond news contents: the role of social context for fake news detection. In: Proceedings of the Twelfth ACM International Conference on Web Search and Data Mining, pp. 312–320 (2019)
26. Shu, K., Zhou, X., Wang, S., Zafarani, R., Liu, H.: The role of user profiles for fake news detection. In: Proceedings of the 2019 IEEE/ACM International Conference on Advances in Social Networks Analysis and Mining, pp. 436–439 (2019)
27. Sitaula, N., Mohan, C.K., Grygiel, J., Zhou, X., Zafarani, R.: Credibility-based fake news detection. In: Shu, K., Wang, S., Lee, D., Liu, H. (eds.) Disinformation, Misinformation, and Fake News in Social Media. LNSN, pp. 163–182. Springer, Cham (2020). https://doi.org/10.1007/978-3-030-42699-6_9
28. Thota, A., Tilak, P., Ahluwalia, S., Lohia, N.: Fake news detection: a deep learning approach. SMU Data Sci. Rev. **1**, 10 (2018)

29. Umer, M., Imtiaz, Z., Ullah, S., Mehmood, A., Choi, G.S., On, B.W.: Fake news stance detection using deep learning architecture (CNN-LSTM). IEEE Access **8**, 156695–156706 (2020)
30. Wang, W.Y.: "Liar, liar pants on fire": a new benchmark dataset for fake news detection. arXiv preprint arXiv:1705.00648 (2017)
31. Yang, S., Shu, K., Wang, S., Gu, R., Wu, F., Liu, H.: Unsupervised fake news detection on social media: a generative approach. In: Proceedings of the AAAI Conference on Artificial Intelligence, vol. 33, pp. 5644–5651 (2019)

# Decision Systems

# A Multi-objective Mathematical Optimization Model and a Mobile Application for Finding Best Pedestrian Routes Considering SARS-CoV-2 Contagions

Juan E. Cantor(✉), Carlos Lozano-Garzón, and Germán A. Montoya

Systems and Computing Engineering Department, Universidad de Los Andes, Bogotá, Colombia
{je.cantor,calozanog,ga.montoya44}@uniandes.edu.co

**Abstract.** Given the spread and pandemic generated by the SARS-CoV-2 coronavirus and the requirements of the Universidad de Los Andes in terms of guaranteeing the health and safety of the students considering possible contagions at the university campus, we propose and design a mobile application to obtain the best route between two places in the campus university for a student. The path obtained by our proposal, reduces the distance traveled by the student as well as a possible contagion of the coronavirus during his journey through the university campus. In this sense, two types of costs were modeled: one cost represents the distance cost to travel the campus by a student, and the second one represents the contagion susceptibility that a student has when he is passing through the campus. In summary, it was developed and validated a solution algorithm that minimizes these two types of costs. The results of our algorithm are compared against a Multi-Objective mathematical optimization solution and interesting findings were found. Finally, a mobile application was created and designed in order to obtain optimal routes to travel between two points in the university campus.

**Keywords:** SARS-CoV-2 · Pedestrian traffic · Multi-objective optimization · Heuristic · Mobile application

## 1 Introduction

The pandemic spread by the SARS-CoV-2 virus, commonly called COVID-19, led multiple countries worldwide to take and enforce social restrictions in order to limit the spread of the virus. The most common contingency measures, taken by several governments, is to limit the movements of its citizens, as well as the social interactions that occur between them. This measure is known as "*social distancing*" and has the purpose of minimizing the exchange of segregations


H. Florez and M. F. Pollo-Cattaneo (Eds.): ICAI 2021, CCIS 1455, pp. 191–206, 2021.
https://doi.org/10.1007/978-3-030-89654-6_14

between one person to another, which is the main event that causes the exponential growth of the contagion of the coronavirus [7].

However, in many scenarios, there are a large number of people spending daily hours walking into multiple pedestrian congestions, where the established social distancing is not accomplished. In 2019, Universidad de Los Andes reported around 14 thousand undergraduate students, 3.000 master's students, and about 364 doctoral students. [22] in a campus area that covers around 11.4 ha [21]. In this sense, this institution has a large volume of pedestrians circulating daily through the different facilities of the campus (multiple corridors and interconnection building routes). This scenario was the common behavior before the implementation of virtual classes by the institution due to the pandemic. Therefore, due to the need to fulfill the social distancing within the campus considering the possible agglomerations, we propose a mathematical model for minimizing the person exposure in places with a risk of contagion (places with high traffic of pedestrians at a specific time). In addition, the mathematical model is proposed to be accessed through a mobile application with the aim of a pedestrian user can determine a route to traverse the campus minimizing the risk of contagion at a specified time.

The remainder of the paper is organized as follows: Sect. 2 presents related works, and the identification and relevance of the problem to be treated. Section 3 shows the descriptions and justifications of the multi-objective mathematical optimization model design process developed for the implementation of the mobile application. Section 4 describes the understanding of the problem modeled in the phases presented in Sect. 3. It also describes the elaboration and validation of the algorithm solution. Section 5 presents the design and development of the mobile application that corresponds to the front-end solution of the proposal. Section 6 analyzes and discusses the findings obtained by our proposal.

## 2 Related Work

The same COVID-19 susceptibility probability equation for an individual in pedestrian congestions was used through these related works. Likewise, these studies presented simulations to represent the population behavior under different types of events in a casual day. In other words, these studies emphasized how displacement or mobility events are related to the spread of the SARS-CoV-2 virus.

In [6], researchers performed simulations of human mobility models to study the dynamics of coronavirus infection in order to get a dynamic model of the virus spread. To capture the nature of the epidemiological dynamics, the study made use of the *SIR* (Susceptible, Infected and Recovered) model. Through contagion rates planned in the model, the behavior of different groups populations was simulated. If a susceptible person and an infected person meet, there is a probability that the susceptible person will become infected. Time after infection, the person typically recovers. Therefore, the study introduced a contact graph constructed from everyday situations such as agglomerations in public trans-port, work, home, and others (scenarios in which the *SIR* model could be applied and studied). Finally, after executing the simulations of the contact graph with the epidemiological dynamics previously stated, the study set out strategies as conclusions, such as the suspension of public transport, total quarantine, and others social decisions, in order to prevent and reduce the spread of the virus.

For instance, one of the researched works created an epidemiological model based on agents (called *PanCitySim*) in which the movement of many individuals were modeled in a city. For this purpose, the model used a graph to represent the contact of many activities performed by the population in the whole city. In this sense, the spatio-temporal properties of an epidemic were studied according to the *SIR* model in order to represent the impact of human interactions to the transmission of the virus [5].

Therefore, it was suitable to study human mobility models in cities in order to understand the spreading behavior of COVID-19. Based on these related works, in our proposal, a solution could apply to the mobility patterns associated with pedestrians and their environment within the Universidad de Los Andes, as well as mathematical models and probabilities associated with the spreading of the virus.

## 3 Multi-objective Mathematical Optimization Model

In this research we propose a mathematical model for minimizing the distance traveled by a student and the risk of contagion when he is walking through campus. Therefore, the model included a transportation cost matrix represented as the distances between each point or node that composes the route of the map to travel. Likewise, it was also established a matrix of interaction costs that represents the pedestrian traffic through the points or nodes in the map. In addition, in the model we defined a minimum coverage radius for each node, which is the maximum distance that must be for each pair of nodes to be a connection.

In order to define the initial nodes network, the *GeoJSON* web tool was used [12], where nodes were placed consecutively on each path of the university campus. Once the coordinates of the different nodes have been exported in a file in *JSON* format, they were loaded as parameters to the mathematical model. To understand the mathematical model, sets, variables, and parameters are presented in Table 1.

Thus, in order to verify the solution of the initial model, the objective function was defined:

**Minimize**

$$\sum_{i \epsilon N} \sum_{j \epsilon N} X_{ij} \cdot (Ct_{ij} + Ci_{ij}) \tag{3.0.1}$$

**Subject to:**

$$\sum_{j \epsilon N} X_{ij} = 1 \qquad \forall i \epsilon N | i = S \tag{3.0.2}$$

$$\sum_{j \epsilon N} X_{ij} = 0 \qquad \forall i \epsilon N | i = S \tag{3.0.3}$$

$$\sum_{i \epsilon N} X_{ij} = 1 \qquad \forall j \epsilon N | i = D \tag{3.0.4}$$

$$\sum_{i \epsilon N} X_{ij} = 0 \qquad \forall j \epsilon N | i = D \tag{3.0.5}$$

$$\sum_{j \epsilon N} X_{ij} - \sum_{j \epsilon N} X_{ji} = 0 \qquad \forall i \epsilon N | i \neq D \wedge i \neq S \tag{3.0.6}$$

$$X_{ij} \cdot Ct_{ij} \leq RC \qquad \forall i, j \epsilon N \tag{3.0.7}$$

**Table 1.** Sets, parameters and variables of the initial mathematical model

| Category | Symbol | Description |
|---|---|---|
| Sets | $N$ | Number of nodes in the network |
| Parameters | $Ci_{ij}$ | Interaction cost between node $i$ and node $j$, which is the number of people that walk through the link (i,j) |
| | $Ct_{ij}$ | Transport cost between node $i$ and node $j$, which is the euclidean distance between these nodes |
| | $S$ | Source node number, in which a student started its walk |
| | $D$ | Destination node number, in which a student finished its walk |
| | $RC$ | Node's Ratio Coverage, which is the maximum distance that two pairs of nodes must have to exist a connection link between them |
| Variables | $X_{ij}$ | Binary decision variable that shows whether the student chooses the link between node $i$ and node $j$ as a part of its path from the source node to the destination node. 0 if otherwise |

The objective function of the mathematical model is subjected to the constraints represented in Eq. 3.0.2 to Eq. 3.0.6. In Eq. 3.0.2 it is indicated that the source node should only have one outflow, that is, the student can only take one path from its point of origin. Equation 3.0.3 ensures that the student cannot be returned to its source node. Equation 3.0.4 ensures that the student reaches its destination node. Equation 3.0.5 establishes that once the student reaches the destination node, he cannot travel to other paths or nodes. For intermediate nodes between the source node and the destination node, Eq. 3.0.6 establishes that if a student reaches an intermediate node, they must leave that node. Finally, to guarantee that the model chose links within the ratio coverage of each node, Eq. 3.0.7 was proposed.

In terms of the mathematical model implementation, we executed the model on the Pyomo optimization framework on an Intel ®Core i5-6200U, 2.3 GHz, 7.7 GiB system.

### 3.1 Interaction Cost Function

The susceptibility probability of contagion proposed by [6] was used in our proposal, which can be observed in Eq. 3.0.7.

$$P_{n,t} = 1 - e^{-\theta \cdot \sum_{m \in M} q_{m,t} \cdot i_{nm,t} \cdot t_{nm,t}} \quad (3.1.1)$$

Where $M$ is the set of people with whom a user was in close contact, with the following variables and parameters: $P_{n,t}$ is the Probability of person $n$ to receive an infectious dose. However, as stated in [8], this probability should not be understood as "probability of contracting the virus", since this depends on the immune system of each person. $\theta$ is the Calibration factor for a particular disease. $q_{nm,t}$ corresponds to the Microbial load between two people, which is the amount of virus that the person $m$ transmits in time $t$. $i_{nm,t}$ is the Contact intensity between the person $n$ and $m$, which corresponds to the distance. Finally, $t_{nm,t}$: Time that person $n$ interacts with person $m$ for a time $t$.

Now, for the simplicity of the proposed model, we assumed that the distance between each pedestrian and each user corresponds to 1 m and that their interaction time was 0.5 s. However, due to the fact that as of the date this project was developed, no calibration parameters and microbial load of COVID-19 were known. Thus, parameters estimated by the previously mentioned studies were used.

First, the value $q_{m,t}$ was assigned as 1 based on estimates made by [6]. Similarly, the calibration parameter $\theta$ was taken as $\frac{1}{20}$ based on studies conducted by [8] in closed spaces such as aircraft interiors during commercial flights.

We considered that the choice of this calibration value should be made on the assumption that all routes in the campus are in a closed space (which is the assumption of a pessimistic scenario). With this decision, it was possible to establish an equal load of micro-organisms in the environment for all the connections between nodes, and therefore, this allowed the model to calculate the best path based on the segregations caused by other students and not on the dispersion caused by other factors.

### 3.2 Pedestrian Traffic Simulation on the University Campus

In order to establish parameters related to the interaction cost of the model in a post COVID-19 scenario, we processed the dataset of the anonymized schedules of undergraduate students in the first semester of 2019, as well as the semester current academic schedule. The data was pre-processed and cleaned through the data analysis software *KNIME* [17]. Using this tool, the courses list was matched to the courses registered by the students. The resulting data set had the following attributes with 65, 536 records: student ID, course ID, department or faculty of the course, name of the course, current academic cycle of the course, building where the course was taught, time and minutes when the course started, time and minutes when the course ended and days on which the course was taught.

Once the final set of data to be used as parameters in the model was got, we loaded the students pedestrian traffic data into a non-relational database [10] using executable scripts [19].

Then, each record of the data set was uploaded as the origin and destination buildings that each student has in a change of classes in an academic cycle, day and specific time.

For a student's first class of the day, we decided in our simulation that a student would arrive at the university campus through the transport nodes identified in *Uniandina Mobility Survey (Encuesta de Movilidad Uniandina)* [23], where each means of transport has an associated percentage, which represents the percentage that uses said means of transport. Then, a random variable was defined that oscillated within the ranges of percentages and that from the value that said variable took, it was assigned within the percentage of the closest mean of transport. Likewise, we defined the nodes in the network equivalent to where the transportation stops are within the university campus.

For non-consecutive classes belonging to the same day, we decided to define random nodes where students would spend their free hours. The choice of these nodes was made arbitrarily.

Once the data had been uploaded to the database, a hierarchical non-relational database was got based on the academic year, day, and class change schedule.

To model the route pattern that each student was going to use when moving around the campus, we decided that for the simulation the students would look for the shortest way to go from their source building to their destination building. For this task, we assumed that each student would move through each node using the Dijkstra Algorithm for minimal cost path (taking the distance between each pair of nodes as cost) [1].

Consequently, the selection of each path by each student assigned the interaction costs of the mathematical model and the solution heuristic. At Table 2 is the terminology used for Algorithm 1, which was used to map the pedestrian traffic parameters to the interaction cost variable:

**Table 2.** Related terminology used for the Algorithm 1

| Symbol | Description | Assumptions |
|---|---|---|
| $N$ | Set of network nodes | |
| $E$ | Set of students that displace in a specific academic cycle, day and schedule | |
| $S(i)$ | Student source node $i$ | $\forall i \in E \quad \forall S(i) \in N$ |
| $D(i)$ | Student destiny node $i$ | $\forall i \in E \quad \forall D(i) \in N$ |
| $Sp(i)$ | Optimal set of links that a student $i$ has to take to displace from $S(i)$ to $D(i)$ | $\forall i \in E \quad \forall S(i), D(i) \in N$ |
| $D_{ij}$ | Optimal links calculated by Dijkstra Algorithm from node $i$ to node $j$ displacement | $\forall i, j \in N$ |

**Algorithm 1:** Algorithm for the simulation of pedestrian traffic and interaction costs assignment to the network nodes' links

```
C_ij = 0        ∀i, j ∈ N;
Sp(e) = {}      ∀e ∈ E;
for e ∈ E do
    Sp(e) = Dijkstra(S(e), D(e));
    for i, j ∈ N do
        if i, j ∈ Sp(e) then
            C_ij ← C_ij + 1;
        end
    end
end
```

### 3.3 Pareto Front Development

Consecutively, we implement a Pareto Front in order to observe the behavior of the transport cost and the interaction cost functions when they are minimized and one function has more importance or weight over the other.

Therefore, the Pareto Front was calculated through the method $\varepsilon$ *-constraint* [2]. For this computation, the previously defined network of nodes was used and the parameters of the transport function were assigned based on the traffic loaded in the database.

Finally, we defined a weights vector in which each vector' element multiply each function on each iteration, in order to give greater weight to one function compared to the other. The mathematical expression was defined:

$$W = [10, 20, \ldots, 50]$$

where for each element of $W$ would represent the upper limit that $f_2$ should have when wanting to minimize $f_2$ (this is reflected in the additional restriction described below). Therefore, each element $W$ was iterated to obtain values close to the limit of $f_2$. It should be noted that the mathematical model of the method also includes the restrictions initially set in the Sect. 3.

The Pareto Front run was performed on an Intel ®Core i5-6200U, 2.3 GHz, 7.7 GiB.

$$\textbf{for each } w \in W :$$

$$f_1 = \sum_{i \epsilon N} \sum_{j \epsilon N} X_{ij} \cdot Cij$$

$$f_2 = \sum_{i \epsilon N} \sum_{j \epsilon N} X_{ij} \cdot Cij$$

**Minimize**

$$f_1$$

**Subject to:**

$$f_2 \leq w \tag{3.3.1}$$

See Eq. 3.0.2–Eq. 3.0.7

# 4 Heuristic

As this problem can be represented as a minimum cost problem, where for each link there are two types of cost - transport cost and interaction cost respectively - we used known heuristic to solve the minimum cost problem by combining in a same scale the two types of cost for each link. Therefore, the Dijkstra Algorithm [1] was used as a solution algorithm, where the user, through the front-end of the project, could choose the criterion to minimize given their preferences and input parameters.

Regarding the unification of costs, the costs of each link were normalized, and then multiplied by a weight represented by a scalar between 1 and 0. Finally, the product of the normalized costs and their scalar was added, resulting in a single cost matrix. It should be noted that the scalar weight is the representation of the path criterion selected by the user (*i.e.* 0 as least importance, 1 as greatest importance). However, we found that given the modeling of pedestrian traffic described in the Subsect. 3.2 and the obtained Pareto Front results in different scenarios, the problem of minimization of the transport cost function and the interaction cost function only had solutions in the limits of each function, which is a solution was not found that minimized the two objectives "as equal", but two optimal solutions were presented that minimize only one function.

## 4.1 Bi-Objective Problem Understanding

To understand the reason for the existence of only two optimal solutions for the minimization problem of $f_1$ and $f_2$ (as conventions, $f_1$ is referred as the transportation cost function and $f_2$ is referred as the interaction cost function.), it must be remembered how the distribution of traffic was defined in the Subsect. 3.2, which was done under the assumption that every student who moved around the campus would choose the shortest route in the network of nodes.

Multiple scenarios were obtained where all the costs associated with $f_2$ are distributed by the links that minimize $f_1$. Therefore, when it is desired to minimize only $f_1$, links are obtained when it is desired to maximize $f_2$. Similarly, when only $f_2$ is minimized, the chosen links have no costs for this function. Those links, since they are not optimal for $f_1$, maximize the value of said function.

To describe the understanding of the problem in more detail, see the terminology described below:

Be

$$D: \text{Set of optimal links selected by Dijkstra's Algorithm}$$
$$C_1: \text{Set of optimal links to minimize} f_1$$
$$C_2: \text{Set of links where there are costs associated with} f_2$$

Where from the student pedestrian traffic displacement model, the following sets relationship occurs:

$$C_1 \in D$$
$$C_2 \in D$$

## 4.2 Single Cost Approach to Solution Heuristic

For the solution heuristic we proceeded to use Dijkstra using as the only cost the weighting of the normalization of the costs of $f_1$ and $f_2$. For this, the scale characteristic normalization method was applied as follows:

$$f_1^{'}(i,j) = \frac{f_1(i,j) - min(f_1)}{max(f_1) - min(f_1)} \tag{4.2.1}$$

$$f_2^{'}(i,j) = \frac{f_2(i,j) - min(f_2)}{max(f_2) - min(f_2)} \tag{4.2.2}$$

Where $f_k^{'}(i,j)$ is the normalization of the scaling characteristic of $f_k$ in the link of node $i$ and node $j$. $max(f_k)$ corresponds to the maximum value in the data set of the function $k$. For the case of both functions, the maximum value was the integer 99999, since that was the default value in the cost matrix where there was no link between a pair of nodes. Finally, $max(f_k)$ is the minimum value in the data set of the function $k$. For $f_1$ it corresponded to the link between nodes of shorter length, and for $f_2$ it was the minimum number of students who traveled on a link given a cycle, day and time.

Finally, the costs of the normalized functions were weighted through the weights that were later defined from the input parameters entered by the user, and they were assigned to a new function. Likewise, from the new function, we defined the heuristic solution that would have as parameters the inputs received by the front-end of the mobile application:

$$F(i,j) = w \cdot f_1^{'}(i,j) + (1-w) \cdot f_2^{'}(i,j) \tag{4.2.3}$$

$$H(s,d) = Dijkstra(F,s,d) \tag{4.2.4}$$

Where $F$ is the function that stores the weight of the two initial normalized functions. $w$ corresponds to the weight that the user implicitly enters as a parameter in the front-end to give a weight to each function. $Dijkstra$ is the function that returns the optimal links given the unified cost function $F$, a source node $s$ and a destination node $d$ given a cycle, day and time. Finally, $H$ is the function that represents the problem's heuristic solution. Once the heuristic solution was modeled, the results of the algorithm were compared with the solutions provided by the Pareto Front in different scenarios.

### 4.3 Results Comparisson and Validation with the Pareto Front

For the validation of the results produced by the heuristic developed with the Pareto Front points, we defined different scenarios to vary the pedestrian traffic given a cycle, day and time with the same origin and destination node in each scenario. This was made with the purpose of comparing how is the variation of the optimal route of two points from the assignment of different costs of $f_2$ (since the costs of $f_1$, being the length of each link between nodes remained fixed in the different executions. However, it should be noted that the accumulated cost of $f_1$ changes from the different routes selected).

The specifications of the different scenarios are observed in Table 3.

For the generation of the Pareto Front, the methods described in Subsect. 3.3 was used.

Regarding the values of $w$ for $F$, different values were iterated in the domain of $w$ ($[0, ..., 1]$) to observe the points got by the heuristic in each iteration.

**Table 3.** Specifications of the scenarios chosen for the comparison of the solutions thrown by the solution heuristic and the Pareto Front.

| Scenario number | Academic cycle | Day | Schedule | Source building | Destiny building |
|---|---|---|---|---|---|
| 1 | 1 | Wednesday | 10:50–11:00 | Julio Mario Santodomingo | Mario Laserna |
| 2 | 1 | Monday | 13:30–14:00 | Julio Mario Santodomingo | Mario Laserna |
| 3 | 1 | Thursday | 13:30–14:00 | Julio Mario Santodomingo | Mario Laserna |

It is observed in Fig. 1a that different points were got from the Pareto front, as well as the heuristic solution from the same source and destiny buildings, but with different pedestrian traffic and assigned weights for each $w \epsilon F$.

f As main evidence, we observed that since it is a problem with only two optimal solutions available the Pareto Front, points were generated particularly at the limits of each function. Likewise, it was obtained that when $w \neq 0$ or $w \neq 1$ the heuristic yielded the same solution in each iteration. Therefore, the solutions thrown by the heuristics behave as if they were only minimize $f_2$, this is due to a single cost function where the costs of $f_2$ associated with the optimal $f_1$ paths made the heuristic look for paths that were not optimal for $f_1$ in order to minimize weight associated with $f_2$. For such reason it can be seen that the algorithm solution does not generate optimal results to minimize $f_1$.

Thus, in the iteration when $w$ = 1-which is when the heuristics were only considered minimizing $f_1$- a solution was got at one end of the Pareto Front, where the associated cost of $f_1$ is minimized and the cost associated with $f_2$

is maximized. However, on iteration when $w = 0$, which is the case where the heuristic only minimized $f_2$. We found that although that the solutions obtained minimize the accumulated cost of $f_2$ with respect to the iteration scenario when $w = 1$, These solutions are solutions dominated by the points that make up the Front, and additionally, it is a non-optimal solution when minimizing $f_2$ compared to the solutions obtained in the iterations when $w \notin \{0, 1\}$

For additional information of the described scenarios, see Fig. 1b, in where the optimal path selected by the heuristic is shown for the different values iterated by $w$ in its domain. Note the similarity of routes chosen when the heuristic only minimizes $f_1$ or $f_2$, and as in a different scenario (Scenario 2) increases the cumulative cost of $f_1$ to minimize the cumulative cost of $f_2$ From the observations derived from the results, and from the naturalness of the modeled problem described in Subsect. 3.2, we decided to only expose two possibilities of paths in the mobile application, where each option corresponds to the minimization of each function. Therefore, to obtain these results, the domain of $w$ was configured to only take the values of 1 and 0.5 (which is a decimal that assigns equal weight to both normalized functions).

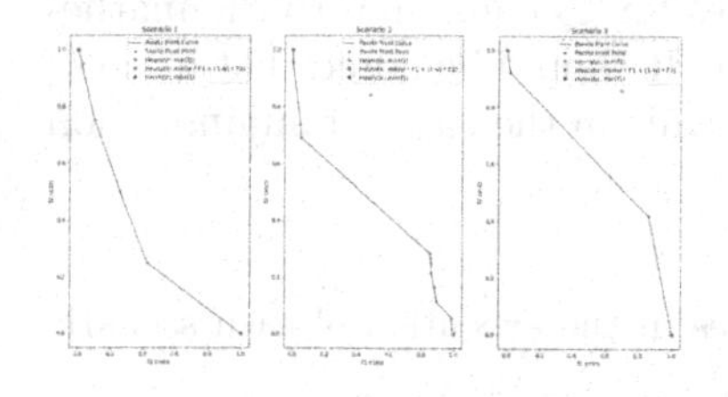

(a) Pareto Fronts and Heuristic solutions for scenarios 1, 2 and 3

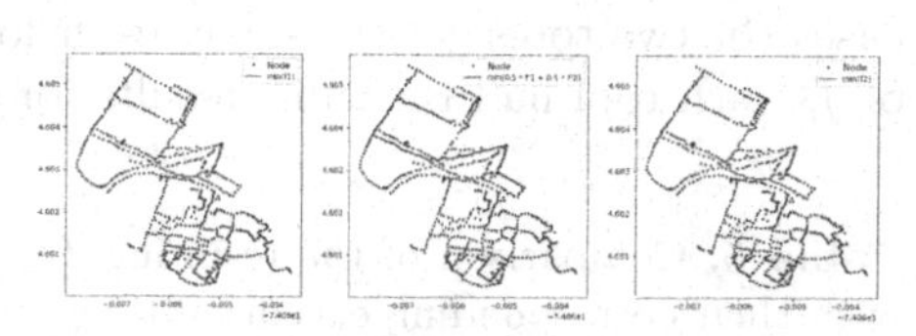

(b) Mapped routes

**Fig. 1.** Scenarios and routes comparisons for different values of $w$ between 0 and 1

## 4.4 Results Comparisons and Validation Based on the Mathematical Model

From the configurations established for the solution heuristic in the Subsect. 3.2 - where $f_1$ and $f_2$ were equally normalized and weighted to minimize $f_2$ - The execution times used by both the mathematical model and the heuristics to minimize $f_1$ and $f_2$ were compared, as well as the values taken by the functions in each scenario.

The mathematical model was configured in the optimization framework *Pyomo* [4] and the heuristic was written and executed in a *script* written in the programming language *Python*.

The specifications of the chosen and executed scenarios can be observed in the Table 4.

**Table 4.** Scenarios specifications.

| Scenario number | Academic cycle | Day | Schedule | Source building | Destiny building |
|---|---|---|---|---|---|
| 1 | 1 | Wednesday | 10:50–11:00 | Julio Mario Santodomingo | Mario Laserna |
| 2 | 1 | Thursday | 13:50–14:00 | Julio Mario Santodomingo | Mario Laserna |
| 3 | 1 | Monday | 13:50–14:00 | Julio Mario Santodomingo | Mario Laserna |

First, the execution times of each scenario were compared, where it was observed that the execution times got by the heuristic are approximately 98.45% less than the times obtained by the mathematical model. The previous result facilitates the response time of the calculation of an optimal route when the heuristic is integrated as a service exposed to the mobile application.

Both the heuristic and the mathematical model were executed on an Intel ®Core i5-6200U, 2.3 GHz, 7.7 GiB. Additionally, in the executions made by *Pyomo* the mathematical model was solved using the *GLPK solver*.

The details of the execution times for each scenario can be observed in the Table 5. Subsequently, the resulting values of $f_1$ and $f_2$ were compared for each tool in the minimization of each function of each scenario. As it is observed in Table 5, the heuristics got relatively lower values for $f_2$ compared to the mathematical model in each case of minimization of $f_1$ (It should be noted that in these cases the two tools got the same result for $f_1$), while in the cases of minimization of $f_2$ each tool had the same results for this function.

**Table 5.** Comparison of the resulting $f_1$, $f_2$ values in the execution of each scenario with their corresponding execution time.

| Scenario number | Function to minimize | Heurístic value for $f_1$ (units) | Mathematical model for $f_1$ (units) | Heurístic value for $f_2$ (units) | Mathematical Model value for $f_2$ (units) | Heuristic Execution Time (sec) | Mathematical model execution time (sec) |
|---|---|---|---|---|---|---|---|
| 1 | $f_1$ | $3.08 \cdot 10^{-3}$ | $3.08 \cdot 10^{-3}$ | 495 | 495 | 2.761 | 182.504 |
| | $f_2$ | $6.58 \cdot 10^{-3}$ | $6.99 \cdot 10^{-3}$ | 5 | 5 | 2.519 | 164.167 |
| 2 | $f_1$ | $3.08 \cdot 10^{-3}$ | $3.08 \cdot 10^{-3}$ | 540 | 540 | 2.522 | 153.907 |
| | $f_2$ | $6.68 \cdot 10^{-3}$ | $7.48 \cdot 10^{-3}$ | 19 | 19 | 2.423 | 171.088 |
| 3 | $f_1$ | $3.08 \cdot 10^{-3}$ | $3.08 \cdot 10^{-3}$ | 682 | 682 | 2.544 | 161.058 |
| | $f_2$ | $6.64 \cdot 10^{-3}$ | $7.11 \cdot 10^{-3}$ | 35 | 35 | 2.695 | 165.645 |

## 5 Mobile Application Design

We implemented *REST API software* architecture style [20] using the web *framework Flask* [18] written in *Python*. Likewise, the steps and instructions for the heuristic solution (see Subsect. 4.2) were developed in the programming language *Python*. The input parameters that the *REST API* received for the solution heuristic to calculate the optimal path were: Text string containing the initials of the building in which the student is located ($s$). Text string containing the acronym of the building the student wants to go to ($d$). Text string that shows the academic cycle in which it is desired to calculate pedestrian traffic. By default, this value corresponds to "*1*" (*cycle*). Text string that indicates the particular day set to calculate pedestrian traffic (*day*). Text string that indicates the schedule for changing classes of pedestrian traffic. Note that class change hours are expressed in the 24-h system (*schedule*). Integer that shows which function it is wanted minimize (1) or if it is wanted to minimize the transport cost function (0), if it is wanted to minimize the interaction cost function ($w$). For more details on what these functions represent, see Sect. 3.

It should be noted that these methods are received by the server through a *POST* request. To test the connection between the *Flask* server and the client, the server was run on a local machine Intel ®Core i5-6200U, 2.3 GHz, 7.7 GiB machine and perform client API requests through the *Postman* [9] software.

For the development of the mobile application, we designed so that it could be executed in the operating system *Android 9* [13]. Likewise, through the programming language *Kotlin* [16] and the graphical interfaces for the design of the *UI* components provided by the *Android Studio* development environment, we implemented the *front end* logic and graphical components to show the probability of susceptibility and the accumulated distance when traveling a route.

For the generation of the university campus map, a *Google Maps* software development kit was used available for *Android* [15].

In Fig. 2 it can be observed the graphical interface of the application with the susceptibility probability and the accumulated distance in the lower bar as well as the optimal route drawn. It should be noted that the icons and figures of the application were taken from the webpage *Flaticon.com* [11].

Regarding the selection of parameters to be sent to the server, a graphical Android component (*AlertDialog*) was used to create a form that would allow the user to select the source building, the destiny building, the opening hours, change of classes available for that day and the route criteria to be minimized. In order to get significant records of the students' trajectory for the generation of pedestrian traffic, we assumed that the default academic cycle parameter was 1. Likewise, the day was automatically assigned according to the current day on which the application was used. If the user uses the application on Sunday - the day on which the university does not hold academic classes - the application assigns Monday as the default parameter. In the Fig. 3 and Fig. 4 it can be observed the parameter selection form.

Once the parameters were selected, when pressing the *Get Path* button of the form, the application - using the *HTTP Android Volley* [14] library - send the request *POST* to the *IP* address of the local machine where the server was deployed.

Once the server response is received, the application updated its interface, showing in the lower bar the susceptibility probability that the user has when traveling the route, as well as the accumulated distance that they have to walk. An example of how the route is drawn in the application can be observed in the Fig. 2.

In Fig. 5 it is observed the final architecture diagram.

Fig. 2. Screenshot of the mobile application.

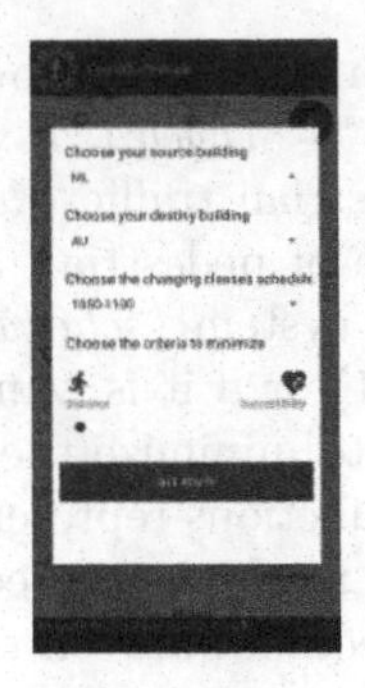


Fig. 3. Parameter selection form with the distance minimization criterion selected.

Fig. 4. Parameter selection form with the susceptibility minimization criteria selected.

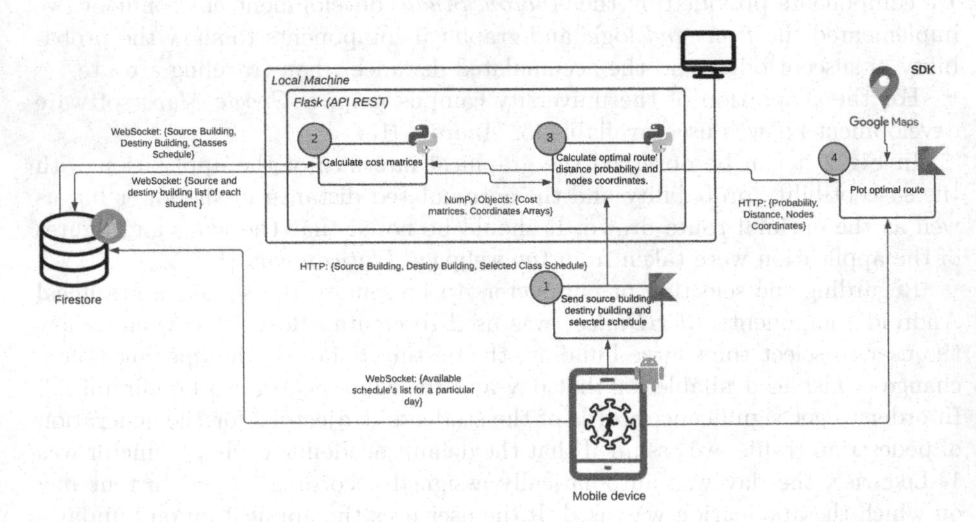


Fig. 5. Diagram of the software architecture built. In the upper right corner of each activity/device it can be observed the programming language, software or platform used. In the upper left corner is the step number of the sequence of steps followed by the process execution. The intermediate texts between the arrows connecting each activity/device show the communication protocols used.

## 5.1 Validations

In order to observe the behavior and functionalities of the final product through its graphical interface, the scenarios presented in the Table 6 were considered to observe the variation between the probability of susceptibility and the accumulated distance when minimizing each criteria, given the different origin-destination buildings and pedestrian traffic on a certain day and class change schedule.

It should be noted that the scenarios presented took Thursday and "1" as the default day and academic cycle respectively, since the tests were made on that specific day of the week.

**Table 6.** Specifications of the tested scenarios on the mobile application, describing the obtained values for each criteria based on the selected criteria to minimize

| Scenario number | Schedule | Source building | Destiny building | Selected criteria | S. probability value (percentage) | A. distance value (units) |
|---|---|---|---|---|---|---|
| 1 | 10:50–11:00 | Mario Laserna | Aulas | Distance | 99% | 0.0016 |
| | | | | Susceptibility | 85% | 0.0025 |
| 2 | 13:30–14:00 | Mario Laserna | City U | Distance | 99% | 0.0023 |
| | | | | Susceptibility | 66% | 0.0037 |
| 3 | 10:50–11:00 | J. M. Santodomingo | Mario Laserna | Distance | 100% | 0.0031 |
| | | | | Susceptibility | 99% | 0.0068 |

# 6 Conclusions

Based on the developed work, a *software* tool has been built in order to optimize the most important criteria to be considered by a pedestrian - a Universidad de Los Andes student- in terms of health effectiveness and safety. Therefore, the developed mobile application allows a pedestrian - related to Universidad de Los Andes - to consult in a easy and quickly way a mobile application that solves the COVID-19 risk contagion problem from a mobility pattern and displacements perspective.

Likewise, based on the pedestrian traffic simulation and the proposed cost functions, it has been possible to generate and understand a pedestrian agglomerations' approximation, given a set of student classes that had time attributes. Moreover, based on the data hierarchy and aggregated pedestrian mobile users data - given the case the developed mobile application is used by multiple users - strategies can be established for data analysis from the different hierarchy data levels (day, academic cycle, etc.). This has been done with the purpose of understanding the mobility patterns that university students generate, and therefore, allows a high-level decision making process that avoids pedestrian agglomerations in a certain period of time.

Additionally, as a future work feature, the validation process must be complemented from the medical domain point of view. In other words, the inclusion and approval of health entities that comprehend the COVID-19 contagion processes could help to calibrate and adjust our work in order to propose a joint solution validated from the computational and medical point of view.

## References


1. Dijkstra, E.W.: A note on two problems in connexion with graphs. Numer. Math. **1**(1), 269–271 (1959)
2. Haimes, Y., Lasdon, L., Wismer, D.: On a bicriterion formulation of the problems of integrated system identification and system optimization. IEEE Trans. Syst. Man Cybern. **SMC-1**(3), 296–297 (1971). https://doi.org/10.1109/TSMC.1971.4308298
3. Harko, T., Lobo, F.S.N., Mak, M.K.: Exact analytical solutions of the Susceptible-Infected-Recovered (SIR) epidemic model and of the SIR model with equal death and birth rates (2014). https://arxiv.org/abs/1403.2160
4. Hart, W., Watson, J.P., Woodruff, D.: Pyomo (2019). https://pyomo.readthedocs.io/en/stable/
5. Kumar, N., Oke, J.B., Nahmias-Biran, B.-h.: Activity-based contact network scaling and epidemic propagation in metropolitan areas (2020). https://arxiv.org/abs/2006.06039
6. Muller, S.A., Balmer, M., Neumann, A., Nagel, K.: Mobility traces and spreading of COVID-19 (2020). https://www.medrxiv.org/content/10.1101/2020.03.27.20045302v1.full.pdf
7. Organización Mundial de la Salud: Q&A. How is COVID-19 transmitted (2020). https://www.who.int/news-room/q-a-detail/q-a-how-iscovid-19-transmitted
8. Schultz, M., Fuchte, J.: Evaluation of aircraft boarding scenarios considering reduced transmissions risks (2020). https://www.mdpi.com/2071-1050/12/13/5329
9. Asthana, A., Kane, A., Sobti, A.: Postman API client (2014). https://www.postman.com/product/api-client/
10. Firebase Inc.: Firebase (2012). https://www.firebase.google.com/
11. Freepik Company: (2020). https://www.flaticon.com/
12. Geographic JSON working group: GeoJSON (2016). https://geojson.org/
13. Google LLC: Android pie (2019). https://www.android.com/versions/pie-9-0/
14. Google LLC: (2020). https://developer.android.com/training/volley
15. Google LLC: Maps SDK for Android (2020). https://developers.google.com/maps/documentation/android-sdk/overview
16. JetBrains: Kotlin (2020). https://github.com/JetBrains/kotlin/releases/latest
17. KNIME: KNIME (2020). https://www.knime.com/
18. Grinberg, M.: Flask Web Development: Developing Web Applications with Python. O'Reilly Media Inc., Sebastopol (2018)
19. Python Software Foundation: Python 3.6 (2020). https://www.python.org/
20. World Wide Web Consortium: Web services architecture (2004)
21. Universidad de Los Andes: Campus en Cifras (2019). https://campusinfo.uniandes.edu.co/es/campusencifras
22. Universidad de Los Andes: Facts and Figures (2019). https://planeacion.uniandes.edu.co/en/statistics/factsand-figures
23. Universidad de Los Andes. Vicerrectoría de Servicios y Sostenibilidad: Reporte de Sostenibilidad 2019 (2019). https://sostenibilidad.uniandes.edu.co/images/Reporte2019/Reporte-de-sostenibilidad-2019W.pdf
24. Zadeh, L.: Optimality and non-scalar-valued performance criteria (1963). https://ieeexplore.ieee.org/abstract/document/1105511

# An Improved Course Recommendation System Based on Historical Grade Data Using Logistic Regression

Idowu Dauda Oladipo[1], Joseph Bamidele Awotunde[1](✉), Muyideen AbdulRaheem[1], Oluwasegun Osemudiame Ige[1], Ghaniyyat Bolanle Balogun[1], Adekola Rasheed Tomori[2], and Fatimoh Abidemi Taofeek-Ibrahim[3]

[1] Department of Computer Science, University of Ilorin, Ilorin, Nigeria
{odidowu,awotunde.jb,muyideen,balogun.gb}@unilorin.edu.ng, 16-52ha043@students.unilorin.edu.ng

[2] Directorate of Computer Sciences and Information Technology, University of Ilorin, Ilorin, Nigeria
tomori@unilorin.edu.ng

[3] Department of Computer Science, Federal Polytechnic, Offa, Nigeria

**Abstract.** Elective course selection is very important to undergraduate students as the right courses could provide a boost to a student's Cumulative Grade Point Average (CGPA) while the wrong courses could cause a drop in CGPA. As a result, institutions of higher learning usually have paid advisers and counsellors to guide students in their choice of courses but this method is limited due to factors such as a high number of students and insufficient time on the part of advisers/counsellors. Another factor that limits advisers/counsellors is the fact that no matter how hard we try, there are patterns in data that are simply impossible to detect by human knowledge alone. While many different methods have been used in an attempt to solve the problem of elective course recommendation, these methods generally ignore student performance in previous courses when recommending courses. Therefore, this paper, proposes an effective course recommendation system for undergraduate students using Python programming language, to solve this problem based on grade data from past students. The logistic regression model alongside a wide and deep recommender were used to classify students based on whether a particular course would be good for them or not and to recommend possible electives to them. The data used for this study was gotten from records of the Department of Computer Science, University of Ilorin only and the courses to be predicted were electives in the department. These models proved to be effective with accuracy scores of 0.84 and 0.76 and a mean-squared error of 0.48.

**Keywords:** Course recommendation · Grade data · Logistic regression · Classification · Wide and deep recommender · Machine learning

## 1 Introduction

The computing environment for learning is changing rapidly, due to the emergence of new information and communication technology such as big data [1]. Learning methods


H. Florez and M. F. Pollo-Cattaneo (Eds.): ICAI 2021, CCIS 1455, pp. 207–221, 2021.
https://doi.org/10.1007/978-3-030-89654-6_15

are also changing every day and so e-learning systems need to develop more techniques and tools to meet the increased need of learners around the world. The choice of elective courses to register is a problem for many undergraduate students in universities as elective courses can help students to improve their Cumulative Grade Point Average (CGPA) if they are chosen wisely; on the other hand, if poor choices are made in the selection of elective courses, students run the risk of a drop in their CGPA. Sometimes, the sheer number of possible electives that can be chosen can cause a student to be confused as to which ones would be best for him/her. Also, the fact that there is no certain way to know how well one would perform in elective courses before selecting them is another factor that makes elective course selection difficult.

Artificial intelligence methods that were developed at the beginning of research are now being applied to information retrieval systems [2]. Recommendation systems provide a promising approach to information filtering as they help users to find the most appropriate items without much effort. Based on the needs of each user, recommendation systems can generate a series of personalized suggestions. Despite the high impact and usefulness of course recommendation systems, they are limited because models based on keywords may not address individual needs. They also do not, usually, in cases of collaborative filtering, association rules, and decision trees, use historical information about the courses. Content-based filtering models also fall short as they are usually based on specific recommendations only rather than more generalized recommendations. They also do not provide comprehensive information about the courses that are most relevant to students.

The proficiency of an algorithm to predict a student's performance in a particular course can be instrumental in aiding students' course selection choices and data mining and machine learning techniques have been seen to be useful in this domain [3]. Singular Value Decomposition (SVD) based algorithms have proved to yield very good results in course recommendation systems but they are limited by the density of the data matrix [4]. CGPA has also been found to be an important measure of course readiness [3]; thus, making it ideal for making predictions. Baseline predictors, Student/Course k-NNs collaborative filtering, Latent factor models - Matrix factorization (MF), Latent factor models - Biased MF (BMF) are other methods that have, so far, been used to build course recommendation systems [5]. Artificial Neural Network models have also been used in an attempt to solve this problem [6].

Recommender systems aim at providing their users with relevant information. A model in a recommender system evaluates the personal information of the user, and a model estimating scores for items not yet seen by the user is also developed [7]. The personalized guidance is based on prior grades of other students, collected in a historical database. Similarities between the elements are evaluated, to find the most suitable elements. The focus of this study is to develop a course recommendation system by training a Logistic Regression model, alongside a Wide and Deep Recommender on collected data of past students' grades to solve this problem. The paper intends to show that the logistic regression model could be effectively used to predict course outcomes for students while the Wide and Deep Recommender could be effectively used to make good recommendations about courses to choose for students after being trained with historical grade data from previous students. A huge amount of grade data can be gathered from school records of students that have passed through the school system previously and more data is known to beat better algorithms [8].

The Python programming language will be used to implement this research due to its available and extensive machine learning, mathematical and statistical packages such as NumPy, Sckit-Learn, MatPlotLib, etc. The course recommendation system to be developed will contribute to making course selection choices easier for undergraduate students. It will also help level advisers to better advice students on what courses they would do better in. Conventional measures of recommendation quality will also be studied in this paper. While some recommender systems have been put in place to solve this problem, they generally struggle due to a lack of sufficient training data or imbalances in data [9]. Also, many recommender systems are based on students' learning styles but these have been proven to change over time [10]. Furthermore, the use of the logistic regression model in course recommendation are still scanty while the effects of the wide and deep recommendation in this area, that is, course recommendation, have not been empirically tested.

In this paper, balanced historical grade data in course recommendations are used due to the huge amount of immutable data that can be gotten from school records of students who have gone through the school. Also, the paper recommends the use of the logistic regression model and the wide and deep recommender in an attempt to solve this problem. Each of these models will be trained and evaluated with historical grade data and implemented using Flask, a minimal web-based framework.

This paper aims to create a course recommendation system for students in undergraduate degree programmes based on collected historical records of past grades. The main contributions of this paper are:

(i) to select the variables used for each course to be predicted;
(ii) to design, implement and train the recommendation models;
(iii) to evaluate the trained models.

## 2 Related Work

One method that has been used in this domain is the use of an objective function to distinguish between courses that are expected to increase or decrease a student's GPA [11]. Morsy and Karypis [11] tried to tackle this problem by combining the grades predicted by grade prediction methods with the classifications generated by course recommendations to improve the final course rankings. In both methods, authors adjusted two commonly-used representation methods to find out the ideal chronological sequence in which courses are to be taken. The representation methods used in this paper were Singular Value Decomposition (SVD) and Course2vec. SVD factorizes a known matrix through getting a solution to $X = U \sum V^T$, where the columns of U and V are the left and right singular vectors respectively, and $\sum$ is a diagonal matrix containing the singular values of X. It was applied on a previous-subsequent co-occurrence frequency matrix F, with Fij being the number of students that had taken course i before taking course j. Having used SVD to estimate the previous and subsequent course sets, and computed each student's inherent profile by averaging over the sets of the courses taken by the said student in all preceding terms, after which, they computed the dot product between said student's profile and the SVD estimates of each course, ranking the courses in non-increasing order corresponding to the dot products and then selecting the top courses for final recommendation. Course2vec, on the other hand, was modelled using a many-to-one, log-linear model, which was inspired by the word2vec Continuous Bag-Of-Word (CBOW) model. While word2vec works on sequences of individual words in a given document where a set of words within a pre-defined window size are used to predict the intended word, course2vec works on sequences of ordered terms taken by each student where each term contains a set of courses and the previous set of courses would be used to predict future courses for each student. The results of the methods showed that course recommendation methods that consider grade information perform better than course recommendation methods which do not consider grade information.

Zabriskie et al. [3] used random forest and logistic regression models to construct early warning models of student success in two physics courses at a university. Combining in-class variables such as homework grades with institutional variables such as CGPA, they were able to predict if students would receive grades less than "B" in the two courses that were considered with 73% accuracy in one and 81% accuracy in the other.

Researchers in [12] approached the course recommendation problem by trying to find relationships between students' activities through the association rules method to enable students to select the best learning materials. The focus of their research was on the analysis of past historical data of course registration or log data. Their article essentially examined the frequent item sets concept to discover the note-worthy rules in the transaction database. Using those rules, the study established a list of more appropriate courses based on the student's behaviours and preferences. Their model was essentially based on the parallel FP-growth algorithm made available by Spark Framework and the Hadoop ecosystem.

The proposed recommendation system involved three major stages: firstly, data collection; secondly, the discovery of connections between user behaviours; thirdly, the recommendation of more appropriate courses for users. FP-growth (frequent pattern growth), being a proficient, scalable, and fast algorithm for extracting items that seem more closely associated, was used as the method of implementing the association rules method to determine the more appealing relationships between items in the database. FP-growth itself can be implemented in several ways. They used a parallel version of the algorithm called parallel FP-growth (PFP) which is based on a new computation distribution scheme, that is, it allocates tasks around a collection of nodes using the MapReduce model, making PFP faster and more scalable than the conventional FP-growth algorithm which is based on single-machine.

Another method that has been considered effective in course recommendation is the k-nearest neighbour algorithm [13]. The system was based on the notion that students who did well in the same previous courses would do well in the same future electives. It comprised of a knowledge base that had gathered prediction proficiency and a collection of rules for applying the knowledge base to individual circumstances. Using a neighbourhood-based approach, the developed system recommended electives as a function of their comparison to the students' grades and predictions for each course were attained by computing a weighted average of the ratings of the chosen electives. The system developed had an accuracy score of 95.65% for correctly classified instances and 4.35% for incorrectly classified instances while it had a mean absolute error of 0.0146 and a root-mean-square error of 0.1178.

It was reported on the examination of a web-based decision assistance application that assists student advising [14]. The tool was evaluated and found to be successful and efficient for academic advising by a substantial percentage of respondents; however, the details of its application were not included in the research.

Techniques for detecting responses [15–17] use a data model based on missing data theory to model NMAR data. To represent NMAR data, the method suggested in [15] modified probabilistic matrix factorization by introducing two modifications. The first variation assumes that the likelihood of detecting a rating is solely determined by its value. The second variation considers that the likelihood is likewise affected by the user and the latent components of the item. None of these strategies take into account the user and item characteristics that influence reaction patterns.

Hana [18] proposed a mechanism-based technique for recommending courses to students by investigating the student's academic record and comparing it to the records of others to determine similarities. The system then determines and advises the course he is good at or interested in taking so that he can pass the course.

The acronym PEL-IRT refers to "Personalized E-Learning System Using Item Response Theory" [19]. It suggests relevant course content to students, taking into account both the difficulty of the course material and the student's competence. Students can utilize PEL-IRT to search for interesting course material by selecting course categories and units and using appropriate keywords.

In [20], educational data was connected to a user/item. The recommendation was generated using the matrix factorization technique, and the approach was validated using logistic regression.

# 3 Methodology

The study addresses the practical problem of course recommendation concerning historical grade data using logistic regression and the wide and deep recommender. This study will be experimental in nature as it would involve running experiments on collected data. The work will be applied in nature to develop a usable technology for course recommendation as a practical problem. It will also be exploratory in nature as the researchers intend on studying the usage of the logistic regression model and the wide and deep recommender in this domain, a position that has not been taken so far in explored literature.

## 3.1 Data Collection

The data that was used for this study was acquired from University of Ilorin. It contained student scores in both core and elective courses for the 2018/2019 session.

### 3.2 Data Pre-processing

After the data had been collected and organized, the classifier undertook data pre-processing operations such as data cleaning, data balancing, etc.

Data cleaning consisted mainly of cleaning missing data from the dataset. Missing data mainly took the form of courses that were not taken by students who came into the school through direct entry rather than the Unified Tertiary Matriculation Examination (UTME) and courses that were taken in some sessions and not taken in others. Machine learning algorithms generally struggle to work with missing data and we tried to avoid this difficulty by filling in the average score of all the students who took such courses.

Usually, when dealing with real-world classification problems, the data tends to be skewed towards a particular class. In elective courses, pass rates are generally positively skewed and this will lead to inaccurate predictions as the model is likely to predict more passes than expected due to finding more passes in the data. To fix this problem, the model will be using the Synthetic Minority Oversampling Technique (SMOTE) to balance the dataset. This will be done by creating "new" instances of failed outcomes using the current failed outcomes in the dataset until the dataset is more equally balanced. This is expected to yield more accurate predictions. Scatterplots showing the observations in the dataset before and after balancing are shown in Figs. 1 and 2.

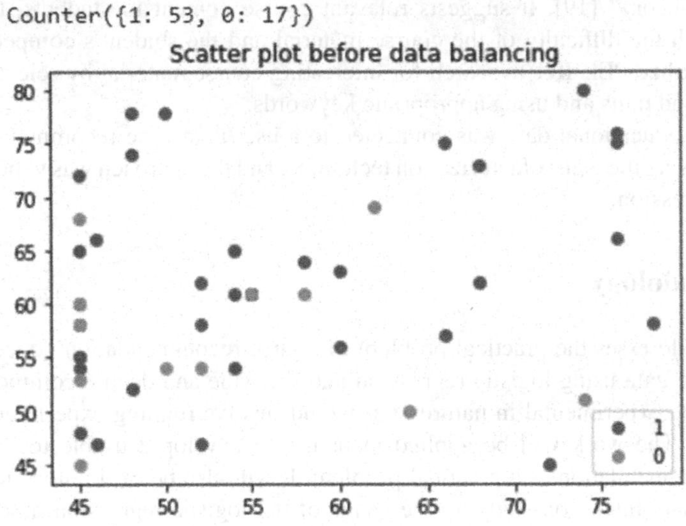

Counter({1: 53, 0: 17})

Scatter plot before data balancing

**Fig. 1.** Scatter plot before data balancing

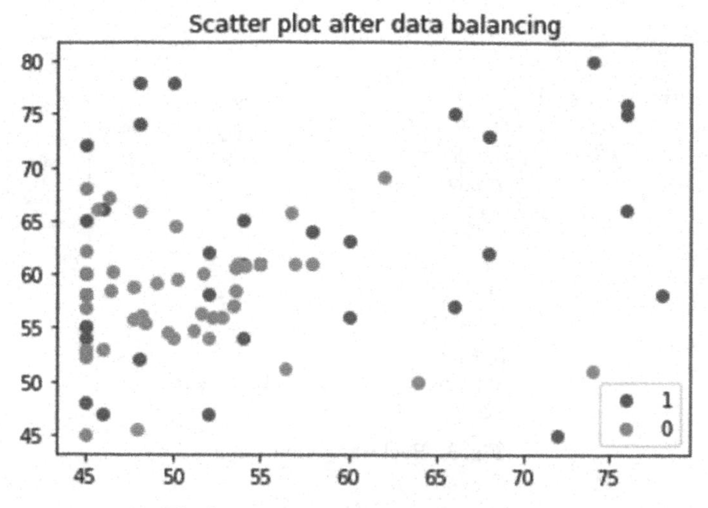


**Fig. 2.** Scatter plot after data balancing

### 3.3 Feature Selection

While the implemented algorithm may be the main focus of any machine learning task, it is also important to concentrate on the data that is being passed into the algorithm when trying to solve real-world problems. Like all other computerized tasks, machine learning algorithms work on the GIGO rule: Garbage In, Garbage Out, and in this case, "Garbage In" refers to noisy, uncorrelated data while "Garbage Out" refers to poor model performance. The goal of feature selection is to pick out only features (independent variables) that show a strong correlation to the output (dependent variable). When feature selection is properly done, it leads to faster model optimization time, reduced model convolution, better model accuracy and performance, and above all, less overfitting.

Several feature selection methods exist including Forward Selection, Recursive Feature Elimination, Bi-directional elimination, and Backward Elimination. For this study, Backward elimination will be used because according to Kleinbaum et al. [15], in models that focus on prediction, backward elimination is one of the more appropriate methods to be used. The algorithm for backward elimination is shown in Fig. 3. Also, the columns in the dataset before feature selection are shown in Fig. 4 while the columns after feature selection are shown in Fig. 5.

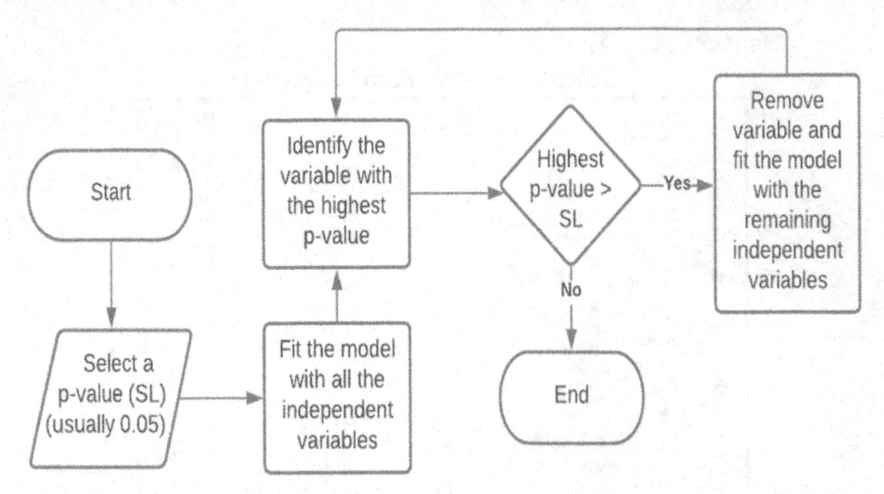


**Fig. 3.** Backward elimination

```
Index(['id', 'CSC111', 'CSC112', 'GNS111', 'GNS112', 'ICS101', 'ICS106',
       'MAT111', 'MAT112', 'MAT113', 'MAT114', 'PHY115', 'PHY152', 'PHY191',
       'PHY192', 'PLB101', 'STA121', 'STA124', 'STA131', 'TCS111', 'TCS112',
       'CSC212', 'CSC214', 'CSC216', 'CSC222', 'CSC224', 'CSC230', 'CSC231',
       'CSC233', 'GNS211', 'GNS212', 'GSE202', 'MAC251', 'MAT201', 'MAT206',
       'MAT211', 'MAT213', 'PHY252', 'STA203', 'STA221', 'CSC311', 'CSC317',
       'CSC321', 'CSC325', 'CSC327', 'CSC331', 'CSC333', 'CSC340', 'GNS311',
       'GNS312', 'GSE301', 'CSC319'],
      dtype='object')
```

**Fig. 4.** Columns before feature selection

```
Index(['CSC111', 'ICS101', 'CSC212', 'CSC216', 'CSC230', 'MAT213', 'CSC325',
       'CSC333', 'CSC340', 'GSE301'],
      dtype='object')
```

**Fig. 5.** Columns after feature selection

## 3.4 Division of Data

The data was then divided into a training dataset and a test dataset. This was done by randomly sampling the whole dataset without replacement. The training dataset (which contained 70% of the original data) was used to construct the model while the test dataset (which contained 30% of the original data) was reserved for model appraisal with "new" data.

**Model Optimization**

*Logistic Regression*

Logistic regression is a mathematical modelling approach that can be used to describe the relationship of several X's to a dichotomous dependent variable, such as D [15] where X represents the independent variable and D represents the dependent variable. It is a very widely used machine learning technique because the logistic function never extends below 0 and above 1 and it provides an attractive S-shaped depiction of the collective influences of the various dependant variables being used in the prediction [15]. The logistic regression model is based on the logistic function, which is given in Eq. 1.

$$f(z) = \frac{1}{1 + e^{-z}} \tag{1}$$

To get the logistic model from the logistic function, we use the formula $z = \alpha + \beta_1 X_1 + \beta_2 X_2 + \cdots + \beta_k X_k$, where the X's are independent variables and $\alpha$ and the βi are constant terms representing unknown parameters. Thus, z becomes an index that pools the X's together, giving us a new logistic function shown in Eq. 2.

$$f(z) = \frac{1}{1 + e^{-(\alpha + \sum \beta_i X_i)}} \tag{2}$$

The logistic model that was employed in this work considers the following framework: Independent variables X1, X2, and so on, up to Xk have been observed on many students for whom elective grades have been determined as either 1 (if "A" or "B") or 0 (if less than "B"). Using this information, the proposed model will attempt to describe the probability that other students will do well (score an "A" or a "B") or not (score less than a "B") in a particular elective course at a particular time having measured the independent variable values X1, X2, up to Xk. The probability being modelled can be denoted by the conditional probability statement in Eq. 3.

$$P(X) = P(D = 1 \vee X_1, X_2, \ldots, X_k) \tag{3}$$

To illustrate how the logistic model was used in this work, we will assume that D is the elective course to be predicted and it is coded 1 if the grade is an "A" or a "B" and it is coded 0 if the grade is less than a "B". For this example, we will consider just three independent variables C1, C2 and C3, each holding continuous values between 0 and 100 that is the student's score in that particular course. Here, the logistic model will be defined as in Eq. 4.

$$P(X) = \frac{1}{1 + e^{-(\alpha + \beta_1 C1 + \beta_2 C2 + \beta_3 C3)}} \tag{4}$$

The proposed system attempts to use the data gathered to approximately guess the unknown factors $\alpha$, $\beta_1$, $\beta_2$, and $\beta_3$.

Logistic regression is very suitable for predicting course outcomes as it is a problem that satisfies some assumptions. One is that the variable to be predicted must follow a binomial distribution [21] and that is the case with a two-class classification problem like this. Also, each outcome must be statistically independent and that is the case with student grades as one student's grade in a course does not affect another student's grade in the same course. Likewise, in educational data mining, logistic regression has been proven to provide more accurate results than some other algorithms [22].

Furthermore, logistic regression is known to perform well even when there is not so much data being used in training [23] and this is important for this study because there are many elective course options for students, thus limiting the number of students who could apply for each course. Besides, logistic regression models tend to suffer in situations where the decision boundary between classes is multiple or non-linear but this isn't such a problem in predicting course outcomes as the decision boundary tends to be quite clear. Lastly, but probably most importantly, logistic regression models can easily be brought up-to-date with fresh data as it becomes available [24].

### 3.5 Wide and Deep Recommender

It has been noted that while generalized linear models can learn associations between wide independent and dependent variables, they generally struggle to take a broad view of features without adequate feature engineering; this is not a problem for deep neural networks as they can generalize more easily but sometimes, too easily when there isn't much interaction between those making the recommendations and the items being recommended, and this leads to poor recommendations [23]. To solve this problem, Cheng et al. [25] proposed Wide and Deep Learning, that is, a recommender that combines wide linear models (beneficial for remembering associations) with deep neural networks (high generalization ability).

The wide and deep recommender is a combination of two separate approaches. The wide element of the recommender is a summarized linear model of the form $y = w^T x + b$ where y is the dependent variable, $x = [x_1, x_2, \ldots, x_d]$ is a vector with d independent variables, $w = [w_1, w_2, \ldots, w_d]$ are the model constraints and $b$ is the bias. On the other hand, the deep element of the recommender is a feed-forward artificial neural network whose categorical features are initially transformed into low-dimensional and dense real-valued vectors (known as embedding vectors) with a dimensionality between $O(10)$ and $O(100)$. These embedding vectors are initially primed arbitrarily after which they are trained to diminish the final loss function while the model is being trained. Subsequently, the embedding vectors are passed into the hidden layers of a neural network in the forward pass, with each hidden layer of the neural network performing the computation in Eq. 5.

$$a^{(l+1)} = f\left(W^{(l)} a^{(l)} + b^{(l)}\right) \tag{5}$$

with l indicating the layer number and f being the activation function (usually rectified linear units – ReLUs); also, a(l), b(l), and W(l) will be the activation, bias and, weights at the l-th layer.

When training a wide and deep recommender, each element, that is, the wide and deep elements, are merged using a weighted sum of their output log odds as the model projection and this is then passed into a shared logistic loss function for both elements to be trained simultaneously.

Three sets of data can be passed into the wide and deep model and each of them is explained below:

**User-Item-Ratings:** This dataset contains three columns only which are identifiers for those making recommendations (in this case, students), the item which is recommended (in this case, elective courses), and a numerical rating is given to the recommended item, with a higher rating indicating that an item comes highly recommended while a lower rating indicates that the item comes unfavourably recommended. In this case, ratings were given to courses based on the grade scored by the student in that course; that is, students who scored an "A" in a particular course gave that course a rating of 5, students who scored a "B" in a particular course gave that course a rating of 4, students who scored a "C" gave the course a rating of 3, and so on. This dataset is the only dataset required for the wide and deep recommender to run.

**User Features:** This dataset contains information about individual users who have provided recommendations. The first column of this dataset (student identifier) must match the identifiers provided in the user-item-ratings dataset. In this study, scores in the required and core courses taken in the student's first three years of study were used to populate the user features dataset. This dataset is not required for the model to provide recommendations but it would lead to better recommendations when provided.

**Item Features:** This dataset contains information about the potential courses that could be recommended to users. It is also not required for the model to provided recommendations and was not used in the study as this could not gather adequate information to necessitate the use of this particular dataset.

The wide and deep model has been proven to be more effective in making recommendations than wide-only and deep-only models [25], thus making it a good choice for this system. Users would supply their grades in required courses as input to the model while the model would produce a set of recommendations as output.

### 3.6 Performance Evaluation Metrics

A number of metrics were used to determine the quality of the trained models. These metrics are described in the table below (Table 1):

**Table 1.** Performance evaluation metrics

| Metric | Description | Formula |
|---|---|---|
| Accuracy | Accuracy measures the correctness of the classification | $\frac{TN+TP}{\text{The total number of test items}}$ |
| Precision | Precision measures the exactness of the classification. It describes the ability of a classification model to return only relevant instances | $\frac{TP}{TP+FP}$ |
| Recall | Recall measures the completeness of classification. It describes the ability of a classification model to identify all relevant instances | $\frac{TP}{TP+FN}$ |
| F1-Measure | F1-measure is a combination of precision and recall. The quality of the model in terms of both precision and recall is combined into a single score, calculated as the standard harmonic mean of precision and recall | $\frac{2 * precision * recall}{precision + recall}$ |
| Receiver operating characteristic curve – Area under curve (ROC-AUC) | ROC curve is a graph showing the performance of a classification model at all classification thresholds. AUC measures the entire two-dimensional area underneath the entire ROC curve | |
| Mean squared error (MSE) | It measures the average of the squares of the errors—that is, the average squared difference between the estimated values and the actual value | $\frac{1}{n}\sum_{i=1}^{n}\left(Y_i - \widehat{Y}_i\right)^2$ |
| Root-Mean-Square error (RMSE) | It represents the square root of the second sample moment of the differences between predicted values and observed values or the quadratic mean of these differences | $\sqrt{\frac{\sum_{i=1}^{n}\left(\widehat{Y}_i - Y_i\right)^2}{n}}$ |

## 4 Results and Discussion

Based on the metrics in Table 2, it can be noted that the models performed effectively in predicting passes in the electives considered and in recommending courses to students. The models also performed better than other models evaluated in Sect. 2 other than the k-nearest neighbour algorithm used in Ogunde and Ajibade [13]. Figure 6 displays the confusion matrix of the proposed classification model.

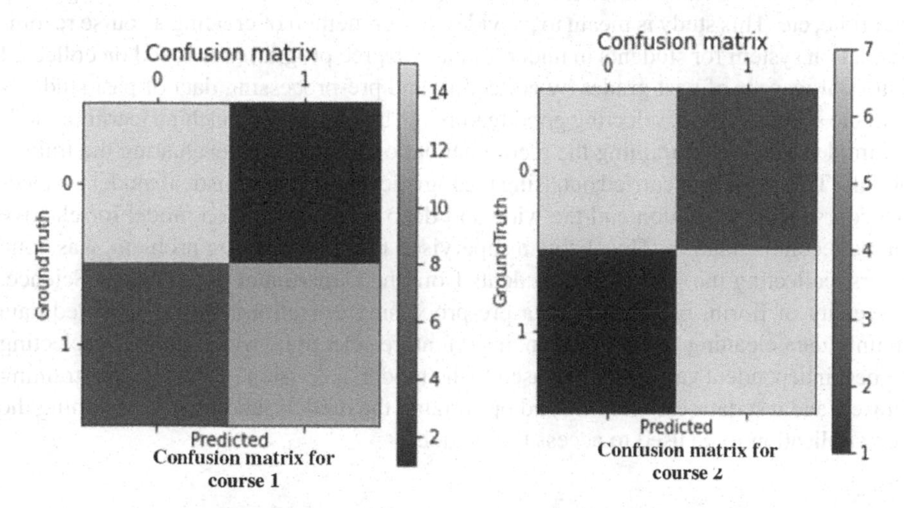


**Fig. 6.** Confusion matrix of the proposed classification model

**Table 2.** The performance metrics for the proposed system

| | Accuracy | Precision | Recall | $F_1$-Measure | ROC AUC | MSE | RMSE |
|---|---|---|---|---|---|---|---|
| Course 1 | 0.84 | 0.92 | 0.75 | 0.83 | 0.84 | 0.48 | 1.63 |
| Course 2 | 0.76 | 0.75 | 0.75 | 0.75 | 0.83 | | |

The perfect precision score is 1.0. The first course predicted was found to have a precision score of 0.92 while the second course had a precision score of 0.75. The first course predicted was found to have a recall score of 0.75 and the second course had a recall score of 0.75. The first course predicted was found to have an $F_1$ score of 0.83 and the second course had an $F_1$ score of 0.75. The constructed model was seen to have a Mean Squared Error value of 0.48. RMSE is always non-negative and a value of zero shows that the model perfectly suits the data; thus, the lower the value, the better the model. The constructed model was seen to have a Root-Mean-Square Error value of 1.63. The first course predicted was found to have an accuracy score of 0.84 while the second course had an accuracy score of 0.76.

## 5 Conclusion

Elective course selection tends to be a problem for many undergraduate students in tertiary institutions. Some reasons for this are the large number of possible electives to be selected, inadequate information about elective courses before selection, insufficient time for level advisers to gather course information, etc. While some recommender systems have been put in place to solve such problems, they tend to struggle because of inadequate training data, imbalances in training data, changes in students' learning styles over time, etc. This study is meant to provide a better method of creating a course recommendation system for students in undergraduate degree programmes based on collected historical records of past grades by collecting and pre-processing data of past students from the school system, selecting good features to be used in the machine learning models, implementing and training the recommendation models, and evaluating the trained models. This study was carried out using the logistic regression statistical model for elective course pass prediction and the wide and deep recommendation model for elective course recommendation. This, being a supervised machine learning problem, was done by first collecting the data of past students from the Department of Computer Science, University of Ilorin, performing data pre-processing operations on the collected data (mainly data cleaning and data balancing), feature selection (which involves selecting the best independent variables to be used in the models), division of the data into training datasets and test datasets, training and optimizing the models, and finally, designing the web application to be used to access the models.

## References

1. Abiodun, M.K., et al.: Cloud and Big Data: A Mutual Benefit for Organization Development. J. Phys. Conf. Ser. **1767**(1), 012020 (2021)
2. Crestani, F.: Application of spreading activation techniques in information retrieval. Artif. Intell. Rev. **11**(6), 453–482 (1997)
3. Zabriskie, C., Yang, J., DeVore, S., Stewart, J.: Using machine learning to predict physics course outcomes. Phys. Rev. Phys. Educ. Res. **15**(2), 020120 (2019). https://doi.org/10.1103/PhysRevPhysEducRes.15.020120
4. Praserttitipong, D., Srisujjalertwaja, W.: Elective course recommendation model for higher education program. Songklanakarin J. Sci. Technol. **40**(6), 1232–1239 (2018). https://doi.org/10.14456/sjst-psu.2018.151
5. Thanh-Nhan, H.L., Nguyen, H.H., Thai-Nghe, N.:. Methods for building course recommendation systems. In: 2016 Eighth International Conference on Knowledge and Systems Engineering (KSE), November 2017, pp. 163–168 (2016). https://doi.org/10.1109/KSE.2016.7758047
6. Jiang, W., Pardos, Z.A., Wei, Q.: Goal-based course recommendation. In: Proceedings of the 9th International Conference on Learning Analytics & Knowledge - LAK19, March, pp. 36–45 (2019). https://doi.org/10.1145/3303772.3303814
7. Kunaver, M., Požrl, T.: Diversity in recommender systems–a survey. Knowl.-Based Syst. **123**, 154–162 (2017)
8. Awotunde, J.B., Adeniyi, A.E., Ogundokun, R.O., Ajamu, G.J., Adebayo, P.O.: MIoT-based big data analytics architecture, opportunities and challenges for enhanced telemedicine systems. Stud. Fuzziness Soft Comput. **2021**(410), 199–220 (2021)

9. Chau, V.T.N., Phung, N.H.: Imbalanced educational data classification: An effective approach with resampling and random forest. In: Proceedings - 2013 RIVF International Conference on Computing and Communication Technologies: Research, Innovation, and Vision for Future, RIVF 2013, November 2013, pp. 135–140 (2013). https://doi.org/10.1109/RIVF.2013.6719882
10. Gurpinar, E., Bati, H., Tetik, C.: Learning styles of medical students change in relation to time. Am. J. Physiol. Adv. Phys. Educ. **35**(3), 307–311 (2011). https://doi.org/10.1152/advan.00047.2011
11. Morsy, S., Karypis, G.: Will this course increase or decrease your gpa? towards grade-aware course recommendation. J. Educ. Data Min. **11**(2), (2019). http://arxiv.org/abs/1904.11798
12. Dahdouh, K., Dakkak, A., Oughdir, L., Ibriz, A.: Large-scale e-learning recommender system based on spark and hadoop. J. Big Data **6**(1), 1–23 (2019). https://doi.org/10.1186/s40537-019-0169-4
13. Ogunde, A.O., Ajibade, E.: A K-nearest neighbour algorithm-based recommender system for the dynamic selection of elective undergraduate courses. Int. J. Data Sci. Anal. **5**(6), 128–135 (2019). https://doi.org/10.11648/j.ijdsa.20190506.14
14. Feghali, T., Zbib, I., Hallal, S.: A web-based decision support tool for academic advising. Educ. Technol. Soc. **14**(1), 82–94 (2011)
15. Ling, G., Yang, H., Lyu, M.R., King, I.: Response aware model-based collaborative filtering, uncertain. In: Artificial Intelligence - Proceedings of 28th Conference UAI 2012, pp. 501–510 (2012)
16. Marlin, B.M., Zemel, R.S.: Collaborative prediction and ranking with non-random missing data. In: Proceedings of 3rd ACM Conference Recommender Systems, RecSys 2009, pp. 5–12 (2009). https://doi.org/10.1145/1639714.1639717
17. J. M. Hernández-Lobato, N. Houlsby, and Z. Ghahramani, "Probabilistic matrix factorization with non-random missing data," 31st Int. Conf. Mach. Learn. ICML 2014, vol. 4, pp. 3394–3436, 2014.
18. Bydžovská, H.: Course enrollment recommender system. In: Proceedings of 9th International Conference Educational Data Mining, EDM 2016, no.10, pp. 312–317 (2016)
19. Chen, C.M., Lee, H.M., Chen, Y.H.: Personalized e-learning system using Item response theory. Comput. Educ. **44**(3), 237–255 (2005). https://doi.org/10.1016/j.compedu.2004.01.006
20. Thai-Nghe, N., Drumond, L., Krohn-Grimberghe, A., Schmidt-Thieme, L.: Recommender system for predicting student performance. Procedia Comput. Sci. **1**(2), 2811–2819 (2010). https://doi.org/10.1016/j.procs.2010.08.006
21. Kleinbaum, D.G., Klein, M.: Logistic Regression A Self-Learning Text. In: Survival (3rd ed.). Springer, New York (2010)
22. Hosmer, D., Lemeshow, S.: Applied Logistic Regression (Issue October). Wiley, Hoboken (2013). https://doi.org/10.1080/00401706.1992.10485291
23. Shaun, R., Baker, J. De, J. E.B., (eds.) Educational Data Mining 2008 The 1st International Conference on Educational Data Mining. Network, January, 187 (2008). http://gdac.uqam.ca/NEWGDAC/proceedingEDM2008.pdf#page=87
24. Sawarkar, N., Raghuwanshi, M.M., Singh, K.R.: Intelligent recommendation system for higher education. Int. J. Future Revolution Comput. Sci. Commun. Eng. **4**(4), 311–320 (2018)
25. Cheng, H.-T., et al.: Wide & Deep Learning for Recommender Systems. In: Proceedings of the 1st Workshop on Deep Learning for Recommender Systems, DLRS 2016, pp. 7–10 (2016)

# Predictive Academic Performance Model to Support, Prevent and Decrease the University Dropout Rate

Diego Bustamante and Olmer Garcia-Bedoya(✉)

Universidad de Bogota Jorge Tadeo Lozano, Carrera 4 22-61, Bogota, Colombia
{diegol.bustamantep,olmer.garciab}@utadeo.edu.co

**Abstract.** One of the biggest problems in higher education is student dropout. Prior to the pandemic, one of the biggest problems for university institutions was the dropout and dropout of many of their students. Today, the situation has become even more critical, as the pandemic has forced many people to drop out of school for a variety of reasons, whether financial or personal. Investigating the causes of dropout with appropriate means to reduce it contributes to decision making within academic management. The objective of this work is to develop a machine learning model that generates early warnings about course loss, which is based on historical data of pupils and students. The model is based on historical data from an undergraduate program that includes, student grades, at various points in time, percentage of course loss in previous semesters, percentage of student loss in previous semesters, subjects passed at the time of evaluating the data, along with student and course average. This would facilitate the identification and internal management of alarms for the early detection of potential dropouts, as well as efficiently display the results found with the execution of these models.

**Keywords:** Higher education · Dropout prevention · Machine learning · Random forest

## 1 Introduction

Student desertion is one of the biggest problems currently faced by educational institutions nationwide. This is very worrying since a large part of the Colombian population is being deprived of this fundamental right, the most worrisome are the reasons why many people have to abandon such studies. In Colombia the highest student dropout rate occurs in higher education, only for 2018, approximately 9% of university level students abandoned their studies for various reasons [11]. This only reviewing the figures for university level, this number becomes even more alarming if the figures for technical and technological levels are taken into account [7]. If the figures at the regional level are reviewed, the situation is even more critical. According to several UNESCO studies in 2015 university dropouts hovered around 50% in Latin America [16]. In most cases the problem


H. Florez and M. F. Pollo-Cattaneo (Eds.): ICAI 2021, CCIS 1455, pp. 222–236, 2021.
https://doi.org/10.1007/978-3-030-89654-6_16

is due to multiple factors, mostly economic and social. Studies conducted by different researchers on this subject identify a variety of causes, ranging from dropout due to poor previous training, frustration for failing multiple times the so-called final exams, social problems and family, wrong choice of career, organizational problems, among others, are the main reasons for the occurrence of the decersion.

At the same time, multiple technological tools have been successfully implemented to solve various problems, this along with various factors in the Colombian market have caused a sudden growth in the ICT sector [10]. Only for 2019 in Colombia a growth of 4% was registered in the first quarter according to DANE data [1]. Moreover, thanks to the pandemic this sector has been forced to accelerate its growth even more, it could be said that it was one of the sectors at national level that was not so crudely hit by the pandemic. Making Colombia even more attractive for foreign investment. One of the most attractive solutions that have entered the market are those based on or using everything related to artificial intelligence and machine learning. Thanks to the fact that these technologies help to optimize some processes, reduce costs and improve effectiveness in many procedures. This has caused an exponential growth in investments in this sector, some media report that the injection of capital earmarked solely for artificial intelligence (AI) is approximately 1.2 billion dollars [2].

The main objective of this project is to implement a Machine Learning model, providing a university tool that profiles students at risk of dropping out, using academic information of the semesters taken such as grades, averages and credits or subjects approved, identifying the risk of loss that a student has at the time of taking a subject. This by means of different models that use the history of grades, the academic progress in the study plan. In this way, universities could implement different strategies to minimize student dropout. In addition to implementing a complete architecture for the loading, evaluation, transformation and visualization of these data along with the results, thus providing a tool that is easy to interpret by the different interested areas of the university entities. Such as the different academic committees and associate professors.

This article is divided as follows. In Sect. 2, we present some context about the notes and desertion in higher education. Section 3 describes several features of the dataset structure and analysis. The following Sect. 4 presented the results of implemented methodologies and we close the paper with Sects. 5 and 6 that consist of concluding comments and provide directions for future work.

## 2 Preliminaries

One of the biggest problems facing the Colombian higher education system is the dropout rates at the undergraduate level. Despite the fact that recent years have been characterized by increases in coverage at the national level, the number of students who complete their higher education studies is much lower than the number of students who entered during the first semester.

### 2.1 Student Desertion

Early diagnosis of students with a high dropout risk enables interventions to be provided early on, which can help these students to complete their studies, graduate, and enhance their future competitiveness in the workplace [17]. There are many approach to use machine learning to improve the student retention. One classification could by the type of data analysed, [12] generate models based on the data available at the time of enrollment, [6] generates clusters based in the logs generated by the student in learning management systems. [8] mix student variable like age, gender with grade point average and other academic variables. [14] presents a review to show different modern techniques widely applied for predicting students' performance finding studies which use Machine Learning, Collaborative Filtering, Recommender Systems, and Artificial Neural Networks, among others. This studies has demonstrated with different approach that Datafication of education has allowed developing automated methods to detect patterns in extensive collections of educational data to estimate unknown information and behavior about the students [4].

It is important to understand that this models help to generate early alarms or understand the students, but are not the final solution to the problem. These strategies give better information to take decision based in data in all levels of the universities (directors, professors, coordinators, students ans parents). When talking about all this, several questions start to become important, such as: What strategies are the institutions taking? What educational alternatives does the institution have? What kind of follow-up is being done on this problem? How effective are these strategies? The answer to these questions may vary among educational institutions, since each one of them has its own mechanisms and alternatives to solve the problems; however, it is difficult to know the effectiveness of these tools or alternatives and even more difficult to quantify the results, since the monitoring of the entire student body of a university is quite complicated due to the number of people and the multiple schedules that are handled.

### 2.2 Functioning of Notes at University and in General at Colombian Universities

In Colombia, university students must complete a certain number of credits or points in order to finish their studies. These credits are used to enroll in subjects of a curriculum that is defined by the faculty in which the student is enrolled. Some subjects have a lower or higher value than others. In order to fulfill the credits required by the course, the student must pass the subjects, thus the amount of credits that this subject has will be completed.

On the other hand, the subjects are divided in 3 moments that have some percentage of the final grade, during the academic period, in order to pass each one of these, a minimum of a 3 over 5 must be obtained. At the end of the academic period, in this article is assumed that the 3 notes are averaged for the final grade obtained.

# 3 Methods

As mentioned in the introduction, this is a software development project using Machine Learning tools to determine the probability of student dropout, however, it is still a software development project.

The CRISP-DM methodology (Cross Industry Standard Process for Data Mining) was used in this project to create and validate the models. CRISP-DM is a framework specifically designed for the implementation of projects oriented to data analysis. Like any common framework, it is focused on results; however, it goes deeper into the analysis stages to identify if the information provided is useful or if it should be refined at some level [5]. This is particularly useful in this project, given the source of the data, which although generated by a university's grade management platform, has a number of inconsistencies from structure to content, an example of which is the lack of control evident in the data when taking failures. Like any CRISP-DM framework, it also has some stages or phases, these are:

- Phase I, Definition of customer needs.
- Phase II, Study and understanding of the data.
- Phase III, Data analysis and feature selection.
- Phase IV, Modeling
- Phase V, Evaluation
- Phase VI, Deployment

## 3.1 Data Preparation

Due to the fact that this project is about student desertion oriented to undergraduate students, support was requested to the faculty of systems engineering and the technology area of the xxxxx University, who provided the academic information since 2017 throughout the university. Always respecting the personal information of students as determined by law (Habeas Data). With the data delivered by the university in an Excel file separated by spreadsheets containing each academic period, a unified matrix was created with all the information of the academic periods delivered. This was done with the help of the multiple tools provided by the PostgreSQL database engine. Although the information provided by the area in charge of the institution is generated by the system that houses the grades, it is important to clarify that these data still do not have the necessary detail to perform a correct analysis, this because the columns provided in the report only have the information corresponding to the strictly academic. The columns provided by the university were the following (Non-Normalized data) in Table 1

**Table 1.** Columns of each register

| Original name | Translated name | Description |
|---|---|---|
| PERIODO_ACADÉMICO | Academic period | cell6 |
| CODIGO_UXXI | Code_UXXI | Academic period to which the qualification belongs |
| PROGRAMA | Program | Academic program or career that belongs |
| CODIGO_ASIG | Code_course | Code of course in university platform |
| ASIGNATURA | Course | Name of course in university |
| NOTA_MOM1 | Qualification#1 | Note of the first moment |
| FALLAS_MOM1 | Absences#1 | Innacistences of the first moment |
| NOTA_MOM2 | Qualification#2 | Note of the Second moment |
| FALLAS_MOM2 | Absences#2 | Innacistences of the second moment |
| NOTA_MOM3 | Qualification#3 | Note of the third moment |
| FALLAS_MOM3 | Absences#3 | Innacistences of the third moment |
| DEFINITIVA | Definitive Qualification | Final grade averaging the 3 moments |
| ORIGEN | Origin | As a course of study |
| ESTADO | State | State of course |
| PROMEDIO_PERIODO | Average period | Student average for that period |
| PROMEDIO_ACUMULADO | Cumulative average | Cumulative student average |

### 3.2 Data Analysis

As can be seen in the previous section, none of the columns specify the name of the student, and although this data is not relevant for the analysis, since there is a unique identifier per person, having the name and surname of each one simplifies the visualization of them. For this purpose, random names and surnames were generated for each of the identifiers. Another variable that had to be added to the data provided was the core to which each of the subjects belonged, in order to better classify the performance of the students by area and subject, crossing this information with the origin. These data were annotated by hand with the information provided by the university on its web page and knowledge of the authors.

With PostgreSQL, a view was created to filter the information and obtain only the data corresponding to the systems engineering students, this facilitated the distribution of the records in the data model. After this, an exploratory data analysis tool called Pandas-Profiling was used [13], which is used to perform the exploratory analysis of the data, facilitate the understanding and description of these, in addition to automating the data analysis.

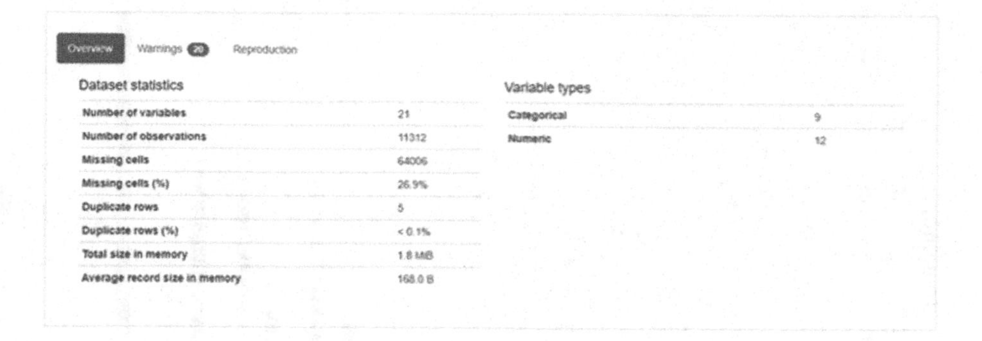


**Fig. 1.** Summary of data analyzed

Figure 1 shows the report returned by the tool, it is possible to identify the basic data of the dataset provided, these data provide a superficial understanding of the behavior of the data.

With this information it can be identified that the data delivered, still have inconsistencies in their content. However, these results are superficial; there is no detail to determine where these inconsistencies are found. By analyzing the information on the variables in more detail, we can identify the variables that are most important, such us:

- Period, because the decrease in the number of students enrolled in
- Each period is evident. Credits per subject, it is possible to identify the number of credits that are addressed per period.
- Origin, because you can see which are the most frequent origins of a subject.
- Core, you can see the different cores during the period.
- Status, general status of the subjects (Fig. 2)

Based on the previous list and Fig. 7, one can see the core, the origin or how the subject was approved, the status of the subject for the period taken and the final grade, respectively. In the analysis of these three graphs it is possible to deduce which group of students have difficulties in taking certain subjects, or which subjects have the most failed students during an academic period.

In Fig. 3, here you can see the relationship between each of the variables with the others found in the data, here you can see that some have a close relationship between them and others do not have any kind of relationship, this taking into account that the report does not have any information on who is dependent and what their dependence is. This means that the tool deduces in the most optimal way the relationship.

Finally, in Fig. 4, you can see one of the possible trees generated, in which you can identify the behavior of the ratings, which is generally positive, however, when looking in detail, it is possible to identify that there are small groups of data, which tend to lower their ratings. It should be noted that these trees are generated with the help of different estimators, however, the behavior can be identified in almost all cases is similar.

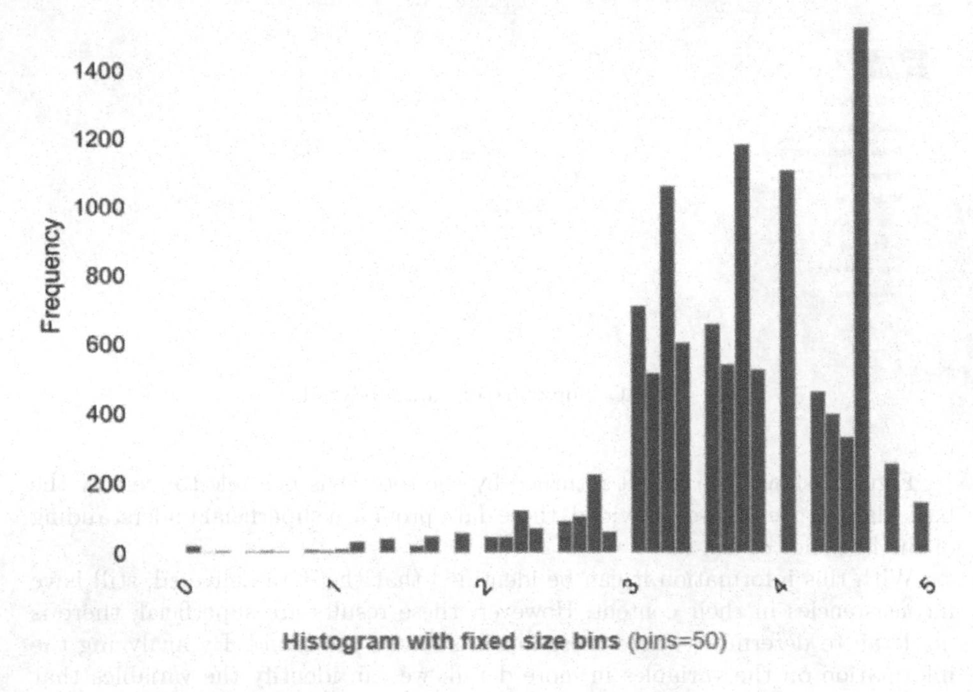


**Fig. 2.** Distribution of the notes over the students analyzed

## 4 Modeling Academic Performance with Machine Learning

This section presents the modeling and testing layer of the CRISP-DM [18] methodology. We used supervised machine learning algorithms presented below. Following that, the training and testing process is presented.

### 4.1 Models Used

In Machine Learning there are a large number of models available, for this case 4 techniques were selected to run, mainly to be able to compare the results and accuracy of each of them.

- Naive-Bayes classifier: A Bayesian classifier is a probabilistic method based on Bayes' theorem and is called naive due to some additional simplifications that determine the hypothesis of independence of the predictor variables [15].
- Decision trees: It is a set of data diagrams of logical constructs, very similar to rule-based prediction systems, that serve to represent and categorize a series of conditions that occur in succession, for the resolution of a problem.
- Logistic regression: is a type of regression analysis used to predict the outcome of a categorical variable as a function of the independent or predictor variables.

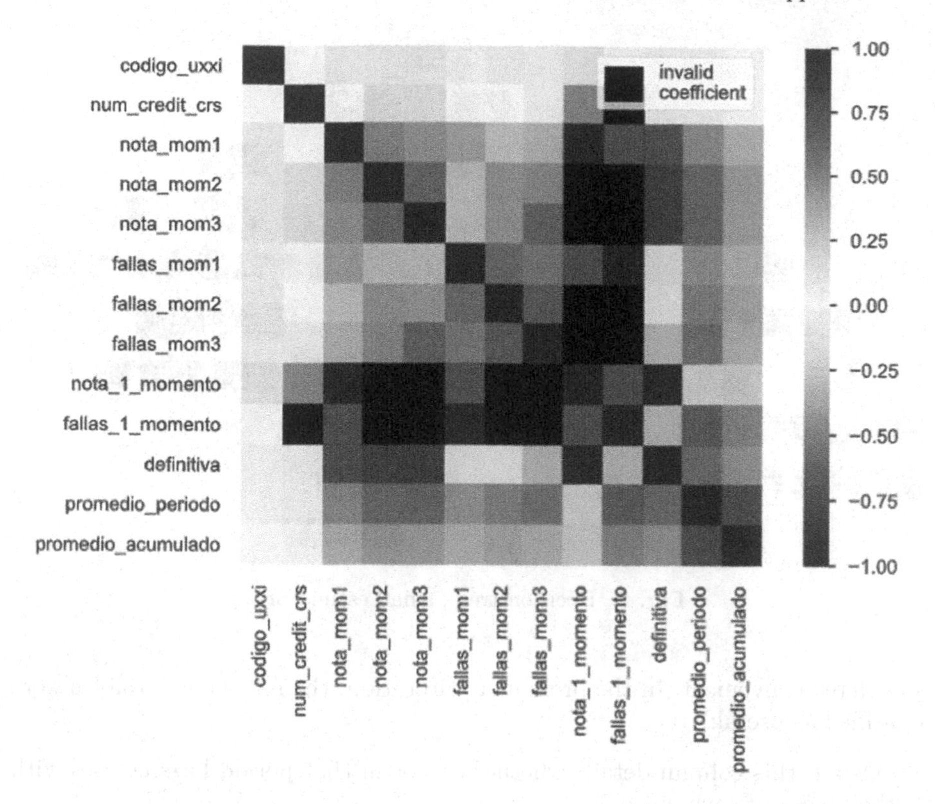


**Fig. 3.** Relationship of variables in the data

- Random forest classifier [9]: is a combination of predictor trees such that each tree depends on the values of a random vector tested independently and with the same distribution for each of these. It is a substantial modification of bagging that builds a large collection of uncorrelated trees and then averages them.

All this techniques were used to solve a classification problem where the output is a categorical variable defined like loss or not loss the assignment which is coursing.

### 4.2 Training Process

After having performed the descriptive analysis or description of the data, the information must be prepared, because the data may contain information that is not relevant for the analysis. In this phase, the data used in the previous stage are taken and the columns that are necessary or useful for the execution of the model are identified. It should be clear that there will be several iterations where the data will be completed and structured, until the expected result is achieved. That is to say that in each iteration variables can enter or leave as it is

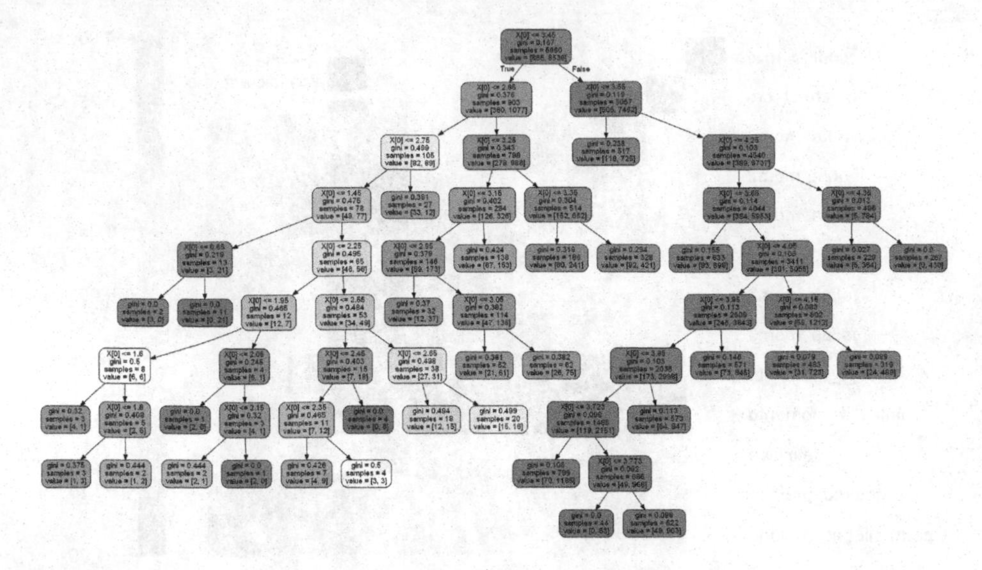

**Fig. 4.** Decision trees, small estimator

considered convenient. In the preliminary iteration, the following variables were identified as useful:

- Output, this column details whether or not in that period he/she loses with the grades of each cut-off.
- Average of the student the previous semester.
- Cumulative average of the student.
- Average of that subject in the previous semester.
- Percentage of loss of the subject in the previous semester
- Approved credits
- Career or faculty
- Subject loss
- Core of the subject
- Subjects taken
- First cut-off grades
- Passing the subject
- Average of failures from previous semesters.
- Origin: This column contains the student's origin information, whether it comes from homologation or was taken.

After performing the initial preparation of the data, it was possible to identify that the failures were not a determining or useful data for this model, this because most of the information recorded in this cell presents many inconsistencies, since a large number of teachers do not record the failures consistently or record them only if the person will lose the subject for exceeding the maximum limit of absences. In other words, the column indicated above "Average number of failures in previous semesters" will not be found in the model.

It is important to define how the algorithm will work, which variables will be evaluated and which variable will be the output. In this implementation, three evaluations were selected.

- Model 1: Probability of loss of subject, without taking into account the cut-off grades 1 and 2.
- Model 2: Probability of loss of subject, only with the grade of the first cut-off.
- Model 3: Probability of loss of subject, with the first and second cut-off grades.

Finally, before performing any evaluation, it must be defined which variables will be marked as labels or categorical, these variables change according to the model to be run, but all have variables that will be taken as a basis. The variables that are marked as labels are:

- Approves subject
- Subject code
- Student code

Where subject code and student code are unique identifiers. The variables that will be taken as categorical are:

- Previous period
- Cumulative average
- Average in the subject previous period
- Approved credits
- Percentage of loss of the subject.

### 4.3 TESTING the MODEL

By performing an evaluation of the selected techniques together with the models built, it is possible to validate the accuracy of each one of them, with the data provided, the result of this evaluation is the Table 2. As can be seen in the table above, the efficiency of each model exceeds 90% with each technique evaluated. Given this level of accuracy, it can be seen that any alternative is useful to evaluate the data. Furthermore, the percentage difference between each technique per model is a maximum of 3%, therefore, in this first iteration the Random forest technique will be used as the basis. In order to validate if the results delivered by this model meet the expectations, we will use the Random forest technique.

**Table 2.** Accuracy of k-fold cross-validated models at 5 using f1-score .

| | Naive Bayes | Decision trees | Logistic regression | Random forest classifier |
|---|---|---|---|---|
| Model 1 | 93% | 94% | 95% | 96% |
| Model 2 | 93% | 93% | 94% | 95% |
| Model 3 | 95% | 95% | 97% | 97% |

## 5 Discussion

The behavior of the evaluated data can be confirmed by validating the confusion matrix for each of the models with the help of Figs. 5, 6 and 7.

```
=====Confusion Matrix (Decision Tree):=====

[[  73   83]
 [  38 1466]]

=====Classification Report (Decision Tree):=====

              precision    recall  f1-score   support

           0       0.66      0.47      0.55       156
           1       0.95      0.97      0.96      1504

    accuracy                           0.93      1660
   macro avg       0.80      0.72      0.75      1660
weighted avg       0.92      0.93      0.92      1660
```

**Fig. 5.** Confusion Matrix, model 1

Where according to the classifier seen in the report, it can be said that:

- Accuracy: Percentage of correctly evaluated data.
- Recall: Students classified in a new category, according to the configured labels.
- F1-score: Corresponds to is the harmonic mean between precision and recall in the data.
- Support: is the number of occurrences in the data set.

Taking into account this information and that the technique with the highest accuracy is Random Forest, we proceed to the execution of the models created supported by this technique. In this technique several estimators can be configured, this is very useful at the moment of evaluating the data since the main characteristic is the partitioning of the data in small groups generating multiple decision trees, being possible to evaluate multiple behaviors in the information. For this case, 10 estimators were configured and 4 different decision trees were generated based on the first model in order to verify the behavior. When evaluating the generated trees, the behavior of the scores is generally positive; however, when looking in detail at each of the trees, it is possible to identify that there are small groups of data that tend to lower their scores. It should be noted that these trees are generated with the help of different estimators.

```
=====Confusion Matrix (Decision Tree):=====

[[ 87  57]
 [ 34 803]]

=====Classification Report (Decision Tree):=====

              precision    recall  f1-score   support

           0       0.72      0.60      0.66       144
           1       0.93      0.96      0.95       837

    accuracy                           0.91       981
   macro avg       0.83      0.78      0.80       981
weighted avg       0.90      0.91      0.90       981

------------------------------------------------------
```

**Fig. 6.** Confusion Matrix, model 2

In order to implement a friendly interface for subsequent executions of these models, we proceed to use the Gradio tool [3], this Python library allows to implement a web server, with a predefined interface according to a given configuration. Where the data to be processed are entered and the library, with the help of the pickle files generated from the execution of the model, proceeds to perform the corresponding calculations.

Gradio also has the possibility of generating a public website, which is managed within its domain, providing a URL so that anyone can run the model without the need for the tools used for this project, i.e. only a file with the data to be evaluated is required. The library displays a simple interface, where you can select the model to be executed, along with a file with csv extension. That will contain the data to be evaluated, depending on the model. As a result, a new file can be generated with the probability that each student has of losing a subject.

This project was designed with the objective of providing a tool to help the university to identify possible dropouts, therefore, it can be scaled in different ways. An example of this is the possibility of identifying the shortcomings of SENA[1] (national learning service in spanish) articulation students, who generally have difficulties in basic areas such as calculus and physics, taking into account that these students are only homologated in the specific subjects of the career. Another possibility is to identify difficulties by academic program, i.e., to identify the shortcomings that students of a given program may have in a given

[1] SENA: The national learning service, SENA, is an entity that offers free training to millions of Colombians who benefit from technical, technological and complementary programmes focused on the economic, technological and social development of the country.

```
=====Confusion Matrix (Decision Tree):=====

[[ 77  41]
 [ 22 791]]

=====Classification Report (Decision Tree):=====

              precision    recall  f1-score   support

           0       0.78      0.65      0.71       118
           1       0.95      0.97      0.96       813

    accuracy                           0.93       931
   macro avg       0.86      0.81      0.84       931
weighted avg       0.93      0.93      0.93       931

-------------------------------------------------
```

**Fig. 7.** Confusion Matrix, model 3

core area, which would help in the configuration of the subject area by career. During the preliminary phase of the project, the implementation of a model based on failures was discarded, this due to the lack of constancy of the professors when registering them, however if the constant entry of this attribute can be guaranteed, it could add precision to the evaluation of the model, in addition, an independent model could be created where it is evaluated what probability a student has of failing certain subjects according to his academic history, this would function as another possible alert in the subjects of the different academic plans. Although this implementation was done with information from systems engineering students, this could be executed with any academic program at the university. The decision was made to use only that career because being a student of this program facilitates the compression of the information as subjects, in addition to knowing from my own experience the core of the subjects along with their academic load or credits. Having said classification or academic core in all the subjects provided by the university, the model could be built or modified to implement greater coherence in the models, that is to say that by seeking a better classification with the cores the effectiveness of each of the models and their results would be better. It is possible to perform an integration with the grade platform of the educational centers, since the data that were used for the training were requested to the technology area, in case of requiring to add a user interface so that teachers or any person of the academic committee can run such models, an interface should be made available where the user selects the minimum data such as the subject and the student that he/she wants to validate. This option should prepare the data to be sent to the model executor, who will be in charge of calculating the probabilities and then deliver the results back to said interface so that it is it who manages the results to the user who requests them, that is to say that this tool will only be in charge of defining and preparing the data with information provided, but who should extract the minimum data required is the university platform, this because it is the only one that has said information due to confidentiality and personal data protection issues.

## 6 Conclusion

This project was developed with the grades of students belonging to the academic program of systems engineering, in which it was possible to identify that at least in this population segment the probability of loss of subjects in general is not so high.

In this type of projects it is important to select the suitable framework or methodology, CRISP-DM is quite comfortable and simple to apply in projects in which multiple iterations must be performed in search of results, and it is optimized for this type of projects.

In this implementation a well-structured data architecture was generated that helps during the multiple iterations and facilitates the access to the data by centralizing all the information. This was developed with models provided by different Python libraries such as SkLearn and Pandas-Profiling for descriptive analysis, thus optimizing resources from the planning, execution and completion of the project.

The classification techniques present a higher performance when most of their variables are categorical; this is due to the subdivision and noise elimination that is performed when multiple estimators are configured, however, this has a significant impact on the performance of the equipment where it is executed.

From the probabilities generated by these reports, multiple internal strategies can be implemented within the university to reduce this rate as much as possible, and by running these models with information from all the careers of the university, the benefit can be further extended and cross-cutting strategies can be generated to help prevent this behavior. It also gives the possibility of building new implementations based on this project, which may eventually implement any other type of prediction technique.

## References

1. Tasa de crecimiento económico del sector de las tic aumentó 4,04 % en los dos primeros trimestres de 2019 - tasa de crecimiento económico del sector de las tic aumentó 4,04 % en los dos primeros trimestres de 2019. MINTIC Colombia 2020. https://www.mintic.gov.co/portal/inicio/Sala-de-Prensa/Noticias/103393:Tasa-de-crecimiento-economico-del-sector-de-las-TIC-aumento-4-04-en-los-dos-primeros-trimestres-de-2019
2. Inteligencia artificial: la región se abre al desarrollo. Conexión Intal, September 2019. https://conexionintal.iadb.org/2018/05/30/ideas-2/

3. Abid, A., Abdalla, A., Abid, A., Khan, D., Alfozan, A., Zou, J.: Gradio: Hassle-free sharing and testing of ml models in the wild. arXiv preprint arXiv:1906.02569 (2019)
4. Bañeres, D., Rodríguez, M.E., Guerrero-Roldán, A.E., Karadeniz, A.: An early warning system to detect at-risk students in online higher education. Appl. Sci. **10**(13) (2020). https://doi.org/10.3390/app10134427, https://www.mdpi.com/2076-3417/10/13/4427
5. Chapman, P., Clinton, J., Kerber, R., Khabaza, T., Reinartz, T., Shearer, C., Wirth, R.: The crisp-dm user guide. In: 4th CRISP-DM SIG Workshop in Brussels in March, vol. 1999. sn (1999)
6. Delgado-Quintero, D., Garcia-Bedoya, O., Aranda-Lozano, D., Munevar-Garcia, P., Diaz, C.O.: Academic behavior analysis in virtual courses using a data mining approach. In: Florez, H., Leon, M., Diaz-Nafria, J.M., Belli, S. (eds.) ICAI 2019. CCIS, vol. 1051, pp. 17–31. Springer, Cham (2019). https://doi.org/10.1007/978-3-030-32475-9_2
7. Fernandes, A., Lima, R., Figueiredo, M., Ribeiro, J., Neves, J., Vicente, H.: Assessing employee satisfaction in the context of covid-19 pandemic. ParadigmPlus **1**(3), 23–43 (2020)
8. Kasthuriarachchi, K., Liyanage, S.: Recommendations for students in higher education: A machine learning approach (2017)
9. Liaw, A., Wiener, M., et al.: Classification and regression by randomforest. R News **2**(3), 18–22 (2002)
10. Mendez, O., Florez, H.: Applying the flipped classroom model using a vle for foreign languages learning. In: International Conference on Applied Informatics, pp. 215–227. Springer (2018)
11. mineducacion: Estadísticas de deserción. Ministerio de educacion nacional 1, 1
12. Nagy, M., Molontay, R.: Predicting dropout in higher education based on secondary school performance. In: 2018 IEEE 22nd International Conference on Intelligent Engineering Systems (INES), pp. 000389–000394 (2018). https://doi.org/10.1109/INES.2018.8523888
13. Peng, J., et al.: Dataprep. eda: task-centric exploratory data analysis for statistical modeling in python. In: Proceedings of the 2021 International Conference on Management of Data, pp. 2271–2280 (2021)
14. Rastrollo-Guerrero, J.L., Gómez-Pulido, J.A., Durán-Domínguez, A.: Analyzing and predicting students' performance by means of machine learning: a review. Appl. Sci. **10**(3) (2020). https://doi.org/10.3390/app10031042, https://www.mdpi.com/2076-3417/10/3/1042
15. Rish, I., et al.: An empirical study of the naive bayes classifier. In: IJCAI 2001 Workshop on Empirical Methods in Artificial Intelligence, vol. 3, pp. 41–46 (2001)
16. Tarsitano, P.: Luiz beltrão, visionário sedutor. Anuário Unesco/Metodista de Comunicação Regional **14**(14), 17–29 (2010). https://doi.org/10.15603/2176-0934/aum.v14n14p17-29
17. Tsai, S.-C., Chen, C.-H., Shiao, Y.-T., Ciou, J.-S., Wu, T.-N.: Precision education with statistical learning and deep learning: a case study in Taiwan. Int. J. Educ. Technol. High. Educ. **17**(1), 1–13 (2020). https://doi.org/10.1186/s41239-020-00186-2
18. Wirth, R., Hipp, J.: Crisp-dm: towards a standard process model for data mining. In: Proceedings of the 4th International Conference on the Practical Applications of Knowledge Discovery and Data Mining, pp. 29–39. Springer, London (2000)

# Proposed Algorithm to Digitize the Root-Cause Analysis and 8Ds of the Automotive Industry

Brenda-Alejandra Mendez-Torres(✉) and Patricia Cano-Olivos(✉)

Universidad Autónoma del Estado de Puebla, 17 Sur 711, Barrio de Santiago, 72410 Puebla, Pueble, México
patricia.cano@upaep.mx

**Abstract.** As we adapt to our new post pandemic reality, many activities are being and have been replaced or modified. Companies need to adapt to digitalization as a new way of interacting with staff and customers. The challenge concerns the management of information since it must be aligned to the trends of Engineering 4.0 to solve everyday problems. This article analyzes the quality tool QRCI (Quick Response Continues Improvement), a solution-based method used by automotive company Faurecia for solving production problems related to 5M (Machine, Raw Material, Labor, Environment and Method) the investigation describes the relationship between QRCI and PPAP (Production Parts Approval Process) a standard with technical information to ensure the product quality. During our investigation we identified a disruptive communication between QRCI & PPAP and we propose to include both tools in the same platform and replace the troubleshooting in QRCI done by humans to artificial intelligence analysis, in order to reduce misperception, improve the analysis time and communicate the results to PPAP in real time, reducing operating costs and increasing productivity and quality since the improvement is continuous. Therefore, the fulfilment of the certifications is ensured through an integrated prevention model. A robust system inspires confidence in suppliers and customers based on the effectiveness and efficiency of the production processes.

**Keywords:** APQP · PPAP · FMEA · Control plan · Quality · Engineering 4.0 · Artificial intelligence random forest · Decision tree · Machine learning

## 1 Introduction

The term automotive is derived from the Greek word auto (by itself) and was proposed by the Society of Automotive Engineers (SAE) [1]. The automotive sector is made up of a group of organizations related to design, development, manufacturing, purchasing, marketing and sales. Each of these areas collaborates significantly in the further development and production process. OEM (Original Equipment Manufacturer) automakers select a template of OES (Original Equipment Supplier) suppliers responsible for the supply of the different parts of the car. In a multilevel hierarchy, the highest ranked suppliers are called first-tier suppliers (FTSs), followed by 2nd and 3rd tier suppliers [2]. The suppliers must be recognized as reliable through the IATF 16949 (International Automotive Task


H. Florez and M. F. Pollo-Cattaneo (Eds.): ICAI 2021, CCIS 1455, pp. 237–249, 2021.
https://doi.org/10.1007/978-3-030-89654-6_17

Force) Certification to prove that companies use and master the essential IATF tools, especially the APQP (Advanced product quality planning) and the PPAP (Production Part Approval Process) that guarantees compliance with the customer's quality.

This paper is organized as follows: the next section examines the related literature and develops the scope of this evaluation study based on four key points: the product quality and its relationship with APQP and PPAP, the QRCI as a production tool resolution method, the communication theories, the digital application development through full stack structure. We then describe our methodology, the problem statement and discuss our change proposal: the Intelligent QRCI with the application of algorithms. The final section concludes the paper.

## 2 Literature Review

The quality and reliability of a vehicle are essential to the end user and therefore to the brand, if a user drives a car with quality problems, whether real or perceived, there is less of a chance to buy products from the same company again [3]. If customers experience four or more issues with a vehicle, they are at risk of becoming detractors and rebranding when they decide to replace their current vehicle [4]. Currently, the quality standards are higher; cars must be safe, reliable, intelligent, easy to maintain, and have a minimum impact on the environment in their manufacture, use and disposal.

Quality problems impact the behavior of the production, sale, import and export of light vehicles. Figure 1 demonstrates that in recent years the Mexican market shows a positive trend in the production of Asian brands because customers perceive these brands as better quality products, while American brands have a decreased, which contrasts with European brands that maintain a constant trend [5].

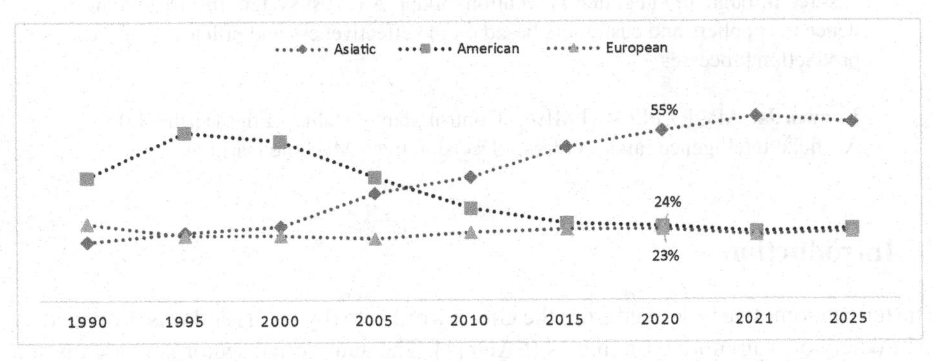


**Fig. 1.** Mexican historic sales automotive market trend by origin.

Product quality is crucial because it guarantees customer satisfaction and loyalty. Therefore, OEM car manufacturers audit suppliers' products and processes through the standard IATF 16949:2016. This standard looks at meeting the needs of customers with quality attributes or non-functional requirements. For products, attributes are technical characteristics, while for services, quality has a human dimension [6]. The certification of the IATF 16949:2016 standard consists of recognizing suppliers as reliable to have a quality management system and continuous improvement to prevent errors and reduce loss in the production phase.

The APQP is a structured method for defining and establishing the steps needed to ensure that a product satisfies the customer [7]. This methodology was agreed upon at three U.S. companies such as Chrysler, General Motors and Ford. The automotive industry's APQP methodology (QS 9000 standard) requirements are widely known and compatible with VDA 6.1, EAQF, AVSQ and ISO 16949 standards [7].

The APQP (Fig. 2) defines the required inputs and outputs of each stage of product development in five phases: 1) Planning: the business plan is carried out, and the objectives of both design, quality and reliability are established; 2) Product design and development: a prototype is manufactured, and the engineering requirements, materials, tools/machines necessary for manufacturing are established through a DFMEA (Failure Mode and Effects Analysis of the design); 3) Design and development of the process: a PFMEA (Failure Mode and Effects Analysis of the process) is carried out for everything related to the plant layout, packaging, quality review of the processes and products; 4) Product and process validation: final tests are performed to validate products and processes, including packaging and measurement systems; 5) Feedback, evaluation and corrective actions: The variation and analysis of delivery times for customer satisfaction are minimized [8].

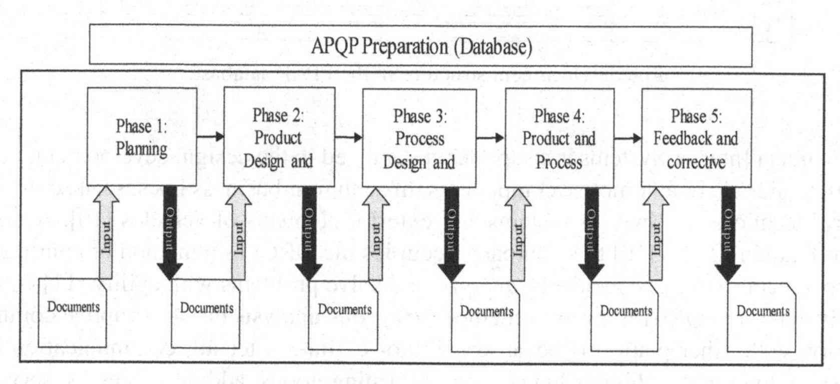


**Fig. 2.** Advanced product quality planning (APQP)

The PPAP (Production Part Approval Process) is a manual responsible for approving the minimum quality requirements through a trial run of the parts produced. The test has at least 300 pieces with the same resources at the same time (machines, equipment and personnel) [8]. This procedure guarantees customers that the final product meets or exceeds the necessary quality requirements and can be reproduced. Figure 3 shows the PPAP approval system for two customers with two products, each with standard information. Global competition, forces the companies to develop products with better performance and quality at a lower cost [9]. Then the key factors of a company's success are reducing time in the product development process, increasing product quality, developing sustainable products, reducing the product's cost in the manufacturing process, and meeting customer requirements [6]. The product development involves a constant evolution of the designed product, as their components continuously change before and during production [9].Technological trends and consumerism have reduced the execution time of the three stages of Product Lifecycle Management (PLM): 1) The beginning of Life (BOL), with the evaluation, concept development, system design and product design takes less than a year; 2) The Middle of Life (MOL) as testing times in the BOL stage were reduced, various quality problems may appear at this stage; and 3) The End of Life (EOL), remanufacturing and recycling [6].

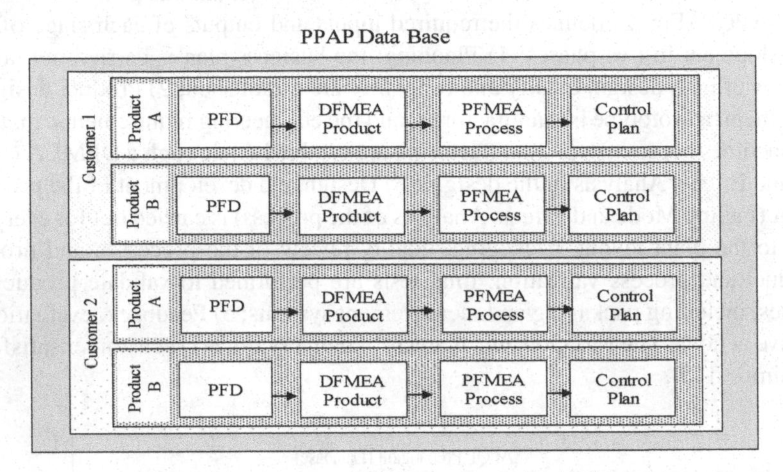


**Fig. 3.** Document structure in the PPAP database

Faurecia Interior Systems is a supplier specialized in the design, development, manufacture and sale of automotive components through four business lines: seats, emission control technologies, interior systems and exterior elements of vehicles [10]. According to Guillén (2012) [10] this company occupies the QRCI as a method of continuous improvement to manage incidents, analyze and solve problems with agility [11].

Even though QRCI is a robust methodology, this analysis has a disruptive communication with other platforms because it is not digital. After all, communication has evolved into complex digital channels, transforming goods, adding processes, services and generating great synergy between different distribution channels, thus obtaining fast responses and constant flows [12]. Value chains have formed within a strong current of globalization, composed of large companies closely related to each other globally and where relocation has played a crucial role in achieving flexibility [13]. A scenario has been reached in which most supply chains are networks [14].

The origins of massive communication date back to 1948 with a mathematical theory of communication [15] where the essential elements are exposed: 1) A source of information that produces a message; 2) A transmitter that operates on the message and sends over a channel; 3) A channel that is the means to carry information; 4) A receiver that transforms the signal; and 5) A destination, which can be a person or a machine. Shannon, C. E (2001) [16] proposes to measure the amount of information in a bit (binary digit) because arbitrarily, the value of the data is one, that is, if the selection of the message has only two different alternatives, true or false, open or closed, black or white, etc.

Information theory is a branch of probability that studies information and its relationship to channels, data compression, and cryptography [16]. The information theory in Internet technology refers to a diversity of independent networks that communicate and connect through a standard TCP/IP protocol (Transmission Control Protocol/Internet Protocol). If networks are not interconnected, two users on different networks cannot exchange information [17]. The TCP is responsible for controlling the flow and congestion of the data transmission rate of both the receiver and the intermediate links in the network path and implements a byte flow over the IP provided data program service, so the TCP is the transport protocol in the TCP/IP protocol set [18].

The main objective of the TCP/IP design is to build a network interconnect called internetwork or internet, which provides universal communication services over heterogeneous physical networks or IP. This is an address or number that logically identifies a network interface on a device that uses the IP protocol [19]. Routers pass IP data programs between different media according to routing information by creating a mesh from an IP Internet, which allows all hosts to communicate with each other over IP. [18]. Each physical network has its technology dependent communication interface in a programming interface that provides a primary (primitive) communication function. To create user facing websites, developers use three main categories: front-end development, back-end development and full-stack.

The development front-end is the interface of a website (or Web or mobile application) built with languages such as HTML, CSS, and JavaScript. It is also the part that the user sees and with which they interact directly [20]. HTML (Hyper Text Markup Language) is the backbone of the Web and is responsible for the entire structure and content [21]. CSS (Cascading Style Sheets) is the language that designates the style or appearance of an HTML on the page [22]. JavaScript allows you to add interactivity, more complex animations and even makes it possible to create fully-featured web applications. However, the effective use of JavaScript requires adjustments [23] and can be used to develop real applications and even system software.

The back-end creates the internal parts of websites that users do not interact with directly. Programming languages include PHP, Rubi, and Piton. The back-end and front-end must be developed together for the code to work within the site's design or application. It is convenient for web developers to use the same language for both front-end and back-end [24].

A full-stack application is a web developer that works with both the front and back of a website or application [20]. Full-stack applications must be written entirely in JavaScript from the client side to the server side. Therefore, you can schedule projects involving data bases, create user oriented websites, or even work with customers during the Project Planning phase.

In 1999 Dorian Pyle published a book called "Data Preparation for Data Mining" in which he proposes a way to use Information Theory to analyze data [25]. Today we must be prepared quality data for pattern recognition, information retrieval, automated learning, data mining, and web intelligence. In practice, data cleansing and preparation requires approximately 80% of the total data engineering effort. Input to data mining algorithms is assumed to be well distributed, i.e., without missing or incorrect values where all characteristics are essential. Which leads to (1) disguising valuable patterns that are hidden in the data, (2) deficient performance, and (3) poor quality outputs [26].

Knowledge Discovery in Data mining is the analysis stage of Knowledge Discovery in Databases (KDD) that attempts to discover patterns in large volumes of datasets. The term Knowledge Discovery understood in the sense of artificial intelligence, refers to the generally true beliefs, ideas, rules and procedures in a particular domain, that is, the interpretation given to the information existing within a specific field [27]. Artificial intelligence (AI), machine learning, statistics, and database systems are used to process data. The overall goal of the data mining process is to extract information from a dataset and transform it into an understandable structure for later use. AI is applied when a machine mimics human cognitive functions, such as perceiving, reasoning, learning and solving problems [28].

Alfred Krupp founded ThyssenKrupp in 1811 and, with more than 200 years of experience in distributing materials and services, released in 2018 the Alfred solution, a Microsoft Azure platform with different applications [19]. AI application helps the company analyze and process more than two million orders a year from 250,000 global customers. The algorithm allows the company to optimize its logistics network, assigning materials to the correct location much faster, minimizing the volume of transportation, and improving its transport capacity.

Digital transformation has been the biggest challenge due to the availability of quality data. Data extraction, processing, and application of artificial intelligence are vital for the automotive sector, both for capturing data generated for the product, process and business integration and the situations and problems in the real-world during production.

## 3 Methodology

This article presents a qualitative analysis of the QRCI and its communication with PPAP. The research method here is exploratory, using direct observation and analysis of primary sources which include data collected from articles published and interviews with two employees from Faurecia company. The investigation is proposed in three stages. 1. Describe how QRCI works and its relationship with PPAP, 2. Identify the main problems that shape the disruptive communication between both standards, 3. Improvement plan, incorporating the architecture of a web site that unifies QRCI and PPAP functions and machine learning for the root cause solution. Stages 1 and 2 are described in the problem statement section, while the stage 3 is addressed in the section: Smart QRCI design. Finally, we respond to the conclusions. The implementation of QRCI with the algorithms and the verification of the results, are not part of the scope of this investigation.

# 4 Problem Statement

## 4.1 Stage 1.

When Faurecia receives a customer claim, the quality team open a QRCI failure analysis in the area, with the process, the defective parts and stakeholders. The QRCI includes methodologies as Deming Wheel or PDCA (Plan, Do, Check, Act/Adjust) and 8D which includes questions as: D1: Why wasn't the problem detected?, D2: Is there a risk of similar problems occurring?, D3: How can the company be protected?, D4: Why was it sent?, D5: What is the root cause of the problem?, D6: What is the solution to the problem?, D7: Tracking the effectiveness of the solution. Is it effective?, D8: What have we learned?. The QRCI also includes the use of the Pareto form and Ishikawa methodology (fish bone diagram). The engineers comparing the current product and process against the PPAP standard and through logical thinking assessment and manual comparison, the team searches the possible cause and test it to confirm or reject it. Once the point of failure or root cause is confirmed the engineers implement strategies to contain or solve the problem. If there is new information that impacts the process or product, this assessment is added to the PPAP in the control plan section.

## 4.2 Stage 2:

The failure analysis QRCI is obtained manually on paper, then QRCI has no application record or web page; therefore, it is detached from the PPAP digital standard. PPAP platforms exist in the network to support automotive vendors to manage product and process standards, however, is just a static database.

The QRCI methodology identifies process failures by comparing problems physically with digital data (PPAP); this causes a communication disruption because the analysis is currently manual. Therefore, the solution is based on experience-intuition and information-reasoning and not on the knowledge-reasoning binomial. Table 1 shows the comparison of the types of information in the QRCI analysis.

**Table 1.** Key points of QRCI methodology and information

| Key points | Description | Information type |
|---|---|---|
| 1. Real workplace | 1. Where and when happens? | Physical |
| 2. Real parts | 2. Cause or situation OK = good vs NOK = bad | Physical |
| 3. Real data | 3. Data compared to standards | Digital |
| 4. Quick response | 4. Respond protect the user first | Physical |
| 5. Logical thinking | 5.What is/are the root cause(s)? | Physical |

Thus, the physically detected problems can be erroneous because the interpretation is from the subjective criteria of the persons responsible for this analysis. Therefore, multidisciplinary participation is required (at least one representative from each area must participate) to consider factors that directly or indirectly affect compliance with customer specifications. The time available to detect quality problems and update information in PPAP standard is minimal. This situation may cause process audits to fail.

Updating the QRCI information outputs is directly related to the information contained in the PPAP, specifically in the FMEA and control plan. Therefore, this research aims to design an electronic algorithm that allows the digitization of the QRCI methodology; it collects 8D and 5M information to identify the cause of the problem through the information contained in the PPAP product and process standard (FMEA and Control Plan).

## 5 Smart QRCI Design

The design of the electronic algorithm to digitize the QRCI uses clustered information, the Random Forest algorithm, and the logical structure of the decision tree through Boolean algebra or logical-numeric relationship [29]. compares information. Thus, the electronic QRCI will compare the current product attributes against the PPAP standard attributes, specifically the Failure Mode and Effects Analysis (FMEA) and Control Plan. The purpose is to solve a wide range of multicriteria problems with qualitative data to be analyzed more easily and independently.

A decision tree evaluates through peer comparisons all input variables hierarchically to select the best one. The most significant variable (root node) is the basis of the algorithm that creates a condition with two responses, one positive and one negative (meets or does not meet the condition). Therefore, the logical-mathematical analysis of the decision tree allows the random forest to identify the most repetitive cause of the problem. So, all problems have the same probability of study because there is no loss of information, therefore, increases the success of identifying the problem and its cause [29]. The result of each comparison is defined as:

An act of choice is independent when class x is chosen from a set of objects and divided into two parts, and then the results are connected in a set conception.

This event is expressed mathematically with Eq. 1.

$$x(u + v) = xu + xv \tag{1}$$

Where $(u + v)$ represents the undivided subject and $(u\ y\ v)$ its constituent parts. The order of successive acts of election is indistinct. For example, if from an animal set, the sheep are chosen and after those with horns; or, vice versa, the result is the same, we reach the sheep with horns. Equation 2 represents the above.

$$xy = yx \tag{2}$$

An act of election realized twice (Eq. 3 and 4) or n times in succession (Eq. 5) gives the same result. If the x objects are chosen from a group, you get a set whose members are all x.

$$xx = 0 \tag{3}$$

$$x^2 = x \tag{4}$$

$$x^n = x \tag{5}$$

The law of the index is characteristic of the symbols because it allows reducing the results to forms suitable for their interpretation. A legitimate operation cannot deny the expression of truth, but it can be limited. That is:

$$y = z \tag{6}$$

It implies that classes $y$ and $z$ are member to member equivalents. Multiply by an x-factor, and you have to:

$$xy = yz \tag{7}$$

The above equation expresses that individual to classes $x$ and $y$ are also common to $x$ and $z$, and vice versa, which is a legitimate inference.

Our proposal is to compare the attributes of the products with the PPAP standards through decision trees. Each tree has a classification, so the random forest algorithm will select the tree with the highest number of matches to find the causes of the most incident problems. The greater the number of trees, the more accurate the random forest results. The proposed algorithm will use two independent "x" variables, which help predict a value, and "y", the dependent or response variable.

## 5.1 Stage 3:

The QRCI, once connected with the PPAP and the web, will be able to send alerts of the quality problems identified to the multidisciplinary team. The objective of digitizing the QRCI is to ensure the massive analysis of all the quality characteristics of the products; that is, when the information is compressed and can be restored to its original form upon reaching their PPAP destination (AMEF and Control Plan). Look at Fig. 4 with the new QRCI system and the integration of the random tree algorithm to search for the root cause of a problem.

Figure 5 shows the classification of the information under different conditions and sub-conditions. At the end of the evaluation, the attribute repeated the most times is selected. Machine learning can analyze the problems stored in the memory. When generating the PPAP (AMEF and Control Plan) of a new product with similar characteristics, the system can send alerts of possible problems to the user.

With this proposal, the organization would have information on quality problems in real-time; therefore, making decisions would be faster and processes could be released as problems are resolved. This helps avoid bottlenecks, reduces costs, increases equipment performance, improves service levels, etc.

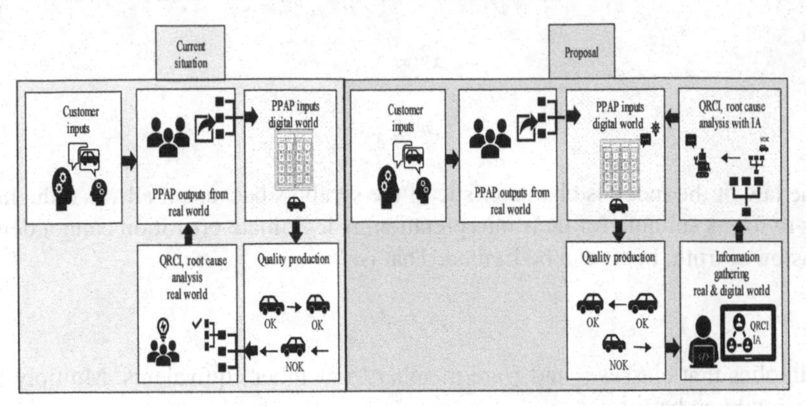


**Fig. 4.** Comparison of a manual QRCI resolution versus electronic QRCI outputs integrated to the electronic PPAP standard.

Failure to meet customer specifications increases returns and decreases sales. Thus, efficient and timely quality management reduces loss of material, reprocessing, downtime, process variation and warranty claims. Traceability of problems can generate reports of damaged production batches; these reports will allow the administrator to analyze the cause – root of non-conforming products and implement solutions in the shortest possible time. The digital analysis creates robust databases, which will prevent the same problems from being reproduced in the design and production of new products. Therefore, traceability and tracking minimize withdrawal costs of incoming material batch codes throughout production operations. The automotive sector is considered one of the most critical strategic industries globally by contributing to each region's economic growth.

However, the sector's importance lies not only in the economy but also in the complexity of its operations. So, the automotive industry must migrate to an industry 4.0 that enables real-time communication through digital channels to collect and analyze data for decision making. Suppose the exchange and analysis of information between machines and administrators are immediate. In that case, problem detection and corrective action implementation are rapid; therefore, Smart QRCI can be avoided defects and non-conforming products. The advantage of applying the random forest algorithm is it can handle a large amount of data (it analyzes thousands of input variables and locates the most significant ones). In addition, it has effective methods for accurately estimating missing data and analyzing nonlinear information. The solution is easy to understand, even for people with a non-analytical background.

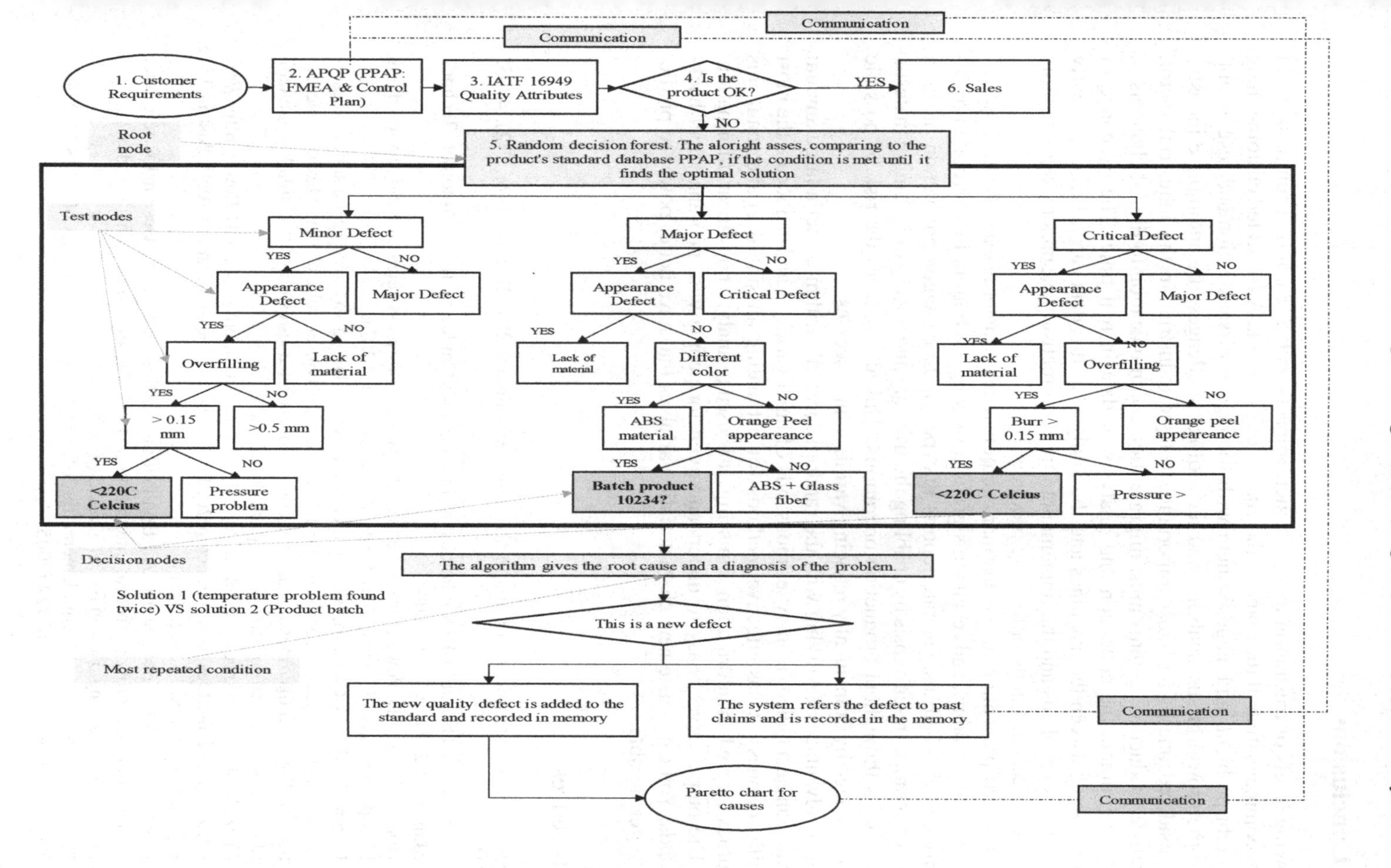


**Fig. 5.** Proposal of electronic architecture of random forest algorithm for QRCI problem solution

## 6 Conclusions

Growing levels of frustration in a customer, impacts on the return of a future sale and the recommendation of the brand in the customer's network. A frustrated customer may highlight problems and magnify them with feelings of dissatisfaction and resentment. With the growth of sustainability and technology challenges, the automotive industry is considered one of the most important strategic industries in the industrial sector, therefore production systems must migrate their communication to digital channels in order to promote the collection and analysis of data in real time. The exchange of information between the machines and the analysis of this data can facilitate the early detection of problems and the implementation of corrective measures by the operators to avoid defects and non-conforming products.

This early prevention and quality management reduces material waste, rework, downtime, warranty claims, and results in significant savings to bottom line. Problem traceability can also help us generate reports that pertain to damaged production batches, which minimizes re-call costs by enabling the tracking and tracing of incoming material batch codes throughout production operations. This also reduces the risk of possible brand damage by dramatically reducing variation in processes.

The advantage of working with the random forest algorithm is that a large amount of data can be handled (it analyzes thousands of input variables and can locate the most significant ones). It has effective methods for estimating missing data and maintains accuracy even when information is missing. It is used to analyze non-linear information.

The output is very easy to understand even for people with non-analytical backgrounds. You can create new characteristic variables that have better power to predict the target variable.

## References

1. Jedlicka, J.: The society of automotive engineers library. Sci. Technol. Libr. **7**(2), 47–52 (1987)
2. Volpato, G.: The OEM-FTS relationship in automotive industry. Int. J. Automot. Technol. Manage. **4**(2–3), 166–197 (2004)
3. Gómez, G.: In: J.D. Power, https://mexico.jdpower.com/es/press-releases/calidad-y-confiabilidad-del-vehiculo-en-mexico-2018-vds. Accessed 03 Jul 2021
4. Duran, F., Álvarez Gallardo, B., Effler, G., In: Power, J.D. (ed.) Mexico El consumo excesivo de combustible afecta negativamente la satisfacción con el vehículo en México, encuentra J.D. Power. https://mexico.jdpower.com/es/press-releases/estudio-de-calidad-y-confiabilidad-del-vehiculo-en-mexico-2020-apeal. Accessed 20 Aug 2021
5. Cruz, S.: In: J.D. Power, Resumen de la Industria Automotriz, por J D Power https://revistamotoryvolante.files.wordpress.com/2021/02/jd-power-mexico-resumen-de-industria-15feb21.pdf. Accessed 15 Aug 2021
6. Benabdellah, A.C., Benghabrit, A., Bouhaddou, I., Benghabrit, O.: Design for relevance concurrent engineering approach: integration of IATF 16949 requirements and design for X techniques. Res. Eng. Design **31**(3), 323–351 (2020)
7. Bobrek, M., Sokovic, M.: Implementation of APQP-concept in design of QMS. J. Mater. Process. Technol. **162**, 718–724 (2005)

8. Mittal, K., Kaushik, P., Khanduja, D.: Evidence of APQP in quality improvement: an SME case study. Int. J. Manage. Sci. Eng. Manage. **7**(1), 20–28 (2012)
9. Shankar, P., Morkos, B., Summers, J.D.: Reasons for change propagation: a case study in an automotive OEM. Res. Eng. Design **23**(4), 291–303 (2012)
10. Guillén Candelas, N.: Implementación de la estrategia en Faurecia Interior Systems (Doctoral dissertation, Universitat Politècnica de València) (2012)
11. Arrondo Durán, S.S.: Análisis e implementación 8 básicos de calidad en la producción de asientos VW270 y VW216, p. 109 (2020)
12. Garay-Rondero, C.L., Martinez-Flores, J.L., Smith, N.R., Caballero Morales, S.O., Aldret-te-Malacara, A.: Digital supply chain model in Industry 4.0. J. Manuf. Technol. Manage. **31**(5), 887–933 (2020)
13. Albors Garrigós, J., Collado Fuentes, A., Dols Ruiz, J.F.: Factores de éxito en la clusterización de la industria del automóvil en España. El rol de los agentes en el clúster. Economía industrial **403**, 125–403 (2017)
14. Chopra, S., Meindl, P., Kalra, D.V.: Supply Chain Management: Strategy, Planning, and Operation, 6th ed., Pearson/Prentice Hall, Upper Saddle River (2016)
15. Shannon, C.E.: A mathematical theory of communication. Bell Syst. Tech. J. **27**(3), 379–423 (1948)
16. Shannon, C.E.: A mathematical theory of communication. ACM SIGMOBILE Mob. Comput. Commun. Rev. **5**(1), 3–55 (2001)
17. Estrada Padilla, N.J., Lorío Rojas, A.M., Ramírez Santana, U.A.: Diseño de puntos de intercambio de internet en entornos virtuales con tecnología Cisco, implementando servicios multimedia (Doctoral dissertation) (2018)
18. Partridge, C., Shepard, T.J.: TCP/IP performance over satellite links. IEEE Netw. **11**(5), 44–49 (1997)
19. Burone Schaffner, N.M., Canessa Michelini, A., Dominguez Acerenza, G.E.: Alfred. (2019)
20. Liu, Z., Gupta, B.: Study of secured full-stack web development. In: CATA, pp. 317–324 (2019)
21. Northwood, C.: The Full Stack Developer: Your Essential Guide to the Everyday Skills Expected of a Modern Full Stack Web Developer. Apress, Berkeley (2018)
22. Alawar, M.W., Abu-Naser, S.S.: CSS-tutor: an intelligent tutoring system for CSS and HTML. Int. J. Acad. Res. Dev. **2**(1), 94–99 (2017)
23. Mikkonen, T., Taivalsaari, A.: Using JavaScript as a real programming language. Citeseer (2007)
24. Alimadadi, S., Mesbah, A., Pattabiraman, K.: Understanding asynchronous interactions in full-stack JavaScript. In: IEEE/ACM 38th International Conference on Software Engineering (ICSE), pp 1169–1180 (2016)
25. Pyle, D.: Data Preparationfor Data Mining, MorganKaufmann, uinlan, JR: Bagging, Boosting, and C4. 5. In: 13th International Conference on Artificial Intelligence, pp. 725–730 (1999)
26. Zhang, S., Zhang, C., Yang, Q.: Data preparation for data mining. Appl. Artif. Intell. **17**(5–6), 375–381 (2003)
27. Moreno-Jimenez, J.M., Mata, E.J.: Nuevos sistemas informáticos de ayuda a la decisión. Sistemas Decisionales Integrales. Actas de la V Reunión ASEPELT-España, Granada **2**, 529–538 (1992)
28. Haenlein, M., Kaplan, A.: A brief history of artificial intelligence: on the past, present, and future of artificial intelligence. Calif. Manage. Rev. **61**(4), 5–14 (2019)
29. Boole, G., Esplugues, J.S.: El análisis matemático de la lógica. Cátedra, Madrid (1984)

# Health Care Information Systems

# Blood Pressure Morphology as a Fingerprint of Cardiovascular Health: A Machine Learning Based Approach

Eugenia Ipar[1(✉)], Nicolás A. Aguirre[1,2], Leandro J. Cymberknop[1], and Ricardo L. Armentano[1]

[1] Bioengineering Research and Development Group (GIBIO), Universidad Tecnológica Nacional, Buenos Aires, Argentina
gibio@frba.utn.edu.ar

[2] Laboratoire de Modélisation et Sûreté des Systèmes, Université de Technologie de Troyes, Troyes, France
https://www.frba.utn.edu.ar/gibio

**Abstract.** Daily exercise and a healthful diet allows the cardiovascular (CV) system to age slower. However, unhealthy life habits and diseases impact negatively on this condition. In consequence, an individual's "chronological age" (CHA) may differ from the "arterial age" (AA). There is proven evidence of correlation between arterial stiffness (AS), highly associated with AA, and central blood pressure (BP) waveform, which combined with other CV properties, define a group of characteristic parameters to assess CV Health (CVH). Such assessment, obtained by means of noninvasive and simple arterial measurements is extremely useful for the prevention of CV diseases. Machine Learning (ML) techniques allow the development of predicting models, but usually a large amount of data is required, often unavailable and/or not fully complete. One-dimensional (1D) CV models become an interesting alternative to overcome this issue, since CV system simulations can be performed. **OBJECTIVE:** A ML model was trained/validated with data from healthy and heterogeneous population (n = 4374), using parameters derived from arterial pressure waveforms (APW) in order to assess CVH. **METHODS:** Sixteen features were extracted from carotid and radial APW. Such were used for prediction of cardiac output (CO), systemic vascular resistance (SVR), systolic BP (SBPc), diastolic BP (DBPc), carotid-femoral pulse wave velocity (PWVcf), aortic augmentation index (AIa) and CHA as a surrogate of AA. **RESULTS:** The normalized-RMSEs for CO, SVR, SBPc, DBPc, PWVcf, AIa and CHA using the best-performing model were 5.4%, 2.2%, 0.1%, 0.06%, 0.6%, 1.2% and 5.5%, respectively. **CONCLUSION:** The acquired results showed that the ML models obtained from measurements in the assessed locations provide an acceptable performance on estimating CVH parameters. Further in-vivo studies are required to validate the application of the outcome.

**Keywords:** One-dimensional modelling · Pulse wave analysis · Cardiovascular health · Machine learning · Risk prediction


H. Florez and M. F. Pollo-Cattaneo (Eds.): ICAI 2021, CCIS 1455, pp. 253–265, 2021.
https://doi.org/10.1007/978-3-030-89654-6_18

# 1 Introduction

Daily exercise and a healthful diet allows the cardiovascular system to age slower. However, unhealthy life habits and diseases impact negatively on this condition, which is reflected by an accelerated "stiffening" of main arteries, acting independently of normal ageing [1]. For that reason, an individuals "chronological age" (CHA) may differ from the "arterial age" (AA). In this sense, "Cardiovascular Health" (CVH) has been defined by the American Heart Association (AHA) as the presence of four health behaviors (non-smoking, regular physical activity at goal levels, body mass index (BMI) under 25 kg/m$^2$ and diet consistent with national recommendations) and three health factors (total cholesterol, blood pressure and fasting blood glucose) at ideal levels [2].

There is proven evidence of correlation between central blood pressure (BP) waveform parameters and arterial stiffness (AS), as they provide important indirect information regarding systemic arterial structure and function allowing for the assessment of wave reflection and amplification [3]. In this sense, it has been demonstrated that a better CVH, comprising both health factors and behaviors, was associated with lower AS [2]. As a result, the assessment of normal ranges of both central hemodynamics and bio-mechanical properties related to CVH such as pulse wave velocity (PWV, the gold standard for estimation of AS), cardiac output (CO), systolic and diastolic central BP (SBPc and DBPc, respectively) and systemic vascular resistance (SVR) becomes extremely useful for the prevention of CV diseases, especially in the young and middle age population. It has to be also noted that chronic disease, functional limitations, and mortality increase exponentially with CHA. Compared with CHA, the concept of AA is a proposed measure that would more accurately reflect structural and functional changes (i.e. vascular stiffness) taking place in the body as it ages [4]. It is possible to show an individual's AA in relation to CHA. If AA is higher than CHA, this indicates that a discrepancy is present and that signs of "early" aging are detected [5]. Furthermore, the use of noninvasive and simple measurements, particularly those obtained from the morphological analysis of BP waveforms belonging to different superficial arteries, constitutes a valuable tool in screening trials [6].

In that behalf, machine learning (ML) techniques involve the study of statistical models and algorithms that can progressively learn from data and achieve desired performance on a specific task. However, it can be challenging and expensive to obtain a large amount of labeled data from specific clinical protocols, annotated by trained clinicians [7]. Consequently, one-dimensional (1D) validated CV models become an interesting alternative to overcome this issue, since CV system simulations can be performed, both in health and disease. 1D computational modeling provides a complementary approach for research, as it allows arterial pressure waves to be simulated under different CV conditions. These waves are influenced by heart and vasculature, with AS and wave reflections acting on their morphology [8]. In this work, ML models were trained and validated through a database based on a 1D model simulations of a healthy and heterogeneous population. Characteristic parameters derived from arterial pressure waveforms (APW) were used for the estimation of central hemodynamic and bio-mechanical parameters (CO, PWVcf, SVR, SBPc, DBPc) in order to act as a "fingerprint" of CVH. Concomitantly, CHA was also estimated, since its potential use as a marker of AA in presence of unhealthy APW.

## 2 Methods

The procedure used in this study is simplified on Fig. 1. The dataset was obtained from the database published by Charlton *et al.* [8], which contains 4,374 healthy virtual subjects aged from 25 to 75 years. Different machine learning models were trained to estimate a series of outputs from a set of input features. Once the models were trained, such were validated using subjects excluded from the training phase. The performance of each model was evaluated in terms of the mean square error (MSE) or accuracy, as appropriate to the model task.

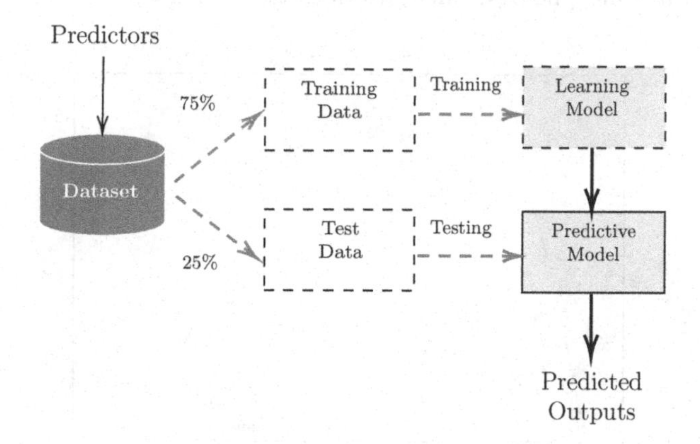


**Fig. 1.** Workflow chart. Each model tested followed the represented workflow.

### 2.1 Signal Analysis

Measurements of the Arterial Pressure Waveforms (APW) of all individuals selected for training were used as input information. During clinical protocols, APW can be easily obtained using the "Applanation Tonometry technique", applied to superficial arteries such as carotid, brachial, radial and femoral, among other arterial sites. The technique employs a high fidelity mechanographic transducer when the surface of the artery is flattened by the sensor, the circumferential stress applied on the vascular wall is balanced and the inner pressure can be registered. The systolic and diastolic values of the pressure wave need to be calibrated using sphygmomanometric values, which are taken previously [9]. The analysis of the APW can provide information about the individual's CV system, and therefore could be used as input for the CVH assessment. Thirty-two features were extracted from each subject's APW, sixteen corresponding to carotid APW and the remaining corresponding to radial APW. A simple visualization of the APW with few features is presented in Fig. 2.

**Arterial Pressure Waveform Analysis.** This section describes the procedure of retrieving features from APW, as follows:

*Amplitude and Time Features.* Nine different amplitude features were retrieved, most presented in Fig. 2. This includes: height at the time of P1 ($P1pk$), height at the time of P1 inflection point ($P1in$), height at the time of P2 ($P2pk$), height at the time of P2 inflection point ($P2in$), height at the systolic peak ($Ps$) and height at the diastolic peak ($Pd$). The above-mentioned include each time interval: the time of P1in ($P1in_t$), the time of P2in ($P2in_t$) and the time of the systolic peak ($Ps_t$).

*Complementary Features.* Seven complementary features were obtained from the APW. These attributes could help to discriminate hidden patterns present in the pulse wave morphology. Such are detailed in Table 1.

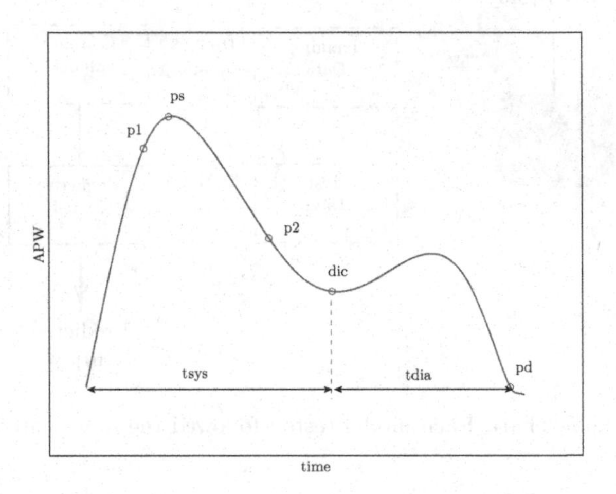


**Fig. 2.** Features from the pressure pulse waveforms. *p1* BP at the time of P1; *ps* systolic BP; *p2* BP at the time of P2; *dic* dicrotic incisura; *pd* diastolic BP; *tsys* systolic time; *tdia* diastolic time.

**Table 1.** Complementary features obtained, and their respective equations. *DT* Duration of the diastolic phase; *HP* Duration of the cardiac cycle; *SP* Systolic phase; *CC* Cardiac Cycle.

| Feature | Formula | Description |
|---|---|---|
| AIa | $(P_2 - P_1)/PP$ | Augmentation Index |
| AP [mmHg] | $P_2 - P_1$ | Augmented Pressure |
| SEVR | $Diastolic_{Area}/Systolic_{Area}$ | Subendocardial viability ratio |
| Area [mmHg · secs] | $\int APW \cdot dt$ | Area under APW curve |
| LVET [secs] | $t_{sys_{end}} - t_{sys_{begin}}$ | Duration of the SP of CC |
| DTF | $DT/HP$ | DT as fraction of the HP |

## 2.2 Models

Individual's CVH was defined by the assessment of PWVcf, CO, SBPc, DBPc, AIa and SVR, since such features have proven to be representative in terms of diagnosis and monitoring of CV diseases. Regression models were used to make estimations of such outputs. Additionally, CHA was also estimated, as a potential surrogate of AA, since a database of healthy population is being used. Since CHA is a discrete value, a classification model was used for this estimation.

Once the features from the APW were obtained, the dataset was subjected through several ML algorithms, since ML techniques have been widely used in many medical applications for clinical diagnosis and insight. Seventy-five percent of the initial dataset was used to train all ML models, while the remaining twenty-five percent was used to test and verify accuracy. Regression analysis was used to estimate CO, SVR, SBPc, DBPc, PWVcf and AIa. As of CHA, a Classification analysis was preferred, since its values were discrete. The implemented models were:

- Neural Networks (NN) [10]: Non-linear model, composed of nodes layers. The NN used a "ReLU" activation function and a 0.01 learning rate. The layer size was of a 100, with a fully-connected architecture. The output layer activation was Softmax for classification, and Linear for regression. Glorot uniform weight initialization was implemented for better performance.
- Gaussian Process (GP): Non-parametric Bayesian model, with the advantage of providing uncertainty measurements on the predictions. Three different kernel functions were tested, choosing the Matern kernel with $\nu = 5/2$. The maximized marginal log likelihood obtained from optimization depends on the output, but its value is usually larger than 900. The optimizer used for parameter estimation was *Quasi-Newton*, which is an approximation to the Hessian.
- Ensemble Methods (EM): Combines several models in order to obtain the highest predictive performance possible. Two categories of ensemble methods are usually distinguished: averaging methods (as Bagging Methods) and boosting methods (AdaBoost, Gradient Tree Boosting). Bagged method was the chosen one to compare for both regression and classification analysis.
- Support Vector Machine (SVM) [11]: Linear model, used both for classification and regression models. Multiple kernels were assessed for both types of estimations, choosing a cubic polynomial kernel. Since a multiclass classification was implemented, a "One vs One" reduction was preferred, and Error-correcting output coding (ECOC) algorithm was applied as to improve accuracy.
- K-Nearest Neighbors (KNN) [12]: Non-parametric model, used only for classification. In a simple way, the decision is made by assessing the near known-data. A Fine KNN was used for classification of CHA.

As to randomly divide the dataset into training and testing set, and prevent over or underfitting, cross-validation with $K = 5$ folds was used. Accompanying cross-validation, grid search hyperparameter optimization was implemented for each learning algorithm. All ML algorithms were performed using the Statistics and Machine Learning Toolbox from Matlab (Mathworks Inc, Massachusetts, USA) [13].

**Model Evaluation.** The correlation coefficient ($r$), root mean squared error (RMSE), normalized RMSE (nRMSE) and mean absolute percentage error (MAPE) were calculated on the predicted and tested values for regression models, whereas for the classification models accuracy was calculated. In both cases, the corresponding Bland-Altman [14,15] analysis was also done.

## 3 Results

The mean ($\mu$) and standard deviation ($\sigma$) of the features extracted from the APW, on each selected location (carotid and radial artery), are represented on Table 2. A comparison of the tested regression models and their respective root mean squared error (RMSE) for each output is shown in Table 3. The lowest RMSE was observed for Gaussian Process model, followed by Neural Networks model. Table 4 compares the classification models tested for CHA estimation,

**Table 2.** Features extracted from the APW. Medium and standard deviation is shown for each feature on both locations. *APW* Arterial Pressure Waveform; $\mu$ medium; $\sigma$ standard deviation

| Predictor | | $\mu$ ($\sigma$) | |
|---|---|---|---|
| Description | Feature | Carotid | Radial |
| BP at the time of P1 | P1pk [mmHg] | 109,59 (10,91) | 127,04 (10,82) |
| BP at the P1 inflection point | P1in [mmHg] | 107,93 (10,31) | 119,80 (11,12) |
| Time of P1in | $P1in_t$ [secs] | 0,13 (0,02) | 0,14 (0,02) |
| BP at the time of P2 | P2pk [mmHg] | 109,46 (11,44) | 101,14 (12,33) |
| BP at the P2 inflection point | P2in [mmHg] | 107,41 (10,33) | 100,86 (12,07) |
| Time of P2in | $P2in_t$ [secs] | 0,27 (0,02) | 0,27 (0,02) |
| Systolic blood pressure | Ps [mmHg] | 112,2 (11,78) | 127,04 (10,83) |
| Time of Ps | $Ps_t$ [secs] | 0,17 (0,05) | 0,11 (0,01) |
| Diastolic blood pressure | Pd [mmHg] | 74,26 (7,29) | 70,76 (7,14) |
| Augmentation index | AIa | 3,94 (12,41) | -34,31 (13,9) |
| Augmented pressure | AP | 1,97 (6,01) | -18,67 (7,24) |
| Time-derivative of the APW | dP/dt [mmHg/s] | 588,62 (6,01) | 588,62 (6,01) |
| Subendocardial viability ratio | SEVR | 2,51 (1,03) | 1,48 (0,66) |
| Area under APW | Area [mmHgs] | 75,18 (9,99) | 74,09 (9,95) |
| Left ventricular ejection time | LVET [secs] | 0,23 (0,07) | 0,29 (0,05) |
| Diastolic time fraction | DTF | 0,71 (0,09) | 0,63 (0,07) |

including both locations. In contrary to the regression analysis, Neural Networks showed the highest accuracy in prediction, followed by SVM.

The comparison of the statistics and errors of both subset inputs (features from carotid and radial arteries) for each of the targeted outputs is shown in Table 5, where best-performing models (Gaussian Process and Neural Networks) are listed. The table reveals that, for regression models, Gaussian Process performs better than Neural Networks, since it led to a smaller RMSE and MAPE. Whilst for classification models, as seen in Table 4 SVM presented lower accuracy percentage than Neural Networks, leaving the last one as the preferred one. Furthermore, when comparing performance between Carotid and Radial input, Carotid results turned out slightly better than Radial.

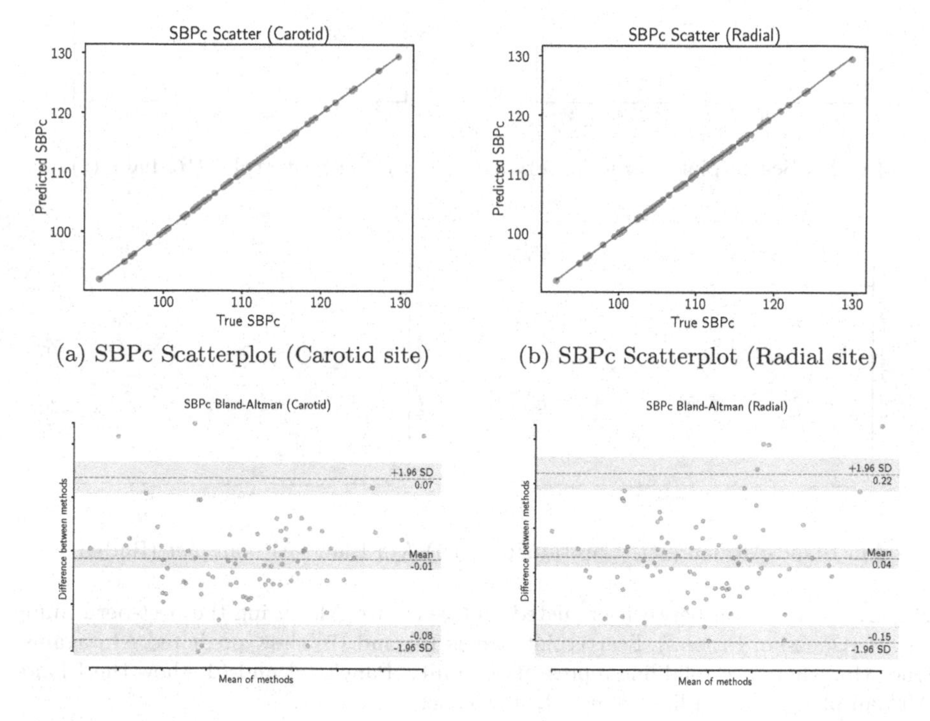


(a) SBPc Scatterplot (Carotid site) (b) SBPc Scatterplot (Radial site)

(c) SBPc Bland-Altman plot (Carotid site) (d) SBPc Bland-Altman plot (Radial site)

**Fig. 3.** Comparison between predicted and reference SBPc, using the best-performing model (Gaussian Process). Scatterplot panels (a) and (b) show predicted SBPc against true SBPc, where the solid line represents equality. Panels (c) and (d) show the Bland-Altman plots, where a limit of $\sigma = 1,96$ was set.

The scatter plots and Bland-Altman graphs for the Gaussian Process model are provided in Fig. 3 for SBPc, Fig. 4 for AIa and Fig. 5 for CO. Similar results were obtained for DBPc, PWVcf, SVR and CHA. In most cases, the mean difference presented a negative value, revealing a trend of predicting under the output's true value.

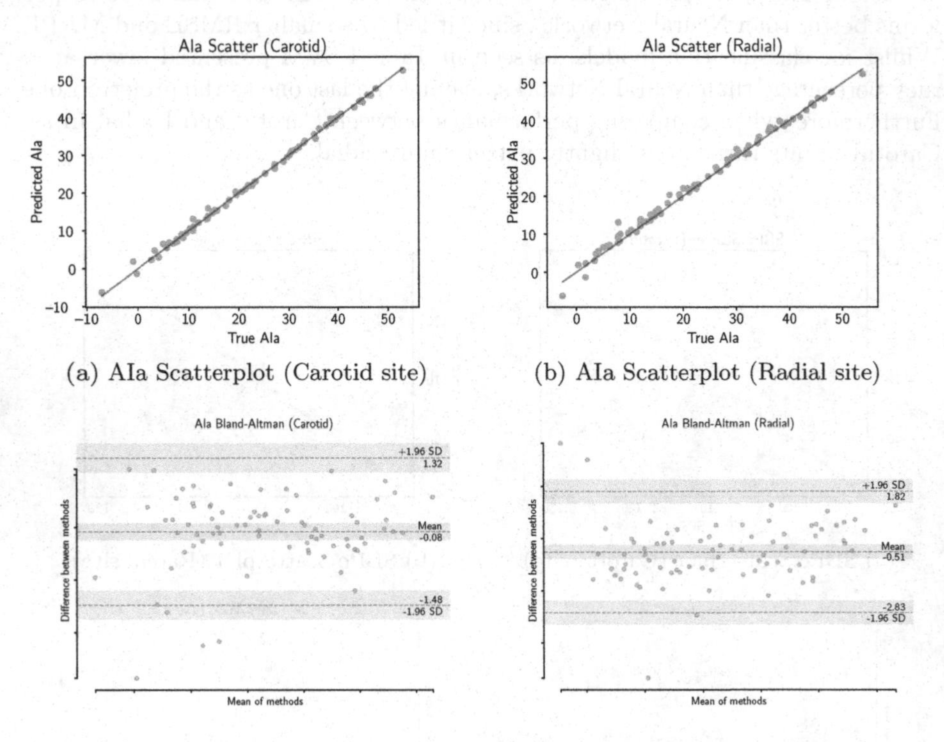


(a) AIa Scatterplot (Carotid site) (b) AIa Scatterplot (Radial site)

(c) AIa Bland-Altman plot (Carotid site) (d) AIa Bland-Altman plot (Radial site)

**Fig. 4.** Comparison between predicted and reference AIa, using the best-performing model (Gaussian Process). Scatterplot panels (a) and (b) show predicted AIa against true AIa, where the solid line represents equality. Panels (c) and (d) show the Bland-Altman plots, where a limit of $\sigma = 1,96$ was set.

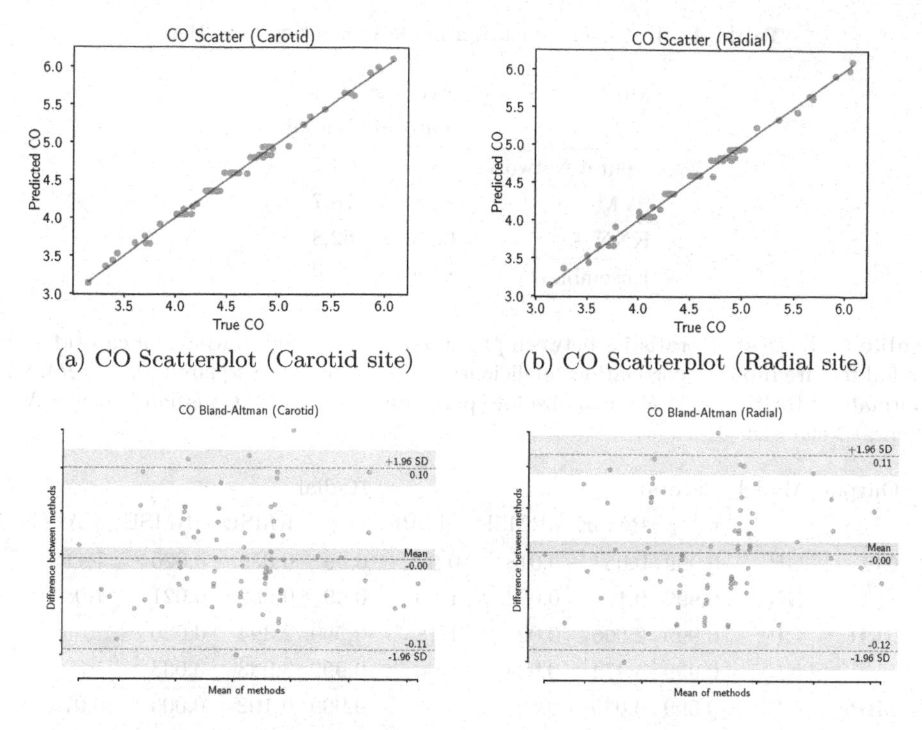


(a) CO Scatterplot (Carotid site) (b) CO Scatterplot (Radial site)

(c) CO Bland-Altman plot (Carotid site) (d) CO Bland-Altman plot (Radial site)

**Fig. 5.** Comparison between predicted and reference CO, using the best-performing model (Gaussian Process). Scatterplot panels (a) and (b) show predicted CO against true CO, where the solid line represents equality. Panels (c) and (d) show the Bland-Altman plots, where a limit of $\sigma = 1,96$ was set.

**Table 3.** Tested regression models RMSE value for each possible output, using APW from both Carotid and Radial Artery. *RMSE* root mean squared error

| Models | Carotid | | | | | | Radial | | | | | |
|---|---|---|---|---|---|---|---|---|---|---|---|---|
| | CO | SVR | SBPc | DBPc | PWVcf | AIa | CO | SVR | SBPc | DBPc | PWVcf | AIa |
| Gaussian | 0,118 | 4,405 | 0,087 | 0,067 | 0,127 | 1,815 | 0,093 | 6,675 | 0,459 | 0,541 | 0,872 | 1,764 |
| Neural Networks | 0,144 | 6,781 | 0,122 | 0,116 | 0,157 | 1,883 | 0,096 | 6,062 | 0,588 | 0,455 | 0,190 | 2,955 |
| SVM | 0,161 | 7,222 | 0,632 | 0,383 | 0,177 | 2,120 | 0,139 | 5,929 | 0,674 | 0,499 | 0,204 | 3,585 |
| Ensemble | 0,326 | 14,507 | 0,518 | 0,479 | 0,379 | 2,681 | 0,326 | 14,347 | 1,323 | 0,815 | 0,410 | 5,016 |

**Table 4.** Tested classification models accuracy for CHA.

| Models | Accuracy [%] | |
|---|---|---|
| | Carotid | Radial |
| Neural Networks | 93,7 | 94,7 |
| SVM | 83,8 | 78,7 |
| KNN | 68,3 | 62,8 |
| Ensemble | 81,4 | 77,6 |

**Table 5.** Regression statistics between predicted and reference output, for carotid and radial feature input. *r* correlation coefficient; *RMSE* root mean squared error; *nRMSE* normalized RMSE; *MAPE* mean absolute percentage error; *GP* Gaussian Process; *NN* Neural Networks.

| Output | Model | Carotid | | | | Radial | | | |
|---|---|---|---|---|---|---|---|---|---|
| | | r | RMSE | nRMSE | MAPE | r | RMSE | nRMSE | MAPE |
| CO | GP | 0,999 | 0,054 | 0,018 | 0,99 | 0,99 | 0,059 | 0,020 | 1,11 |
| | NN | 0,999 | 0,101 | 0,034 | 1,85 | 0,99 | 0,062 | 0,021 | 1,08 |
| SVR | GP | 0,999 | 2,666 | 0,022 | 1,18 | 0,999 | 2,394 | 0,020 | 1,06 |
| | NN | 0,999 | 4,474 | 0,038 | 2,06 | 0,999 | 3,933 | 0,033 | 1,69 |
| SBPc | GP | 0,999 | 0,039 | 0,001 | 0,03 | 0,999 | 0,102 | 0,003 | 0,07 |
| | NN | 0,999 | 0,079 | 0,002 | 0,05 | 0,999 | 0,300 | 0,008 | 0,20 |
| DBPc | GP | 0,999 | 0,014 | 0,0006 | 0,014 | 0,999 | 0,070 | 0,003 | 0,05 |
| | NN | 0,999 | 0,039 | 0,0016 | 0,04 | 0,999 | 0,114 | 0,005 | 0,11 |
| PWVcf | GP | 0,999 | 0,054 | 0,006 | 0,35 | 0,999 | 0,072 | 0,008 | 0,43 |
| | NN | 0,998 | 0,105 | 0,012 | 0,86 | 0,999 | 0,106 | 0,013 | 0,83 |
| AIa | GP | 0,993 | 0,717 | 0,012 | 4,12 | 0,997 | 1,290 | 0,022 | 3,16 |
| | NN | 0,999 | 0,838 | 0,014 | 2,83 | 0,993 | 2,194 | 0,037 | 9,57 |

## 4 Discussion

In the present study, accurate estimations of parameters related to CVH were obtained, based on APW (from different arterial sites) information and ML techniques. Our data shows that the APW, from both locations, provides valuable information for such prediction.

The obtained results suggest that Gaussian Regression Process was the best-performing model for all continuous domain outputs, as it presented the lowest mean absolute percentage error (MAPE). Such model was followed by Neural Networks, accomplishing as well high accuracy in prediction. Said model outperformed other classification models, being the best one to predict CHA. The validation of the ML models was carried out based on a percentage of the entire database, constituted by a large amount of records, since it was generated from a 1D cardiovascular model simulation.

Concerning the comparison between different arterial sites, both locations presented satisfactory results, being carotid APW better in terms of lower RMSE and higher accuracy. Nevertheless, radial APW outcomes suggest that accurate predictions are possible to be obtained from that vascular site, since radial measurements can be performed in a more secure and repeatable way [16]. In addition, Radial APW measurement methods represents a simpler, more comfortable and readily-available measurement, in comparison to traditional carotid tonometry. This is an important issue, in terms of feasibility, to be included in consumer-available devices, such as smart watches and fitness wristbands. In consequence, daily life devices have the potential to give an insight in the individual's CVH.

This study is in agreement with previous results [1,17–20]. Although an $RMSE = 0,1$ was obtained in [Jin 2021] for the estimation of PWVcf with Neural Networks using radial APW as an input, a lower RMSE was obtained from the implemented algorithms of this work. This difference may be due, not only to using distinct predictors, but also to the fact that [Jin 2021] used in-vivo measurements, in contrast to our study that used a 1D model. Nevertheless, said previous studies reported to have use either Gaussian Process or Neural Networks models, corroborating our results. In addition, [Tavelli 2018] proposed the use of neural networks through bootstrap averaging to estimate PWVcf, resulting in a $RMSE = 1,12$. This result differ both from [Jin 2021] and the present study, basically due to the type of dataset considered, as it mainly focus on the Intrinsic Frequencies (IF) concept obtained from the carotid APW and other features constructed from IF methodology. In [Bikia 2020] a nRMSE of 7.6% was reported for CO estimation, and a nRMSE of 3.36% for SBPc. Our results appear to corroborate their claim, in fact our nRMSE were a bit lower for each output. This slight improvement may correspond to the amount of used predictors. In our study, sixteen predictors were used, in contrast to [Bikia 2020] where they included six predictors. Regarding the chosen ML models, the one reported by [Bikia 2020] as the golden-result was SVM. In our study, we found two other models (Gaussian Process and Neural Networks) which performed better. Such were not reported to have been analyzed by [Bikia 2020].

Regarding the age assessments, it was proposed the estimation of CHA as a potential surrogate of AA in relation to the presence of pathological waves [3,21]. It is well known that cardiovascular risk increase with age, being accelerated in presence of risk factors, such as chronic diseases (like hypertension or diabetes) and unhealthy habits [22]. In this sense, hemodynamic parameters are consequently affected by the presence of such conditions, which can be directly reflected on APW morphology, acting as a fingerprint. This implication turns of interest when CHA is estimated with a pathological wave, as it could give a sign of "early aging". Since the used database was created from healthy, heterogeneous wide age ranged subjects, it would allow to detect an individual with an altered APW due to the presence of cardiovascular alterations, as it would predict a CHA greater than the actual individual's CHA.

To summarize, the evidence obtained from this study points towards the idea of obtaining APW information from daily life, allowing an earlier identification of possible CV alterations, using predictive, personalized and preventive methodologies. The obtained results constitute the basis for a future implementation of a clinical decision support system, where a set of features derived from non-invasive measurements of the arterial pressure waveform could provide valuable information regarding CV health.

## 5 Conclusion

This paper presented a ML approach for the estimation of several hemodynamic and bio-mechanical parameters (CO, PWVcf, SVR, SBPc, DBPc), as well as CHA, derived from APW in carotid and radial arteries, in order to act as a "fingerprint" of CVH. The Gaussian Regression Process model proved to be the most suitable for continuous domain outputs, while the Neural Network model achieved good performance in CHA classification. In addition, carotid artery presented the best results, when comparing to radial's. Consequently, APW provides powerful information for the assessment of CVH. The presented methodology has the underlying capability to detect pathologies or potential CV events in low/medium CV risk individuals. The latter is proposed based on our results with simulated APW from a 1D CV model. Nonetheless, further in-vivo studies are required to validate the application of the outcome.

## References

1. Bikia, V., Papaiaoannou, T.G., Pagoulatou, S. et al.: Noninvasive estimation of aortic hemodynamics and cardiac contractility using machine learning. Sci. Rep. (10), 15015 (2020). https://doi.org/10.1038/s41598-020-72147-8
2. Crichton, G., Elias, M., Robbins, M.: Cardiovascular health and arterial stiffness: the Maiane-Syracuse Longitudinal Study. J. Hum. Hypertens. **28**, 444–449 (2014). https://doi.org/10.1038/jhh.2013.131
3. Kucharska-Newton, A.M., Stoner, L., Meyer, M.L.: Determinants of vascular age: an epidemiological perspective. Clin. Chemistry **1**(65) (2019). https://doi.org/10.1373/clinchem.2018.287623
4. Wolsk, E., Bakkestrøm, R., Thomsen, J.H., Balling, L., Andersen, M.J., Dahl, J.S., Hassager, C., Møller, J.E., Gustafsson, F.: The influence of age on hemodynamic parameters during rest and exercise in healthy individuals. JACC Heart Faial **5**, 337–346 (2017). https://doi.org/10.1016/j.jchf.2016.10.012
5. Nilsson, P.: Vascular age: how can it be determined? What are its clinical applications. Medicographia **37**(4), 454–60 (2016)
6. Kullo, I.J., Malik, A.R.: Arterial ultrasonography and tonometry as adjuncts to cardiovascular risk stratification. J. Am. College Cardiology **49**(13), 1413–26 (2007). https://doi.org/10.1016/j.jacc.2006.11.039
7. Kumar, N.K., Sindhu, G.S., Prashanthi, D.K., Sulthana, A.S.: Analysis and prediction of cardio vascular disease using machine learning classifiers. In: 6th International Conference on Advanced Computing and Communication Systems (ICACCS), pp. 15–21 (2020). https://doi.org/10.1109/ICACCS48705.2020.9074183

8. Charlton, P.H., Mariscal Harana, J., Vennin, S., Li, Y., Chowienczyk, P., Alastruey, J.: Modeling arterial pulse waves in healthy aging: a database for in silico evaluation of hemodynamics and pulse wave indexes. Am. J. Physiol. Heart Circulatory Physiol. **317**(5) (2019). https://doi.org/10.1152/ajpheart.00218.2019
9. Santana, D.B., Zócalo, Y.A., Ventura, I.F., Arrosa, J.F., Florio, L., Lluberas, R., Armentano, R.L.: Health informatics design for assisted diagnosis of subclinical atherosclerosis, structural, and functional arterial age calculus and patient-specific cardiovascular risk evaluation. IEEE Trans. Inf. Technol. Biomed. **16**(5), 943–51 (2012). https://doi.org/10.1109/TITB.2012.2190990
10. McCulloch, W., Pitts, W.: A logical calculus of ideas immanent in nervous activity. Bull. Math. Biophysi. **5**(4), 115–133 (1943). https://doi.org/10.1007/BF02478259
11. Boser, B.E., Guyon, I.M., Vapnik, V.N.: A training algorithm for optimal margin classifiers. In: Fifth Annual Workshop on Computational Learning Theory, p. 144 (1992). 10.1.1.21.3818
12. Fix, E., Hodges, J.L.: Discriminatory Analysis. Consistency Properties. USAF School of Aviation Medicine, Randolph Field, Texas, Non parametric Discrimination (1951)
13. MATLAB. (2021). 9.10.0.1684407 (R2021a). Natick, Massachusetts: The MathWorks Inc. https://www.mathworks.com/. Accessed 16 July 2021
14. Altman, D.G., Bland, J.M.: Measurement in medicine: the analysis of method comparison studies. J. Roy. Stat. Soc. **3**(32), 307–317 (1983). https://doi.org/10.2307/2987937
15. Altman, D.G., Bland, J.M.: Measuring agreement in method comparison studies. Stat. Methods Med. Res. (8)2, 135–160 (1999). https://doi.org/10.1177/096228029900800204
16. Armentano, R.L., Cymberknop, L.J.: Quantitative vascular evaluation: from laboratory experiments to point-of-care patient (experimental approach). Current Hypertension Rev. (14)2, 76–85 (2018). https://doi.org/10.2174/1573402114666180423110658
17. Tavallali, P., Razavi, M., Pahlevan, N.M.: Artificial intelligence estimation of carotid-femoral pulse wave velocity using carotid waveform. Sci. Rep. (8)1014 (2018). https://doi.org/10.1038/s41598-018-19457-0
18. Jin, W., Chowienczyk, P., Alastruey, J.: Estimating pulse wave velocity from the radial pressure wave using machine learning algorithms. PLoS ONE **16**(6), e0245026 (2021). https://doi.org/10.1371/journal.pone.0245026
19. Sorelli, M., Perrella, A., Bocchi, L.: Detecting vascular age using the analysis of peripheral pulse. IEEE Trans Biomed Eng. **65**(12), 2742–2750 (2018). https://doi.org/10.1109/TBME.2018.2814630
20. Pereira, T., Santos, I., Oliveira, T., Vaz, P., Pereira, T., Santos et al.: Local PWV and other hemodynamic parameters assessment - validation of a new optical technique in an healthy population. In: Proceedings of the International Conference on Bio-inspired Systems and Signal Processing 2013, Barcelona, Spaian (2013)
21. Dakik, H.A.: Vascular age for predicting cardiovascular risk: a novel clinical marker or just a mathematical permutation. J. Nucl. Cardiol. **26**(4), 1356–1357 (2018). https://doi.org/10.1007/s12350-018-1223-x
22. Groenewegen, K.A., den Ruijter, H.M., Pasterkamp, G., Polak, J.F., Bots, M.L., Peters, S.A.: Vascular age to determine cardiovascular disease risk: a systematic review of its concepts, definitions, and clinical applications. Eur. J. Preventive Cardiol. **23**(3), 264–74 (2016). https://doi.org/10.1177/2047487314566999

# Characterizing Musculoskeletal Disorders. A Case Study Involving Kindergarten Employees

Gonzalo Yepes Calderon[2,3], Julio Perea Sandoval[3], Julietha Oviedo-Correa[3,4], Fredy Linero-Moreno[4], and Fernando Yepes-Calderon[1,3,5](✉)

[1] Science Based Platforms, 405 Beact CT, Fort Pierce, FL, USA
fernando.yepes@strategicbp.net

[2] GYM Group SA, Cra 78A No. 6-58, Cali, Colombia

[3] Universidad ECCI, Cra. 19 No. 49-20, Bogotá, Colombia
{gyepesc,joviedoc,jpereas}@ecci.edu.co

[4] Jardín Infantil Eco Kids, Av. Boyacá #175-75, Bogotá, Colombia
gerencia@jardininfantilecokids.com

[5] Universidad Distrital Francisco Jose de Caldas, Ak. 7 No. 40b-53, Bogotá, Colombia
fyepesc@udistrital.edu.co

http://www.gym-group.org, http://www.strategicbp.net, https://www.udistrital.edu.co/inicio

**Abstract.** This article presents the results of applying the Nordic Kuorinka questionnaire in a group of educators to infer musculoskeletal healthiness. The Kuorinka instrument queries the workers about postural pains and muscular afflictions that might have appeared due to systematic effort in the workplace. The sample, n = 42, included administrative workers, teachers, general services members, and kitchen individuals a kindergarten. These people execute different activities with diverse workloads. The results show that individuals with general discomfort in the last 12 months pointed at the neck as the source of the problem. In 19% of the workers, the discomfort lasted around 30 days, but the employees decided not to report the health-related event to the human resources office. 25% of the participants indicate that each episode of pain or discomfort lasted from one to four weeks. Other results indicate that among the workers incapacitated for 1 to 7 days, only 17% received medical treatment. According to Liberty Mutual, the largest workers' compensation insurance provider in the United States, musculoskeletal diseases cost employers 13.4 billion every year the United States. Nevertheless, and perhaps more disturbing is that workers do not use the health systems after work-related discomforts. Also, corrective actions are few or non-existent in Latin American countries, perpetuating work-related diseases and increasing the burden of losing health and productivity.

**Keywords:** Work related injuries · Ergonomics in preschool teachers · Musculoskeletal discomfort


H. Florez and M. F. Pollo-Cattaneo (Eds.): ICAI 2021, CCIS 1455, pp. 266–277, 2021.
https://doi.org/10.1007/978-3-030-89654-6_19

## 1 Introduction

Musculoskeletal disorders (MSD) originated by work impact employees' health and precluded their capacity to work in all economic sectors. Authors in diverse countries have studied the topic along with the evolution of occupational safety and health. An excellent example of the subject's relevance happened in Colombia between 2001–2002 when the general direction of occupational risks attached to the Ministry of Social Protection evidenced carpal tunnel syndrome as the leading cause of occupational disease. Likewise, lumbar pain, painful shoulder syndrome, Quervain's tenosynovitis, medial and lateral epicondylitis, and herniated discs increased during 2003 and 2004. In 2007 – also in Colombia – the results of the first national survey of health and work conditions in the general system of occupational risks claimed that agents with the highest exposure to positional conditions during work were related to ergonomic conditions. Those positional conditions include but are not limited to non-natural movements, repetitive tasks, keeping posture for extended periods, and lifting or moving heavy loads without mechanical assistance [16].

In the United States – 1995 – a study evidenced the annoyances of educators in a Montessori school, indicating that these are related to actions such as kneeling, sitting on the floor, squatting or bending at waist level. The operators performed these actions while cleaning, note-taking, serving food, executing bathroom activities, and direct work with children [9].

In another study, researchers identified some problems that were considered generalizable in these work settings. Among these are the incorrect lifting of children, toys, and materials by workers; the inadequate height of work tables, sitting on the floor without support on the back, reaching above the shoulders, making sudden movements combined with awkward postures when lifting objects, carrying garbage bags [11].

The authors in [10] coincide with previous claims saying that physical demands such as feeding children, changing diapers, providing learning activities, lifting and carrying children and materials, lowering and raising children in chairs, squatting to interact with children, and moving furniture, among several others, put workers of the studied sector in muscular discomfort.

In the study named Prevalence and risk factors of neck pain and pain in the upper extremities among school teachers in Hong Kong, researchers highlighted that working head down constitutes a risk factor for neck and upper limb pain (69.3%). The same study in Brazil yielded 41.1% of primary and secondary teachers who reported back pain associated with the workload and the positional constraints within the workplace [2,4].

Finally, in Colombia, in the study of Osteomuscular Pain in Teachers of an Educational Institution for Technical and Technological Training, it was shown that 65% of teachers had musculoskeletal pain in the last year due to executing work-related activities [17].

These antecedents suggest that MSD and sequelae are a silent pandemic; therefore, our interest investigating the subject in different environments. In this opportunity, we provide insights into the problem in a kindergarten, aiming to identify the sources MSD and associated discomfort suffered by the staff. In further deliveries of this project, we will provide scientifically-based mitigation strategies.

## 2 Materials and Methods

### 2.1 Evalu@ Parametrization

To promptly deliver the questionnaire to several people and yet, make the querying tool available for continuous data gathering even out of the scope of the current exploration, we employed the on-cloud data centralizer Evalu@ http://www.evalualos.com/test/. The Fig. 1, shows a section of the configuration file used and the part of the web form created by the system.

### 2.2 The Kuourika Questionnaire (KQ) Administration

The Nordic KQ is a standardized instrument for the analysis of musculoskeletal symptoms in the workplace [14].

The questionnaire can be administered as part of an interview or in a self-administered format. The questionnaire evaluates the neck, shoulder, thoracic spine, elbow, hand/wrist, and lumbar spine, aiming to identify the annoyances in the early stages and thus, improving well-being in people and working conditions.

Considering the characteristics of the questionnaire described above, we guide each of the 42 participants in accessing the Evalu@ tool application for handheld devices. Once we ended the training, each worker fulfilled the 11 main questions of the Nordic questionnaire taking five minutes on average in a self-administered manner. In Evalu@, 462 yes/no answers were obtained, characterizing the annoyances of the kindergarten workers [18].

## 3 Results

### 3.1 Discomfort in the Upper Body

A total of 42 questionnaires were processed using the Evalu@ tool. An overview of the results showed that 58% of the workers evaluated presented some discomfort. Among the 42 workers, 40% rated the annoyance with a rating of 4; on the scale, zero (0) is without annoyance, and five (5) is very strong. Among the employees suffering pain, 28% pointed at activities not associated with their work as the annoyance's origin.

In the detailed results, starting with the discomfort presented in the upper part of the body, it is evidenced that 58% of the workers have had neck discomfort. The previous percentage coincides with the general result, which indicates that the same percentage of workers have had at least one of the listed sources of discomfort. In their order, the next affectation in the upper part of the body is the wrist, with 46%. Finally, 42% of the workers presented discomfort in the lumbar spine. (See Table 1).

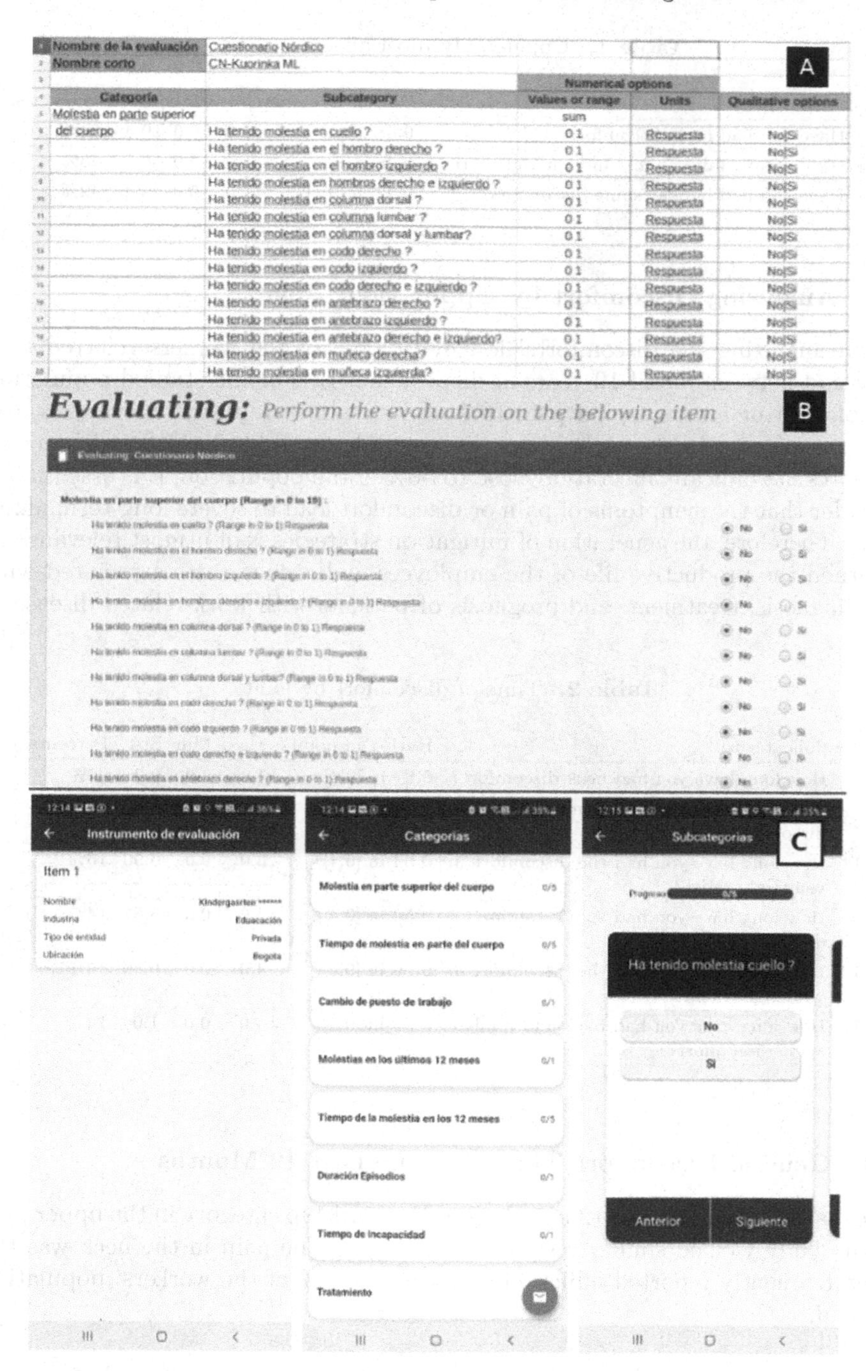


**Fig. 1.** In panel A, the top part of the excel file to teach Evalu@ how to perform the Kuorinka Questionaire. In panel B, the tracking instrument ready for web usage. In panel C, the cell phone interfaces that are synchronized in real-time and allow the final users to register the data from any place in the world. The configuring forms follow the official language spoken at the studied institution.

**Table 1.** Upper body most affected joints

| No. | Subcategory | Rating in points | Max | Min | Std | Percentage |
|---|---|---|---|---|---|---|
| 1.1 | Have you had neck discomfort? | 0.58 in [0,1] | 1.0 | 0.0 | 0.49 | 58% |
| 1.14 | Have you had discomfort in your right wrist? | 0.46 in [0,1] | 1.0 | 0.0 | 0.50 | 46% |
| 1.6 | Have you had lumbar spine discomfort? | 0.42 in [0,1] | 1.0 | 0.0 | 0.49 | 42% |

### 3.2 Analysing Discomfort by Upper Body Joint

When analyzing the discomforts' longevity, 20% of the workers referred that they had between 6 and 10 years of displeasure. 19% of the studied population complained of lumbar spine pain-16% of individuals reported between 6 and 10 years of pain in the right wrist. See records on Table 2. Although none of these results indicate affectation close to 50% of the population, it is essential to consider that the symptoms of pain or discomfort lead to severe long-term afflictions. Therefore, the generation of mitigation strategies is of utmost relevance to lengthen the productive life of the employees and reduce costs associated with the diagnosis, treatment, and prognosis of patients with work-related diseases.

**Table 2.** Time of discomfort by joint

| No. | Subcategory | Rating in points | Max | Min | Std | Percentage |
|---|---|---|---|---|---|---|
| 2.1 | How long have you had neck discomfort? | 0.81 in [0,4] | 4.0 | 0.0 | 0.91 | 20% |
| 2.6 | How long have you had lumbar spine discomfort? | 0.75 in [0,4] | 4.0 | 0.0 | 1.02 | 19% |
| 2.14 | How long have you had the discomfort in your right wrist? | 0.64 in [0,4] | 3.0 | 0.0 | 0.90 | 16% |
| 2.5 | How long have you had spinal column discomfort? | 0.54 in [0,4] | 4.0 | 0.0 | 0.87 | 14% |
| 2.17 | How long have you had the discomfort in your right hand? | 0.56 in [0,4]] | 4.0 | 0.0 | 0.89 | 14% |
| 2.7 | How long have you had back and lumbar spine discomfort? | 0.56 in [0,4] | 4.0 | 0.0 | 1.00 | 14% |

### 3.3 General Discomfort Timing in the Last 12 Months

The discomfort in the last 12 months and the hardship category in the upper part of the body yielded similar results (see Table 3). The pain in the neck was the most frequently reported affliction going up to 58% of the workers' population studied.

**Table 3.** General discomfort in the past 12 months

| No. | Subcategory | Rating in points | Max | Min | Std | Percentage |
|---|---|---|---|---|---|---|
| 4.1 | You have had discomfort in the last 12 months? | 0.58 en [0,1] | 1.0 | 0.0 | 0.49 | 58% |

Regarding discomfort in the last 12 months (see Table 4), 19% of the workers reported non-continuous pain for more than 30 days. 20% indicated discomfort in the lumbar spine, and 17% reported non-continuous discomfort in the dorsal and lumbar spine for more than 30 days. It is essential to mention that the discomfort in the right wrist has not occurred in the last 12 months among the studied population. The discomfort in the dorsal and lumbar spine is more relevant than the discomfort in the right wrist.

### 3.4 Upper Body Discomfort Timing in the Last 12 Months

**Table 4.** Localized discomfort timing in the last 12 months

| No. | Subcategory | Rating in points | Max | Min | Std | Percentage |
|---|---|---|---|---|---|---|
| 5.6 | How long have you had lumbar spine discomfort in the last 12 months? | 0.81 in [0,4] | 4.0 | 0.0 | 1.14 | 20% |
| 5.1 | How long have you had neck discomfort in the last 12 months? | 0.75 in [0,4] | 4.0 | 0.0 | 0.93 | 19% |
| 5.7 | How long have you had discomfort in the last 12 months in the thoracic and lumbar spine? | 0.66 in [0,4] | 0.0 | 1.10 | 0.90 | 17% |
| 5.17 | How long have you had discomfort in the last 12 months on your right hand? | 0.59 in [0,4] | 4.0 | 0.0 | 0.99 | 15% |
| 5.5 | How long have you had back discomfort in the last 12 months? | 0.54 in [0,4]] | 4.0 | 0.0 | 1.03 | 14% |

### 3.5 Discomfort in the Last Seven Days

Regarding having discomfort in the last seven days (see Table 5), 34% of the workers indicated the neck as the source of displeasure. This same percentage of workers have presented discomfort in the lumbar spine. 29% presented discomfort in the thoracic spine; and 24% in the dorsal and lumbar spine.

**Table 5.** Discomfort in the last 7 days

| No. | Subcategory | Rating in points | Max | Min | Std | Percentage |
|---|---|---|---|---|---|---|
| 9.1 | Suffered neck discomfort in the last 7 days? | 0.34 in [0,4] | 1.0 | 0.0 | 0.47 | 34% |
| 9.6 | Suffered lumbar discomfort in the last 7 days? | 0.34 en [0,1] | 1.0 | 0.0 | 0.47 | 34% |
| 9.5 | Suffered thoracic spine discomfort in the last 7 days? | 0.29 in [0,1] | 1.0 | 0.0 | 0.45 | 29% |
| 9.17 | Suffered right hand discomfort in the last 7 days? | 0.24 in [0,1] | 1.0 | 0.0 | 0.43 | 24% |
| 9.7 | Suffered discomfort thoracic and lumbar spines in the last 7 days? | 0.24 in [0,1] | 1.0 | 0.0 | 0.43 | 24% |

### 3.6 Duration of the Discomfort Episodes

The duration of the episodes in the study population coincides with the previous categories in terms of neck discomfort (see Table 6). A 25% of workers in the study indicated that the episode of neck discomfort lasted from one to four weeks. Also, 23% of the studied population reported episodes of discomfort in the right hand lasting from one to four weeks. Finally, 20% of the workers reported discomfort in the right hand and wrist lasting from one to four weeks.

**Table 6.** Duration of episodes

| No. | Subcategory | Rating in points | Max | Min | Std | Percentage |
|---|---|---|---|---|---|---|
| 6.1 | How long is each episode in the neck? | 1.27 in [0,5] | 5.0 | 0.0 | 1.38 | 25% |
| 6.17 | How long is each episode on the right hand? | 1.14 in [0,5] | 5.0 | 0.0 | 1.51 | 23% |
| 6.6 | How long does each episode in the lumbar spine last? | 1.00 in [0,5] | 4.0 | 0.0 | 1.30 | 20% |
| 6.14 | How long is each episode on the right wrist? | 1.02 in [0,5] | 5.0 | 0.0 | 1.46 | 20% |

Of high significance, no worker reported pain episodes lasting more than a month. Additionally, none of the episodes lasted less than an hour. Only 6% of the workers were incapacitated, indicating that the period of incapacity lasted from 1 to 7 days. 40% of the studied population assigned a level of four (4) to the qualification of the annoyance in the scale zero (0) without annoyance and five (5) very strong annoyances; 28% of the population defined activities other than work and sports activities as the cause of their discomfort. As for receiving treatment, only 17% of workers have received any intervention for these complaints. Although the disability does not necessarily indicate that the worker requires a specific treatment, it is necessary to clarify that rest (not treatment) is a valid alternative intervention. Finally, 7% of the population refers that it has been necessary to change their job.

### 3.7 Annoyance Vs. Disability Time

Regardless of the moments in which the discomfort occurs, the workers did not go to a medical service; see Figure . Recall that this study pointed to disability time in two-time schemes of discomfort – twelve months and seven days –, showing that the annoyance persisted without being attended-the Figures and present evidence for the mentioned reluctance to visit the health providers.

Regarding the time of disability versus the discomfort's presence in 12 months, In Fig. 2 although 58% of the workers reported displeasure only 6% received medical attention.

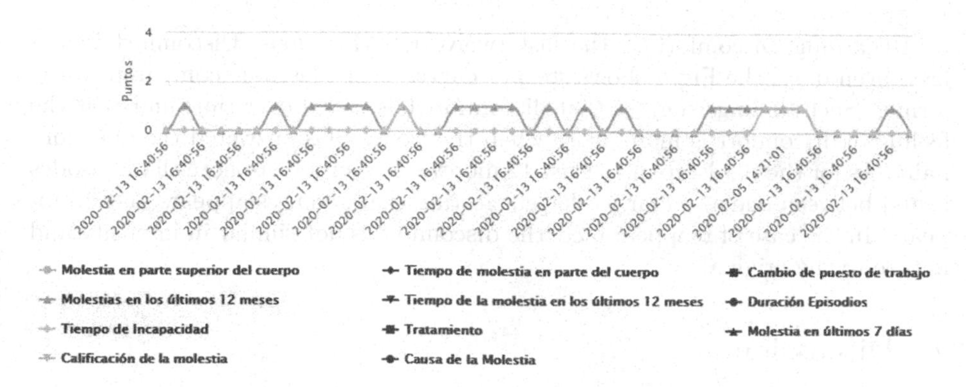


**Fig. 2.** The discomfort remains without receiving medical attention. Only a small part of the working population made the medical consultation.

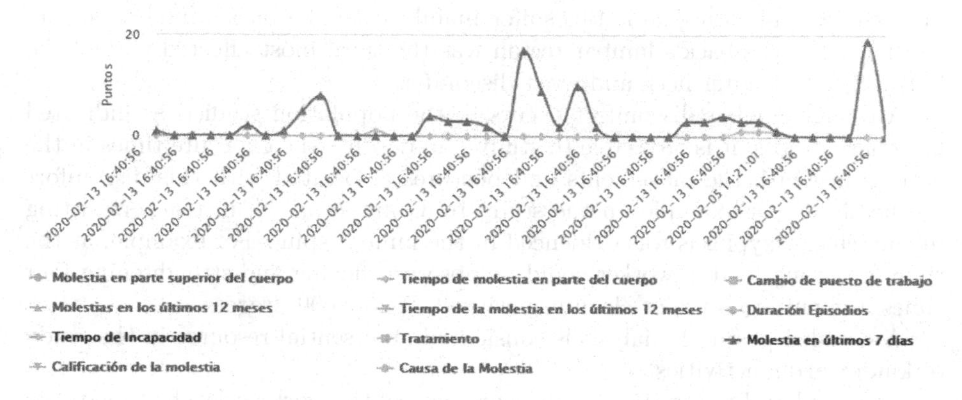


**Fig. 3.** As the previous graph, the worker continues in his activities even with discomfort.

In the case of discomfort in the last seven days vs. Time of disability, In Fig. 3 indicates that the workers, although presenting the discomfort, did not visit the doctor. In this sense, the workers tend to continue their work even while suffering pain.

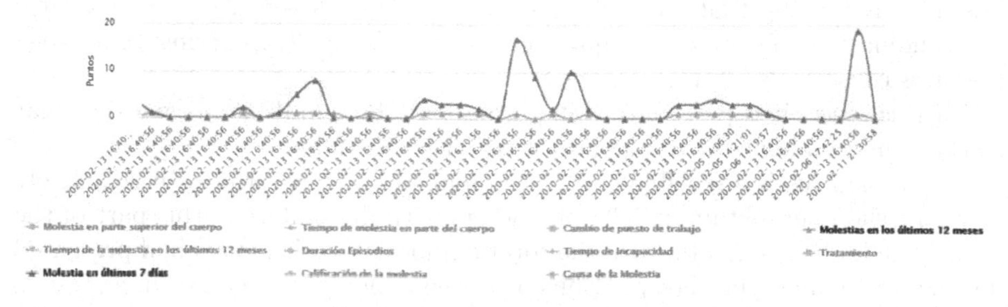


**Fig. 4.** The red line indicates that the discomfort was more acute in the last seven days, compared to the last 12 months. However the annoyance persists. (Color figure online)

Regarding Discomfort in the last twelve months versus Discomfort in the last seven days, the Fig. 4 shows steeper curves in the last one compared to the former. Such findings suggest that discomfort has lasted over time and that the feeling of discomfort is more acute when the worker has answered the questionnaire. As for the hardship in the last 12 months, it is crucial to note that episodes lasted between one and four weeks, suggesting that soreness appears and disappears. In the case of reappearance, the discomfort is not similar in intensity and nature to the original.

## 4 Discussion

Several authors claim that kindergarten teachers would suffer pain and discomfort originating in the body's upper part. At the same time, people who work in nurseries and caring activities suffer mainly from the back's lumbar region. In this study, the back's lumbar region was the third most affected part of the body with 42%, after neck and wrist discomfort.

Although lumbar discomfort occurs in the population studied as indicated by other authors, it is advisable to study – in-depth – the exposure times in the actions in which the movements and postures associated with this discomfort are involved. For example, in the study by Grant *et al.* [9] in teachers sitting in tiny chairs, kyphosis was evidenced in the lumbar spine. For example, at the time of serving food, a worker could be observed flexing and straightening four times a minute to serve or clean in positions of 45 to 90 degrees.

A detailed postural analysis is considered an essential resource in the study of kindergarten activities.

As stated by Punnet [15], the measurement of the work cycle, the time taken in each position, and the time that the actions last within the cycle, are variables to take into account for future evaluation proposals.

The types of postures assumed by the workers are related to the activities and gestures learn during childhood and later through adaptation. As reported Kumagai [13], in activities with children from 0 to 1 years, the most frequent postures were "sitting on the floor" and "kneeling," while in activities with children from 4 to 5 years, the most frequent postures were "standing" and "sitting in a chair." In the study above, he draws attention to the non-neutral postures, indicating that 35% of the shift time was employed in postures such as standing, leaning forward, squatting, and kneeling, considering that these postures caused a load on the lower back.

In this sense, it is pertinent to inquire about the population of workers that in this study remained standing or sitting with discomfort.

Since the neck is a sensible part of the human body, we studied it separately from the shoulder, obtaining 58% of workers with discomfort in this part of the body. The finding coincides with the total number of workers who had presented discomfort in any upper body regions: neck, shoulder, spine, elbow, hand/wrist, or lumbar spine. In the reviewed studies, reports of discomfort in this part of the body are evidenced in secondary school teachers more than preschool teachers.

In the study carried out by Chiu *et al.* [4], 69.3% reported neck pain in secondary school teachers. Other studies indicate the same results exceeding 50% of the population. In the study carried out by Chong*et al.* [5], 68.9% had neck pain, and in the study carried out by Korkmaz*et al.* [12] 42.5% of neck pain was reported in male and female teachers. Similar results to those obtained in this study were shown by Damayanti *et al.* [7] , indicating that neck pain occurred in 53.52% of the respondents.

Regarding neck discomfort, it is vital to indicate that the symptoms may persist even after diagnosis. In the study by Converso [6], 75.6% of diagnosed workers (238) reported neck pain even while being treated.

Given that in this population of workers, 58% of them presented neck discomfort. We suggest two postural aspects to consider in later studies and mitigation strategies to apply in kindergartens. The first is the use of a "head down" posture; the second referring to the posture of "pushing the chin" both associated with neck pain, reported in the study by Chiu *et al.* [4].

Regarding the discomfort in the right wrist, in contrast to the results obtained in this study, Grant *et al.* [9] evidenced discomfort in 11%, while Korkmaz et al. (2010) reported 13.4%. Erik *et al.* [8] reported 30.7% discomfort. Although these results are well below those obtained in the present study, it is necessary to consider the specific activities performed by kindergarten workers, including extension, abduction, and wrist adduction. Also, the posture assumed by the teachers and the exposure time when carrying out the writing activities, given that the teachers write and describe the tasks and the observable advances in the children's development daily.

As for the discomfort in the last 12 months, our experiments yielded numbers above the findings of Alias [1], who stated the neck discomfort affliction in 22.6% and 22.2% for male and female teachers, respectively. Our results in this regard are similar to the ones obtained by Erick [8] claiming that neck discomfort affects 50.8% of teachers.

Concerning the discomfort category in the last seven days, 34% of the teachers had neck discomfort and 29% in the spine. Alias [1] reported that teachers suffer pain, 22.2% in the neck, 27.4% in the back, and 26.4% in the lower back.

In results obtained in research similar to this study, cheng [3] showed that 16.1% of workers had neck discomfort greater than 30 days and 14.2% of workers presented lumbar spine discomfort greater than 30 days.

Regarding the right wrist, in the present study the workers did not present discomfort in the last 12 months, which coincides with the study by cheng [3]. However, they pointed out that when these episodes occur, they last from one to four weeks in 23% of them, which indicates again the relevance of carrying out a detailed analysis of the movements of the right wrist, the exposure time and the tasks in which these movements are executed.

Concerning disabilities, cheng [3] showed that 19.4% caused sick leave, while in this study, only 6% were incapacitated for 1 to 7 days. Similarly, cheng [3] recorded 25% suggestions for work changing in contrast to 7% of workers considering work change in the current work. The reasons for the workers to continue carrying out their work despite the discomforts might be diverse. Since the

Kuorinka test does not provide mechanisms to discern this question, we might slightly modify the test to gain the required resolution and deliver responses to this and other crucial details.

Regarding the above, Damayanti [7], reported 17.7% of teachers visiting the doctor due to neck and shoulder discomfort; whereas, 22.3% did visit the doctor complaining in back discomfort. In the present study, 17% of the teachers have received some intervention to their discomfort.

The data used to derive the results presented in this work were gathered by Evalu@, which enables the constant monitoring of variables, accessible data capturing (by computer or handheld devices), and the rapid management of the information, facilitating the analysis. The data gathered in the kindergarten is still available as a baseline to support intrinsic monitoring and interinstitutional analysis.

## 5 Conclusion

This study presents the discomfort analysis in teachers of a Kindergarten. The affected regions in their order were: neck, wrist, and lumbar spine. The analysis corroborated some of the findings presented by other authors in similar methodologic approaches. Nevertheless, the studied subjects switched the incidence of discomfort per body zone, being the neck the most prevalent source of pain. Although already reported as a significant finding, the workers' reluctance to visit the health care system was not as marginal as this investigation present it. The objection to visiting the health care providers and, thus, unfeasibility to threatening the work-related discomforts enlights substantial flaws in the risk management systems. Consequently, we may venture into predicting a significant number of current employees falling into chronic afflictions that considerably and progressively deteriorate the quality of life. Further work involves increasing the sample by diagnosing other companies and using the artificial-intelligence-based prediction capabilities offered by Evalu@ to sensibilize companies and the risk management entities.

## References

1. Alias, A.N., Karuppiah, K., How, V., Perumal, V.: Prevalence of musculoskeletal disorders (msds) among primary school female teachers in Terengganu, Malaysia. Int. J. Ind. Ergon. **77**, 102957 (2020)
2. Cardoso, J.P., Ribeiro, I.D.Q.B., Araújo, T.M.D., Carvalho, F.M., Reis, E.J.F.B.D.: Prevalence of musculoskeletal pain among teachers. Revista Brasileira de Epidemiologia **12**, 604–614 (2009)
3. Cheng, H.Y.K., Cheng, C.Y., Ju, Y.Y.: Work-related musculoskeletal disorders and ergonomic risk factors in early intervention educators. Appl. Ergon. **44**(1), 134–141 (2013)
4. Chiu, T.T., Lam, P.K.: The prevalence of and risk factors for neck pain and upper limb pain among secondary school teachers in Hong Kong. J. Occup. Rehabil. **17**(1), 19–32 (2007)

5. Chong, E.Y., Chan, A.H.: Subjective health complaints of teachers from primary and secondary schools in Hong Kong. Int. J. Occup. Saf. Ergon. **16**(1), 23–39 (2010)
6. Converso, D., Viotti, S., Sottimano, I., Cascio, V., Guidetti, G.: Musculoskeletal disorders among preschool teachers: analyzing the relationships among relational demands, work meaning, and intention to leave the job. BMC Musculoskel. Disord. **19**(1), 1–8 (2018)
7. Damayanti, S., Zorem, M., Pankaj, B.: Occurrence of work related musculoskeletal disorders among school teachers in eastern and northeastern part of india. Int. J. Musculoskel. Pain Prev **2**(1), 187–192 (2017)
8. Erick, P.N., Smith, D.R.: The prevalence and risk factors for musculoskeletal disorders among school teachers in botswana. Occup. Med. Health Affairs., 1–13 (2014)
9. Grant, K.A., Habes, D.J., Tepper, A.L.: Work activities and musculoskeletal complaints among preschool workers. Appl. Ergon. **26**(6), 405–410 (1995)
10. Gratz, R.R., Claffey, A., King, P., Scheuer, G.: The physical demands and ergonomics of working with young children. Early Child Dev. Care **172**(6), 531–537 (2002)
11. King, P.M., Gratz, R., Scheuer, G., Claffey, A.: The ergonomics of child care: conducting worksite analyses. Work **6**(1), 25–32 (1996)
12. Korkmaz, N.C., Cavlak, U., Telci, E.A.: Musculoskeletal pain, associated risk factors and coping strategies in school teachers. Sci. Res. Essays **6**(3), 649–657 (2011)
13. Kumagai, S., et al.: Load on the low back of teachers in nursery schools. Int. Arch. Occup. Environ. Health **68**(1), 52–57 (1996)
14. Kuorinka, I., et al.: Standardised nordic questionnaires for the analysis of musculoskeletal symptoms. Appl. Ergon. **18**(3), 233–237 (1987)
15. Punnett, L., Fine, L.J., Keyserling, W.M., Herrin, G.D., Chaffin, D.B.: Back disorders and nonneutral trunk postures of automobile assembly workers. Scand. J. Work Environ. Health **17**, 337–346 (1991)
16. Social, M.D.: Primera encuesta nacional de condiciones de salud y trabajo en el sistema general de riesgos profesionales. Primera Encuesta Nacional de 157 (2007)
17. Vélez, D.F.G., Terranova, O.E.L., Valderrama-Aguirre, A.: Dolor osteomuscular en docentes de una institución educativa de formación técnica y tecnológica. Revista Colombiana de Salud Ocupacional **5**(4), 27–31 (2015)
18. Yepes-Calderon, F., Yepes Zuluaga, J.F., Yepes Calderon, G.E.: Evalu@: an agnostic web-based tool for consistent and constant evaluation used as a data gatherer for artificial intelligence implementations. In: Florez, H., Leon, M., Diaz-Nafria, J.M., Belli, S. (eds.) ICAI 2019. CCIS, vol. 1051, pp. 73–84. Springer, Cham (2019). https://doi.org/10.1007/978-3-030-32475-9_6

# Comparison of Heart Rate Variability Analysis with Empirical Mode Decomposition and Fourier Transform

Javier Zelechower[1], Fernando Pose[2], Francisco Redelico[2], and Marcelo Risk[1,2(✉)]

[1] Universidad Tecnológica Nacional (UTN, FRBA), Buenos Aires, Argentina
[2] Instituto de Medicina Traslacional e Ingeniería Biomédica (IMTIB), CONICET - Instituto Universitario del Hospital Italiano - Hospital Italiano de Buenos Aires, Buenos Aires, Argentina
marcelo.risk@hospitalitaliano.org.ar

**Abstract.** The heart rate variability (HRV) analysis allows the study of the regulation mechanisms of the cardiovascular system, in both normal and pathological conditions, and the power spectral density analysis of the short-term HRV was adopted as a tool for the evaluation of the autonomic function. The Ensemble Empirical Mode Decomposition (EEMD) is an adaptive method generally used to analyze non-stationary signals from non-linear systems. In this work, the performance of the EEMD in the decomposition of the HRV signal in the main spectral components is studied, in a first instance to a synthesized series to calibrate the method and achieve confidence and then to a real HRV database. In conclusion, the results of this work propose the EEMD as useful method for analysis HRV data. The ability of decomposes the main spectral bands and the capability to deal with non-linear and non-stationary behaviors makes the EEMD a powerful method for tracking frequency changes and amplitude modulations in HRV signals generated by autonomic regulation.

**Keywords:** Heart rate variability · Comparison of methods · EEMD · Fourier analysis · Spectrum decomposition

## 1 Introduction

The HRV analysis allows the study of the regulation mechanisms of the cardiovascular system, not only under normal conditions, but also when these are altered to produce pathological conditions, for example high blood pressure, heart failure and diabetes among others [1,2].

It is feasible to involve the autonomic control mechanisms in cardiac function, these influence the short-term fluctuations of the time interval between consecutive heart beats (RR) [3,4]. Indeed, the power spectral density analysis of the short-term HRV was adopted as a tool for the evaluation of the autonomic


H. Florez and M. F. Pollo-Cattaneo (Eds.): ICAI 2021, CCIS 1455, pp. 278–289, 2021.
https://doi.org/10.1007/978-3-030-89654-6_20

function [3,4]. Three main spectral components can be highlighted, very low frequency component (VLF), from 0.003 Hz to 0.04 Hz; low frequency component (LF), from 0.04 Hz to 0.15 Hz; and high frequency component (HF), from 0.15 Hz to 0.4 Hz [3,5].

Numerous techniques have been applied in the frequency domain, among the most important measures, some linear models that generate comparable results can be included. Nonparametric models, such as windowed fast Fourier transform (FFT) and Blackman-Tukey spectral estimation, and parametric models, such as autoregressive (AR) and moving average autoregressive (ARMA). After calculating the spectrum with any of the above methods, the energies within each band can be calculated [1,2].

These linear methods must assume stationary conditions that are difficult to achieve, even in short-term records under controlled conditions. To correctly attribute spectral components to specific physiological conditions, the heart rate modulation mechanisms must not make any changes during the measurement process [3]. Due to the nature of HRV signals, non-linear techniques appears as attractive methods for their analysis in order to solve the difficulty of achieving the conditions of strict stationarity and correctly reflecting the non-linear content of the data.

The Empirical Mode Decomposition (EMD) is an adaptive method generally used to analyze non-stationary signals from non-linear systems [6]. The algorithm produces a decomposition of the time series into a finite quantity of oscillating functions and a residue. These zero local mean functions are modulated amplitude / frequency signals called intrinsic mode functions (IMF).

In the EMD process a problem called mode mixing occurs, oscillations with very different scales can exist in one mode. To reduce this effect, a new method called ensemble empirical mode decomposition [7] was proposed. The decomposition is performed from a set of noisy copies of the original signal, obtaining the final results by averaging whereas the noise converges to zero. These decomposition methods have proven their competence in different applications, for heart rate variability analysis [8], assessment of cardiovascular autonomic control [9], automated identification of congestive heart failure[10], classification of ECG heartbeats [11], early detection of sudden cardiac death [12].

In this work, the performance of the EEMD in the decomposition of the HRV signal in the main spectral components is studied, in a first instance to a synthesized series to calibrate the method and achieve confidence and then to a real HRV database. Looking to generate a stationary behavior for the measurement, we used 5 min short-term recordings of 14 subjects before and during the application of a pharmacological autonomic blockade in combination with posture changes during controlled breathing [13]. The energy of the three main spectral bands acquired by the windowed FFT method and the energy of each IMF obtained using the EEMD was calculated to obtain their correlation. The objetive was to validate the correlation between them and verify the effectiveness of the EEMD method applied to HRV signals.

## 2 Materials and Methods

### 2.1 Simulated Signal

A typical 5-minutes short-term HRV segment with zero mean was produced to evaluate the performance of the EEMD. As shown in the Fig. 1, the synthesized HRV series $s(t)$ is composed of $s_1(t)$,$s_2(t)$ and $s_3(t)$, three sinusoidal signals whose oscillation frequency was located in the center of each main spectral band [3].

The $s_1$ component recreates the HF with an angular frequency $\omega_{HF} = 2\pi(0.275)t$ , $s_2$ represents the LF with an angular frequency $\omega_{LF} = 2\pi(0.095)t$, and $s_3$ simulate the VLF with an angular frequency $\omega_{VLF} = 2\pi(0.0215)t$. The signal was divided into 3 different segments in order to recreate a non-stationary environment, amplitude changes of the LF and HF components were made every 100 s. For the sampling frequency a rate of 7 hz was used [14]. The resultant series $s(t) = s_1(t) + s_2(t) + s_3(t)$ can be described as follows:

$$s(t) = \begin{cases} 30sin(\omega_{VLF}) + 15sin(\omega_{LF}) + 5sin(\omega_{HF}) & 0 \leq t < 100 \\ 30sin(\omega_{VLF}) + 20sin(\omega_{LF}) + 2sin(\omega_{HF}) & 100 \leq t < 200 \\ 30sin(\omega_{VLF}) + 10sin(\omega_{LF}) + 10sin(\omega_{HF}) & 200 \leq t \leq 300 \end{cases} \quad (1)$$

### 2.2 Dataset

We used recordings from a database developed by Harvard Medical School (HMS, Children's Hospital), Massachusetts Institute of Technology (MIT), and the Favaloro Foundation School of Medicine (FFMS). The HMS-MIT-FFMS database was designed to perform training and comparison of a variety of methodologies used to analyze the cardiac function. A total of 82 segments of 5-minute short-term recordings of 14 subjects were used [13].

The signal measurement process begins with each subject in a supine position where they perform a breathing protocol, then they move to the standing position and after 5 min for hemodynamic balance they repeat the breathing protocol. After that, the subject is then returned to the supine position and atropine (0.03 mg/kg, n = 7) or propranolol (0.2 mg/kg, n = 7) is administered, after 10 min, it is performed the breathing protocol with the subject in the supine position and then standing. Finally, the subject is placed back in the supine position to administer the other autonomous blocking agent, and the measurement process is repeated for the supine and standing positions. The doses chosen for complete parasympathetic blockade (atropine) and complete P-adrenergic blockade (propranolol) were based on previous studies [15]. Instances of pure sympathetic or parasympathetic modulation before and during the application of a pharmacological autonomic block in combination with posture changes during controlled breathing were generated. The RR interval signal were acquired from an ECG using a peak detection program, where time series of HR smoothed 3 Hz [16].

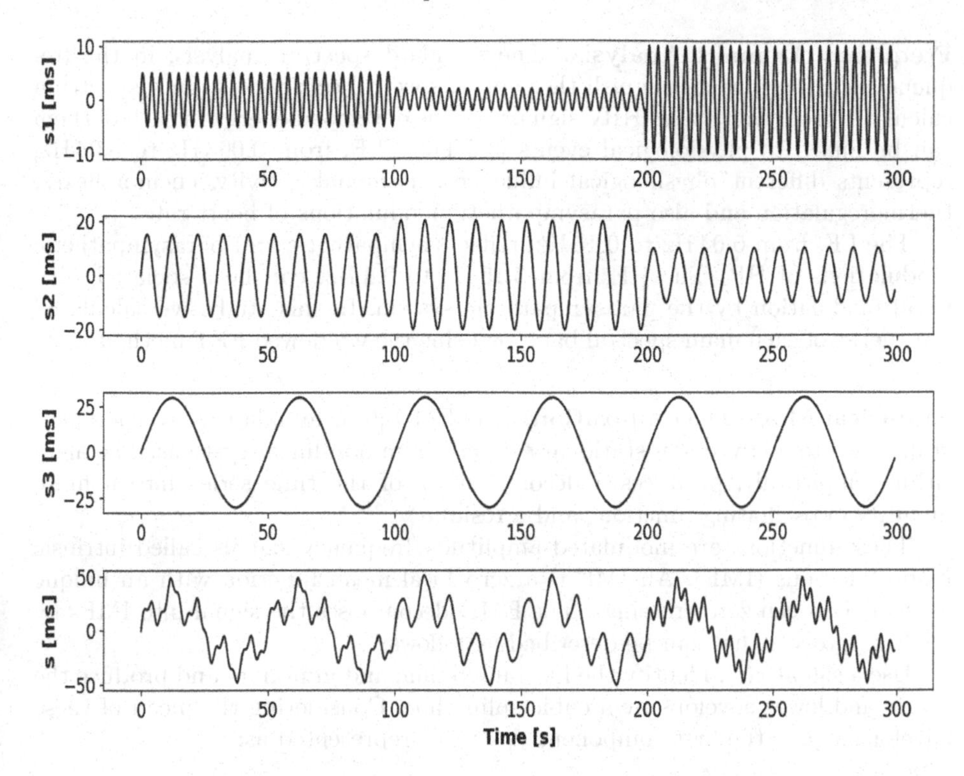


**Fig. 1.** Composition of the synthesized typical HRV 5-min segment. $s_1(t)$ represents the HF, $s_2(t)$ the LF and $s_3(t)$ the VLF. The resultant segment can be described as: $s(t) = s_1(t) + s_2(t) + s_3(t)$.

## 2.3 Resampling

It is necessary to convert the signal with an interpolation step, HRV data has an uneven sampling nature, whereas most feature extraction methods require uniform sampling [17]. Among the many resampling methods already defined in the literature, the Cubic Spline method is possibly the most widely used due to the minimal interference it has on frequency domain measurements [18]. The upper frequency of HRV data is 0.5 Hz [3] and a rate of 1 10 Hz is normally used for interpolation [14]. In this case, we applied the Cubic Spline resampling method with a sampling rate 7 Hz to HRV data.

## 2.4 HRV Signal Analysis

As suggested by HRV guidelines [19], standard frequency domain features for short-term HRV were extracted from 5-minute segments. In addition, an EEMD was applied in these HRV segments to analyze them based on a non-linear model [20].

**Frequency Domain Analysis.** The standard spectral analysis, in the frequency domain, can distinguish three main spectral components in a spectrum calculated from short-term HRV signals [5], the developed energy in each of them can be linked to physiological events [3]. The VLF, from 0.003 Hz to 0.04 Hz, represents different physiological influences, hormonal activity, chemoreflexes, thermoregulation and also parasympathetic modulations of heart rate.

The LF, from 0.04 Hz to 0.15 Hz, reflects sympathetic and parasympathetic modulations of HR. The HF, from 0.15 Hz to 0.4 Hz, can be associated with vagal modulation by the parasympathetic system. In this study, we calculated the energy of each main spectral band by using the windowed FFT method [1,2].

**Empirical Mode Decomposition.** The EMD [6] is an adaptive method generally used to analyze non-stationary signals from non-linear systems. The algorithm, empirically, produces a decomposition of the time series into a finite quantity of oscillating functions and a residue.

These functions are modulated amplitude/frequency signals called intrinsic mode functions (IMF). An IMF is a zero local mean function with an unique extreme between zero crossings. The EMD decomposes the signal into IMFs by a sifting process that can be described as follows:

Use a signal $x(t)$, identify the local maximum and minimum and produce the upper and lower envelops by a cubic spline line. Considering the mean of these envelops as $m_1$, the first component $f_1$ can be represented as:

$$f_1 = x(t) - m_1 \tag{2}$$

Take $f_1$ as the original signal and replicate the sifting process considering $m_{11}$ as the mean of its upper and lower envelopes. Then, get $f_{11}$ as:

$$f_{11} = f_1 - m_{11} \tag{3}$$

Repeat $k$ times the sifting process until $f_{1k}$ becomes itself to an IMF

$$f_{1(k-1)} - m_{1k} = f_{1k} \tag{4}$$

Then, obtain the first IMF component from the data:

$$IMF_1 = f_{1k} \tag{5}$$

The $IMF_1$ must contain the fastest, or the highest frequency, components of the signal. Once determined, get the residue $r_1$ separating this $IMF_1$ from the rest of the data

$$r_1 = x(t) - IMF_1 \tag{6}$$

Take the residue $r_1$ as a new signal, repeat the sifting process steps performed previously and obtain the second IMF component $IMF_2$. This procedure can be repeated resulting in:

$$r_1 - IMF_2 = r_2, ..., r_{n-1} - IMF_n = r_n \tag{7}$$

The stop process criteria are two: when the $IMF_n$ or the $r_n$ becomes less than a predetermined value or when the $r_n$ becomes a monotonic function. Finally, the signal $x(t)$ can then be represented as a sum of the IMFs and a residue:

$$x(t) = \sum_{i=1}^{n} IMF_i + r_n \tag{8}$$

In the EMD process a problem called mode mixing occurs, oscillations with very different scales can exist in one mode. To reduce this effect, the new method EEMD [7] was proposed. The decomposition is performed from a set of noisy copies of the original signal, obtaining the final results by averaging whereas the noise converges to zero.

The addition of Gaussian white noise reduces the mixing of modes by populating the entire time-frequency space, taking advantage of the behavior of the EMD's dyadic filter bank [21] and obtaining more regular modes. Therefore, EMD mode mixing problem is solved effectively by using EEMD [20]. The EEMD algorithm can be computed as follows:

1. Add to the original signal $x(t)$, $J$ different series of Gaussian white noise with a defined standard deviation $n^j$ ($j = 1, ..., J$). Then, return the observations:

$$x^j_{(t)} = x(t) + n^j_{(t)} \tag{9}$$

2. Decompose each $x^j_{(t)}$ into IMFs by the EMD process, obtaining a $J$ sets of IMFs per observation $IMF^j_i$.
3. Compute the ensemble mean of each IMF as the final $i^{th}$ IMF:

$$\overline{IMF_i} = \frac{1}{J} \sum_{j=1}^{J} IMF^j_i \tag{10}$$

### 2.5 Software Tools

In this experiment, the Python language was used to developed all the scripts and they were run in a JupyterLab notebook as integrated development environment. Pandas 1.0.3, Numpy 1.19.4, Scipy 1.5.4 and emd 0.4.0 packages were used.

## 3 Results

The EEMD analysis on a synthesized signal as shown in Fig. 3 produces the decomposition in five IMF, therefore is possible to see that the $IMF_1$ is mainly composed of the remaining noise used to average the assembly, the $IMF_2$ and $IMF_3$ are identified as the HF component of the signal, the $IMF_4$ is associated with the LF component and the $IMF_5$ with the VLF component.

Also, it can be observed that the amplitude modulations carried out in the LF and VLF components during the simulation are correctly represented by each corresponding IMF.

As suggested [7], a noise ratio of 0.2 of the signal standard deviation and a ensemble number of 100 was used. With this configuration, a set of 5 IMFs of each HRV segment was obtained both in the simulated case and with the real signals as shown in Fig. 2 and Fig. 3 respectively.

In the case of real RR signals, as shown in Fig. 3, the EEMD also produces the decomposition of five IMFs. A total of 410 IMFs from 82 short-term HRV segment were obtained, the correlation between their energies with the energies of the main spectral bands was calculated in order to validate their link.

In the Fig. 4 it can be observed a 0.96 correlation between the HF energy ($Ehf$) with the energy of the $IMF_1$ ($Eimf_1$), a correlation of 0.98 and 0.97 between the LF energy ($Elf$) with the energies of $IMF_2$ and $IMF_3$ respectively ($Eimf_2, Eimf_3$) and a correlation of 0.93 and 0.88 between the VLF energy ($Evlf$) with the energies of $IMF_4$ and $IMF_5$ ($Eimf_4, Eimf_5$). It is also possible to see that exist a correlation of 0.94 between the energies of $IMF_2$ and $IMF_3$ ($Eimf_2, Eimf_3$), and a 0.91 correlation with the energies of $IMF_4$ and $IMF_5$ ($Eimf_4, Eimf_5$).

## 4 Discussion

In this paper, the EEMD is presented as an adaptive method suitable for the non-linear and non-stationary behavior of HRV signals. This multi-resolution decomposition technique can evaluate the intrinsic characteristics of signals even with non-stationary components [14]. In the study of the congestive heart failure was used to improve the accuracy of an automated identification system [10], and to produce new features and indices that could serve as a new way of evaluating [22]. Also has been used to extract the feto-maternal heart rate from the abdominal ECG signal where they obtained a mean accuracy of approximately 95% in the quantification [23] and to classify ECG heartbeats with a mean accuracy greater than 95% [11]. The EEMD has been demonstrated that is a very useful technique for tracking frequency changes and amplitude modulations generated by autonomic regulation [8].

Autonomic function, through the sympathetic and parasympathetic nervous systems, controls the functioning of vital organs [24]. The autonomic control mechanisms are involved in cardiac function influencing short-term fluctuations of the RR time interval [3,4]. Indeed, the power spectral density analysis of the short-term HRV was adopted as an evaluator of the autonomic function [3,4]

The EEMD has the issue of decomposing the series analyzed in a set of IMFs according to the nature of the signal analyzed [25]. In this experiment, the numbers of the modes obtained by analyzing the synthesized signal do not coincide with those obtained in the real data set, a problem that requires prior knowledge of the data to correctly attribute the information to an IMF.

The method has shown that it can decompose HRV signals into four functions associated with four frequency bands, functions even validated by Hilbert-Huang transform [26]. It seems that, using the IMFs obtained through the EEMD in real short-term HRV series, it is possible to reconstruct separately the main spectral

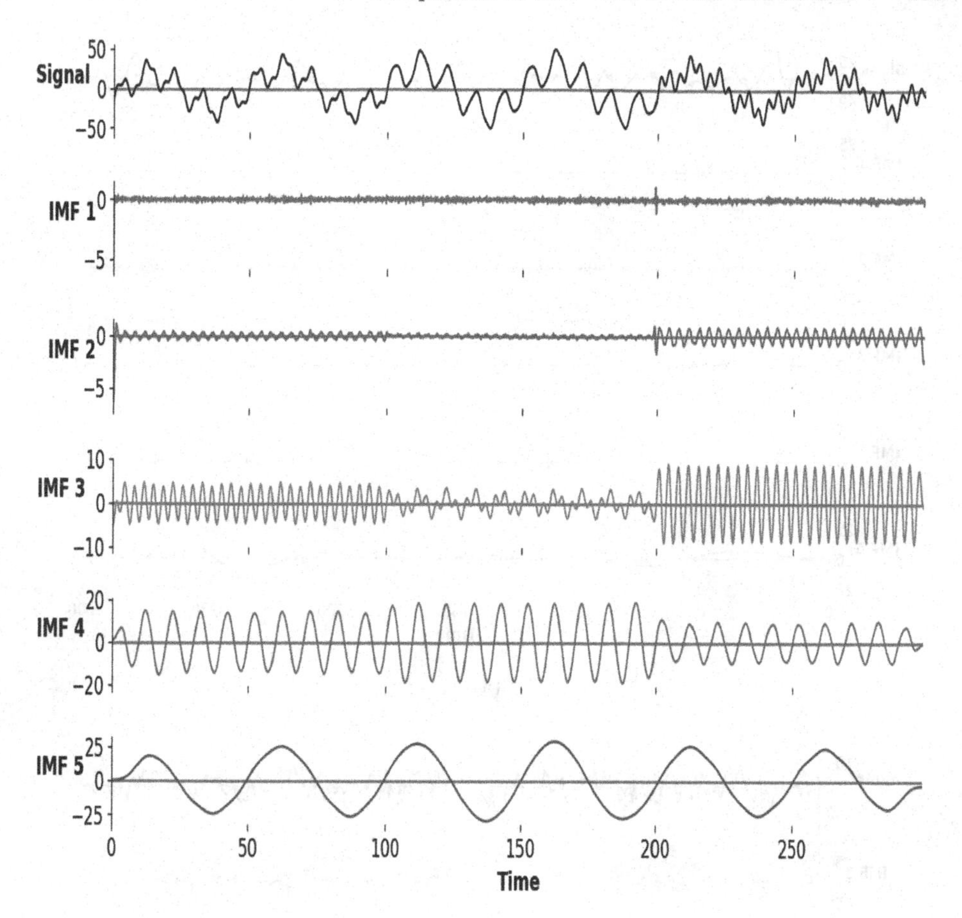


**Fig. 2.** Results of EEMD on synthesized typical HRV 5-minutes segment s(t).

bands avoiding their interference and allowing to analyze each one of them in particular. In the results it can be observed a 0.96 correlation between the $IMF1$ and the HF band and correlations of 0.98 and 0.97 between the $IMF2$ and the $IMF3$ with the LF band achieving a correct discrimination without using any type of fixed filtering. If this correlation is sufficient to validate the link between the IMFs with the autonomic function, it is possible to say that IMFs describe their temporal activity.

In future works, with the objective to analyze the mode mixing level that exists in the different IMFs, will be to use a spectral or time-frequency method, as the Hilbert-Huang transform [6], in order to validate the correspondence of the IMFs with the main spectral bands. Also, to confirm the correlation between the IMFs with the autonomic function, we will construct a classification system where the IMFs will be evaluated as classification features.

One of the most significant limitations is the computational cost attributed by its iterative nature that slows down the method [27], it can be solved by

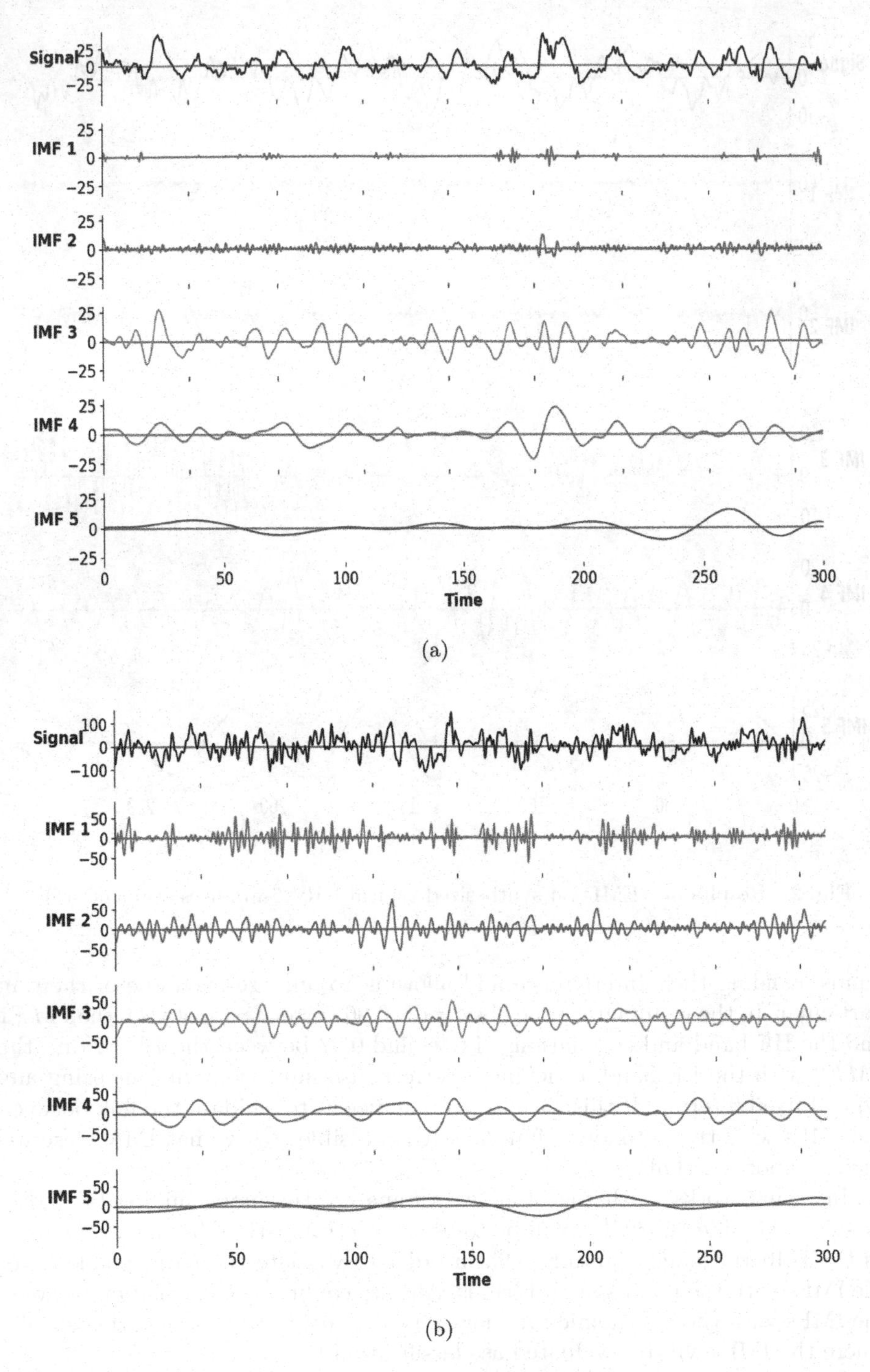


**Fig. 3.** Results of EEMD on two real HRV segments: (a) Subject in a pure sympathetic modulation state, (b) another subject in a pure parasympathetic modulation state.

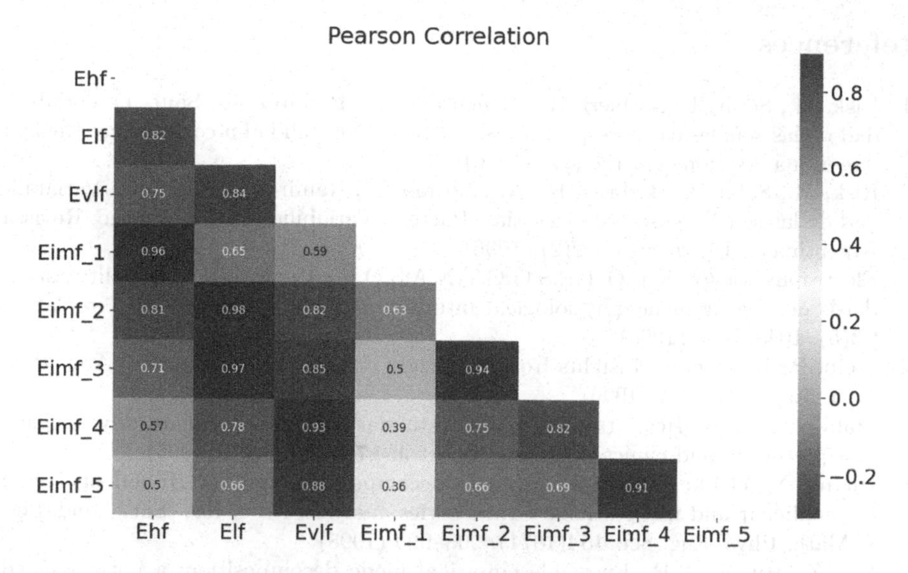


**Fig. 4.** Correlation between the energies of the HRV main spectral bands with the energies of the 5 IMFs obtained with the EEMD.

computing the EEMD in parallel processing threads, making real-time implementation possible. Another limitation was that we have used only data from 14 subjects, the analysis must be applied to a larger dataset to validate the results. Finally, we can mention that although the FFT is taken as the goal standard measurement, it is based on a linear method that requires stationarity for its correct operation.

### 4.1 Conclusions

In this work we have presented the EEMD as an algorithm that can solve the problem of analyzing non-linear and non-stationary signals such as the HRV series. The method was successfully tested on artificial and real signals. It is possible to say that the EEMD solves the mode mixing phenomenon by adding white noise to the original signal, the synthetized signal shows the capability to decompose the signal into IMFs.

Regarding the performance using the real data set, it can be observed a 0.96 correlation between the $IMF1$ and the HF band and correlations of 0.98 and 0.97 between the $IMF2$ and the $IMF3$ with the LF band, we can achieve a correct discrimination without using any type of fixed filtering.

In conclusion, the results of this work propose the EEMD as useful method for analysis HRV data. The ability of decompose the main spectral bands and the capability to deal with non-linear and non-stationary behaviors makes the EEMD a powerful method for tracking frequency changes and amplitude modulations in HRV signals generated by autonomic regulation.

## References


1. Risk, M., Sobh, J., Barbieri, R., Armentano, R., Ramirez, A., Saul, J.: Variabilidad de las señales cardiorespiratorias. Parte 1: Variabilidad a corto plazo. Revista Argentina de Bioingenieria **2**(1) (1996)
2. Risk, M., Sobh, J., Barbieri, R., Armentano, R., Ramirez, A., Saul, J.: Variabilidad de las señales cardiorespiratorias. Parte 2: Variabilidad a largo plazo. Revista Argentina de Bioingenieria **2**(2) (1996)
3. Electrophysiology, T.F.O.T.E.S.O.C.T.N.A.S.O.P.: Heart rate variability: standards of measurement, physiological interpretation, and clinical use. Circulation **93**(5), 1043–1065 (1996)
4. Stein, P., Kleiger, R.: Insights from the study of heart rate variability. Ann. Rev. Med. **50**(1), 249–261 (1999)
5. Malik, M., et al.: Heart rate variability: standards of measurement, physiological interpretation, and clinical use. Eur. Heart J. **17**(3), 354–381 (1996)
6. Huang, N., et al.: The empirical mode decomposition and the Hilbert spectrum for nonlinear and non-stationary time series analysis. Proc. Roy. Soc. Lond. Ser. A Math. Phys. Eng. Sci. **454**(1971), 903–995 (1998)
7. Wu, Z., Huang, N.E.: Ensemble empirical mode decomposition: a noise-assisted data analysis method. Adv. Adapt. Data Anal. **1**(01), 1–41 (2009)
8. Echeverria, J., Crowe, J., Woolfson, M., Hayes-Gill, B.: Application of empirical mode decomposition to heart rate variability analysis. Med. Biol. Eng. Comput. **39**(4), 471–479 (2001)
9. Neto, E.S., et al.: Assessment of cardiovascular autonomic control by the empirical mode decomposition. Methods Inf. Med. **43**(1), 60–65 (2004)
10. Acharya, U.R., et al.: Application of empirical mode decomposition (emd) for automated identification of congestive heart failure using heart rate signals. Neural Comput. Appl. **28**(10), 3073–3094 (2017)
11. Rajesh, K.N., Dhuli, R.: Classification of ECG heartbeats using nonlinear decomposition methods and support vector machine. Comput. Biol. Med. **87**, 271–284 (2017)
12. Shi, M., et al.: Early detection of sudden cardiac death by using ensemble empirical mode decomposition-based entropy and classical linear features from heart rate variability signals. Front. Physiol. **11**, 118 (2020)
13. Sobh, J.F., Risk, M., Barbieri, R., Saul, J.P.: Database for ecg, arterial blood pressure, and respiration signal analysis: feature extraction, spectral estimation, and parameter quantification. In: Proceedings of 17th International Conference of the Engineering in Medicine and Biology Society, vol. 2, pp. 955–956 (1995). https://doi.org/10.1109/IEMBS.1995.579378
14. Clifford, G.D., Azuaje, F., McSharry, P., et al.: Advanced methods and tools for ECG data analysis. Artech house, Boston (2006)
15. Jose, A.D., Taylor, R.R., et al.: Autonomic blockade by propranolol and atropine to study intrinsic myocardial function in man. J. Clin. Invest. **48**(11), 2019–2031 (1969)
16. Berger, R.D., Akselrod, S., Gordon, D., Cohen, R.J.: An efficient algorithm for spectral analysis of heart rate variability. IEEE Trans. Biomed. Eng. **9**, 900–904 (1986)
17. Laguna, P., Moody, G.B., Mark, R.G.: Power spectral density of unevenly sampled data by least-square analysis: performance and application to heart rate signals. IEEE Trans. Biomed. Eng. **45**(6), 698–715 (1998)

18. Clifford, G.D., Tarassenko, L.: Quantifying errors in spectral estimates of HRV due to beat replacement and resampling. IEEE Trans. Biomed. Eng. **52**(4), 630–638 (2005)
19. Sassi, R., et al.: Advances in heart rate variability signal analysis: joint position statement by the e-Cardiology ESC working group and the European heart rhythm association co-endorsed by the Asia pacific heart rhythm society. EP Europace **17**(9), 1341–1353 (2015). https://doi.org/10.1093/europace/euv015
20. Zhao, Z., Yang, L., Chen, D., Luo, Y.: A human ECG identification system based on ensemble empirical mode decomposition. Sensors **13**(5), 6832–6864 (2013)
21. Flandrin, P., Rilling, G., Goncalves, P.: Empirical mode decomposition as a filter bank. IEEE Signal Process. Lett. **11**(2), 112–114 (2004)
22. Chen, M., He, A., Feng, K., Liu, G., Wang, Q.: Empirical mode decomposition as a novel approach to study heart rate variability in congestive heart failure assessment. Entropy **21**(12), 1169 (2019)
23. Bin Queyam, A., Kumar Pahuja, S., Singh, D.: Quantification of feto-maternal heart rate from abdominal ECG signal using empirical mode decomposition for heart rate variability analysis. Technologies **5**(4), 68 (2017)
24. Lahiri, M.K., Kannankeril, P.J., Goldberger, J.J.: Assessment of autonomic function in cardiovascular disease: physiological basis and prognostic implications. J. Am. Coll. Cardiol **51**(18), 1725–1733 (2008)
25. Martis, R.J., Acharya, U.R., Min, L.C.: ECG beat classification using PCA, LDA, ICA and discrete wavelet transform. Biomed. Signal Process. Control **8**(5), 437–448 (2013)
26. Pan, W., He, A., Feng, K., Li, Y., Wu, D., Liu, G.: Multi-frequency components entropy as novel heart rate variability indices in congestive heart failure assessment. IEEE Access **7**, 37708–37717 (2019)
27. Torres, M.E., Colominas, M.A., Schlotthauer, G., Flandrin, P.: A complete ensemble empirical mode decomposition with adaptive noise. In: 2011 IEEE International Conference on Acoustics, Speech and Signal Processing (ICASSP), pp. 4144–4147. IEEE (2011)

# Evalu@ + Sports. Creatine Phosphokinase and Urea in High-Performance Athletes During Competition. a Framework for Predicting Injuries Caused by Fatigue

Juan Fernando Yepes Zuluaga[3], Alvin David Gregory Tatis[2], Daniel Santiago Forero Arévalo[2], and Fernando Yepes-Calderon[1,3](✉)

[1] GYM Group SA, Carrera 78 A No. 6-58, Cali, Colombia
fernando.yepes@strategicbp.net

[2] Universidad de los Andes, Cra. 1 18a 12, Bogotá, Colombia
{ad.gregory,ds.foreroa}@uniandes.edu.co

[3] SBP LLC, 604 Beach CT, Fort Pierce, FL 34950, USA
juan.yepes@strategicbp.net

**Abstract.** Elite athletes follow a strict regime of physical training that forces muscle deterioration - reconstruction cycles and specific energy generation patterns. One can monitor metabolic functions in blood for further training planning and optimization. The creatine phosphokinase (CPK) and the Urea appear in the serum-blood with higher average values in elite-athletes in comparison with sedentary subjects. In this manuscript, CPK and Urea recorded in professional soccer players are studied along a full season to create a framework where training sessions could be customized. Preliminary results set the foundation for building a platform capable of anticipating fatigue-induced injuries and a detailed recovery follow-up of lesions.

**Keywords:** Elite athletes fatigue · Fatigue analysis · Training planning · Evalu@ data centralizer · Prediction of injures by fatigue

## 1 Introduction

Medical departments specialized in sports medicine use the serum-Urea (SU) and serum creatine kinase (CK) as indexes of physical effort and training adjustment variables in soccer and several other disciplines [1,12]. Another primary aim of continuously reading these biomarkers is to reduce the probability of lesions by fatigue, injuries, rhabdomyolysis, and muscle malfunction [2,4,16].

The Urea is a final residual of Adenosine Triphosphate (ATP) generation that transforms the highly toxic intermediate step called ammonia into a molecule that the body can metabolize and excrete through urination [9].


H. Florez and M. F. Pollo-Cattaneo (Eds.): ICAI 2021, CCIS 1455, pp. 290–302, 2021.
https://doi.org/10.1007/978-3-030-89654-6_21

On their side, the CK enzyme is present in the fast-energy supply of the muscle contraction scheme. The CK serves as energy storage when phosphorylated (CPK), and its contribution lasts few seconds after the muscle reaches exhaustion [3,13,15,17]

Urea's production is associated with the metabolic process of energy generation called the Kreb's cycle. Hartmann and Mester [10] reported urea's levels in the range $[5-7] mmol.L^{-1}$ for 75% of 717 male athletes in the study where 6981 samples were gathered. The maximum urea reading in this study was $8.3 : .L^{-1}$. However, other authors reported different values for exigent sportive activities. E.g., Haralambie et al. [8] reported a range of $[5.69 - 9.61] : mmol.L^{-1}$ in the groping-test with the biggest dispersion among six groups athletes .

One can associate high levels of Urea with prolific production of ammonia. In sedentary subjects, ammonia's abundance could lead to edema, increased intracranial pressure, dementia, and coma due to specific receptors present in the brain that boost water demand [6,7].

As for the CK enzyme, the reading spam is higher than those of urea. Some authors report athletes' levels of CK in the range $[82 - 1083] : U.L^{-1}$ [13] while Hartmann and Mester reported low-entries in the range $[100, 350] : U.L^{-1}$ and high entries in $[1000 - 2000] : U.L^{-1}$.

High levels of CK in sedentary subjects are associated with tissue damage, disruption of muscle fibers or muscle disease, and myocardial and brain tissue abnormality [2].

Urea's increased reading levels in athletes are due to the abundant protein ingestion that covers their greater energetic demands. Although contradictory, human metabolism can yield high levels of Urea when dealing with starvation due to autolysis, a mechanism where the body destroys muscle fibers to obtain the needed protein that supports energy creation [14].

Concerning CPK, athletes' high levels are due to muscle cell disruption or changes in cell permeability [15] that leak the enzyme into plasma after continuous muscle contraction. Several authors coincide in determining eccentric contractions as the most prominent factor for serum CK proliferation[2].

Since these two biomarkers are metabolically separated and only found a commonplace while being transported by blood, one can study them together, create a new analyzing and visualization framework that will provide insights never explored before.

In the current work, Evalu@, an online tool designed to assist companies in assuring quality through constant and coherent evaluation in time that also provides artificial intelligence functionality [18], is used to gather creatinine phosphokinase (CPK) and Urea data.

In Evalu@, user-defined indexes and raw variables monitored over time are employed to create players' individualized active-recovering records. The methods used here apply to a myriad of analyses. Coaches could use the platform's flexibility to test and validate their training initiatives that might depend on other observable variables.

For the particular development, we gather CPK and Urea using Evalu@'s dynamic inputting methods. The software presents analytical instruments where coaches can see the players' performance in stand-alone fashion and comparison bases.

## 2 Materials and Methods

In-Blood saturation of CPK and Urea content in urination were recorded periodically before and after every soccer practice session during the pre-season and, before and after every official match, while attending two tournaments.

We configured Evalu@ software http://www.evalualos.com/test/ to perform evaluations of CPK and Urea over 47 professional soccer players along the 2017 season. The software proposes a simplistic methodology that antagonizes the customized software tendency. Evalu@ does not perform any operation before we configured it to do so through excel or comma separated values (CSV) files. Fortunately, Evalu@ is rendered fully operational after configuring two functional elements: the EV-Box and the EV-Surv. The EV-box is a class that instantiates aggrupation boxes with variable descriptors in name and number; therefore, anything can be evaluated and tracked over time within the software. The EV-Surv is a class that operates as a template for all sorts of evaluations, regardless of their grading scheme, qualitative or quantitative nature, the number of questions, and whether we organize the questions in endlessly nested structures holding categories and subcategories.

Since we have only two tracking variables in the accumulated-fatigue analysis are proposed, the software configuration takes no more than two minutes. Then, Evalu@ is ready to register and track the CK and Urea readings before and after the competition. The platform administrator grants access to specific services, including index creation, average analysis, rankings, versus mode to compare players, and automatic reporting. Hence, operators receive benefits for using the software.

The following steps end by configuring Evalu@ to gather CPK and Urea specific records and ensure reproducibility.

### 2.1 EV-box Configuration

After completing the procedure depicted in Fig. 1, the system can allocate the evaluating items or, in this case, the elite soccer players.

After adding the item (elite soccer player), data bout CPK and Urea can be gathered using the Evalu@ tool, as shown in Fig. 2.

### 2.2 EV-Surv Configuration

Like the process in Sect. 2.1, the software provides a template that should be customized according to the test's needs. The "m" character in the column "numerical values" tells the online interpreting system that evaluation will not

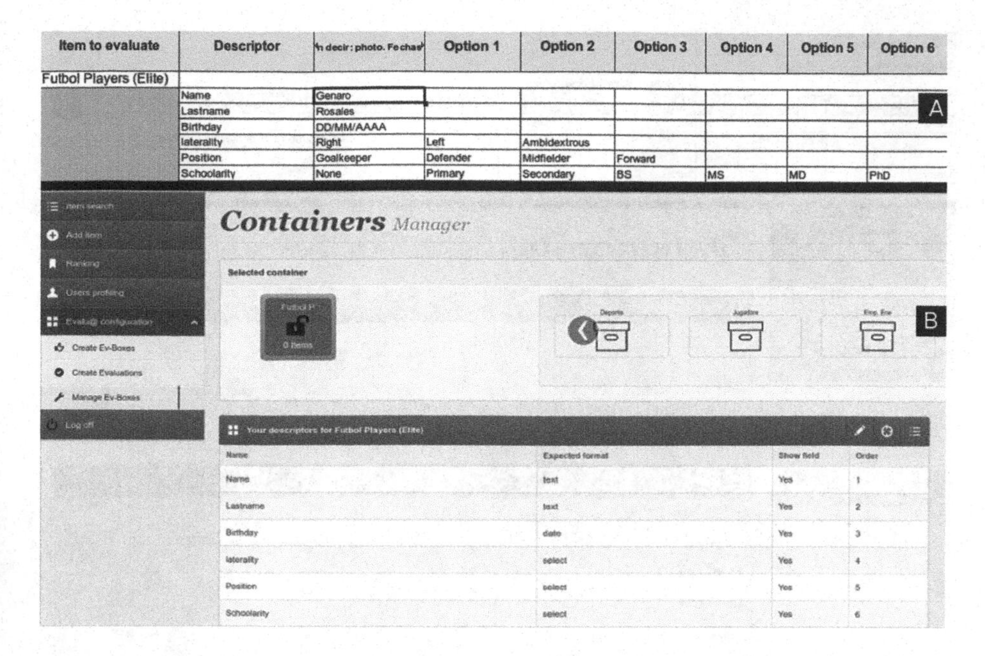

Item to evaluate
Descriptor
Option 1
Option 2
Option 3
Option 4
Option 5
Option 6
Futbol Players (Elite)
Name
Genaro
Lastname
Rosales
Birthday
DD/MM/AAAA
laterality
Right
Left
Ambidextrous
Position
Goalkeeper
Defender
Midfielder
Forward
Schoolarity
None
Primary
Secondary
BS
MS
MD
PhD
A
Containers Manager
Selected container
B
Create Ev-Boxes
Create Evaluations
Manage Ev-Boxes
Name
Expected format
Show field
Order
Name
text
Yes
1
Lastname
text
Yes
2
Birthday
date
Yes
3
laterality
select
Yes
4
Position
select
Yes
5
Schoolarity
select
Yes
6


**Fig. 1.** Configure the data gethering platform – step 1 –. In frame A, the excel sheet used as a template for creating the EV-Box where the elite players will be grouped. In frame B, the EV-Box manager where all the grouping boxes are listed.

use select controls with pre-defined value; instead, the HTML input control will be of type text. This configuration is useful when the variable range is too open, and a select control would have too many options making it unpractical by the moment of evaluating.

Readings of CPK and Urea from 47 professional soccer players were registered in Evalu@. The online software takes care of the dates in format DD-MM-YYY HH:MM:SS, who performed the test, and whether the information was inserted from a PC, a tablet, or a cell phone.

An initial glance variables' evolution along the season, allows the medical department to determine how each athlete reacts to the training. Recall that CPK and Urea readings are performed before and after high-intensity sessions (such as official matches). Therefore, the average of the exhaustion-recovering cycles is expected to trace a horizontal line that is characteristic of zero accumulated fatigue. See Fig. 9.

Since the two analyzed variables are independent and do not metabolically interfere (see Fig. 3 and 4), one can combine them along the time. A scatter of the two variables allows for a complete energy management profile, and a linear regression defines the metabolic influence of a variable over the other.

Finally, it is of interest to the medical department to create a fatigue signature for each player. This request is covered by creating a probability map. Further

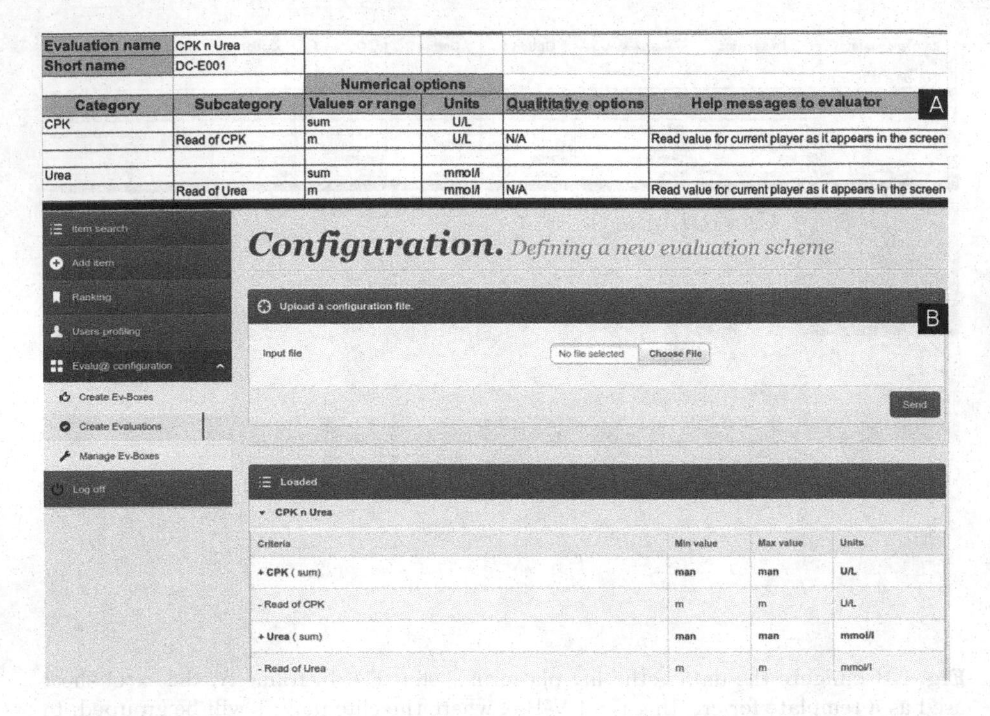


**Fig. 2.** Configure the data gathering platform – step 2 –. In frame A, the excel sheet used as template for creating the EV-Surv instance. In frame B, the EV-Surv in the evaluations manager where all the evaluations are listed and editions can be made.

intakes that displace or abruptly modify the probability map are the basements for anticipating events such as injuries by fatigue.

## 2.3 Variable Independence

After a detailed look at the metabolic pathways of CPK and urea creation, one can conclude that the two variables enzymes are independent and only have commonplace in the bloodstream. The reactors used in the detection are specific for each enzyme; therefore, the measuring methods keep the variable independence.

The creatine phosphate can be ingested or endogenously created by the liver and kidneys.

### 2.4 Studied Indexes

The club's medical department has been interested in constructing robust evidence of accumulated fatigue in every player; therefore, the active-resting sessions can be customized to every player's needs.

In every moment of the season, the software provides a meaningful analysis of the read values, and plotting capabilities; consequently, the medical department and the player, are aware of events regarding their immediate storage of energy and their long-term energy creation.

Due to the enzymatic independence in the pathway, a conjoint scattering and central tendency over the two variables provides the user with information about the accumulated fatigue; moreover, a probability map sets the bases for injures anticipation.

## 3 Results

### 3.1 Tracking CPK and Urea

Since the software controls all aspects of dates and evaluators, the operators can focus on the data registration tasks. The CPK tracking by itself does not provide meaningful information. However, the medical department can detect abrupt changes when analyzing a player in the current state compared to his historical records.

Recall that readings are done before and after matches or high-intensity sessions. That is the reason for a large number of peaks in the CPK profile in the shown player. Seasonal periods might favor prediction by machine learning implementations, but despite not obvious in the Fig. 6, seasonal patterns are unlikely due to the variable amount of minutes-per-match in the field that influences the management of energy differently in each athlete.

The urea is flat most of the time despite the fact that it was recorded in the same before-and-after matches fashion. However, peaks are expected in high-energy-demand periods. In the Fig. 7, the regions of higher variability in urea readings coincide with higher CPK readings, indicative of high physical exigence when lesions are not reported, as in the case of this midfielder (Fig. 8).

o comply with the legacy reporting format, we provide the group statistics that assume a Gaussian distribution. The numbers registered here are copied from the group statistics bar in Fig 5.

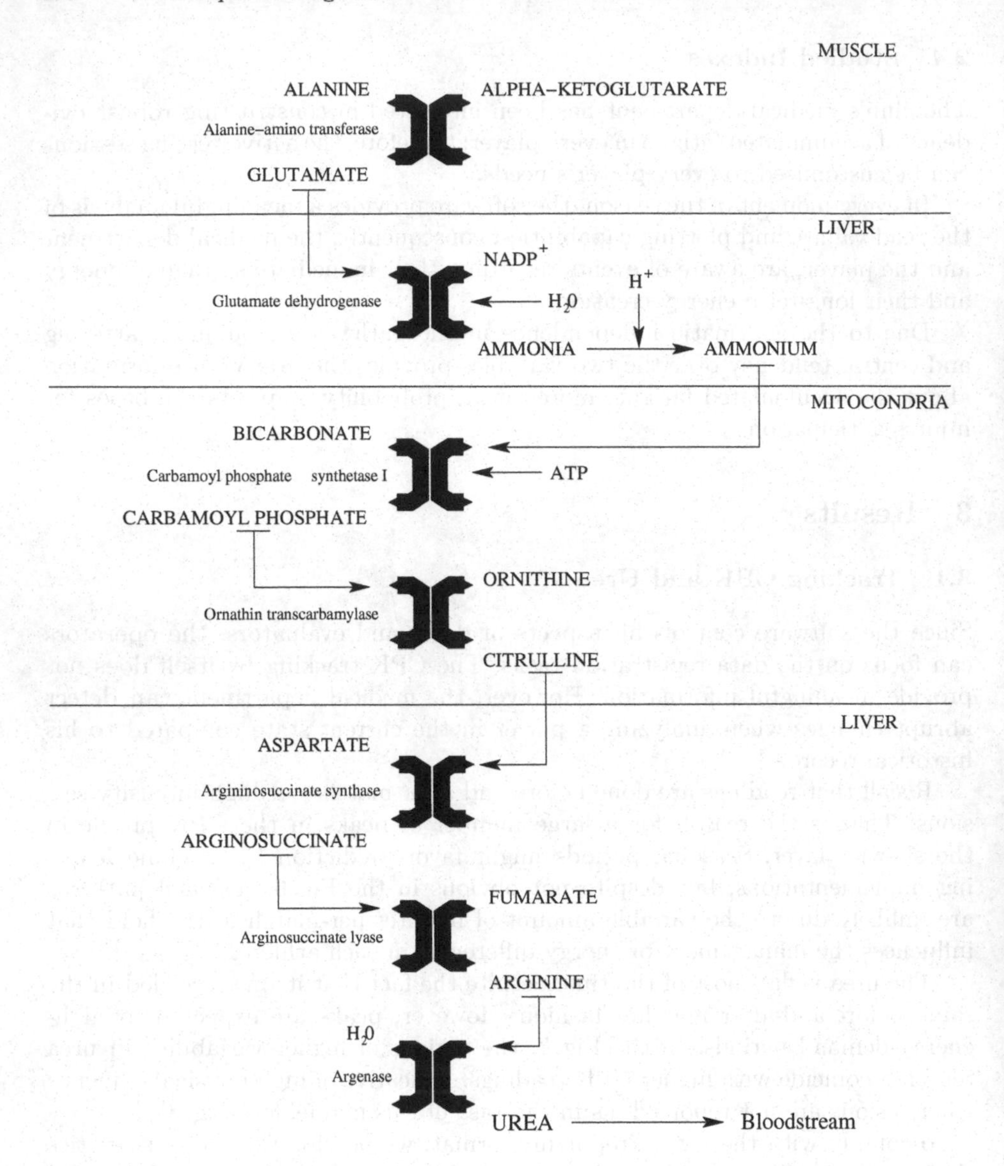


**Fig. 3.** Urea cycle. The urea cycle is intended to transform ammonia into the less toxic urea and dispose the waste product through urination

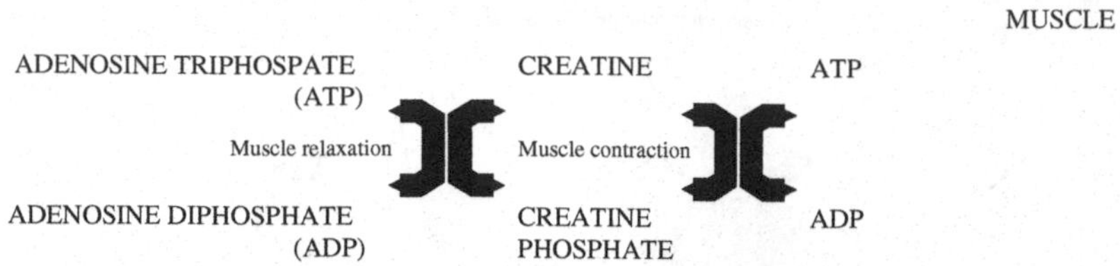

MUSCLE

ADENOSINE TRIPHOSPATE (ATP) — Muscle relaxation — ADENOSINE DIPHOSPHATE (ADP)

CREATINE — Muscle contraction — CREATINE PHOSPHATE

ATP — ADP

**Fig. 4.** The CPK provides the muscle with instant energy by releasing phosphates that turn ADP into ATP.

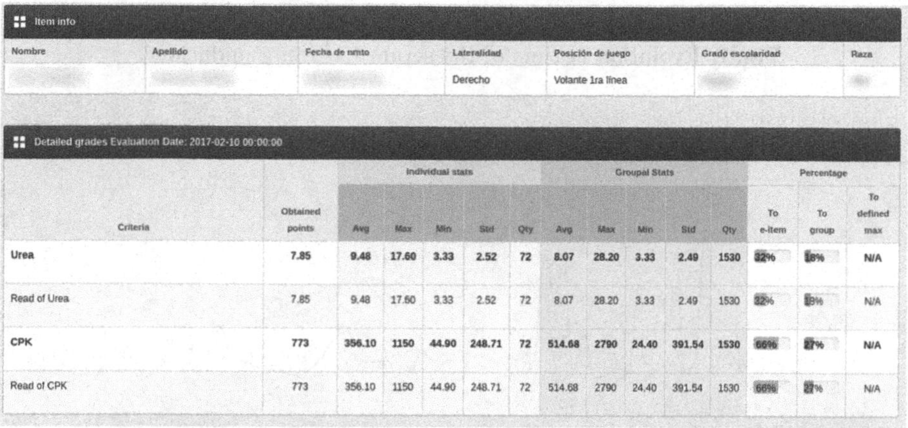

Item info

| Nombre | Apellido | Fecha de nmto | Lateralidad | Posición de juego | Grado escolaridad | Raza |
|---|---|---|---|---|---|---|
| | | | Derecho | Volante 1ra línea | | |

Detailed grades Evaluation Date: 2017-02-10 00:00:00

| | | Individual stats | | | | | Groupal Stats | | | | | Percentage | | |
|---|---|---|---|---|---|---|---|---|---|---|---|---|---|---|
| Criteria | Obtained points | Avg | Max | Min | Std | Qty | Avg | Max | Min | Std | Qty | To e-item | To group | To defined max |
| **Urea** | **7.85** | **9.48** | **17.60** | **3.33** | **2.52** | **72** | **8.07** | **28.20** | **3.33** | **2.49** | **1530** | **32%** | **18%** | **N/A** |
| Read of Urea | 7.85 | 9.48 | 17.60 | 3.33 | 2.52 | 72 | 8.07 | 28.20 | 3.33 | 2.49 | 1530 | 32% | 18% | N/A |
| **CPK** | **773** | **356.10** | **1150** | **44.90** | **248.71** | **72** | **514.68** | **2790** | **24.40** | **391.54** | **1530** | **66%** | **27%** | **N/A** |
| Read of CPK | 773 | 356.10 | 1150 | 44.90 | 248.71 | 72 | 514.68 | 2790 | 24.40 | 391.54 | 1530 | 66% | 27% | N/A |

**Fig. 5.** Instant respond of the system after pushing an evaluation. Note how individual and group statistics are created in real time. The records belong to a midfielder.

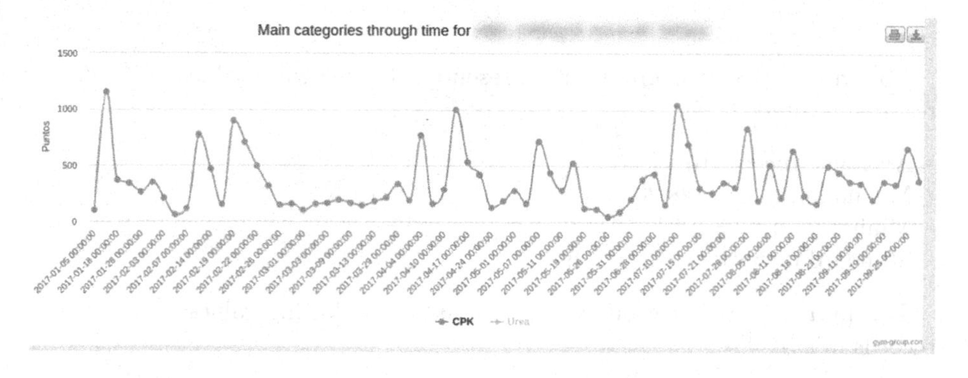


**Fig. 6.** Evolution in time of the serum CPK in a midfielder.

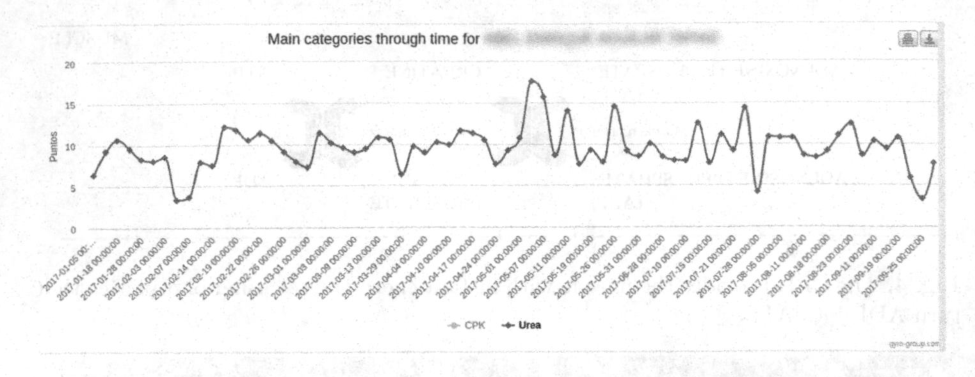


**Fig. 7.** Evolution in time of the serum Urea in a midfielder.

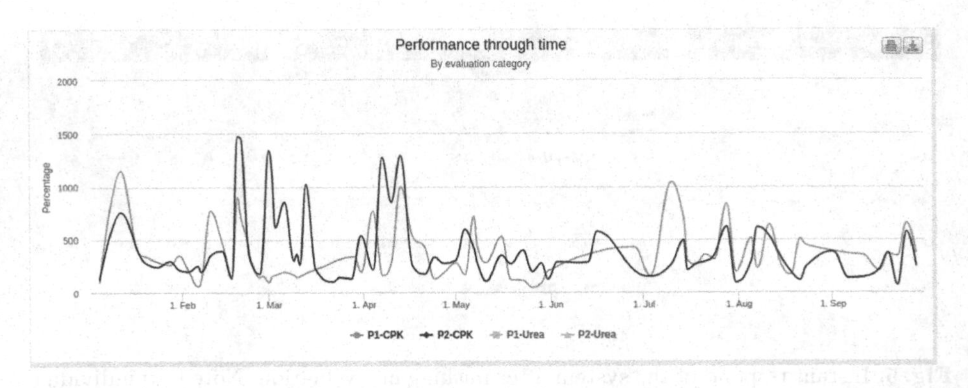


**Fig. 8.** Comparison scheme. Two midfielders compared in their CPK readings along time. Urea records are also available for comparison

For an $n = 1530$, the group urea presented the following values (records in $mmol.L^{-1}$):

- Average. $avg = 8.07$
- Maximum. $max = 28.20$
- Minimum. $min = 3.33$
- Standard deviation. $std = 2.49$

For an $n = 1530$, the group CK presented the following values (records in $U.L^{-1}$):

- Average. $avg = 514.68$
- Maximum. $max = 2790.00$
- Minimum. $min = 24.20$
- Standard deviation. $std = 391.54$

### 3.2 Joint Analysis of the Variables

As explained in Sect. 2.3, the two variables are orthogonal and, thus, can be combined in the joint analysis.

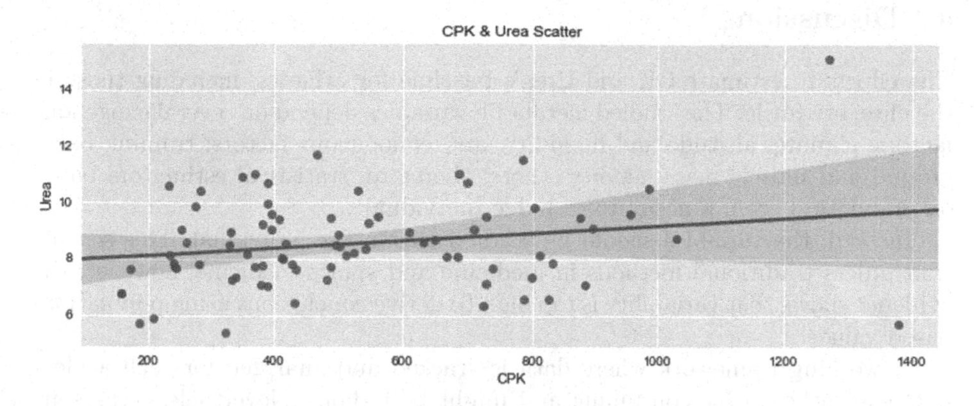


**Fig. 9.** Co-joint analysis of CPK and Urea for a defensive midfielder.

In this case, the player presents a positive slope in the linear regression, which is indicative of fatigue accumulation. Players with fewer minutes in the matches tend to present negative slopes in this analysis.

In general, the midfielders and flankers presented the higher slopes ($0.32 \pm 0.08$) in the co-joint curve for fatigue accumulation. This is coherent with the distances covered by athletes in the field according to their position. The slope in attackers was found to be $0.23 \pm 0.9$, $0.12 \pm 0.08$ for central defenders, and $0.10 \pm 0.02$ for goalkeepers. Some players into the position exercise presented negative slopes, indicative of sub-training. Several other grouping exercises could be performed, but these grouping approaches reverse the individualizing intention set in the conception of Evalu@ (Fig. 10).

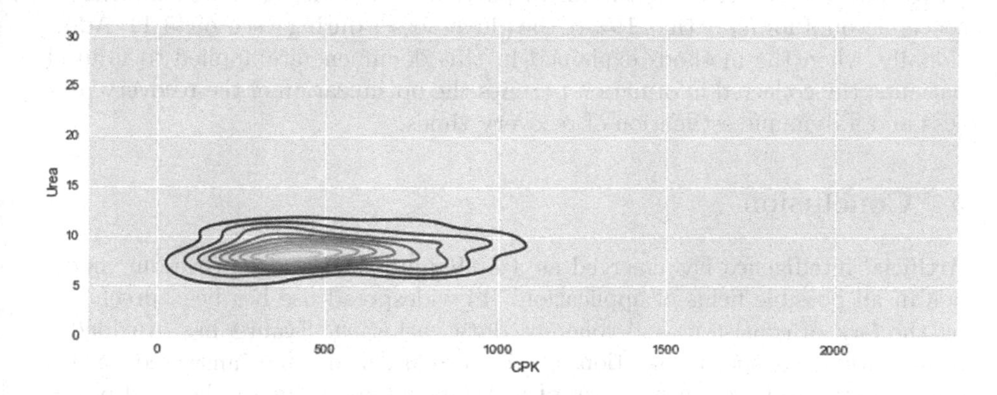


**Fig. 10.** Probabilistic signature created by a defensive midfielder during the season.

The probabilistic signature of fatigue is unique for each player and serves as a ground for inter-season comparisons, within the same player and with players that perform in the same position in the field.

## 4 Discussion

The efforts to estimate CK and Urea's baseline for athletes, including those in the elite, are futile. The studied metabolic variables depend on several exogenous factors: climate, altitude and humidity, and endogenous factors: training background and muscle mass, among others; therefore, statistical estimators based on population are not accurate to judge individuals.

Instead, the variables should be studied within the individual. This scheme contradicts traditional methods in medicine and sports medicine; however, the evidence shows that variability is too high to derive conclusions using population-based values.

A working framework where data is tracked and analyzed for each athlete independently is time-consuming and might be tedious. Nevertheless, the software presented in this document performs most of the repetitive tasks leaving to professionals the activities that give added value to their duties.

Training is the core of performance in elite competition. Historically, training in collective disciplines is done in groups without considering the individuals' inherent metabolic variability. Such a scheme could favor some athletes while over-train or sub-train others. With instruments like the one described in this development, coaches and physical trainers could customize the training sessions according to each player's fatigue profile and guarantee that athletes are in the best possible shape.

Thanks to the structural storage of variables and the possibility of querying historical results provided by the gathering platform, retrospective studies can be performed on injured individuals to anticipate injuries by fatigue. The use of retrospective data empowers artificial intelligence strategies. Losing players represent one of the most significant budget-draining factors affecting both, the player and the sportive institution. In particular, muscle associated injuries in soccer account for more than 15% of the player losses during a season [5,11] Additionally, when the methods explained in this document are applied to injured patients, the collected information permits the optimization of the recovery process and a dynamic estimation of recovery times.

## 5 Conclusion

Artificial intelligence has emerged as a technique that can improve our methods in all possible fields of application. Its widespread use has been precluded by the lack of consistent and coherent data gathering. Evalu@ has provided a framework where specific questions attained to any field can be answered. In this opportunity, we have created a set of instruments useful for coaches and physical trainers, where athletes' CPK and urea are analyzed to determine individual fatigue and management of long term energy. Since Evalu@ is configurable and indexes are unlimited, we have added other variables to the current analysis, and soon we will have sufficient features to implement a machine learning analysis to accurately anticipate injuries. The results of such an implementation will be part of a further delivery of the Evalu@ + Sports project.

## References


1. Aristotelis, G.: Biochemical changes from preparation to competitive period in soccer. International J. Sci. Cult. Sport **4**(June), 150–161 (2016). https://doi.org/10.14486/IntJSCS495, www.iscsjournal.com
2. Baird, M.F., Graham, S.M., Baker, J.S., Bickerstaff, G.F.: Creatine-kinase- and exercise-related muscle damage implications for muscle performance and recovery. J. Nutr. Metab. **2012** (2012). https://doi.org/10.1155/2012/960363
3. Brancaccio, P., Lippi, G., Maffulli, N.: Biochemical markers of muscular damage. Clin. Chem. Lab. Med. **48**(6), 757–767 (2010). https://doi.org/10.1515/CCLM.2010.179
4. Clarkson, P.M., Nosaka, K., Braun, B.: Muscle function after exercise-induced muscle damage and rapid adaptation. Med. Sci. Sports Exerc. **24**(5), 512–520 (1992). https://doi.org/10.1249/00005768-199205000-00004
5. Ekstrand, J., Hägglund, M., Waldén, M.: Epidemiology of muscle injuries in professional football (soccer). Am. J. Sports Med. **39**(6), 1226–1232 (2011). https://doi.org/10.1177/0363546510395879
6. Hagel, C., Krasemann, S., Löffler, J., Püschel, K., Magnus, T., Glatzel, M.: Upregulation of shiga toxin receptor CD77/Gb3 and interleukin-1$\beta$ expression in the brain of EHEC patients with hemolytic uremic syndrome and neurologic symptoms. Brain Pathol. **25**(2), 146–156 (2015). https://doi.org/10.1111/bpa.12166
7. Handley, R.R., et al.: Brain urea increase is an early Huntington's disease pathogenic event observed in a prodromal transgenic sheep model and HD cases. Proc. Natl. Acad. Sci. United States Am. **114**(52), E11293–E11302 (2017). https://doi.org/10.1073/pnas.1711243115
8. Haralambie, G., Berg, A.: Serum urea and amino nitrogen changes with exercise duration. Eur. J. Appl. Physiol. Occup. Physiol. **36**(1), 39–48 (1976). https://doi.org/10.1007/BF00421632
9. Hartmann, U., Mester, J.: Selected sport events. Medicine **32**, 209–215 (2000). https://doi.org/10.1097/00005768-200001000-00031, http://www.msse.org
10. Hartmann, U., Mester, J.: Training and overtraining markers in selected sport events. Med. Sci. Sports Exerc. **32**(1), 209–215 (2000). https://doi.org/10.1097/00005768-200001000-00031
11. Junge, A., Dvořák, J.: Football injuries during the 2014 FIFA World Cup. Brit. J. Sports Med. **49**(9), 599–602 (2015). https://doi.org/10.1136/bjsports-2014-094469
12. Majumdar, P.: Physiological analysis to quantify training load in badminton. Brit. J. Sports Med. **31**(4), 342–345 (1997). https://doi.org/10.1136/bjsm.31.4.342
13. Mougios, V.: Reference intervals for serum creatine kinase in athletes. Brit. J. Sports Med. **41**(10), 674–678 (2007). https://doi.org/10.1136/bjsm.2006.034041
14. Myers, V.C., Riger, M., Benson, O.O.: The formation of urea in autolysis. Proc. Soc. Exp. Biol. Med. **23**(6), 474–476 (1926). https://doi.org/10.3181/00379727-23-3021

15. Shen, Y.Q., Tang, L., Zhou, H.M., Lin, Z.J.: Structure of human muscle creatine kinase. Acta Crystallographica Sect. D Biol. Crystallogr **57**(8), 1196–1200 (2001). https://doi.org/10.1107/S0907444901007703
16. Silva, J.R., et al.: Acute and residual soccer match-related fatigue: a systematic review and meta-analysis. Sports Med. **48**(3), 539–583 (2017). https://doi.org/10.1007/s40279-017-0798-8
17. Wallimann, T., et al.: Some new aspects of creatine kinase (CK): compartmentation, structure, function and regulation for cellular and mitochondrial bioenergetics and physiology. BioFactors **8**(3–4), 229–234 (1998). https://doi.org/10.1002/biof.5520080310
18. Yepes-Calderon, F., Yepes Zuluaga, J.F., Yepes Calderon, G.E.: Evalu@: an agnostic web-based tool for consistent and constant evaluation used as a data gatherer for artificial intelligence implementations. In: Florez, H., Leon, M., Diaz-Nafria, J.M., Belli, S. (eds.) ICAI 2019. CCIS, vol. 1051, pp. 73–84. Springer, Cham (2019). https://doi.org/10.1007/978-3-030-32475-9_6

# Heart Rate Variability: Validity of Autonomic Balance Indicators in Ultra-Short Recordings

Jose Gallardo[1,4](✉), Giannina Bellone[2,4], and Marcelo Risk[1,3,4]

[1] Universidad Tecnológica Nacional (UTN, FRBA), Buenos Aires, Argentina
jgallardo@frba.utn.edu.ar

[2] Research (BIOMED), Catholic University of Argentina (UCA), Buenos Aires, Argentina

[3] Instituto de Medicina Traslacional e Ingeniería Biomédica (IMTIB), Buenos Aires, Argentina

[4] Consejo Nacional de Investigaciones Científicas y Técnicas (CONICET), Buenos Aires, Argentina

**Abstract.** This work aims to find the minimum recording times of ultra-short heart rate variability (HRV) that enable the analysis of autonomic activity indexes. Samples covering 5 min are employed to extract SS and S/PS from the Poincaré diagram from a group of 23 subjects. The RR series, extracted from the electrocardiogram signal, were recorded for 300 s at rest – used as the gold standard – and at intervals of 60, 90, 120, 180 and 240 s to perform the concordance analysis with the gold standard derived indexes. We used four different techniques of concordance: Spearman, Bland, and Altman correlation, and Cliff's Delta.

The SS times within records of 120 s were equivalent to those of short-term HRV and S/PS of 90 s. Also, ultra-short HRV indexes were similar to those obtained for the short-term HRV analysis. Such a reduction in measurement times will allow expanding the use of HRV to monitor the state of health and well-being and help physical trainers achieve better performance in the registration and processing of the information obtained.

The results motivate the conduct of new studies to analyze the behavior of these indicators in different populations, using different pre-processing methods than the RR series.

**Keywords:** HRV · ultra-short HRV · Poincaré · SS · S/SP

## 1 Introduction

Heart rate variability (HRV) is a physiological phenomenon, where the regulatory action of the autonomic nervous system (ANS) on the sinus node produces beat-to-beat changes. HRV is analyzed non-invasively by recording the electrocardiogram (ECG), in whose tracing R waves are identified. The time distance


H. Florez and M. F. Pollo-Cattaneo (Eds.): ICAI 2021, CCIS 1455, pp. 303–315, 2021.
https://doi.org/10.1007/978-3-030-89654-6_22

between two consecutive R waves (cardiac period) is defined as the duration of the RR interval [1]. Another proposed method for HRV, is the measurement of the photoplethysmographic arterial pulse wave (PPG), in this non-invasive method, changes in blood volume in peripheral tissue, which are caused by heart rate, are recorded [2,3].

HRV is used as an indicator parameter of the level of cardiovascular health, since it is possible to study the changes in the sympathetic-vagal balance of the cardiac response, resulting in a very useful tool in the study and analysis of the evaluation of individuals who perform sports activities [4,5]. Depending on the duration of the records, HRV studies are classified into long-term studies (24-hour records) [6] and short-term studies (5 min records) [1,7,8].

Recently, short-term HRV analyzes for monitoring the health and well-being of people increased due to the use of portable devices (smart watches and bracelets) and applications for smartphones, a record of 5 min is very long in comparison with those required for measurements of blood glucose, blood pressure, body temperature, heart rate, oxygen saturation and body weight. Therefore, there is a trend towards the use of ultra-short-term records for HRV, with times less than 5 min. [9], This would allow, on the one hand, practically real-time monitoring and, on the other hand, increase the use of HRV in the daily clinic [10]. Although there are still no clear guidelines or a standard method to analyze ultra-short-term HRV, the methods currently used in short-term HRV are; time domain analysis using a statistical approach, frequency domain analysis, by power spectral estimation, geometric and nonlinear analysis [1].

Poincaré analysis is a geometric method that allows to analyze, visually or through calculations of the plot parameters, beat-by-beat information on the behavior of the heart and can be a better way to monitor dynamic change in function. autonomic [11], Fig. 1. Cross axis (*SD1*) reflects short-term changes in RR and is directly related to parasympathetic activity. However, the longitudinal axis (*SD2*) not as well defined, but appears to be inversely proportional to sympathetic activity [12]. The relationship *SD2*/*SD1*, is typically used to assess the interaction between sympathetic-parasympathetic activity (*SD2*/*SD1*) [13]. However, its interpretation is still unclear, when both terms of the relationship increase or decrease simultaneously, so [12] proposed new indices of stress level (SS) (stress score) and the sympathetic-parasympathetic balance (S/PS) (sympathetic-parasympathetic ratio). The use of SS as a direct index for the evaluation of sympathetic activity and of S/PS (SPS) as an indicator of autonomous balance used in athletes, improve the physiological interpretation of HRV [12].

Ultra-short HRV is a new methodology for HRV analysis because of this, recent studies suggest minimum lengths of the RR series depending on the index to be calculated. [9], but we still have not found studies where SS and SPS are analyzed. Therefore, the objective of this work is to find the minimum recording times for ultra-short HRV, required for the SS and SPS indices to be equivalent to those obtained using HRV of 5 min.

The reduction of the recording times of the SS and SPS indicators, on the one hand, will increase their application in the evaluation of sports performance in athletes and, on the other, improves the knowledge of the physiological meaning of HRV.

## 2 Materials and Methods

### 2.1 Subjects

Twenty-three subjects, twelve women, between the ages of 21 and 25 voluntarily participated in the experiment. Subjects were healthy and medication free. Each subject provided their written consent before starting the experiment and all subjects were free to terminate the experiment at any time. This study was approved by the Ethics and Research Committee of the University of Quilmes, Buenos Aires, Argentina (04/07/2016/No.3).

### 2.2 HRV Analysis

The ECG signal of each subject was obtained in the supine resting position, for 5 min (300 s) and 2 min of stabilization time, with a Holter, by means of the DI lead, at a sampling frequency 225 Hz. ECG signal was processed, 50 Hz notch filter was applied to remove interferences from the electrical power supply network, the non-linear trend of the ECG was eliminated by means of its difference with a polynomial of order 6 previously adjusted to the signal and then oversampled at 1 KHz, to improve the determination of the maximum point of the R wave. The modified Pan-Tompkins algorithm was used [14] to detect R peaks, in the QRS complex of the ECG, followed by visual inspection, then the RR time interval series was calculated. In this series, the linear trend and the continuous component were eliminated, in addition to being able to detect atypical beats, an impulse rejection filter was used [15,16]. Atypical beats were replaced by interpolation by another interval whose value was the average of the six RR intervals neighboring the one considered artifact. [17,18]. Subsequently, for the analysis of ultra-short HRV, to each series of 300 s (the "Gold Standard" (GS) series) windows of different time lengths (60, 90, 120, 180 and 240 s) were extracted, starting from the beginning, considering the natural way of making the measurements.

### 2.3 Poincaré Geometric Domain

The Poincare plot, Fig. 1, describes the nature of the fluctuations of the RR interval when plotting a certain interval *RRn* versus the next *RRn+1*. To extract HRV parameters, fit the graph to an ellipse where the width relative to the axis $y = x$, *SD1* (Standard deviation short-term RR intervals) can be considered as an indicator of parasympathetic activity. The length of the ellipse on the axis is *SD2* (Standard Deviation RR intervals in the long term) is an inverse function of sympathetic activity, (1), (2):

$$SD1 = \sqrt{Var(\frac{RR_n - RR_{n+1}}{\sqrt{2}})} \quad (1)$$

$$SD2 = \sqrt{Var(\frac{RR_n + RR_{n+1}}{\sqrt{2}})} \quad (2)$$

where *RRn* is the nth interval, *RRn+1* is the interval following the nth interval and *Var* represents the variance.

It was calculated from the indixes extracted from the Poincaré diagram [12]; $SS = 1000 * 1/SD2$ and $SPS = SS/SD1$. For the series of 300 s, GS, of short-term HRV and the ultra-short-term, of 60, 90, 120, 180 and 240 s.

All signal processing and HRV calculations were performed by the GNU/Linux environment, using algorithms developed in GNU Octave, version 4.2.2, Copyright (C) 2018, John W. Eaton and others.

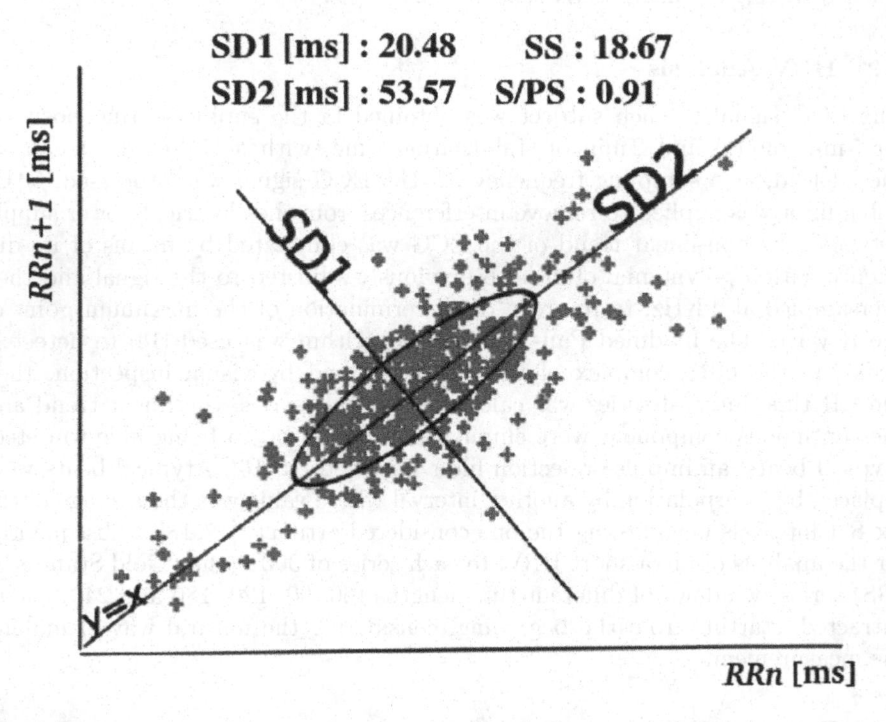


**Fig. 1.** Poincaré graph, used to determine HRV parameters. The abscissa represents the time interval *RRn*, the ordinate the next interval $RR_{n+1}$. The fit to an ellipse is represented, the identity line $y = x$, $SD1$ width of the ellipse (perpendicular dispersion of RR intervals), $SD2$ length of the ellipse (longitudinal dispersion of RR intervals) with respect to the identity line $y = x$.

### 2.4 Statistical Analysis

To analyze the equivalence between the GS series and each of the ultra-short series, the algorithm proposed by [19] for the rest condition and also the concurrent validity criterion similar to that used in the work of [9], because it is an analysis between series of the same sample.

In addition, non-parametric statistics were applied, given the lack of normality in some series [19]. The Spearman correlation coefficients (*rho*) [20] of the

SS series of the GS and each of the SS indicators corresponding to the HRV-US series of 60, 90, 120, 180 were calculated and 240 s, in the same way we proceeded with the SPS indicator.

To establish the degree of agreement, the Bland-Altman (BA) [21] method was used between the GS series and each of the ultra-short series for each index. The bias was calculated for each index as the median of the difference between the HRV measurements of the GS and that of each ultra-short series. The test of Wilcoxon was used to analyze bias.

In addition, the Cliff Delta, ($d$) [22], was calculated to quantify the effect size, a quantitative measure of the magnitude of a phenomenon, in this case, the bias of the parameter measurements of HRV of the different series time lengths and the GS. The value of $d$ can be between 1 and –1, being the threshold proposed by CLiff for $|d| < 0.15$ a "negligible" effect size, $|d| < 0.33$ "small", $|d| < 0.47$ "medium" and "large", for another situation.

In all the tests, values of $p < 0.05$ were considered statistically significant.

The relative error percentage ($e$) between the GS series and each of the ultra-short RR series was also calculated for each SS and SPS index, using (3) and (4) [23]:

$$e(\%) = \frac{median(GS_{SS}) - median(serieX_{SS})}{median(GS_{SS})} * 100 \tag{3}$$

$$e(\%) = \frac{median(GS_{SPS}) - median(serieX_{SPS})}{median(GS_{SPS})} * 100 \tag{4}$$

with X = 60, 90, 120, 180 and 240. All algorithms for statistical calculations were developed with free software R version 3.6.3, Copyright (C) 2020. The R Foundation for Statistical Computing. Platform: x86_64-pc-linux-gnu (64-bit).

### 2.5 Results

The ECG signals of the 23 subjects in a resting situation were processed and as a result the series of RR time intervals were obtained, which prior to the HRV analysis using the Poincaré plot, editing and artifact correction methods were applied.

Table 1 shows the values of the medians and quartiles (1st; 3rd) of the SS and SPS indicators for the different lengths of the series of time intervals RR, 60, 90, 120, 180, 240 and 300 s for both indices the values are practically constant.

**Table 1.** HRV indexes calculated after pre-processing RR series. Values for each index expressed in median and [1st; 3rd] quartile.

| RR series | SS | SPS |
|---|---|---|
| 60 | 13.61 (10.63 21.12) | 0.39 (0.21 1.26) |
| 90 | 14.10 (11.48 17.36) | 0.48 (0.23 1.07) |
| 120 | 13.48 (10.53 17.16) | 0.47 (0.24 1.05) |
| 180 | 13.18 (9.92 17.62) | 0.47 (0.23 1.09) |
| 240 | 13.47 (10.07 17.88) | 0.45 (0.24 1.01) |
| 300 | 12.87 (10.10 16.85) | 0.43 (0.25 0.96) |

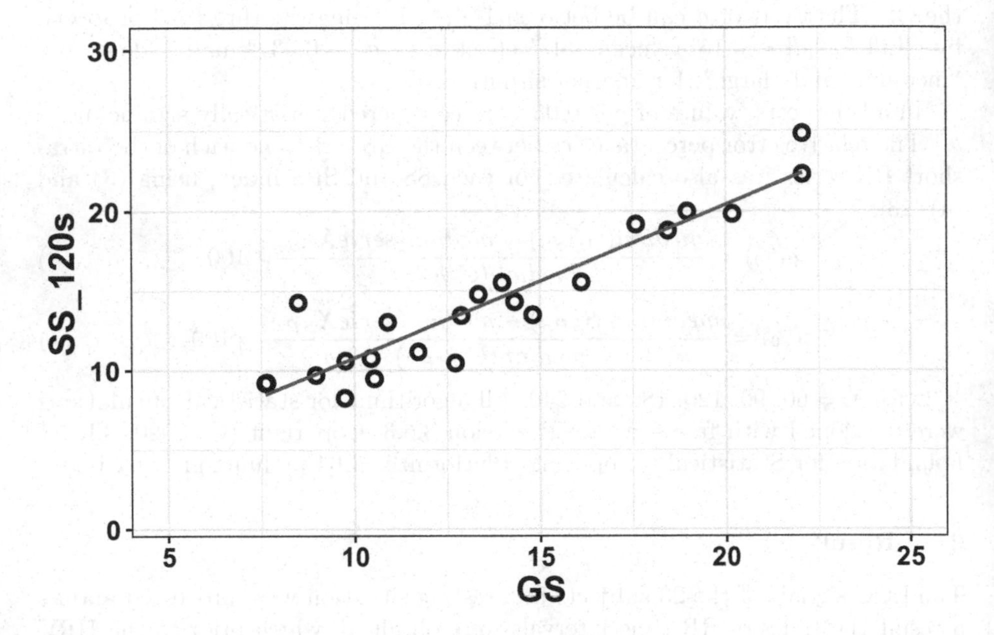


**Fig. 2.** Spearman correlation GS vs SS 120 s of the SS index, $rho = 0.90$, $p = 3.28e\text{-}06$.

Figure 2 is the plot corresponding to the analysis of the Spearman correlation of the SS index, between the GS series and the 120 s series.

Figure 3 shows the Bland-Altman plot corresponding to the concordance analysis between the 120 s series and the GS of the SPS index, showing the lines corresponding to the Bias, the median of the difference between the series, and the LoA 95%, estimated using the 2.5 and 97.5 percentiles. Note that GS is represented on the abscissa axis instead of the average of the measurements [24].

In Table 2, the results of the different methods used in the comparison between the GS series and each of the RR series for the SS indicator are presented.

In this table, a positive relationship is observed between the variables that as the length of the series increases, the values of ($rho$) the correlation coefficient of Spearman increase with a $rho = 0.90$, $p = 3.28\text{e-}06$, from the series of 120 s and the values corresponding to $d$ the Delta of Cliff decrease, being negligible for the 60 s series and small for the 90 s series and negligible for the 120, 180 and 240 s series, while the % error ($e$) is less than 10 % in each series.

The results of the Wilcoxon test, and the Bias values show that it is equal to 0, only, for the series of 120 s. The negative values obtained from the Bias show an overestimation of the GS series. The LoA of 95 % decreased with the temporary increase in the series, evidencing the decrease in LoA.

Figure 4 shows the graph of the Spearman correlation between GS and the series of 90 s corresponding to the SPS index. An accumulation of values below 0.5 can be seen in the graph.

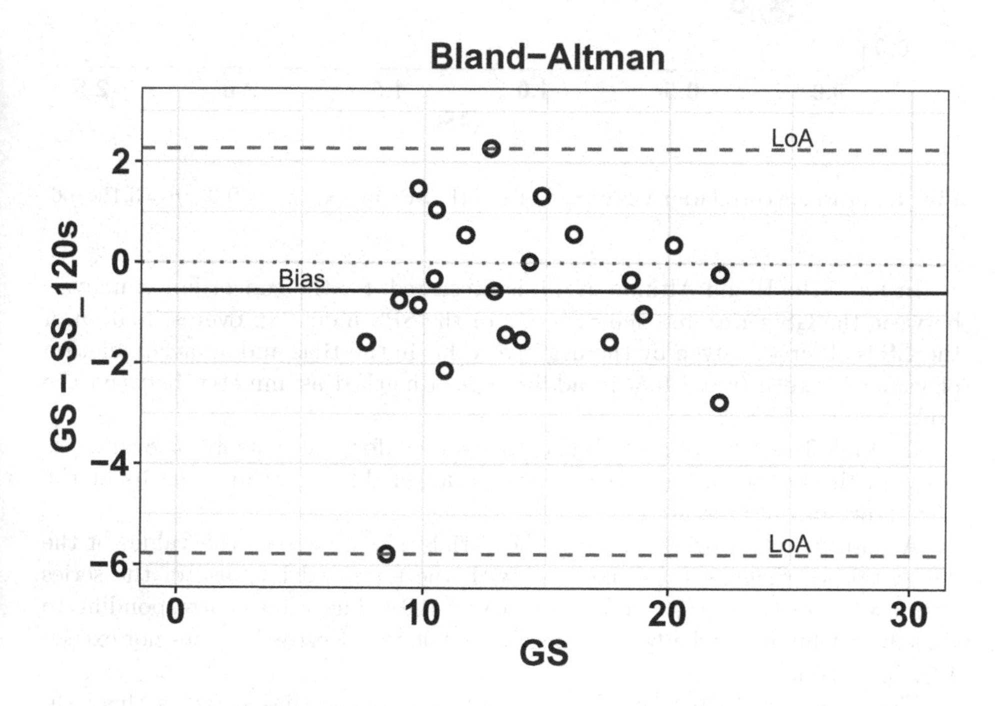


**Fig. 3.** Bland-Altman concordance analysis of GS vs SS 120 s series, of the SS index.

**Table 2.** Table SS.

| SS RR series (s) | Correlation($rho$) ($p$) | Cliff ($d$) (c.i. 95%) | % error ($e$) | Wilcoxon ($p$) | Bias (1st; 3rd) | LoA 95% |
|---|---|---|---|---|---|---|
| 300 vs 60 | 0.71 (1.98e–04) | 0.14 (–0.20 0.45) | 5.74 | 0.04 | –1.51 (–3.90 0.57) | -14.65 5.10 |
| 300 vs 90 | 0.84 (1.71e–06) | 0.16 (–0.18 0.48) | 9.48 | 0.02 | –1.04 (–2.56 0.35) | -6.67 1.89 |
| 300 vs 120 | 0.90 (3.28e–06) | 0.09 (–0.25 0.41) | 4.66 | 0.09 | –0.56 (–1.54 0.46) | -5.78 2.27 |
| 300 vs 180 | 0.96 (2.45e–06) | 0.06 (–0.28 0.38) | 2.37 | 0.03 | –0.58 (–1.36 0.35) | -3.90 1.61 |
| 300 vs 240 | 0.98 (1.99e–06) | 0.06 (–0.28 0.38) | 4.61 | 0.01 | –0.48 (–0.95 –0.07) | -3.23 1.40 |

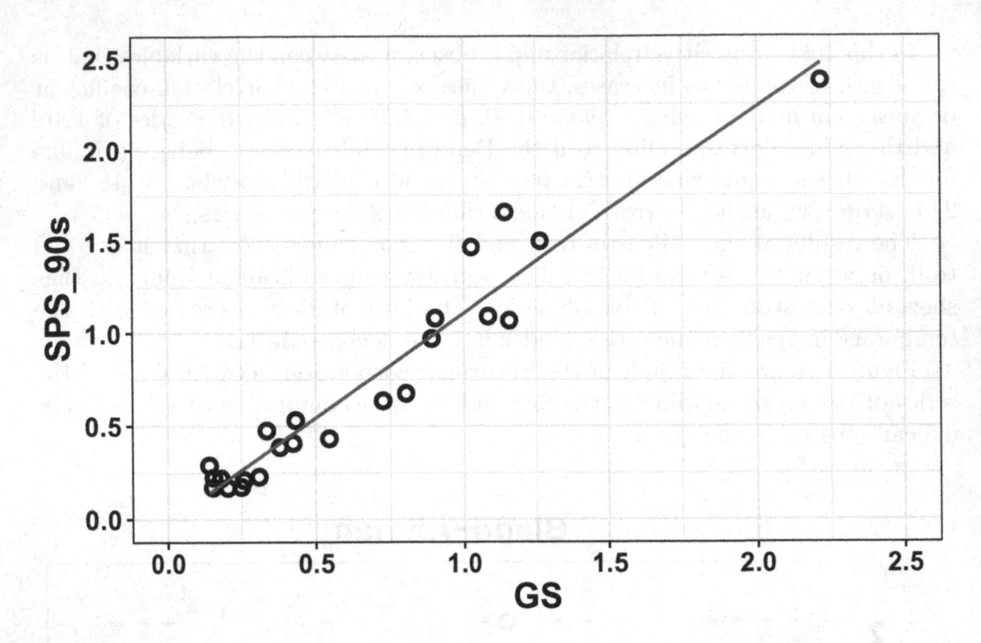


**Fig. 4.** Spearman correlation GS vs SPS 90 s of the SPS index, $rho = 0.93$, $p = 3.15\text{e–}06$.

In Fig. 5 the Bland-Altman graph is presented, for the concordance analysis between the GS series and the SPS 90 s of the SPS index, an overestimation of the GS is observed, given by the negative value in the Bias and an accumulation of values between 0 and 0.5, in addition to a marked asymmetry between the LoA.

In Table 3, the results of the methods used for the concordance analysis between the GS series and each of the series of different time lengths of the SPS indicator are presented.

A positive relationship between the variables is observed, the values of the correlation coefficients ($rho$) increase with the temporal increase of the series with a value of $rho = 0.90$ from the series of 120 s. The values corresponding to $d$ Delta of Cliff are negligible for all series, while the % error ($e$) does not exceed 10% in all series.

The results of the test of Wilcoxon and those of the Bias show for this indicator that it is possible that the Bias is 0 for the different series and the negative values show an overestimation of the GS series. The values corresponding to LoA 95 % decrease with the temporal increase of the series and improve symmetry from the 90 s series.

Table 3 shows the results of the comparison between the series corresponding to the SSP indicator, a high correlation value is observed between the GS series and the 60 s series with a value of $rho = 0.95$, $p = 2.79\text{e–}06$, in coefficient of then correlation of Spearman ($rho$) that is incremented until arriving at $rho = 0.99$, $p = 1.71\text{e–}06$, for the series of 240 s. The values corresponding to the Cliff Delta

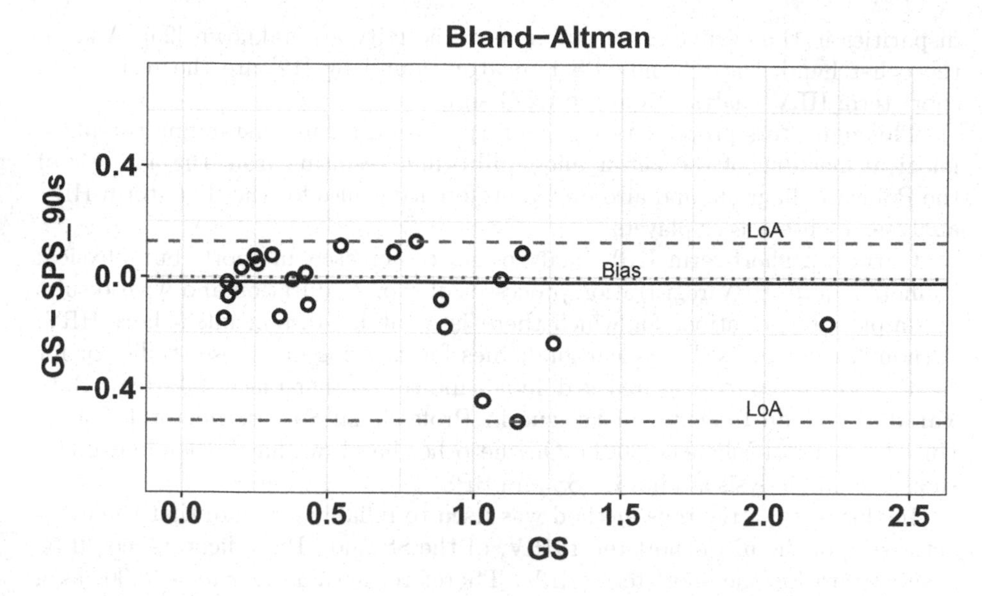


**Fig. 5.** Bland-Altman concordance analysis of GS vs SPS 90 s series, of the SPS index.

($d$) denote negligible differences between the GS series and each of the 90, 120, 180 and 240 s series. The % error ($e$) is less than 10% in all comparison cases, and their absolute values are similar for all comparisons between series. The values corresponding to bias are also negative, showing an overestimation of the GS series in each comparison, although the Bias values are lower than in the case of SS and in the Student test it is observed that the bias could take the value null in any of the comparisons, the values extracted from the Bland-Altman LoA 95% graphs show a decrease with the temporal increase of the series, evidencing a better level of agreement between each time series and the GS.

### 2.6 Discussion

The signal from the ECG was used to perform the HRV analysis, this is because although both signals, ECG and PPG, have been used, in different methods, to evaluate autonomic activity and their results are considered equivalent in subjects healthy [25]. But, on the one hand, differences in their results have been observed under various physiological conditions and the reasons for their

**Table 3.** Table SPS.

| SPS RR series (s) | Correlation($rho$) ($p$) | Cliff ($d$) (c.i. 95%) | % error ($e$) | Wilcoxon ($p$) | Bias (1st; 3rd) | LoA 95% |
|---|---|---|---|---|---|---|
| 300 vs 60 | 0.80 (7.02e–06) | 0.01 (–0.33 0.34) | 9.63 | 0.52 | –0.02 (–0.15 0.08) | –1.04 0.74 |
| 300 vs 90 | 0.93 (3.15e–06) | 0.05 (–0.28 0.38) | 10.15 | 0.15 | –0.02 (–0.14 0.06) | –0.51 0.13 |
| 300 vs 120 | 0.95 (2.79e–06) | 0.03 (–0.30 0.35) | 9.48 | 0.71 | 0.00 (–0.09 0.06) | –0.22 0.18 |
| 300 vs 180 | 0.98 (2.05e–06) | 0.05 (–0.29 0.372) | 8.85 | 0.29 | –0.02 (–0.08 0.03) | –0.32 0.22 |
| 300 vs 240 | 0.99 (1.71e–06) | 0.06 (–0.28 0.38) | 4.91 | 0.20 | –0.03 (–0.06 0.01) | –0.26 0.23 |

disparities in the evaluation of autonomous activity are unknown [26]. And on the other hand, the SS and SPS indicators found by [12] are the result of a short-term HRV analysis from the ECG signals.

This indicators propose a clearer interpretation for understanding the physiological meaning of the autonomic equilibrium resulting from the analysis of the Poincaré diagram, and also provide reference values for the short-term HRV analysis in elite soccer players.

Currently, short-term HRV analyzes have increased in sport, but physical trainers require HRV registration processes of shorter duration and with results of simple interpretation, for which there is great interest in ultra-short HRV. Although there are still no clear guidelines for carrying out these studies or for the times required to measure the different indices (Forner-Llácer FJ. Heart Rate Variability as an Indicator of Fatigue in Professional Soccer. Doctoral Thesis. Univ Catholic of Valencia, 2021). On the other hand, we have not found ultra-short record analysis studies for SS and SPS.

In this work, a rigorous method was used to reliably evaluate that the characteristics of the ultra-short-term HRV, of the SS and SPS indicators, could be a substitute for the short-term HRV. Therefore, a value of $rho = 0.90$, as a minimum, was adopted as the cutoff criterion. As stated by [9], this threshold ensures that an acceptable measurement of HRV-US represents at least 81 % of the variability ($rho^2 = 0.81$) in the value corresponding to 5 min and a extremely low risk of making a Type 1 error.

The analysis of the results shows a similarity in the values of the medians in most of the series of each index with differences in the series of 60 s, this is probably due to the fact that a time window of 60 s is small for the measurement of these indices [27], in addition an increase in general is observed in all the indicators of the levels of agreement calculated as the length of the series increases, both for SS and SPS.

The ultra-short series of the SS indicator, whose correlation value meets the cut-off criterion is given from the 120 s series, in addition a negligible difference value is observed for the Delta of Cliff, very low values of the error ($e$) and this behavior is similar in the series of 180 and 240 s. The values extracted from the Bland-Altman graphs reflect a high level of agreement between the GS series and the 120 s series, similar to GS with 180 s and GS with the 240 s time length, showing the similarity between the series. This allows the exchange of one series for another, that is, the 120 s series of the SS indicator is equivalent to the GS series, of 5 min, of the same indicator.

A similar analysis for the SPS indicator allows us to establish that the 90 s series meets the established cut-off condition given that $rho = 0.93$, $p = 3.15e\text{-}06$, the values of the Delta of Cliff show negligible differences with the GS, so it is concluded that the series are similar to the GS from the time-length series of 90 s, that is, the series of 90 s is equivalent to the GS series of the SPS indicator.

This study is preliminary and has limitations, on the one hand, the sample size of 23 subjects, which should be larger to investigate the effect of the variables associated with HRV, such as age, sex, physical activity, etc. On the other hand, different stabilization times prior to the records were not analyzed. In addition, for this analysis, no other position than the supine was used, nor was another segment of the signal different at the beginning, although we consider it possible to keep the ECG signal stable, avoiding artifacts and using robust methods for pre-processing. and edition of the series RR, GS, the results of the comparison between the series present similar levels of agreement. However, all these limitations require further studies.

### 2.7 Conclusions

Ultra-short recordings of 120 s were obtained for SS, an index for the evaluation of sympathetic activity and 90 s for SPS, an indicator of autonomic balance, equivalent to those obtained for short-term HRV analysis (5 min). This reduction in measurement times will allow an increase in the use of HRV to monitor people's health and well-being. Enabling physical trainers to obtain better performance in the registration and processing of the information obtained.

The results found motivate the carrying out of new studies to analyze the behavior of these indicators in different populations and with the use of different methods of pre-processing and editing of the RR series.

## References


1. Malik, M.: Heart Rate Variability, 1st edn. Futura Publishing Company Inc., New York (1995)
2. Girčys, R., Kazanavičius, E., Maskeliūnas, R., Damaševičius, R., Woźniak, M.: Wearable system for real-time monitoring of hemodynamic parameters: implementation and evaluation. Biomed. Signal Process. Control **59** (2020). https://doi.org/10.1016/j.bspc.2020.101873
3. Lee, M.S., Lee, Y.K., Lim, M.T., Kang, T.K.: Emotion recognition using convolutional neural network with selected statistical photoplethysmogram features. Appl. Sci. **10**(10) (2020). https://doi.org/10.3390/app10103501
4. Bellenger, C.R., Fuller, J.T., Thomson, R.L., Davison, K., Robertson, E.Y., Buckley, J.D.: Monitoring athletic training status through autonomic heart rate regulation: a systematic review and meta-analysis. Sports Med. **46**(10), 1461–1486 (2016). https://doi.org/10.1007/s40279-016-0484-2
5. Jiménez Morgan, S., Molina Mora, J.A.: Effect of heart rate variability biofeedback on sport performance, a systematic review. Appl. Psychophysiol. Biofeedback **42**(3), 235–245 (2017). https://doi.org/10.1007/s10484-017-9364-2
6. Risk, M., Sobh, J., Barbieri, R., Armentano, R., Ramirez, A., Saul, J.: Variabilidad de las señales cardiorespiratorias. Parte 2: Variabilidad a largo plazo. Rev. Argentina Bioingenieria **2**(2) (1996)
7. Risk, M., Sobh, J., Barbieri, R., Armentano, R., Ramirez, A., Saul, J.: Variabilidad de las señales cardiorespiratorias. Parte 1: Variabilidad a corto plazo. Rev. Argentina Bioingenieria **2**(1) (1996)
8. Gamero, L.G., Risk, M., Sobh, J.F., Ramirez, A.J., Saul, J.P.: Heart rate variability analysis using wavelet transform. Comput. Cardiol. 177–180 (1996). https://doi.org/10.1109/cic.1996.542502

9. Shaffer, F., Shearman, S., Meehan, Z.M.: The promise of ultra-short-term (UST) heart rate variability measurements. Biofeedback **44**(4), 229–233 (2016). https://doi.org/10.5298/1081-5937-44.3.09, http://www.aapb-biofeedback.com/doi/10.5298/1081-5937-44.3.09
10. Massaro, S., Pecchia, L.: Heart rate variability (HRV) analysis: a methodology for organizational neuroscience, vol. 22 (2019). https://doi.org/10.1177/1094428116681072
11. Bracale, U., Rovani, M., Bracale, M., Pignata, G., Corcione, F., Pecchia, L.: Totally laparoscopic gastrectomy for gastric cancer: Meta-analysis of short-term outcomes. Minim. Invasive Ther. Allied Technol. **21**(3), 150–160 (2012). https://doi.org/10.3109/13645706.2011.588712
12. Orellana, N., Orellana, J.N., De La Cruz Torres, B., Cachadiña, E.S., De Hoyo, M., Cobo, S.D.: Two new indexes for the assessment of autonomic balance in elite soccer players. Int. J. Sport. Physiol. Perform., 452–457 (2014). https://doi.org/10.1123/ijspp.2014-0235
13. Gallardo, J., Bellone, G., Plano, S., Vigo, D., Risk, M.: Heart rate variability: influence of pre-processing methods in identifying single-night sleep-deprived subjects. J. Med. Biol. Eng. **41**(2), 224–230 (2021). https://doi.org/10.1007/s40846-020-00595-8
14. Sathyapriya, L., Murali, L., Manigandan, T.: Analysis and detection R-peak detection using Modified Pan-Tompkins algorithm. In: Proceedings 2014 IEEE International Conference Advanced Communication Control Computer Technology, ICACCCT 2014, no. 978, pp. 483–487 (2015). https://doi.org/10.1109/ICACCCT.2014.7019490
15. McNames, J., Thong, T., Aboy, M.: Impulse rejection filter for artifact removal in spectral analysis of biomedical signals. In: Proceedings 26th Annual International Conference IEEE EMBS, San Fransico, CA, USA, vol. 1, 145–148 (2004). https://doi.org/10.1109/IEMBS.2004.1403112, http://www.ncbi.nlm.nih.gov/pubmed/17271626
16. Thuraisingham, R.A.: Preprocessing RR interval time series for heart rate variability analysis and estimates of standard deviation of RR intervals. Comput. Methods Progr. Biomed. **83**(1), 78–82 (2006). https://doi.org/10.1016/j.cmpb.2006.05.002
17. Wejer, D., Makowiec, D., et. al. Struzik, Z., Żarczyńska-Buchowiecka, M.: Impact of the editing of patterns with abnormal RR intervals on the assessment of heart rate variability. Acta Phys. Pol. B **45**(11), 2103 (2014). https://doi.org/10.5506/APhysPolB.45.2103, http://www.actaphys.uj.edu.pl/vol45/abs/v45p2103
18. Aubert, A.E., Ramaekers, D., Beckers, F.: Analysis of heart rate variability in unrestrained rats: assessment of method and results. Med. Biol. Eng. Comp. **60**, 197–213 (1999)
19. Pecchia, L., Castaldo, R., Montesinos, L., Melillo, P.: Are ultra-short heart rate variability features good surrogates of short-term ones? state-of-the-art review and recommendations. Healthc. Technol. Lett. **5**(3), 94–100 (2018). https://doi.org/10.1049/htl.2017.0090, http://digital-library.theiet.org/content/journals/10.1049/htl.2017.0090
20. Guzik, P., et al.: Correlations between the poincaré plot and conventional heart rate variability parameters assessed during paced breathing. J. Physiol. Sci. **57**(1), 63–71 (2007). https://doi.org/10.2170/physiolsci.RP005506, http://joi.jlc.jst.go.jp/JST.JSTAGE/physiolsci/RP005506?from=CrossRef
21. Bland, J.M., Altman, D.G.: Statistical methods for assessing agreement between two methods of clinical measurement. Lancet **1**(8476), 307–310 (1986). https://www-users.york.ac.uk/~mb55/meas/ba.pdf

22. Cliff, N.: Ordinal Methods for Behavioral Data Analysis, 1st edn. Psychology Press, Hove (2014). https://doi.org/10.4324/9781315806730, https://www.taylorfrancis.com/books/9781315806730
23. Nardelli, M., Greco, A., Bolea, J., Valenza, G., Scilingo, E.P., Bailon, R.: Reliability of lagged poincaré plot parameters in ultra-short heart rate variability series: application on affective sounds. IEEE J. Biomed. Health Inf. **2194**(c), 1–1 (2017). https://doi.org/10.1109/JBHI.2017.2694999, http://ieeexplore.ieee.org/document/7903605/
24. Munoz, M.L., et al.: Validity of (Ultra-)Short recordings for heart rate variability measurements. PLoS One **10**(9), 1–15 (2015). https://doi.org/10.1371/journal.pone.0138921
25. Lu, G., Yang, F., Taylor, J.A., Stein, J.F.: A comparison of photoplethysmography and ECG recording to analyse heart rate variability in healthy subjects. J. Med. Eng. Technol. **33**(8), 634–641 (2009). https://doi.org/10.3109/03091900903150998
26. Mejía-Mejía, E., Budidha, K., Abay, T.Y., May, J.M., Kyriacou, P.A.: Heart rate variability (HRV) and pulse rate variability (PRV) for the assessment of autonomic responses. Front. Physiol. **11**(July), 1–17 (2020). https://doi.org/10.3389/fphys.2020.00779
27. Shaffer, F., Ginsberg, J.P.: An overview of heart rate variability metrics and norms. Front. Public Health **5**(September), 1–17 (2017). https://doi.org/10.3389/fpubh.2017.00258, http://journal.frontiersin.org/article/10.3389/fpubh.2017.00258/full

# Image Processing

# An Improved Machine Learnings Diagnosis Technique for COVID-19 Pandemic Using Chest X-ray Images

Joseph Bamidele Awotunde[1], Sunday Adeola Ajagbe[2](✉), Matthew A. Oladipupo[3], Jimmisayo A. Awokola[4], Olakunle S. Afolabi[5], Timothy O. Mathew[6], and Yetunde J. Oguns[7]

[1] Department of Computer Science, University of Ilorin, Ilorin, Nigeria
awotunde.jb@unilorin.edu.ng

[2] Department of Computer Engineering, Ladoke Akintola University of Technology LAUTECH, Ogbomoso, Nigeria
saajagbe@pgschool.lautech.edu.ng

[3] Salford University, The Greater Manchester, Manchester, UK
m.a.oladipupo@edu.salford.ac.uk

[4] Mobile and e-Computing Research Group, Faculty of Computing and Informatics, LAUTECH, Oyo State, Ogbomoso, Nigeria
jaawokola@lautech.edu.ng

[5] Department of Computer Science, University of Abuja, Abuja, Nigeria

[6] The Federal Polytechnic, Ilaro, Nigeria
timothy.mathew@federalpolyilaro.edu.ng

[7] Department of Computer Studies, The Polytechnic Ibadan, Ibadan, Nigeria

**Abstract.** The pandemic produced by coronavirus2 (COVID-19) has confined the world, and avoiding close human contact is still suggested to combat the outbreak although the vaccination campaigns. It is expectable that emerging technologies have prominent roles to play during this pandemic, and the use of Artificial Intelligence (AI) has been proved useful in this direction. The use of AI by researchers in developing novel models for diagnosis, classification, and prediction of COVID-19 has really assist reduce the spread of the outbreak. Therefore, this paper proposes a machine learning diagnostic system to combat the spread of COVID-19. Four machine learning algorithms: Random Forest (RF), XGBoost, and Light Gradient Boosting Machine (LGBM) were used for quick and better identification of potential COVID-19 cases. The dataset used contains COVID-19 symptoms and selects the relevant symptoms of the diagnosis of a suspicious individual. The experiments yielded the LGBM leading with an accuracy of 0.97, recall of 0.96, precision of 0.97, F1-Score of 0.96, and ROC of 0.97 respectively. The real-time data capture would effectively diagnose and monitor COVID-19 patients, as revealed by the results.

**Keywords:** Chest X-ray · Coronavirus · Machine learning · Random forest · XGBoost · Extra trees · Light gradient boosting machine


H. Florez and M. F. Pollo-Cattaneo (Eds.): ICAI 2021, CCIS 1455, pp. 319–330, 2021.
https://doi.org/10.1007/978-3-030-89654-6_23

## 1 Introduction

The emerging of coronavirus a family of novel severe contagious respiratory syndrome called (COVID-19) has caused the greatest public health challenge globally, after the pandemic of the influenza outbreak of 1918. According to the World Health Organization, as of 5:02 pm CEST, 23rd November 2020, 58,425,681 confirmed, and 1,385,218 (2.4%) death cases globally [1]. Globally, people spend much of their time indoor to contain or avoid people infected with the virus. Until now there has been a rapid increase in diverts research works to find a lasting solution to this worldwide threat.

Around ten billion people from various cities and towns are facing serious problems with self-quarantine themselves at home with lockdown measures. In recent time, the medical tools and supplies are of increases due to the pandemic, and almost all the existing one vitally needs necessary replenishment. To seek medical help, potential citizens have to leave their homes, and that post real risks for the unaffected citizens and a hole on the efforts of isolation and quarantine put in place to contain the outbreak globally [2, 3]. Also, the shortage of medical personnel, lack of proper medical equipment, and the shortage of isolation clinics have prompted the policymakers to encourage those with suspected symptoms, or mild stages to stay at home. Hence, there is an urgent need for a home-based diagnostic test to serve as an alternative method that will be cost-effective, flexible, and can deliver invaluable solutions for the self-isolated patient [4].

Fast and accurate COVID-19 detection is becoming increasingly crucial for preventing infection and supporting patients in preventing disease progression. Soft Computing (SC) techniques such as fuzzy logic, neural networks, and genomics have been shown to be helpful in identifying diseases [5, 6]. They will aid decision-making by enabling for immediate patient seclusion and care [7]. Several approaches for detecting COVID-19 infections have been proposed, but none have yet attained the requisite detection sensitivity. Several approaches for diagnosing COVID-19 have been developed recently, but none of them consider the impact of feature weight on classifier judgment [8–10]. Because both techniques treat all functions in the same way, consistency suffers. If COVID-19 cases are misdiagnosed, the epidemic will spread to the rest of the world [11]. Giving each piece a weight or a rank, on the other hand, would assist the classifier in making accurate decisions, enhancing diagnostic accuracy.

Researchers now have enough data to anticipate the pandemic's rapid growth tendency using learning and prediction methods, and they're doing a good job of it. This sort of forecasting allows us to comprehend the potential implications, which could have a significant impact on socioeconomic growth. The initial imaging modality is a chest X-ray, which plays an important role in the detection of COVID-19. The retrieval of similar images based on the visual content of data is known as content-based image retrieval (CBIR). It is a modern and active research field. Content-based image retrieval is used to handle ever-growing visual data in the digital imaging era. The retrieval system aids in the development of an image-based diagnostic system that is automated.

Furthermore, the retrieval procedure of COVID-19 affected Chest X-Ray images is more significant because to the abundance of chest X-Ray images compared to other modalities and the restricted availability of adequately annotated data. Based on similarity measurement, the technique extracts similar images from a huge database of collected chest X-Ray images. Manually retrieving similar photos from a large database takes a long time and demands a lot of human resources.

Given COVID-19's X-ray image improvements, the study plans to apply an improved ML technique for extracting COVID-19's imagery features to give a medical study prior to the pathogenic examination, so saving a critical period for ailment control. In general, the goal of this work was to find the most appropriate algorithm for early COVID-19 identification using chest X-ray datasets. The dataset includes 372 patients, and screening data was evaluated to identify patients with COVID-19 or other respiratory infectious disorders from the Kaggle database repository. The study employed four ML classification techniques, the results of the classifiers were compared in this study.

The proposed method has the following contributions:

(i) The study use ML algorithms for the diagnosis, and prediction based on COVID-19 dataset.
(ii) The significant of ML are highlighted in smart healthcare system to solve the real-time diagnosis, and prediction during the COVID-19 pandemic.
(iii) The study validates the Four ML Extra Trees, Random Forest (RF), XGBoost, and Light Gradient Boosting Machine (LGBM) on COVID-19 diagnosis results based on normalization of image data.
(iv) The performance evaluation of the proposed techniques was carried out using accuracy, precision, F1-score, Receiver operating characteristic (ROC), and recall, and the techniques were compared using state-of-the-art existing technique.

This research is organized as follows. Section 2 presents review of related work. Section 3 presents the methodology, sample dataset and proposed flowchart for the system. Section 4 experimentation of the proposed model, data preprocessing and diagnosis model, parameter and metrics were discussed. Section 5 presents the results and discussion, and Sect. 6 conclusion and future works for the realization of efficient uses of ML-based technique in fighting the COVID-19 pandemic.

## 2 Review of Related Work

Classification of biomedical images is an emerged topic of study to form the care sector additional promising. Amid the aftermath of the outbreak of COVID-19, many current international research round the world have utilized subtle ML algorithms and AI based mainly technologies to higher perceive the pattern of infective agent transmission, more increase diagnostic speed and accuracy. Not only that, but it would benefit health-care workers by limiting their contact with COVID-19 patients. Furthermore, ML/AI-based systems aid in the development of novel, effective treatment approaches as well as the theoretical classification of the most vulnerable individuals. Infectious and non-infectious illness screening and detection of COVID-19 on chest X-ray scans are predicted to improve with the use of ML/AI-based approaches. In the medical field, AI is used to guide clinicians in making decisions about what they model, rather than to replace human experiences.

Following the International Health Regulations (IHR) (2005) [12], WHO scheduled an emergency conference on the 30th of January, 2020, to discuss the new coronavirus 2019 epidemic in China. According to them, China acted quickly to assist other countries in combating COVID-19. According to a mathematical model published in The Lancet [13], the epidemic's expansion can be halted if the transmission rate drops to 0.25. Santosh [14] proposes using AI-driven techniques to predict the type of COVID-19 outbreaks. Deep learning is currently being used extensively in the medical picture arena.

In the medical imaging sector, the authors in [15] and [16] are examples of the use of deep learning applications in healthcare sectors. Deep learning network architectures are trained on a significant amount of data. The training is frequently computationally hard and time-consuming, requiring expensive resources. Pereira et al. used a multi-class and hierarchical classification strategy to classify COVID-19 damaged chest X-ray pictures in [17]. To solve the problem of data imbalance, they used a resampling algorithm. They used texture descriptors to extract features, which were then fed through a CNN model that had already been trained. They created a dataset for work evaluation called RYDLS-20.

Attempts are currently being made to develop novel diagnostic procedures using ML algorithms. In this context, research shows the promise of AI and machine learning techniques by providing novel models for quick and accurate SARS-CoV-2 diagnosis, such as: Advanced convolutional neural network structures were utilized to recognize CXR pictures by author in reference [18]. To deal with a number of gift abnormalities in the dataset, transfer learning was implemented. With a 97% accuracy, two datasets from somewhat different repositories are employed to research images in two classes: COVID-19 and the traditional state (sensitivity of 98.66% and specificity of 96.46%). For COVID-19 identification, the authors in [19] presented a CXR-b intelligent system. A vector support machine (SVM) classifier was employed to identify deep characteristics of the CXRs. The study concluded that ResNet50 paired with SVM produced the highest results (accuracy, F1 score, Matthew's correlation coefficient (MCC), and Kappa, respectively, of 95.38%, 95.52%, 91.41%, and 90.76%).

In [20], the authors employed generative adversarial networks (GAN) to detect respiratory illness from CXRs in another study. They overcame the issue of over-processing and asserted their dominance by creating more GAN images. There were 5863 CXR images in the data set for both natural and nuclear uses. To diagnose pneumonia, deep learning structures such as GoogLeNet, AlexNet, Squeznet, and ResNet were employed. ResNet18 and GAN outperformed other deep transport models, according to their research. Instead of using a typical CNN network, the authors in [21] recommended using a capsule network called COVID-CAPS to communicate with a data collection. COVID-CAPS has been tuned to achieve 95.7% accuracy, 90% sensitivity, and 95.8% specificity.

## 3 Materials and Methods

The ML models play a significant role in the decision-making process, even when the data volume is very large [22, 23]. The method of implementing data processing techniques for particular fields requires specifying types of data such as velocity, variety, and volume. Normal data analysis modeling involves the model of the neural network, the model of classification and the process of clustering, and the implementation of efficient algorithms as well [24, 25]. Data can be generated from different sources with specific types of data, and it is also important for the development of methods capable of handling data characteristics.

### 3.1 Dataset

The dataset is a chest X-ray image that was retrieved from the Kaggle database. https://www.kaggle.com/bachrr/covid-chest-xray is the link to the dataset. Sample datasets are shown in Figs. 1 and 2. Figure 1 shows a sample of COVID-19-positive chest X-ray images, while Fig. 2 shows a sample of COVID-19-negative chest X-ray images. 357 chest X-ray datasets were used in the preparation phase, while 126 chest X-ray datasets were used in the testing phase. The strategy employed a 80:20 dataset split, meaning that 65% of the information was utilized for preparation and 35% was used for the research.

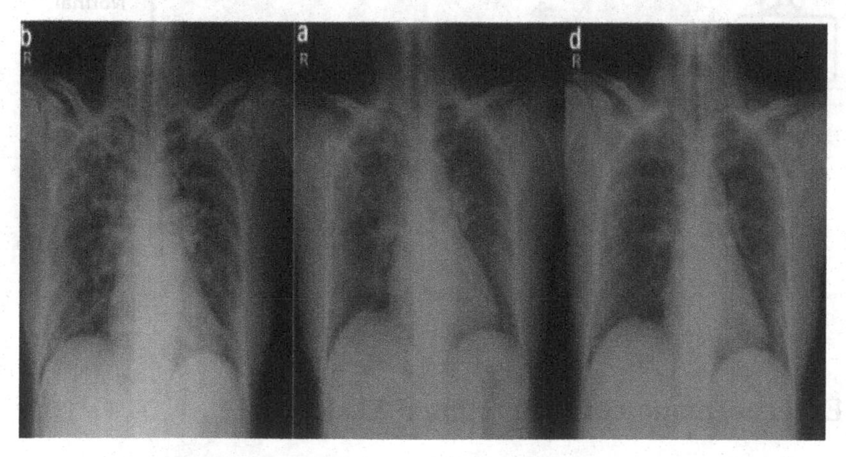


**Fig. 1.** Sample of X-ray image for an infected person

### 3.2 The Flowchart for the Propose Model

As shown in Fig. 2, the proposed architecture for completing the diagnostic procedure and detection of the presence of COVID-19 using X-ray pictures consists of three phases: (i) preprocessing; (ii) feature extraction; and (iii) ML classification phase (Fig. 3).

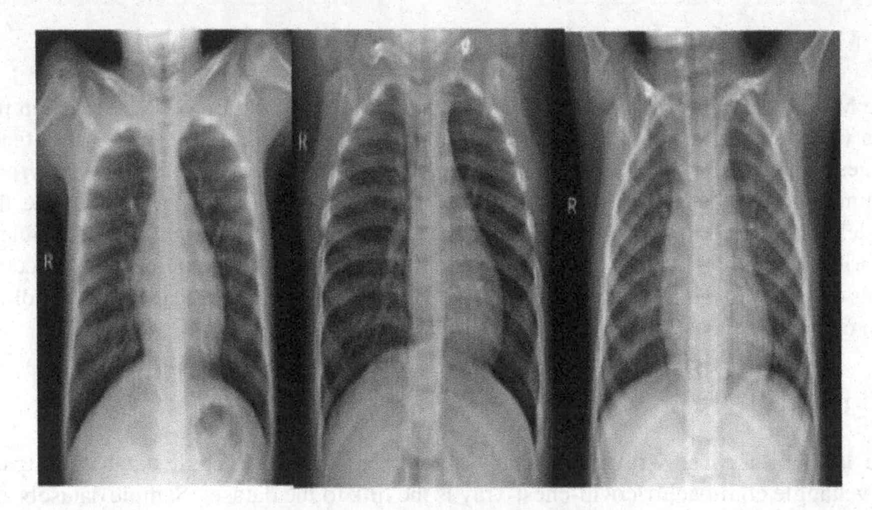

Fig. 2. Sample of X-ray of Normal person

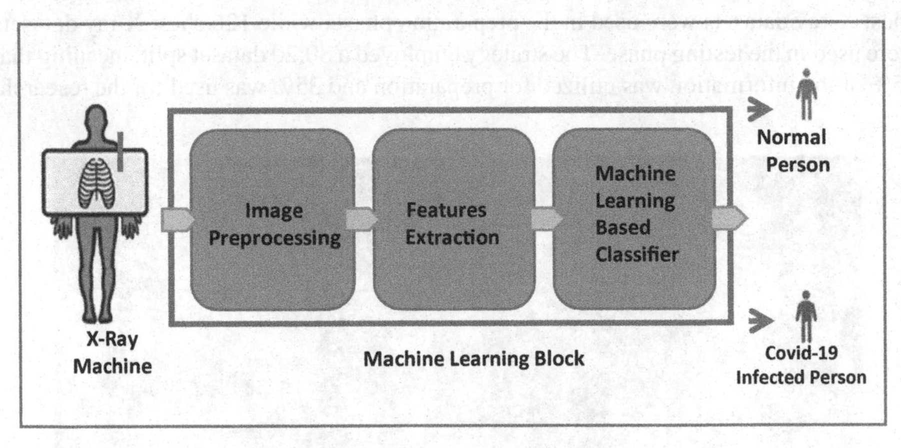


Fig. 3. Flowchart of the Proposed system

## 4 Experimentation of the Proposed Model

The experimentation of the proposed model basically entails: 1) pre-processing of the dataset, and 2) diagnosis using the dataset, and the performance metrics was used to evaluate the proposed model.

### 4.1 Pre-processing

Presenting the study as a classification problem: Let $S = \{(x_1, y_1), (x_2, y_2), \cdots (x_n, y_n)\}$ be the set of training instances of dimension $d$. $Y = \{y_1, y_2 \cdots, y_n\}$ be the set of labels (COVID-19, PTB, and Normal) where $x_i$ is a feature with corresponding $y_i$ label is a set of features according to classification task definition. The initial step taken in the image classification model is the extraction of features. This is pertinent when the features extracted which is also the input data is extremely large and difficult to process in its raw form. The selection of important features will resolve this problem; this was done according to [26, 27].

### 4.2 Diagnosis Models, Parameters and Metrics

This section describes the models, their parameters, and metrics used for the multi-class classification experiment. The task was performed with four classifiers namely: LGBM, Extra Trees, RF, XGBoost. Table 1 presents the parameters of the model used in this study. A confusion matrix is a table representing the prediction performance of a model. The row and column represent the predicted and the actual class respectively as shown in Table 2. The formula for metrics computed from the confusion matrix is presented by Eqs. (1)–(5).

**Table 1.** Model parameters applied in this study

| Models | Parameters |
|---|---|
| *LGBM* | *n_estimators = 100, random_state = 10* |
| *Extra Trees* | *random_state = 10* |
| *RF* | *n _ estimators = 700, max _ depth = 3* |
| *XGBoost* | *Learning _ rate = 0.05, max _ depth = 40, max _features = 1.0, min _ samples _ leaf = 4, n _ estimators = 100, random _ state = 10, subsample = 0.8* |

Basic evaluation criteria and criterion based on classification accuracy utilizing True Negative (TN), True Positive (TP), False Negative (FN), and False Positive (FP) (FP). Table 2 show the confusion matrix.

**Table 2.** Confusion matrix

| | |
|---|---|
| *TP* | *FP* |
| *TN* | *FN* |

$$Accuracy = \frac{TP + TN}{TP + TN + FP + FN} \tag{1}$$

$$Precision = \frac{TP}{TP + FP} \tag{2}$$

$$Recall(TPRate) = \frac{TP}{TP + FN} \tag{3}$$

$$(FPRate) = \frac{FP}{FP + TN} \tag{4}$$

$$F1 - Score = 2 \times \frac{Precision \times Recall}{Precision + Recall} \tag{5}$$

# 5 Results and Discussion

This section presents the results and discusses the discoveries in the study. The experiment was performed with Jupyter notebook with sklearn [28]. libraries on the Anaconda platform. All experiments were performed on an Intel® core™ i5–7200 CPU @ 2.50 GHz to 2.70 GHz Pentium Windows computer with 8 GB RAM. The images were manually cropped to remove some unwanted background images and noise. Then, they were resized to 128 × 128, flattened, and converted to greyscale before their features were extracted. Firstly, Fig. 1 presents a sample CXR image and their corresponding transformed HOG images for the 3 different classes (COVID-19, NORMAL, and PTB). The feature vectors were divided into 80:20 train-test split ratio. As discussed in Sects. 5.1, the results of the comparison of the performance of the 4 different machine learning models on the extracted and reduced dataset were presented. The values for all metrics range from between 0 and 1. The closer the value of the metrics to 1, the better the model. Table 1 displayed the results of the ML models used on the dataset.

**Table 3.** The summary performance of the machine learning algorithms

| Model | Accuracy | Precision | Recall | F1-score | ROC |
|---|---|---|---|---|---|
| XGBoost | 0.90 | 0.91 | 0.90 | 0.90 | 0.90 |
| LGBM | 0.97 | 0.97 | 0.96 | 0.96 | 0.97 |
| Random forest | 0.75 | 0.87 | 0.72 | 0.72 | 0.75 |
| Extra trees | 0.86 | 0.89 | 0.85 | 0.86 | 0.86 |

Table 3 presents the performance comparison of the four different classifiers based on precision, recall, F1-Score, and accuracy. The result is based on the test set which is 20% of the dataset. It is observed that LGBM achieved the highest classification report values of 0.91 across all metrics. Its performance value is more consistent than with other models across all metrics. Using F1- score as for comparing all models, Random Forest obtained the least value of 0.72 following closely by Extra Tree with the value of 0.86. LGBM outperformed all other models with a value of 0.97. Random Forest obtained the least recall rate and accuracy values of 0.85 and 0.86 respectively. Therefore, making it the least performing model.

### 5.1 ROC Curves

The ROC curves for the four models are displayed in Fig. 4 which is the trade-off between true positive and false positive rate having established that it performed best of the models as shown by Fig. 4. The ROC values for all metrics are close to 1 showing a very good classification performance. For example, for LGBM model the ROC values 0.96, 1.00 and 0.97 for COVID-19, NORMAL and PTB respectively. The overall average is 0.98 showing a good trade-off between recall and precision, and the RF also has ROC of 0.98.

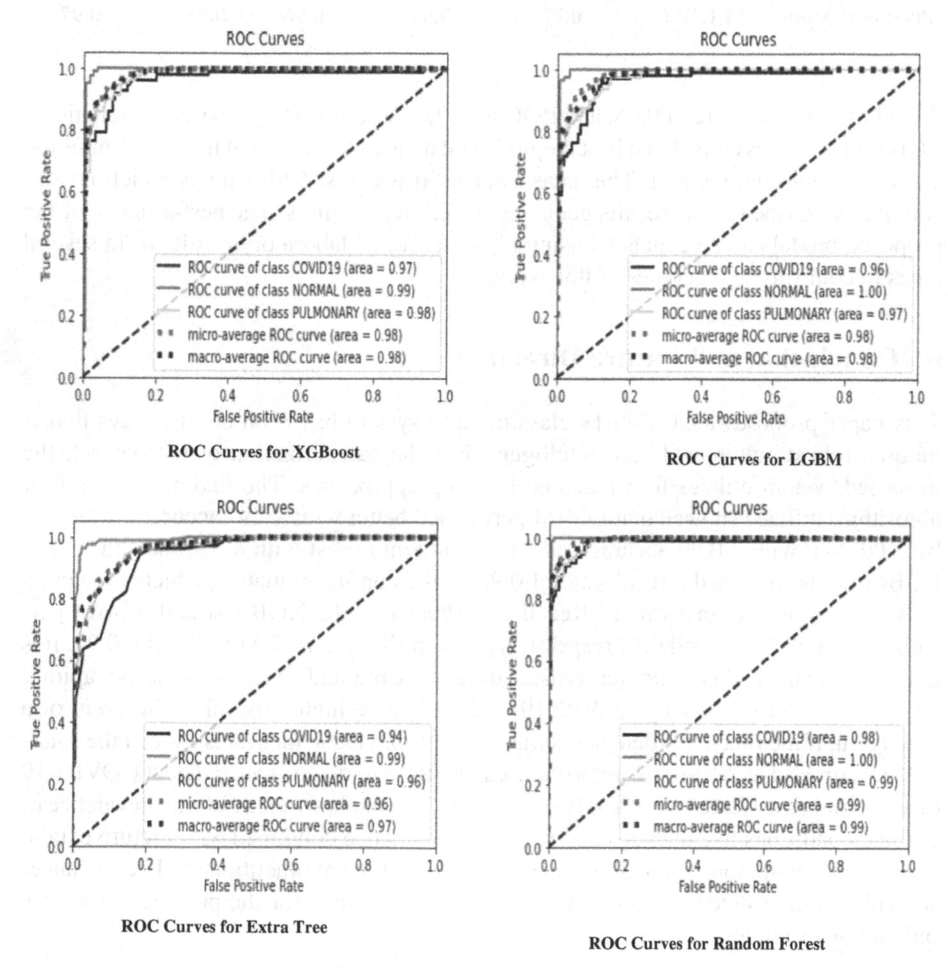


**Fig. 4.** ROC for all the models

Table 4 show the result of various models has compared with the proposed model. The existing model used the same dataset with several conventional ML and DL algorithms. Based on the results obtained, the proposed LGBM model performed better using various measurements. The LGBM classifier yield an accuracy of 0.97, Precision of 0.97,

**Table 4.** Performance evaluation of the proposed system with the existing state-of-the-art method for COVID-19 pandemic

| Models | Method | Accuracy | Precision | Recall | F1-Score | ROC |
|---|---|---|---|---|---|---|
| [29] | CNNLSTM | 0.93 | | | 0.93 | 0.90 |
| [30] | SVM, RF | | | | 0.72 | 0.87 |
| [31] | XGB | | | | | 0.66 |
| [32] | SVM | 0.95 | 0.95 | 0.95 | 0.94 | 0.95 |
| **Proposed Model** | **LGBM** | **0.97** | **0.97** | **0.96** | **0.96** | **0.97** |

Recall of 0.96, F1-score of 0.96, and ROC of 0.97 Therefore, the proposed system using LGBM classifier is considered best method. The dataset size is one of the most limitation associated with the method. The dataset set used was just 520 patients which is very small, this ascribed to the results get using different classifiers. The performance of the proposed model can be enhanced using big dataset and laboratory results from several areas to confirm the outcomes of this work.

## 6 Conclusion and Future Direction

This paper proposes a COVID-19 classification system based on chest X-rays that is unique, robust, automated, and intelligent. For the categorization of the dataset, the proposed system utilizes four machine learning approaches. The findings of the four algorithms utilized showed that LGBM performed better with a 0.97 accuracy, followed by XGBoost with a 0.90 accuracy, and the Random Forest with a 0.96 accuracy. The LGBM model obtained a recall rate of 0.96 in the confusion matrix, which showing a good detection rate. In terms of Recall and Precision, the XGBoost and LGBM performed best with 0.90 and 0.91 respectively, but in F-score, LGBM leads with 0.91. It is assumed that using this computer-assisted diagnostic method would increase the rapidity and precision of identifying COVID 19. This might be highly useful in the event of a contagion, if the illness burden and demand for prophylactic measures exceed the available resources. Deep learning algorithms can be used to diagnose and detect COVID-19 disease, and other features such as body temperature and knowledge of the prevalence of chronic conditions like diabetes and heart disease can be combined with picture data to make the system more efficient and accurate for healthcare practitioners. Breast cancer screening, tumor detection, and other medical applications for the proposed work are only a few examples.

## References

1. Ogundokun, R.O., Lukman, A.F., Kibria, G.B., Awotunde, J.B., Aladeitan, B.B.: Predictive modelling of COVID-19 confirmed cases in Nigeria. Infect. Dis Model. **5**, 543–548 (2020)
2. Asai, A., et al.: COVID-19 drug discovery using intensive approaches. Int. J. Mol. Sci. **21**(8), 2839 (2020)

3. Awotunde, J.B., Jimoh, R.G., AbdulRaheem, M., Oladipo, I.D., Folorunso, S.O., Ajamu, G.J.: IoT-based wearable body sensor network for COVID-19 pandemic. Stud. Syst. Decis. Control **2022**(378), 253–275 (2022)
4. Devi, A., Nayyar, A.: Perspectives on the definition of data visualization: a mapping study and discussion on coronavirus (COVID-19) dataset. In: Al-Turjman, F., Devi, A., Nayyar, A. (eds.) Emerging Technologies for Battling Covid-19. SSDC, vol. 324, pp. 223–240. Springer, Cham (2021). https://doi.org/10.1007/978-3-030-60039-6_11
5. Aydin, N., Yurdakul, G.: Assessing countries' performances against COVID-19 via WSIDEA and machine learning algorithms. Appl. Soft Comput. **97**, 106792 (2020)
6. Gayathri, G.V., Satapathy, S.C.: A Survey on techniques for prediction of asthma. In: Satapathy, S.C., Vikrant Bhateja, J.R., Mohanty, S.K., Udgata, (eds.) Smart Intelligent Computing and Applications. SIST, vol. 159, pp. 751–758. Springer, Singapore (2020). https://doi.org/10.1007/978-981-13-9282-5_72
7. Awotunde, J.B., Bhoi, A.K., Barsocchi, P.: Hybrid cloud/Fog environment for healthcare: an exploratory study, opportunities, challenges, and future prospects. In: Bhoi, A.K., Mallick, P.K., Mohanty, M.N., de Victor Hugo, C., Albuquerque, (eds.) Hybrid Artificial Intelligence and IoT in Healthcare. ISRL, vol. 209, pp. 1–20. Springer, Singapore (2021). https://doi.org/10.1007/978-981-16-2972-3_1
8. Brunese, L., Mercaldo, F., Reginelli, A., Santone, A.: Explainable deep learning for pulmonary disease and coronavirus COVID-19 detection from X-rays. Comput. Methods Program. Biomed. **196**, 105608 (2020)
9. Islam, M.Z., Islam, M.M., Asraf, A.: A combined deep CNN-LSTM network for the detection of novel coronavirus (COVID-19) using X-ray images. Inf. Med. Unlocked **20**, 100412 (2020)
10. Salehi, A.W., Baglat, P., Gupta, G.: Review on machine and deep learning models for the detection and prediction of Coronavirus. Mater. Today Proc. **33**, 3896–3901 (2020)
11. Guo, L., et al.: Profiling early humoral response to diagnose novel coronavirus disease (COVID-19). Clin. Infect. Dis. **71**(15), 778–785 (2020)
12. Statement on the second meeting of the International Health Regulations, Emergency committee regarding the outbreak of novel coronavirus (2019-nCoV). World Health Organization (WHO). 30 Jan 2020 Switzerland Geneva (2005)
13. Wu, J.T., Leung, K., Leung, G.M.: Nowcasting and forecasting the potential domestic and international spread of the 2019-nCoV outbreak originating in Wuhan, China: a modelling study. The Lancet **395**(10225), 689–697 (2020)
14. Santosh, K.C.: AI-driven tools for coronavirus outbreak: need of active learning and cross-population train/test models on multitudinal/multimodal data. J. Med. Syst. **44**(5), 1–5 (2020)
15. Fu, H., Xu, Y., Wong, D. W. K., Liu, J.: Retinal vessel segmentation via deep learning network and fully-connected conditional random fields. In: 2016 IEEE 13th international symposium on biomedical imaging (ISBI), pp. 698–701. IEEE (April 2016)
16. Folorunso, S.O., Awotunde, J.B., Adeboye, N.O., Matiluko, O.E.: Data classification model for COVID-19 pandemic. Stud. Syst. Decis. Control **2022**(378), 93–118 (2022)
17. Pereira, R.M., Bertolini, D., Teixeira, L.O., Silla Jr, C.N., Costa, Y.M.: COVID-19 identification in chest X-ray images on flat and hierarchical classification scenarios. Comput. Methods Program. Biomed. **194**, 105532 (2020)
18. Apostolopoulos, I.D., Mpesiana, T.A.: Covid-19: automatic detection from x-ray images utilizing transfer learning with convolutional neural networks. Phys. Eng. Sci. Med. **43**(2), 635–640 (2020)
19. Sethy, P.K., Behera, S.K.: Detection of coronavirus disease (covid-19) based on deep features (2020)

20. Khalifa, N.E.M., Taha, M.H.N., Hassanien, A.E., Elghamrawy, S.: Detection of coronavirus (COVID-19) associated pneumonia based on generative adversarial networks and a fine-tuned deep transfer learning model using chest X-ray dataset. arXiv preprint arXiv:2004.01184 (2020)
21. Afshar, P., Heidarian, S., Naderkhani, F., Oikonomou, A., Plataniotis, K.N., Mohammadi, A.: Covid-caps: a capsule network-based framework for identification of covid-19 cases from x-ray images. Pattern Recogn. Lett. **138**, 638–643 (2020)
22. Ayo, F.E., Awotunde, J.B., Ogundokun, R.O., Folorunso, S.O., Adekunle, A.O.: A decision support system for multi-target disease diagnosis: A bioinformatics approach. Heliyon **6**(3), e03657 (2020)
23. Ayo, F.E., Ogundokun, R.O., Awotunde, J.B., Adebiyi, M.O., Adeniyi, A.E. (2020) Severe Acne Skin Disease: A Fuzzy-Based Method for Diagnosis. In: Gervasi, O., et al. (eds.) Computational Science and Its Applications – ICCSA 2020. ICCSA 2020. Lecture Notes in Computer Science, vol. 12254. Springer, Cham. https://doi.org/10.1007/978-3-030-58817-5_25
24. Oladele T.O., Ogundokun, R.O., Awotunde, J.B., Adebiyi, M.O., Adeniyi, J.K.: Diagmal: A malaria coactive neuro-fuzzy expert system. In: Gervasi, O., et al. (eds.) Computational Science and Its Applications – ICCSA 2020. ICCSA 2020. Lecture Notes in Computer Science, vol. 12254. Springer, Cham (2020). https://doi.org/10.1007/978-3-030-58817-5_32
25. Folorunso, S.O., Awotunde, J.B., Ayo, F.E., Abdullah, K.K.A.: RADIoT: the unifying framework for iot, radiomics and deep learning modeling. Intell. Syst. Ref. Libr. **2021**(209), 109–128 (2021)
26. Ajagbe, S.A., Amuda, K.A., Oladipupo, M.A., Oluwaseyi, F., Okesola, K.I.: Multi-classification of alzheimer disease on magnetic resonance images (MRI) using deep convolutional neural network (DCNN) approaches. Int. J. Adv. Comput. Res. **11**(53), 51–60 (2021). https://doi.org/10.19101/IJACR.2021.1152001
27. Awotunde, J.B., Folorunso, S.O., Bhoi, A.K., Adebayo, P.O., Ijaz, M.F.: disease diagnosis system for IoT-based wearable body sensors with machine learning algorithm. In: Bhoi, A.K., Mallick, P.K., Mohanty, M.N., de Victor Hugo, C., Albuquerque, (eds.) Hybrid Artificial Intelligence and IoT in Healthcare. ISRL, vol. 209, pp. 201–222. Springer, Singapore (2021). https://doi.org/10.1007/978-981-16-2972-3_10
28. Pedregosa, F., et al.: Scikit-learn: Machine learning in Python. J. Mach. Learn. Res. **12**, 2825–2830 (2011)
29. Mohammed, M.A., et al.: Benchmarking methodology for selection of optimal COVID-19 diagnostic model based on entropy and TOPSIS methods. IEEE Access **8**, 99115–99131 (2020)
30. de Moraes Batista, A.F., Miraglia, J.L., Donato, T.H.R., Chiavegatto Filho, A.D.P.: COVID-19 diagnosis prediction in emergency care patients: a machine learning approach. medRxiv (2020)
31. Schwab, P., Schütte, A.D., Dietz, B., Bauer, S.: predcovid-19: a systematic study of clinical predictive models for coronavirus disease 2019. arXiv preprint arXiv:2005.08302 (2020)
32. Abdulkareem, K. H., et al.: Realizing an Effective COVID-19 Diagnosis System Based on Machine Learning and IOT in Smart Hosp (2021)

# Evaluation of Local Thresholding Algorithms for Segmentation of White Matter Hyperintensities in Magnetic Resonance Images of the Brain

Adam Piórkowski(✉) and Julia Lasek

Department of Biocybernetics and Biomedical Engineering,
AGH University of Science and Technology, Mickiewicza 30 Av.,
30–059 Krakow, Poland
pioro@agh.edu.pl

**Abstract.** White matter hyperintensities are distinguished in magnetic resonance images as areas of abnormal signal intensity. In clinical research, determining the region and position of these hyperintensities in brain MRIs is critical; it is believed this will find applications in clinical practice and will support the diagnosis, prognosis, and therapy monitoring of neurodegenerative diseases. The properties of hyperintensities vary greatly, thus segmenting them is a challenging task. A substantial amount of time and effort has gone into developing satisfactory automatic segmentation systems.

In this work, a wide range of local thresholding algorithms has been evaluated for the segmentation of white matter hyperintensities. Nine local thresholding approaches implemented in ImageJ software are considered: Bernsen, Contrast, Mean, Median, MidGrey, Niblack, Otsu, Phansalkar, Sauvola. Additionally, the use of other local algorithms (Local Normalization and Statistical Dominance Algorithm) with global thresholding was evaluated. The segmentation accuracy results for all algorithms, and the parameter spaces of the best algorithms are presented.

**Keywords:** White matter hyperintensities segmentation · White matter lesions · Plaques · Local thresholding

## 1 Introduction

White matter hyperintensities (WMH, also known as white matter lesions) are distinguished in T2-weighted MR imaging as areas of abnormal signal intensity. WMHs are related to various geriatric and cardiac disorders, e.g., cerebrovascular diseases, cardiovascular diseases, dementia, and psychiatric disorders [10]. WHMs exhibit high variability in their characteristics, thus making their segmentation a very demanding task [9,26]. Evaluation of a WMH area and position


H. Florez and M. F. Pollo-Cattaneo (Eds.): ICAI 2021, CCIS 1455, pp. 331–345, 2021.
https://doi.org/10.1007/978-3-030-89654-6_24

is critical in clinical research investigations; it is anticipated that this will make its way into clinical practice in terms of assisting in neurodegenerative disease diagnosis, prognosis, and therapy monitoring [6]. As a result, a lot of effort has gone into designing appropriate automatic segmentation systems.

The segmentation methods used a decade ago were based on classic methods of image processing and analysis [9], such as probability functions whose segmentation threshold is determined by the minimum probability between distributions [8]; a mean intensity-based region-growing algorithm started from seeds [5]; and k-NN [1,7] and Naive Bayesian [13] classifiers. The current trend [2] is primarily the use of artificial intelligence for segmentation, e.g., [3,12,19]. Some of the classic segmentation methods are based on thresholding algorithms. However, due to the high variability of lesions and of images captured with different equipment or settings, the use of global thresholding (GT) is inefficient. Using GT frequently leads to the exclusion of deep lesions (resulting in lower contrast) or the overestimation of periventricular lesions (which are brighter). As a result, the sensitivity and specificity of lesion identification are frequently compromised [26].

In contrast to GT, which uses only one value across the whole process, a local thresholding (LT) technique would allow each pixel in the image to have its own threshold value set separately based on the intensities of pixels in a square or round neighborhood centered on the pixel in question.

This work investigated how local thresholding methods handle lesion segmentation. The most essential and widely used local thresholding algorithms were tested, including Bernsen, Contrast, Mean, Median, MidGrey, Niblack, Otsu, Sauvola, and Phansalkar Thresholding. Additionally, two local algorithms with global thresholding were tested for comparison with LT. A similar paper comparing the performance results of local algorithms has been presented recently for the microbial cell counting problem [17], in which the Bernsen algorithm appeared to be the optimum selection method.

As a follow-up to previous work [14], in which local thresholding methods were shown to be a promising research direction, this work studied how they handle lesion segmentation. More of these types of algorithms have been tested, and the results have been collected into a coherent whole.

## 2 Materials and Methods

### 2.1 The Dataset

The dataset contains 4 examples of DICOM images of brain MRI in TIRM mode (Fig. 1). For the analysis, the images were reduced by scaling from 12-bit to 8-bit depth. The image resolution and the types of scanners used to acquire the images are presented in (Table 1). The white matter in the images was segmented using semi-automatic methods, and the plaques were segmented manually.

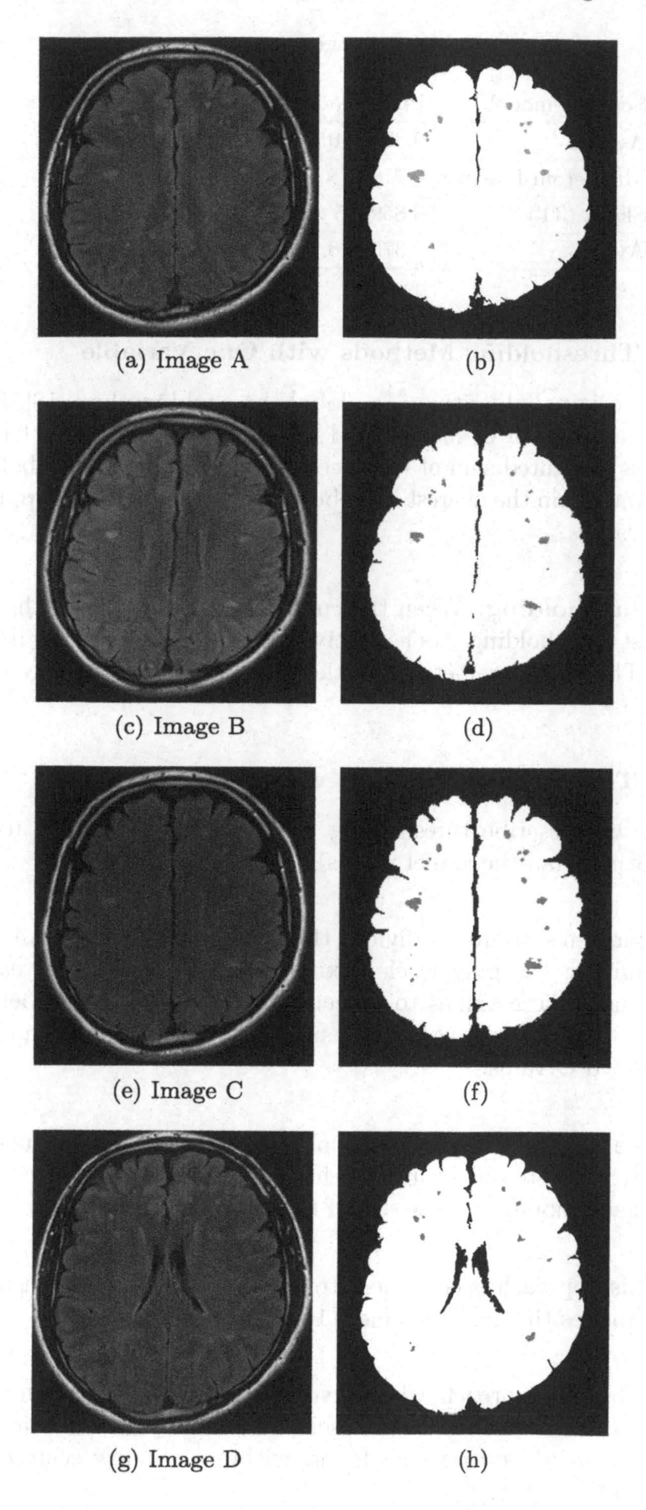


**Fig. 1.** Images used in the paper, their ROIs (white) and WMHs (blue).

**Table 1.** Image properties.

| ID | Scanner model | Pixel spacing | Original matrix (px) |
|---|---|---|---|
| A | Avanto | 0.375 × 0.375 | 540 × 640 |
| B | Magnetom Essenza | 0.71875 × 0.71875 | 252 × 320 |
| C | Skyra (3T) | 0.859375 × 0.859375 | 256 × 256 |
| D | Avanto | 0.375 × 0.375 | 540 × 640 |

### 2.2 Local Thresholding Methods with One Variable

**Otsu.** Otsu's original algorithm is a global threshold method [18]. ImageJ contains a local variation of Otsu's method which selects threshold that minimizes the two classes' weighted sum of variances. Rather than using global Otsu, local pixels are examined in the nearest neighborhood. The only user-supplied variable is the kernel radius.

**Contrast Thresholding.** When the current value is closer to the local maximum, contrast thresholding sets a positive pixel value; otherwise, it sets a value of zero [25]. The only parameter that the operator should modify is the kernel radius.

### 2.3 Local Thresholding Methods with Two Variables

This section will describe thresholding methods that have one user-provided parameter (other than the kernel radius).

**Bernsen.** Bernsen's approach divides the image into predetermined segments before determining the gray level maxima and minima within each segment [4]. Then, it uses these values to detect whether the segment belongs to the foreground or background layer by measuring the local contrast and comparing it to a predefined $C$ value.

**Mean.** In the Mean method of thresholding, the gray-scale values within the window are averaged to determine the threshold. Pixels that are larger than the mean by a fixed amount $C$ are assigned to the foreground [11].

**Median.** This approach is equivalent to the previous one, but instead of using the mean it applies the median value [11].

**MidGrey.** The last approach, which involves two variables, is different from the median and mean methods in that it uses the mid-gray value, which is obtained by dividing the $min + max$ values found within a window centered about the pixel by two [11].

## 2.4 Local Thresholding Methods with Three Variables

The thresholding methods described in this section have two variables in addition to the kernel radius.

**Niblack.** The Niblack method works by calculating the binarization $T(x, y)$ threshold using the local mean $m(x, y)$ and the standard deviation $s(x, y)$ of the intensity values of neighboring pixels [16]:

$$T(x, y) = m(x, y) + k \cdot s(x, y) \tag{1}$$

The $k$ parameter adjusts the value of the threshold in the local window. An additional argument has been added to ImageJ's implementation [24]: the $C$ value, which is an offset that was not included in the algorithm's initial implementation. With the addition of this extra parameter, the formula is as follows:

$$T(x, y) = m(x, y) + k \cdot s(x, y) - C \tag{2}$$

**Sauvola.** The Sauvola Algorithm is a variation of Niblack's method [23]. The dynamic range of standard deviation, $r-value$, is used to compute the threshold in this modification. In addition, the local mean is used to multiply the term $r - value$ and a constant value $k$:

$$T(x, y) = m(x, y) \left[1 + k \left(\frac{s(x, y)}{r} - 1\right)\right] \tag{3}$$

where $m(x, y)$ stands for the local mean; $s(x, y)$ stands for the standard deviation, as in the Niblack Algorithm; $r-value$ is the dynamic range of standard deviation; $k$ is a constant.

**Phansalkar.** To deal with low-contrast images, the Phansalkar binarization approach [20] is a variation of Sauvola's thresholding method [23]. The equation for obtaining the threshold value is as follows:

$$T(x, y) = m(x, y) \left[1 + pe^{-g \cdot m(x,y)} + k \left(\frac{s(x, y)}{r} - 1\right)\right] \tag{4}$$

where $m(x, y)$, $s(x, y)$, $r-value$ and $k$ implies the same variables and constant as in the previous Niblack and Sauvola algorithms. In the ImageJ implementation of this algorithm, the $p$ and $q$ values are fixed and amount to $p = 2$ and $q = 10$.

## 2.5 Local Algorithms with Global Thresholding

In a different approach, we tested how the two algorithms perform in combination with Global Thresholding.

**Local Normalization.** Local Normalization Algorithm [22] is computed as follows:

$$g(x,y) = \frac{f(x,y) - m(x,y)}{\sigma(x,y)} \tag{5}$$

where $f(x,y)$ is the original image; $m(x,y)$ stands for an estimation of a local mean ($sigma1$) of $f(x,y)$; $\sigma(x,y)$ is an estimation of the local variance ($sigma2$); $g(x,y)$ is the output image.

**Statistical Dominance Algorithm.** A promising method for medical image segmentation is the Statistical Dominance Algorithm (SDA) [21]. After applying SDA, when the algorithm is applied the image is translated from the brightness domain to the statistical information domain. SDA calculates the number of neighbors with a specified radius in a given region.

## 3 Results

The test scripts were processed in ImageJ (v. 1.53j) [15,24]. Most of the algorithms were implemented in the Auto Local Threshold plug-in (v. 1.17.2). The highest Dice score for each approach has been collated and presented in Table 2, along with all settings required to achieve this score. Density plots of parameter spaces for Local Normalization and the Niblack and Phansalkar algorithms for all images are presented in Fig. 2, Fig. 3, Fig. 4. Figure 7 shows a chart of maximum Dice coefficients, depending on the radius for the Niblack, Phansalkar, and Sauvola algorithms. Visual comparison of the segmentation is shown in Fig. 6.

The best results were achieved by the Local Thresholding algorithms that were controlled by three parameters, including the radius of the surrounding pixel. This group is based on the Niblack algorithm and the algorithms which were created on the basis of modifications of the Niblack, Sauvola and Phansalkar algorithms. In this group, the accuracy results according to the Dice coefficient are in the range of 0.5-0.6. An interesting aspect of this group is the repeatability of the best results of these algorithms for a similar radius value, depending on the image. Unfortunately, despite the initial observed relationship between the pixel spacing of these images and the radius for the best result, such a relationship cannot be relied on because the ratio of these two parameters is different even for images acquired with the same equipment. A bigger problem is the choice of the other parameters of the algorithms. No repeatability of these parameters for different images was shown for this group, except for the Phansalkar algorithm.

In combination with the use of global thresholding, the second group, i.e., algorithms for enhancing the brightness of small objects (Fig. 5), gave results that were about 10% worse. The algorithms in the first group, which had only one variable, gave very inaccurate segmentations.

The algorithms which use only the radius value are not suitable for the segmentation of white matter hyperintensities.

**Table 2.** Results of tested algorithms for A, B, C and D images.

| Image | Algorithm | Settings | Best Dice |
|---|---|---|---|
| A | Otsu (Local) | r = 69 | 0.0365 |
| | Contrast | r = 63 | 0.1505 |
| | Bernsen | r = 79, c = 122 | 0.1603 |
| | Midgray | r = 66, c = −31 | 0.2567 |
| | Mean | r = 48, c = −25 | 0.2583 |
| | Median | r = 16, c = −21 | 0.3194 |
| | SDA + GT | r = 25, t = 14, gl_thr = 198 | 0.4865 |
| | Local Norm. + GT | sigma1 = 16.9, sigma2 = 20.9, gl_thr = 162 | 0.5507 |
| | Phansalkar | r = 38, kv = 0.21, rv = 0.02 | 0.5784 |
| | Niblack | r = 51, kv = 3.7, Cv = 10 | 0.6084 |
| | Sauvola | r = 51, kv = 0.15, rv = 2.45 | 0.6158 |
| B | Otsu (Local) | r = 21 | 0.0284 |
| | Contrast | r = 38 | 0.0434 |
| | Mean | r = 26, c = −34 | 0.1655 |
| | Bernsen | r = 45, c = 164 | 0.2756 |
| | Median | r = 23, c = −34 | 0.2897 |
| | Midgray | r = 37, c = −36 | 0.3636 |
| | SDA + GT | r = 14, t = 17, gl_thr = 217 | 0.4532 |
| | Sauvola | r = 26, kv = 0.25, rv = 3.95 | 0.5185 |
| | Local Norm. + GT | sigma1 = 6.4, sigma2 = 10.2, gl_thr = 190 | 0.5283 |
| | Phansalkar | r = 26, kv = 0.2, rv = 0.02 | 0.5294 |
| | Niblack | r = 26, kv = 3, Cv = 7 | 0.5478 |
| C | Contrast | r = 24 | 0.0459 |
| | Otsu (Local) | r = 35 | 0.0612 |
| | Mean | r = 24, c = −17 | 0.2091 |
| | Bernsen | r = 33, c = 75 | 0.2484 |
| | Median | r = 6, c = −9 | 0.3030 |
| | Midgray | r = 29, c = −19 | 0.3904 |
| | SDA + GT | r = 18, t = 5, gl_thr = 234 | 0.5566 |
| | Local Norm. + GT | sigma1 = 28, sigma2 = 8.1, gl_thr = 185 | 0.5969 |
| | Sauvola | r = 15, kv = −0.05, rv = −1.15 | 0.6097 |
| | Phansalkar | r = 16, kv = 0.45, rv = 0.03 | 0.6400 |
| | Niblack | r = 16, kv = 1.8, Cv = −1 | 0.6437 |
| D | Otsu (Local) | r = 19 | 0.0179 |
| | Mean | r = 48, c = −28 | 0.1368 |
| | Contrast | r = 64 | 0.1669 |
| | Median | r = 9, c = −21 | 0.2192 |
| | Midgray | r = 65, c = −29 | 0.2243 |
| | Bernsen | r = 77, c = 106 | 0.3417 |
| | SDA + GT | r = 29, t = 8, gl_thr = 240 | 0.4401 |
| | Local Norm. + GT | sigma1 = 50, sigma2 = 17, gl_thr = 191 | 0.5191 |
| | Phansalkar | r = 33, kv = 0.24, rv = 0.02 | 0.5232 |
| | Sauvola | r = 27, kv = 0.1, rv = 1.45 | 0.5466 |
| | Niblack | r = 30, kv = 2, Cv = 0 | 0.5583 |

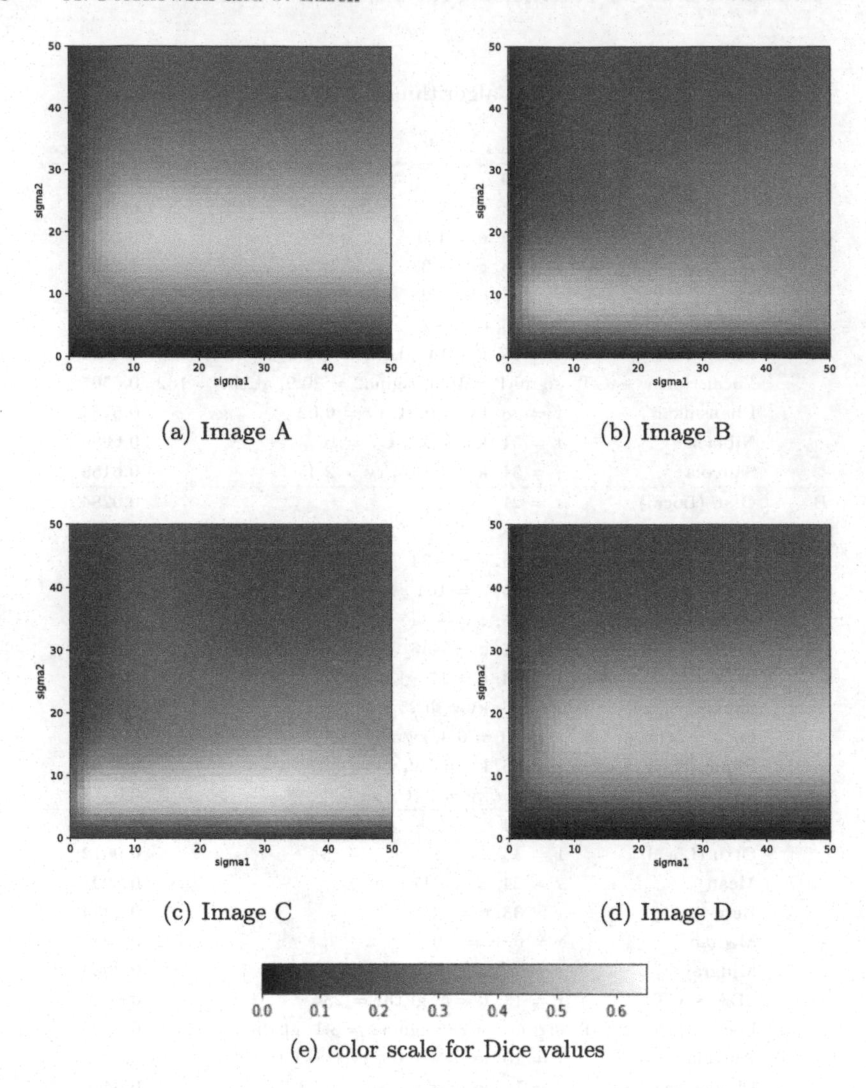


(a) Image A (b) Image B

(c) Image C (d) Image D

(e) color scale for Dice values

**Fig. 2.** Parameter space of the Local Normalization algorithm for images A, B, C and D.

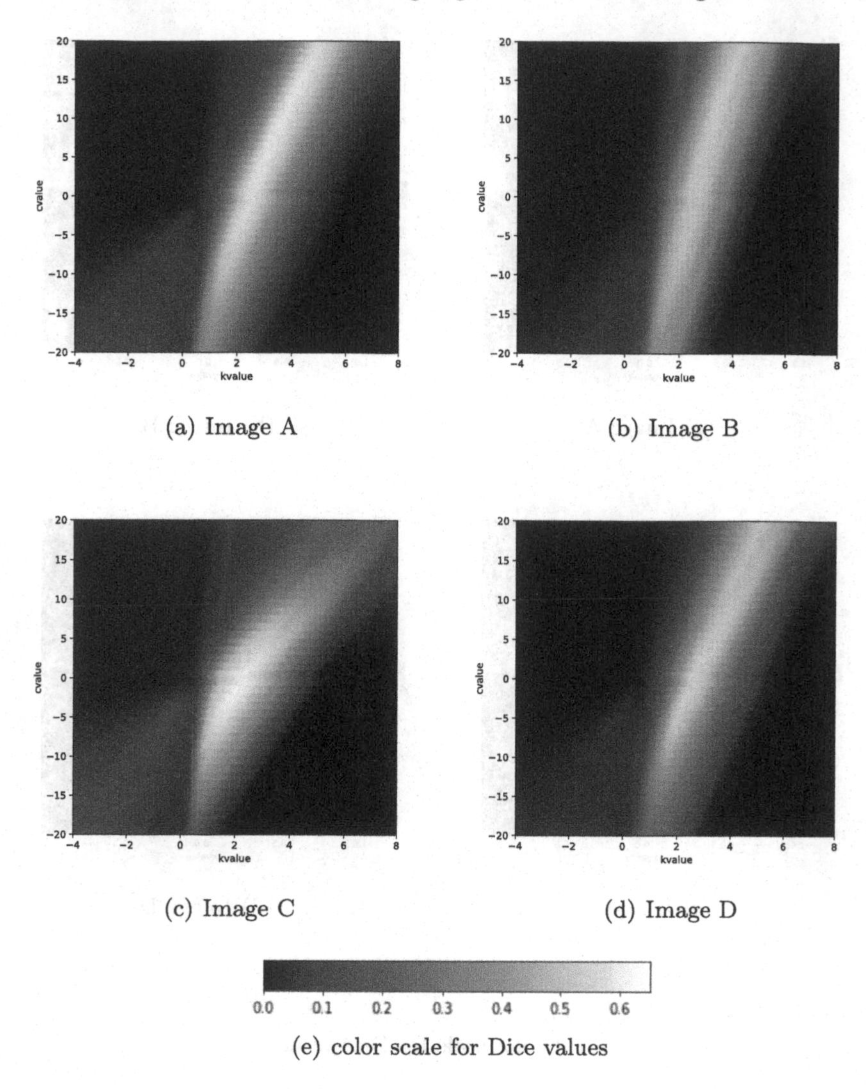


**Fig. 3.** Parameter space of the Niblack algorithm for images A, B, C and D.

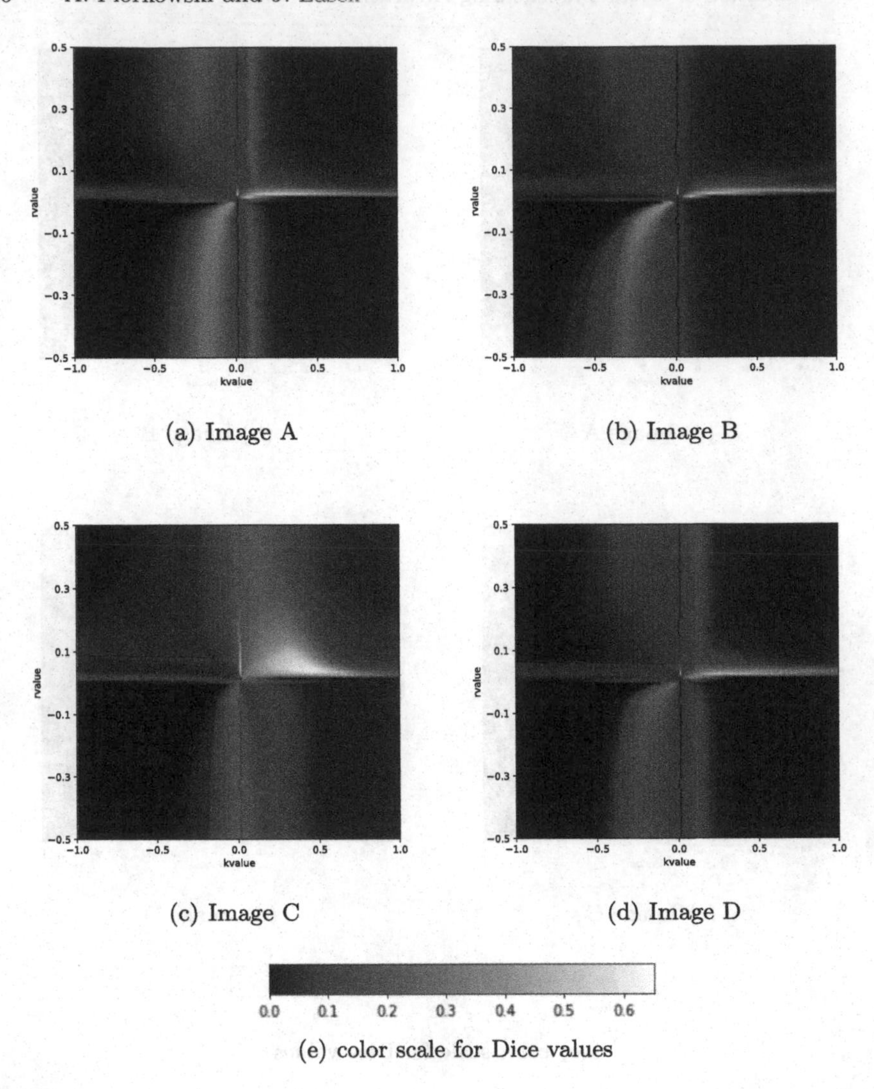


(a) Image A

(b) Image B

(c) Image C

(d) Image D

(e) color scale for Dice values

**Fig. 4.** Parameter space of the Phansalkar algorithm for images A, B, C and D.

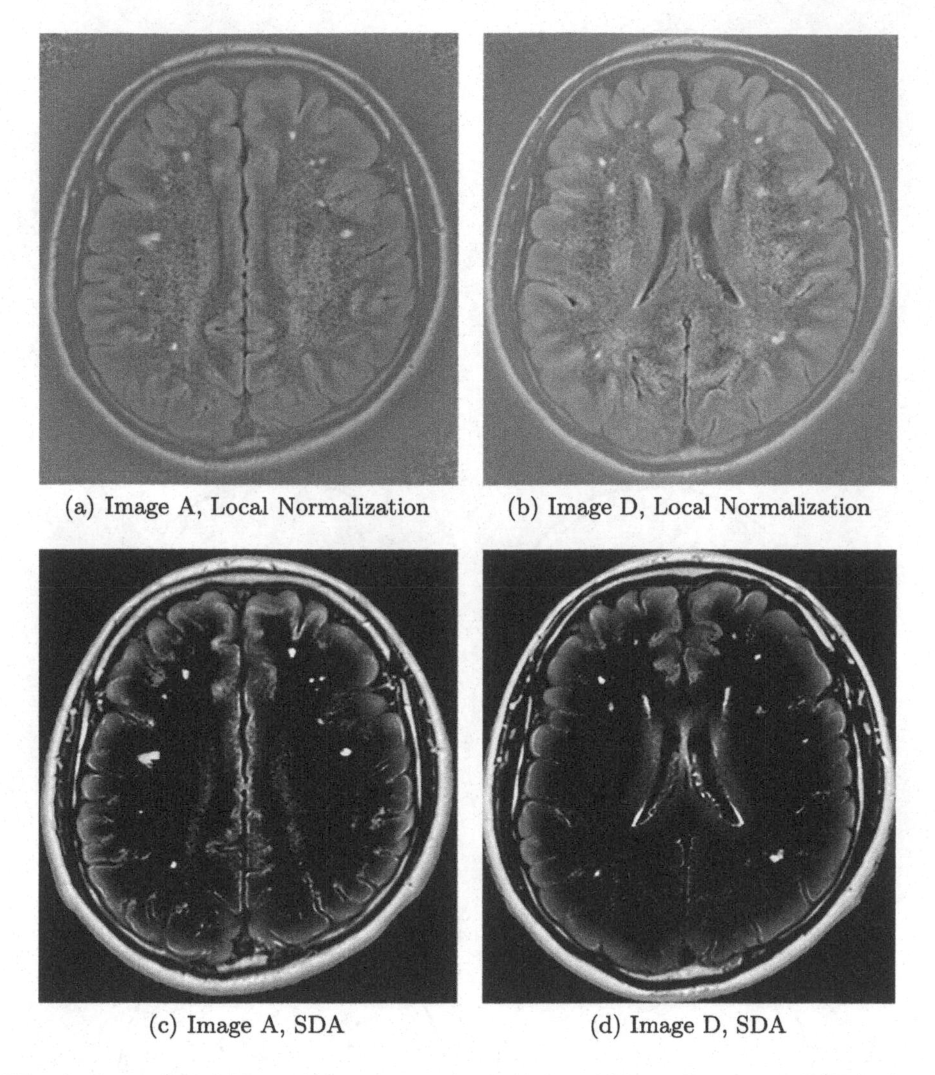

(a) Image A, Local Normalization (b) Image D, Local Normalization

(c) Image A, SDA (d) Image D, SDA

**Fig. 5.** Examples of images after processing with Local Normalization and SDA algorithms (both without global thresholding).

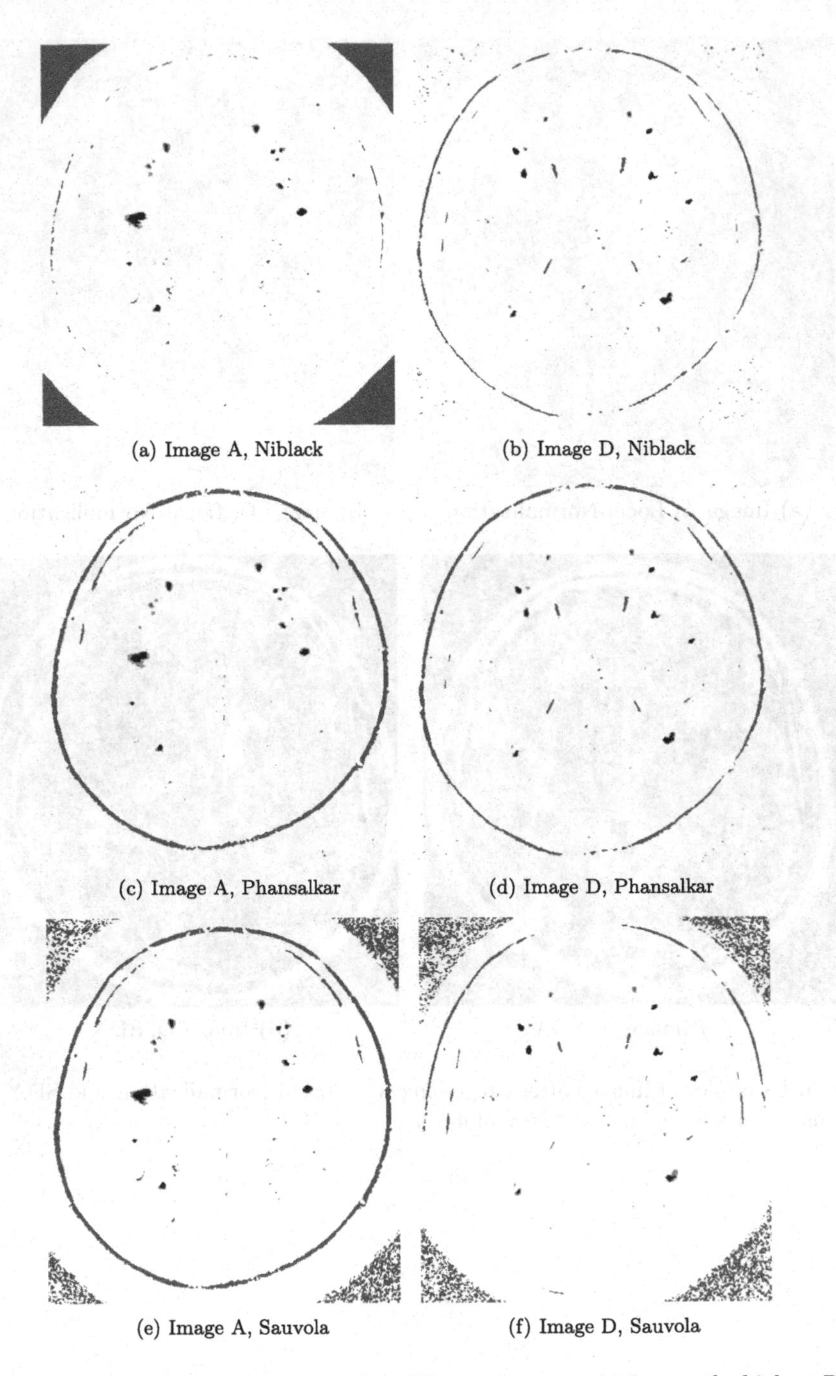

(a) Image A, Niblack
(b) Image D, Niblack
(c) Image A, Phansalkar
(d) Image D, Phansalkar
(e) Image A, Sauvola
(f) Image D, Sauvola

**Fig. 6.** Visual comparison of segmentation. The parameters which gave the highest Dice for each algorithm were selected according to the Table 2. Black color of segmentation indicates true positive; blue indicates false negative; white indicates true negative; red indicates false positive. (Color figure online)

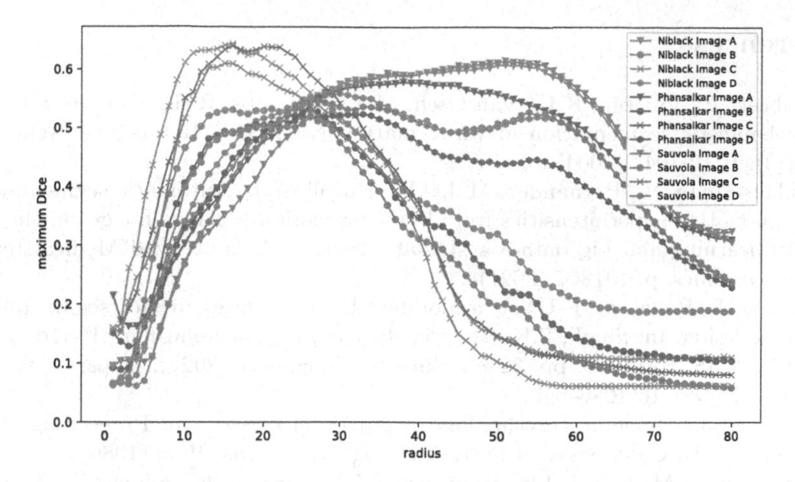


**Fig. 7.** Maximum Dice coefficients depending on radius for the Niblack, Phansalkar and Sauvola algorithms.

# 4 Conclusions

The paper investigates the applicability of classic algorithms for segmentation of white matter hyperintensities in medical resonance imaging of the brain. It is shown that the algorithms controlled only by the kernel radius value are totally unsuitable for the investigated task, but algorithms controlled by two parameters also do not offer the required efficiency. Slightly better results are given by algorithms that enhance the contrast of small objects (LN, SDA), but even they do not reach acceptable values.

The recommended approach to plaque segmentation is the use of algorithms from the Niblack group (Niblack, Sauvola, Phansalkar). These algorithms allow the accurate detection of white matter hyperintensities in MR images. Their disadvantage is that they require the selection of as many as three parameters. Unfortunately, it has been shown that these parameters depend on the characteristics of the image in question. The range of values of these parameters within which the optimal values should be found has been indicated.

Further work will consist in the search for further segmentation possibilities of white matter hyperintensities, including algorithms based on artificial intelligence.

**Acknowledgement.** This publication was funded by AGH University of Science and Technology, Faculty of Electrical Engineering, Automatics, Computer Science and Biomedical Engineering, KBIB no 16.16.120.773.

## References

1. Anbeek, P., Vincken, K.L., Van Osch, M.J., Bisschops, R.H., Van Der Grond, J.: Probabilistic segmentation of white matter lesions in MR imaging. Neuroimage **21**(3), 1037–1044 (2004)
2. Balakrishnan, R., Hernández, M.d.C.V., Farrall, A.J.: Automatic segmentation of white matter hyperintensities from brain magnetic resonance images in the era of deep learning and big data-a systematic review. Computerized Medical Imaging and Graphics, p. 101867 (2021)
3. Basak, H., Rana, A.: F-UNet: a modified U-Net architecture for segmentation of stroke lesion. In: Singh, S.K., Roy, P., Raman, B., Nagabhushan, P. (eds.) CVIP 2020. CCIS, vol. 1376, pp. 32–43. Springer, Singapore (2021). https://doi.org/10.1007/978-981-16-1086-8_4
4. Bernsen, J.: Dynamic thresholding of gray-level images. In: Proceedings Eighth International Conference on Pattern Recognition, Paris, 1986 (1986)
5. Brickman, A.M., Sneed, J.R., Provenzano, F.A., Garcon, E., Johnert, L., Muraskin, J., Yeung, L.K., Zimmerman, M.E., Roose, S.P.: Quantitative approaches for assessment of white matter hyperintensities in elderly populations. Psychiatry Res. Neuroimaging **193**(2), 101–106 (2011)
6. Caligiuri, M., Perrotta, P., Augimeri, A., Rocca, F., Quattrone, A., Cherubini, A.: Automatic detection of white matter hyperintensities in healthy aging and pathology using magnetic resonance imaging: a review. Neuroinformatics **13**, 261–276 (2015)
7. De Boer, R., Vrooman, H.A., Van Der Lijn, F., Vernooij, M.W., Ikram, M.A., Van Der Lugt, A., Breteler, M.M., Niessen, W.J.: White matter lesion extension to automatic brain tissue segmentation on MRI. Neuroimage **45**(4), 1151–1161 (2009)
8. DeCarli, C., et al.: Measures of brain morphology and infarction in the framingham heart study: establishing what is normal. Neurobiol. Aging **26**(4), 491–510 (2005)
9. Frey, B.M., Petersen, M., Mayer, C., Schulz, M., Cheng, B., Thomalla, G.: Characterization of white matter hyperintensities in large-scale MRI-studies. Front. Neurol. **10**, 238 (2019)
10. Kim, K.W., MacFall, J.R., Payne, M.E.: Classification of white matter lesions on magnetic resonance imaging in elderly persons. Biol. Psychiatry **64**(4), 273–280 (2008). https://doi.org/10.1016/j.biopsych.2008.03.024. Stress and Synaptic Plasticity
11. Krig, S.: Computer Vision Metrics: Survey, Taxonomy, Analysis. Apress Open (2014). https://doi.org/10.1007/978-1-4302-5930-5
12. Liu, L., Chen, S., Zhu, X., Zhao, X.M., Wu, F.X., Wang, J.: Deep convolutional neural network for accurate segmentation and quantification of white matter hyperintensities. Neurocomputing **384**, 231–242 (2020)
13. Maillard, P., Delcroix, N., Crivello, F., Dufouil, C., Gicquel, S., Joliot, M., Tzourio-Mazoyer, N., Alpérovitch, A., Tzourio, C., Mazoyer, B.: An automated procedure for the assessment of white matter hyperintensities by multispectral (t1, t2, pd) MRI and an evaluation of its between-centre reproducibility based on two large community databases. Neuroradiology **50**(1), 31–42 (2008)
14. Milewska, K., Obuchowicz, R., Piorkowski, A.: A preliminary approach to plaque detection in MRI brain images. In: Innovations and Developments of Technologies in Medicine, Biology amd Healthcare - Proceedings of the IEEE EMB International Student Conference 2020. AISC. Springer (2022)

15. Mutterer, J., Rasband, W.: Imagej macro language programmers reference guide v1. 46d. RSB Homepage, pp. 1–45 (2012)
16. Niblack, W.: An Introduction to Digital Image Processing, 115–116 Prentice Hall. Englewood Cliffs, New Jersey (1986)
17. Nichele, L., Persichetti, V., Lucidi, M., Cincotti, G.: Quantitative evaluation of imagej thresholding algorithms for microbial cell counting. OSA Continuum **3**(6), 1417–1427 (2020). https://doi.org/10.1364/OSAC.393971
18. Otsu, N.: A threshold selection method from gray-level histograms. IEEE Trans. Syst. Man Cybern. **9**(1), 62–66 (1979)
19. Park, G., Hong, J., Duffy, B.A., Lee, J.M., Kim, H.: White matter hyperintensities segmentation using the ensemble u-net with multi-scale highlighting foregrounds. Neuroimage **237**, 118140 (2021)
20. Phansalkar, N., More, S., Sabale, A., Joshi, M.: Adaptive local thresholding for detection of nuclei in diversity stained cytology images. In: 2011 International Conference on Communications and Signal Processing, pp. 218–220. IEEE (2011)
21. Piórkowski, A.: A statistical dominance algorithm for edge detection and segmentation of medical images. In: Piętka, E., Badura, P., Kawa, J., Wieclawek, W. (eds.) Information Technologies in Medicine. AISC, vol. 471, pp. 3–14. Springer, Cham (2016). https://doi.org/10.1007/978-3-319-39796-2_1
22. Sage, D., Unser, M.: Easy Java programming for teaching image-processing. In: Proceedings of 2001 International Conference on Image Processing. vol. 3, pp. 298–301. IEEE (2001)
23. Sauvola, J., Pietikäinen, M.: Adaptive document image binarization. Pattern Recogn. **33**(2), 225–236 (2000)
24. Schneider, C.A., Rasband, W.S., Eliceiri, K.W.: NIH Image to ImageJ: 25 years of image analysis. Nat. Methods **9**(7), 671–675 (2012)
25. Soille, P.: Morphological Image Analysis. Springer (2004)
26. Sundaresan, V., et al.: Automated lesion segmentation with bianca: Impact of population-level features, classification algorithm and locally adaptive thresholding. NeuroImage **202**, 116056 (2019). https://doi.org/10.1016/j.neuroimage.2019.116056

# Super-Resolution Algorithm Applied in the Zoning of Aerial Images

J. A. Baldion(✉), E. Cascavita, and C. H. Rodriguez-Garavito

Automation Engineering Program, La Salle University, Bogotá, Colombia
{jbaldion06,ecascavita17,cerodriguez}@unisalle.edu.co

**Abstract.** Nowadays, multiple applications based on images and unmanned aerial vehicles (UAVs), such as autonomous flying, precision agriculture, and zoning for territorial planning, are possible thanks to the growing development of machine learning and the evolution of convolutional and adversarial networks. Nevertheless, this type of application implies a significant challenge because even though the images taken by a high-end drone are very accurate, it is not enough since the level of detail required for most precision agriculture and zoning applications is very high. So, it is necessary to further improve the images by implementing different techniques to recognize small details. Hence, an alternative to follow is the super-resolution method, which allows constructing an image with the information from multiple images. An efficient tool can be obtained by combining drones' advantages with different image processing techniques. This article proposes a method to improve the quality of images taken on board in a drone by increasing information obtained from multiple images that present noise, vibration-induced displacements, and illumination changes. These higher resolution images, called super-resolution images, allow supervised training processes to perform different zoning methods better. In this study, GAN-type networks show the best results to recognize visually differentiated ones on an aerial image automatically. The quality measure of the super-resolution image obtained by different methods was defined using sharpness and entropy metrics, and a semantic confusion matrix measures the accuracy of the following semantic segmentation network. Finally, the results show that the super-resolution algorithm's implementation and the automatic segmentation provide an acceptable accuracy according to the defined metrics.

**Keywords:** Super-resolution · Zoning · Modified linear interpolation · Semantic segmentation

## 1 Introduction

Technological development has allowed the advance in Unmanned Aerial Vehicles (UAV) and their use in a wide range of applications [1], such as precision agriculture, search and rescue, remote sensing, and infrastructure inspections. Likewise,

Supported by Universidad de La Salle Bogotá-Colombia.


H. Florez and M. F. Pollo-Cattaneo (Eds.): ICAI 2021, CCIS 1455, pp. 346–359, 2021.
https://doi.org/10.1007/978-3-030-89654-6_25

UAVs have advantages when capturing images because they can cover extensive zones and take bursts of overlapped images quickly. Nevertheless, these features are not enough for most practical applications related to precision agriculture, for instance, crop health analysis base on leaf plant symptoms. One approach for improving details in regions of interest on UAV images consists of creating from multiple low-resolution overlapped images a suitable Super-resolution image using SISR (Single Image Super-Resolution) [16,18] or MIRS (Multiple Images Super Resolution) [2,14,15,17] algorithms.

Different approaches based on interpolation [19], reconstruction [20], and learning [21,22] tackle the lack of details observed in images taken from UAVs. Several works focus on using Convolutional Neural Networks to decompose, hierarchically, intrinsic and context features from a low-resolution image to generate an image with more information expressed by its level of details.

Improving the quality and resolution of the images captured from drones on flying opens an enormous range of applications like the diagnosis of crops [23], control production in agriculture [24], fault detection [25,26], analysis of human settlements [27], among others.

A super-resolution image in the context of zoning of a geographical area can provide more visual information than a single UAV image, let it recognize water bodies, rural extension, and delimitation of crop areas. At this point, the authors propose a variant of the traditional linear interpolation method for making a super-resolution image, comparing this with super-resolution images generated by other methods base on neural networks. Finally, the best super-resolution method is determined, and automatic zoning from the PIX2PIX [3] network performs a semantic segmentation with the super-resolution image. The improved zoning is measured to check if the precision grows up using the methodology proposed.

## 2 Generation of Super-Resolution Images

Some applications like zoning for development territorial planning use very high-resolution images. Since UAV's camera captures its field of view from a high height, image resolution is not enough for trim details, so super-resolution methods offer a way to increase visual information. This section describes different techniques of super-resolution, such as a modified Linear interpolation method proposed by the authors, and other techniques found in the literature, such as ESRGAN [4] and PIX2PIX [3].

### 2.1 Modified Linear Interpolation

Modified linear interpolation is dependent on linear interpolation, but with some modifications made by the authors. In the first instance, Linear interpolation is a MISR method; therefore, it needs several images from the same geographic area to generate a super-resolution image. One hundred ninety-six captures of the same scene form an image bank (Available in: https://github.com/juanbaldion/

SuperResolutionImages.git), where the first capture is the reference image, and each one of the others is selected as a test image to implement this method. The image shown in Fig. 1 has a size of 3648 × 5472 pixels, but since the interpolation of this scene has a high computational cost, only a section of this image will be treated with this technique; this zone of interest will have a size of 912 × 1513 pixels. Afterward, each image will be loaded by trimming the section around a specific coordinate point, but due to changes in camera pose caused by the drone overflight, it is necessary to transform every capture to a standard top perspective. Therefore, relevant points in the images, reference and test, are detected using SIFT features [5] to extract local descriptors and then return matching feature points. Then, to calculate the change of perspective between each pair of images, $I_{ref}$ and $I_{test_i}$, a parametric estimator RANSAC (Random Sample Consensus) [6] is implemented to obtain the Homography matrix ${}^{i}H_0$. The reference image must be enlarged in this method, considering that a higher amplitude factor will cause a higher computational cost to continue with the super-resolution process, the expansion of the base image will only be three times its original size, introducing intermediate black pixels within the pixels of the base image. The filling process is done by taking one by one of the black pixels of the enlarged image $I_{ref}$ located in coordinates $[u, v]$. Then, multiplying by the homography matrix ${}^{0}H_i$ and the scale factor $\mathcal{K}_s$, it is possible to know the coordinates in which each pixel $[u, v]$ is projected on $I_{test_i}$, $[x, y]$. Later, the linear interpolation is made on the corresponding pixel $[x, y]$, so that it is possible to obtain the intensity $I_{test_i}(x, y)$ according to the current pixel and its $j$ neighbours $(x_j, y_j)$. Hence, already having the intensity $I_{test_i}(x, y)$, it will be replaced in the black pixel in question ${}^{i}I_{ref}(u, v)$. This process is performed for each of the pixels that are in black Fig. 2. Finally, in order to improve accuracy in interpolated information added to super-resolution image $I_{ref}$, the process is repeated with all burst of images taken around the zone of interest, and the $I_{ref}(u, v)$ are averaged across interpolated intensities extracted from all test images $i^{th}$ ${}^{i}I_{ref}(u, v)$.

$$[x, y] = {}^{i}H_0\mathcal{K}_s \begin{bmatrix} u \\ v \end{bmatrix}; {}^{i}I_{test}(x, y) = \sum_{n=1}^{j} I_{test}(x_j, y_j)\frac{|(x, y) - (x_j, y_j)|}{\sum_{n=1}^{j} |(x, y) - (x_j, y_j)|} \quad (1)$$

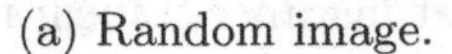

(a) Random image. (b) Zone of Interest.

**Fig. 1.** (a) Random image of the 196 captures, (b) Interest zone magnified three times compared to the base image.

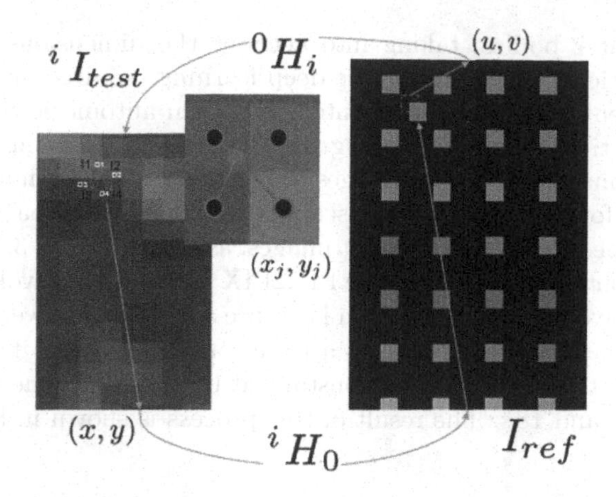


**Fig. 2.** Interpolation based on homography matrix.

The process develops each pixel's average in all the channels, resulting in a super-resolution image containing combined information of all the processed images. At this point, the result is a super-resolution image, but some sections of the image encounter strange intensity patterns. The method called blind deconvolution is used to improve regions that contain this distortion [7].

### 2.2 ESRGAN

It is a GAN (Generative Adversarial Network) type architecture. This type of network is composed of 2 convolutional neural networks that solve two different tasks. A U-Net type generator network which create the desired content. Additionally, there is the discriminating network. This network is used to know whether the output image is real or false, i.e., whether an image was taken from the set of images or is generated from the generator network. These networks are in opposition. One always seeks to deceive the other and in this way both networks improve over time.

### 2.3 PIX2PIX

As in ESRGAN, this is also a GAN-based architecture, the difference is that it is conditional, which is known as cGAN.

This architecture stands for "picture to picture", which means that this network is going to be in charge of creating images, not modifying the ones that already exist. The architecture is in charge of translating images from one domain to another domain.

## 3 Automatic Image Zoning

Firstly, an image storage bank with 250 samples from free databases was collected constrained to exhibit the following geographical areas: Rural zone, Green zone,

Roads, and Water bodies, taking into account the dimensions of each image for standardization. However, because deep learning required an extensive data set, data augmentation was implemented, and then automatic zoning was done based on semantic segmentation using the PIX2PIX network mentioned above. To do this, we must have output images to which the network must be adjusted. The tool used for labelling was Inkscape which helped to paint the areas of interest mentioned above on the input images, as shown in Fig. 3. The automatic image zoning continues modifying the PIX2PIX network; a convolutional layer, a normalization layer, and an activation layer are added to use over the segmented images. Since the training process is a time-consuming task, it halts after 267 epochs, because the error remains constant at 0.17%. The time needed for this was 35 h, 6 min, and 18 s. The result of this process is shown in Fig. 4.

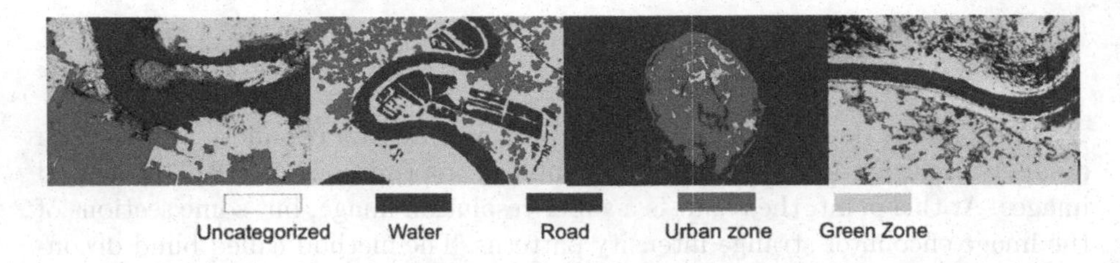


**Fig. 3.** Manual segmentation of the input images.

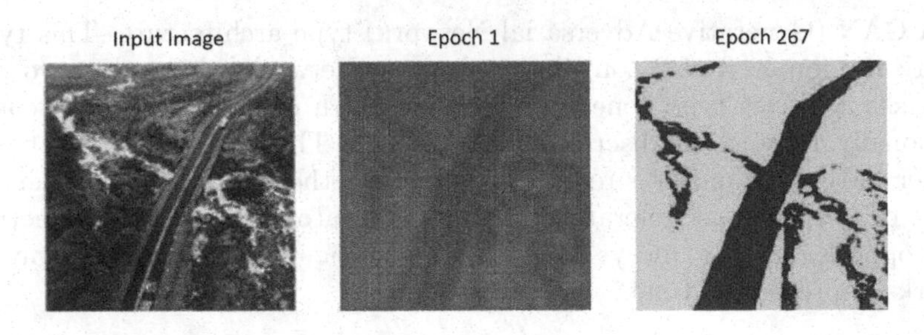


**Fig. 4.** Segmentation training.

## 4 Summary Results

### 4.1 Super-Resolution

By implementing ESRGAN, PIX2PIX, and the Modified Linear Interpolation methods over Fig. 1 and set a ROI, the results are shown in Fig. 5.

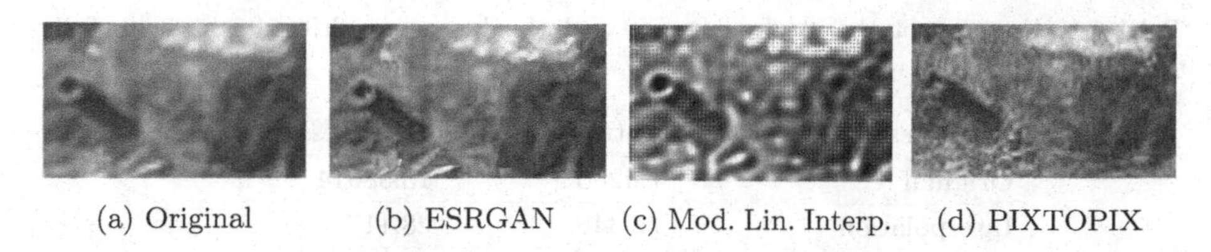

(a) Original (b) ESRGAN (c) Mod. Lin. Interp. (d) PIXTOPIX

**Fig. 5.** Prediction of the implemented techniques, (a) Original Image, (b) ESRGAN Network, (c) Technique, (d) PIX2PIX Network

**Metric in Super-Resolution Algorithms.** Two metrics are used to measure each of the resulting images in the super-resolution process, allowing them to determine with greater precision which yields the most accurate super-resolution image and is closest to reality.

*Entropy Metric.* Entropy is a statistical measure of randomness that can be used to characterize the texture of an image. It is defined as the corresponding states of the level of intensity at which individual pixels can adapt. It is used in the details of the super-resolution image under test, as it provides a better comparison of image details. A higher entropy value means more precise information, i.e., an image that offers more detail [8]. Equation 2 defines the entropy used in this study. Where: $H(S_m)$ is a numerical value representing the entropy of the image in the gray scale, and$P_n(S_m)$ is the probability density, which is calculated from the image's histogram. The quantitative results for this metric are shown in Table 1. The image obtained by modified linear interpolation offers more details than the original image; for this reason, it is determined that this method successfully generates an image of super-resolution.

$$H(S_m) = -\sum_{n=1}^{256} P_n(S_m) * log_2(P_n(S_m) \tag{2}$$

*Sharpness Metric.* Sharpness is defined as the clarity of image detail [9]. There are different techniques for calculating this metric, but the magnitude of the average gradient is used in this case. This metric allows the calculation rate of intensity change at pixel level [10]. In this study, sharpness is mathematically defined by Eq. 3. Where: $SP$ means the sharpness level of the image, a value close to 0 means that the image is completely blurred, and a value above 20 indicates that the image has good sharpness [11]. $\Delta Fx$ and $\Delta Fy$ are the gradients in the X and Y directions. This metric calculates the overall gradient for the whole test image.

$$SP = \sqrt{\Delta Fx^2 + \Delta Fy^2} \tag{3}$$

The result of applying all metrics in a test image is shown in Table 1. As can be seen, interpolation shows a higher rate of intensity change at the pixel level, which allows us to determine that this method has greater clarity in its details than other methods.

**Table 1.** Entropy of the original image and the image resulting from each of the methods.

| Super-resolution method | Entropy metric | Sharpness metric |
|---|---|---|
| Original | 7.4256 | 13.58014 |
| Interpolation | 7.7345 | 28.117 |
| ESRGAN | 7.4188 | 14.5721 |
| PIX2PIX | 7.4201 | 13.2368 |

### 4.2 Zoning

The PIX2PIX network is tested by entering images of the training set observing the behaviour of the network as it is shown in Fig. 6.

**Metrics for Zoned Images.** Different metrics are used, such as pixel by pixel accuracy and (accuracy, sensitivity, and precision of the image) from the confusion matrix to know the margin of error for the implemented method. That will be done with three images from the test set, see Fig. 7.

*Pixel by Pixel Metric.* This metric consists of comparing the pixels of two images in the value of their intensity. The input image must be painted manually since the output image is already segmented. Then, the size of the images is adjusted to $1500 \times 1500$ pixels. Both images are subtracted pixel by pixel, taking into account the pixel position to be measured, where the result will be a 3-colour image; white and gray are pixels that had differences, and black are areas where

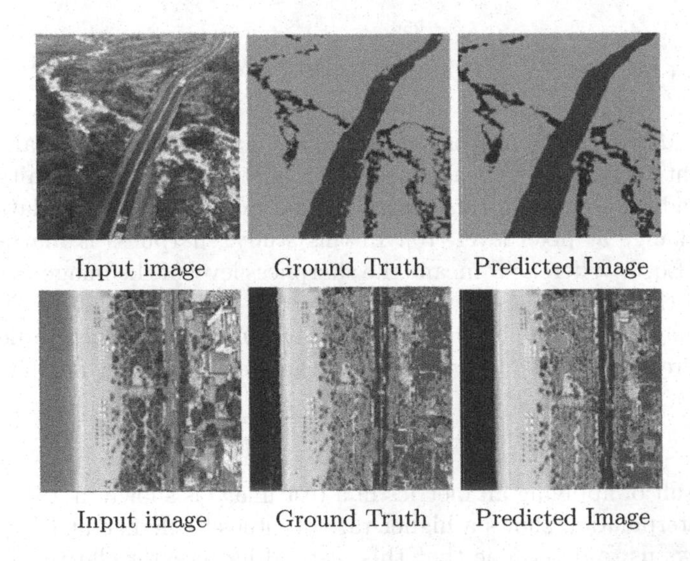


**Fig. 6.** Result of the network trained after 267 epochs.

the images were painted the same way. This process is repeated for each of the three images with which the network is tested. When there are pixels of all three colours, it must be calculated how many pixels did not match each of the methods' input image and output image. That is done using Eq. 4.

$$PA_c = 100\% - \frac{K_c * 100}{s} \quad (4)$$

where: $PA_c$ corresponds to the success percentage per channel, $K_c$ is the number of pixels where there were differences between the two images, and $s$ is the number of pixels of the corresponding colour in the image. One example of applying this metric can be seen in Fig. 8. Where averaging each of the percentages, and the result is 79.765% accuracy per pixel. Figure 9 shows the results of applying pixel to pixel metric over the three test images.

*Confusion Matrix.* It shows the segmentation's performance, describing how many categories of the network were correctly predicted, taking into account the natural values of the manually painted image and the one that was predicted by the network [12]. It was necessary to count in each image how many pixels were correctly classified and how many were not to create this matrix. Finally, the metrics of accuracy, precision, and sensitivity are implemented [13]. These are measured for all the zoned images and are given by Eqs. 5–7. Where: $Ex$ is accuracy, $Pr$ is precision, and $Se$ is sensitivity. $M$ represents the number of classes. $Total$ represents the number of samples submitted for classification. $TP$ represents the values marked adequately as positive. $FP$ is the values wrongly marked as positive, and $FN$ the values erroneously Super-resolution algorithm applied in the zoning of aerial images marked as negative. Tables 2, 3, 4 and 5 summarize the results applying segmentation metrics based on Confusion Matrix.

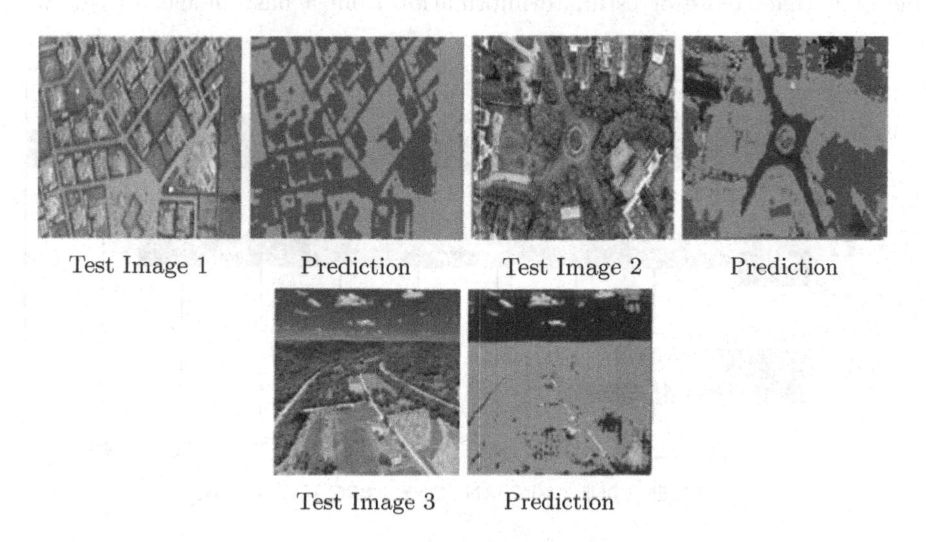


**Fig. 7.** Zoning Results.

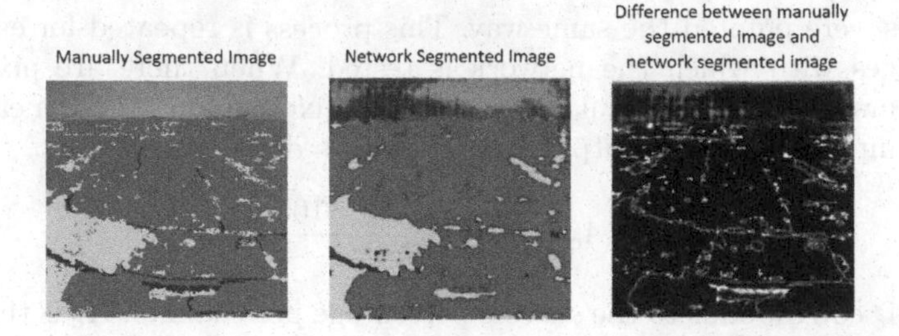


**Fig. 8.** Result between manually segmented image and network segmented image.

$$Ex = \frac{\sum_{i=1}^{M} TPi}{Total} \tag{5}$$

$$Pr = \frac{\sum_{i=1}^{M} \frac{TPi}{TPi+FPi}}{M} \tag{6}$$

$$Se = \frac{\sum_{i=1}^{M} \frac{TPi}{TPi+FNi}}{M} \tag{7}$$

## 5 Analysis of Results

After evaluating all super-resolution methods throughout the metrics described above, it can be observed that the best result is the modified linear interpolation because it shows more detailed information within the zone of analysis. Furthermore, this technique resembles a closer reconstruction of reality than the other methods that create or estimate information from a base image. In this way,

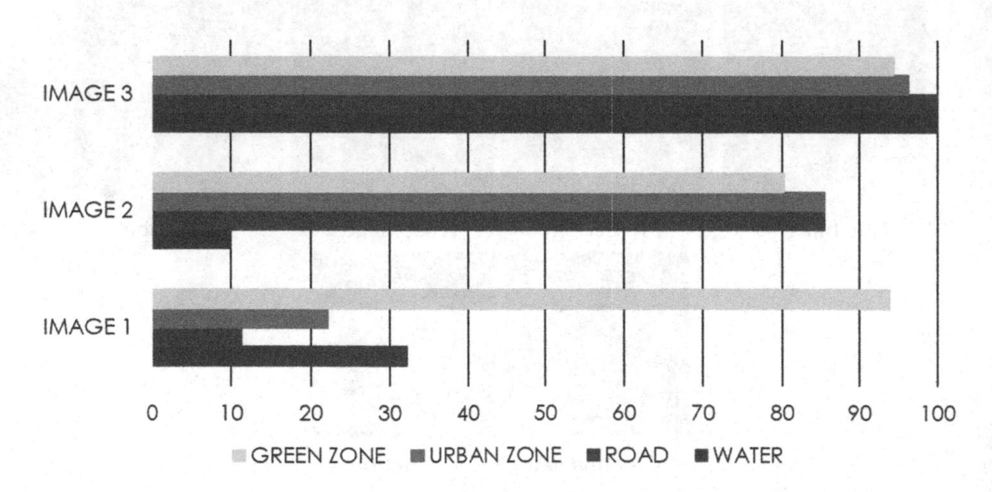


**Fig. 9.** Zoning Results.

**Table 2.** Confusion matrix for images 1.

| Class | Water | Road | Urban zone | Green zone | Total |
|---|---|---|---|---|---|
| Water | 51364 | 236 | 189 | 2105 | 53894 |
| Road | 429 | 539 | 198 | 246 | 1412 |
| Urban zone | 47219 | 3289 | 1169821 | 216945 | 1537274 |
| Green zone | 1513 | 1647 | 128784 | 408674 | 440618 |
| Total | 100525 | 5711 | 1398992 | 527970 | 2033198 |

modified linear interpolation shows a higher entropy value than the PIX2PIX and ESRGAN images, corresponding to greater detail at the pixel level. Moreover, it can be analysed that the best method, according to the values of the metric sharpness, is the modified linear interpolation too, where high values indicate the clarity of details.

## 6 Implementation

It consists of taking the image in Fig. 1 as a reference and considering the results obtained by the different metrics, applying the modified linear interpolation output to the PIX2PIX network for zoning by automatic segmentation. The entropy metric measures the level of noise of each image, having that in the original image is 0.70388, while the image in super-resolution is 0.70722. Although the difference is minimal, at the pixel level, it is equivalent to a higher resolution. The accuracy of zoned super-resolution is verified pixel by pixel, and in this way, the percentage of the successful zones is quantified. The visual result of this process can be seen in Fig. 10, and the performance of the super-resolution zoned image is shown in Fig. 11.

**Table 3.** Confusion matrix for images 2.

| Class | Water | Road | Urban zone | Green zone | Total |
|---|---|---|---|---|---|
| Water | 31577 | 9354 | 28662 | 5334 | 74927 |
| Road | 293 | 395478 | 12347 | 2183 | 410301 |
| Urban zone | 27465 | 4184 | 518859 | 10635 | 561143 |
| Green zone | 1783 | 2149 | 19572 | 1115371 | 1138875 |
| Total | 61118 | 411165 | 579440 | 1133523 | 2185246 |

**Table 4.** Confusion matrix for images 3.

| Class | Water | Road | Urban zone | Green zone | Total |
|---|---|---|---|---|---|
| Water | 0 | 0 | 0 | 0 | 0 |
| Road | 293 | 395478 | 12347 | 2183 | 410301 |
| Urban zone | 0 | 0 | 1329416 | 47894 | 1377310 |
| Green zone | 0 | 0 | 89175 | 706136 | 795311 |
| Total | 0 | 0 | 1418591 | 754030 | 2172621 |

**Table 5.** General metrics data for zoning

| Metric | Image 1 | Image 2 | Image 3 |
|---|---|---|---|
| Accuracy | 0.7411 | 0.7824 | 0.9252 |
| Precision | 0.5391 | 0.7391 | 0.9684 |
| Sensitivity | 0.7261 | 0.7223 | 0.9632 |

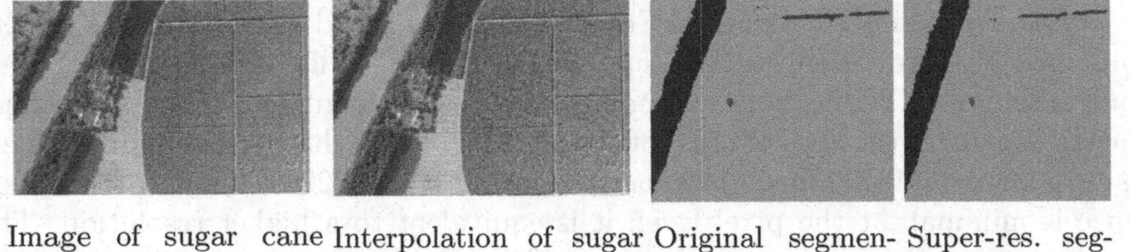

Image of sugar cane field) Interpolation of sugar cane field Original segmentation Super-res. segmentation

**Fig. 10.** Original and interpolated image; Original and super-resolution zoned image

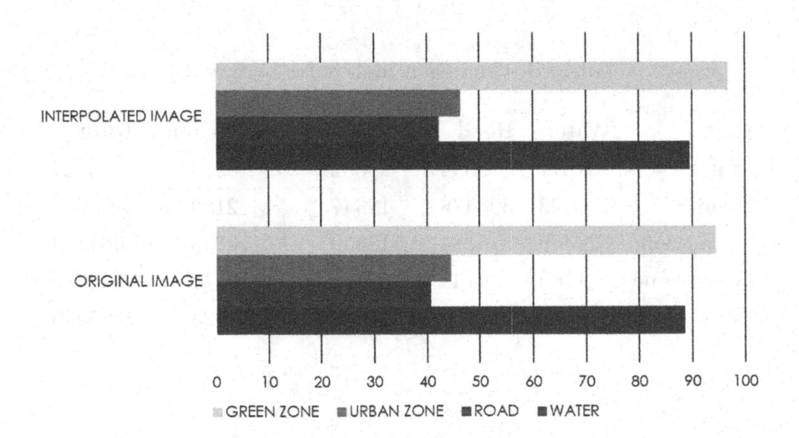


**Fig. 11.** Pixel-to-pixel comparison between the zoned original image and the super-resolution zoned image.

## 7 Conclusions

In this work, several ways of addressing two problems related to images captured from drones have been presented, super-resolution and zoning. As for the super-resolution, a considerable visual improvement was obtained by using the modified linear interpolation method, but this method is the one that takes more time for its execution, considering the available comprehensive set of images of the same scene with slight pose and noise variations among the images. The output image from the modified linear interpolation method gives the best results since the final image was obtained from multiple images that are slightly rotated or displaced, which allows the scene to be explored in greater detail. Simultaneously, using convolutional neural network methods, information is estimated from a base image, the intensity of the pixels in each channel. However, it is not real. It was not obtained from many samples but only from one. That is why the modified linear interpolation method proposed by the authors gives the best results because it resembles a closer reconstruction of reality. However, when ESRGAN was used, an improvement could be seen in the objects' textures in the images. As future work, a new methodology could involve the three methods used in this study. We recommend improving the resolution of an input image by first using ESRGAN to obtain better textures. Then implement the modified linear interpolation to provide much more detail to the image, and finally, as this image will have a very high quality, use that image as the target image with which to train the PIX2PIX network so that the network will learn the patterns of an image with much better information. Although the PIX2PIX architecture gave good results in zoning, it was not so in super-resolution. Showing that, even though it is an excellent and versatile model, it does not necessarily give good results for all cases. That is why there are so many types of neural network architectures since each one specializes in a different task and using it in a task for which it was not adequately designed can lead to inefficient performance. Based on the results obtained, it can be determined that the category with the best segmentation was the green area, then the urban area, followed by water, and finally roads and highways. That happens because the green zone tones are very characteristic; it is difficult to find shades of green that are not green zone. Reddish and white tones characterize the urban zone. Water takes mainly blue tones, but water can be white if the sun shines on it or yellowish if the water is dirty. Finally, the roads and highways take only gray and white tones on different scales, confused with water when there are shadows or urban areas when it has much light. One of the advantages of using convolutional neural networks is that they learn in a global rather than local context. Meaning that, in this specific case of zoning, the network does not learn to associate an intensity of one pixel with another pixel's intensity, but rather the network analyses a pixel and its neighbours and, based on that analysis, it makes the prediction. That is an enormous advantage, since, if it were not so, the predicted image would be a mixture of many colours everywhere, since there are many tonalities of the pixel that can correspond to different categories, as it happens with the different tonalities of the water, of the trees or the buildings.

## References


1. Shakhatreh, H., et al.: Unmanned aerial vehicles (UAVs): a survey on civil applications and key research challenges. IEEE Access **7**, 48572–48634 (2019)
2. Del Gallego, N.P., Ilao, J.: Multiple-image super-resolution on mobile devices: an image warping approach. EURASIP J. Image Video Process. **2017**(1), 1–15 (2017)
3. Isola, P., Zhu, J.Y., Zhou, T., Efros, A.A.: Image-to-image translation with conditional adversarial networks. In: Proceedings of the IEEE Conference on Computer Vision and Pattern Recognition, pp. 1125–1134 (2017)
4. Wang, X., et al.: Esrgan: enhanced super-resolution generative adversarial networks. In: Proceedings of the European Conference on Computer Vision (ECCV) Workshops, pp. 63–79 (2018)
5. Lowe, D.G.: Distinctive image features from scale-invariant keypoints. Int. J. Comput. Vision **60**(2), 91–110 (2004)
6. Fischler, M.A., Bolles, R.C.: Random sample consensus: a paradigm for model fitting with applications to image analysis and automated cartography. Commun. ACM **24**(6), 381–395 (1981)
7. Shaw, P., Rawlins, D.: The point-spread function of a confocal microscope: its measurement and use in deconvolution of 3-D data. J. Microscopy **163**, 151–165 (1991)
8. Kopriva, I., Ju, W., Zhang, B., Xiang, D.: Single-channel sparse nonnegative blind source separation method for automatic 3d delineation of lung tumor in PET images. IEEE J. Biomed. Health Inf **21**, November 2017
9. Panetta, K., Gao, C., Agaian, S.: No reference color image contrast and quality measures. IEEE Trans. Consum. Electron. **59**(3), 643–651 (2013)
10. Mlnsa, P., Rodríguez, J.: The Essential Guide to Image Processing. Academic Press, Estados Unidos (2009)
11. Atkins, B.: Sharpness: What is it and How it is Measured (2006). www.imatest.com/docs/sharpness/#:~:text=Image%20sharpness%20can%20be%20measured,distance%3B%20see%20Figure%203
12. Jalife, A., Calderón, G., Fierro, A., Nakano, M.: Clasificación de Imágenes Urbanas Aéreas: Comparación entre Descriptores de Bajo Nivel y Aprendizaje Profundo. Información tecnológica **28**(3), 209–224 (2017)
13. Tharwat, A.: Classification assessment methods. Applied Computing and Informatics (2020)
14. Ur, H., Gross, D.: Improved resolution from subpixel shifted pictures. CVGIP: Graph. Models Image Process. **54**(2), 181–186 (1992)
15. Farsiu, S., Robinson, M.D., Elad, M., Milanfar, P.: Fast and robust multiframe super resolution. IEEE Trans. Image Process. **13**(10), 1327–1344 (2004)
16. Li, K., Yang, S., Dong, R., Wang, X., Huang, J.: Survey of single image super-resolution reconstruction. IET Image Proc. **14**(11), 2273–2290 (2020)
17. Protter, M., Elad, M., Takeda, H., Milanfar, P.: Generalizing the nonlocal-means to super-resolution reconstruction. IEEE Trans. Image Process. **18**(1), 36–51 (2008)
18. Yang, J., Wright, J., Huang, T.S., Ma, Y.: Image super-resolution via sparse representation. IEEE Trans. Image Process. **19**(11), 2861–2873 (2010)
19. Zhang, L., Wu, X.: An edge-guided image interpolation algorithm via directional filtering and data fusion. IEEE Trans. Image Process. **15**(8), 2226–2238 (2006)
20. Zhang, K., Gao, X., Tao, D., Li, X.: Single image super-resolution with non-local means and steering kernel regression. IEEE Trans. Image Process. **21**(11), 4544–4556 (2012)

21. Lim, B., Son, S., Kim, H., Nah, S., Mu Lee, K.: Enhanced deep residual networks for single image super-resolution. In Proceedings of the IEEE Conference on Computer Vision and Pattern Recognition Workshops, pp. 136–144 (2017)
22. Ledig, C., et al.: Photo-realistic single image super-resolution using a generative adversarial network. In: Proceedings of the IEEE Conference on Computer Vision and Pattern Recognition, pp. 4681–4690 (2017)
23. Garcia-Ruiz, F., Sankaran, S., Maja, J.M., Lee, W.S., Rasmussen, J., Ehsani, R.: Comparison of two aerial imaging platforms for identification of Huanglongbing-infected citrus trees. Comput. Electron. Agric. **91**, 106–115 (2013)
24. Huang, Y., Thomson, S.J., Hoffmann, W.C., Lan, Y., Fritz, B.K.: Development and prospect of unmanned aerial vehicle technologies for agricultural production management. Int. J. Agric. Biol. Eng. **6**(3), 1–10 (2013)
25. Mohamadi, F.: U.S. Patent No. 8,880,241. Washington, DC: U.S. Patent and Trademark Office (2014)
26. Sampedro, C., Martinez, C., Chauhan, A., Campoy, P.: A supervised approach to electric tower detection and classification for power line inspection. In 2014 International Joint Conference on Neural Networks (IJCNN), pp. 1970–1977. IEEE, July 2014
27. Santos, M.J., Disney, M., Chave, J.: Detecting human presence and influence on neotropical forests with remote sensing. Remote Sens. **10**(10), 1593 (2018)

# Security Services

# An Enhanced Lightweight Speck System for Cloud-Based Smart Healthcare

Muyideen AbdulRaheem[1], Ghaniyyat Bolanle Balogun[1], Moses Kazeem Abiodun[2], Fatimoh Abidemi Taofeek-Ibrahim[3], Adekola Rasheed Tomori[4], Idowu Dauda Oladipo[1], and Joseph Bamidele Awotunde[1](✉)

[1] Department of Computer Science, University of Ilorin, Ilorin, Nigeria
{muyideen,balogun.gb,odidowu,awotunde.jb}@unilorin.edu.ng
[2] Department of Computer Science, Landmark University, Omu Aran, Nigeria
moses.abiodun@lmu.edu.ng
[3] Department of Computer Science, Federal Polytechnic, Offa, Nigeria
[4] Directorate of Computer Sciences and Information Technology, University of Ilorin, Ilorin, Nigeria
tomori@unilorin.edu.ng

**Abstract.** In the realm of information and communication sciences, the Internet of Things (IoT) is a new technology with sensors in the healthcare sector. Sensors are critical IoT devices that receive and send crucial bodily characteristics like blood pressure, temperature, heart rate, and breathing rate to and from cloud repositories for healthcare specialists. As a result of technical advancements, the usage of these devices, referred to as smart sensors, is becoming acceptable in smart healthcare for illness diagnosis and treatment. Data generated from these devices is huge and intrinsically tied to every sphere of daily life including healthcare domain. This information must be safeguarded and processed in a safe location. The term "cloud computing" refers to the type of innovation that is employed to safe keep such tremendous volume of information. As a result, it has become critical to protect healthcare data from hackers in order to maintain its protection, privacy, confidentiality, integrity, and its processing mode. This research suggested a New Lightweight Speck Cryptographic Algorithm to Improve High - Performance Computing Security for Healthcare Data. In contrasted to the cryptographic methods commonly employed in cloud computing, the investigational results of the proposed methodology showed a high level of security and an evident improvement in terms of the time it takes to encrypt data and the security obtainable.

**Keywords:** Cloud computing · Lightweight · Speck encryption · Smart healthcare system · Medical data · Security and privacy

## 1 Introduction

Technology and System for cloud applications have in the recent past evolved dramatically. A high number of architectural and infrastructural distributed models have


H. Florez and M. F. Pollo-Cattaneo (Eds.): ICAI 2021, CCIS 1455, pp. 363–376, 2021.
https://doi.org/10.1007/978-3-030-89654-6_26

emanated from these technologies. Cloud computing as one of the emerging technologies also called computing network, is classically connected utilizing the online and shared a large number of services distributed around the region to meet the need of users [1]. The National Institute of Science and Technology of India (NIST) idea of computational cloud is a context that offer a standardized collection of reconfigurable cloud computing with quick access, and ubiquitous access to the internet as needed. The purpose of using cloud computing is not limited to its manageability, scalability, and affordability but also on-demand storage facility that is characterized by uniformity, simplicity, leased variety, dependability, and flexibility [2].

A cloud client like Smart Healthcare System can use the cloud tools on request to store its data and can be accessed anywhere, and on any device at any time. The NIST concept on cloud identified three service models which are made available by Cloud Service Provider (CSP) to various cloud users. It also recognizes four models for deployment which are based on Public, Hybrid, Private, and Community Cloud, highlighting the framework for distributed computing resources easy organization and management cloud [3].

With these models available on the cloud infrastructure, the most critical issue facing the cloud is the exposure of its data to vulnerability and a range of hazards being a technology that uses network connectivity for its communication and transmission. Cloud computing's widespread adoption is hampered by security concerns [4]. In reality, the sharing of cloud computing services poses the difficulty of keeping these services secure and secured against unauthorized access or usage. Mostly the cloud client data outsourced to it faces this challenge [5]. One of the most important security challenges in cloud computing is Smart Healthcare data security, which involves both direct and indirect threats. Providing a safe connection between Smart Healthcare and cloud providers has been created to protect the safety of data transmissions on the cloud network. Lightweight Cryptography is the most effective for data security of Smart Healthcare System. The transformation to ciphertext from plaintext is involved in the process and its tools are always send contents safely by guaranteeing that only the authorized parties are able to retrieve them [6].

Similar to Information Technology Management Systems, Cloud Computing requires fundamental security apparatus such as confidentiality, availability and integrity, authorization, accountability, authentication and privacy as part of essential cloud protection measures. In this paper, an Enhanced Lightweight Speck System for Cloud-based Smart Healthcare was presented. The proposed method called Enhanced Lightweight Speck System (ELSS) improve cloud computing data security for smart healthcare systems. It uses encryption set of rules according to lightweight symmetric cryptography.

The paper is structured as follows. Section one is the introduction to the study. Section to examines background to the study while section three is the methodology used. Lastly, section is the discussion of the result.

## 2 Background

Many researches have explored the security challenges facing cloud computing. This part, however, highlights a few current findings that looked into security of cloud computing. Some categories of symmetric algorithms such as Advanced Encryption Standard

(AES) algorithm, Data Encryption Standard (DES), and International Data Encryption Algorithm (IDEA) were employed to secure cloud computing [7]. A comparison of these symmetric algorithms and asymmetric algorithms such as Rivest-Shamir-Adleman were introduced based on Key size, Rounds, Degree of Safety, and Execution Time. In a cloud setting, the results are relatively efficient [8].

Blowfish, AES, RSA, DSA, and Eclipse IDA were used in a hybrid encryption technique to improve the security of data stored in multiple cloud servers. The research focuses on giving clients the power to choose how their data is to be encrypted rather than relying on third parties to do it [9]. To reduce latency and processing time, an efficient cloud computing architecture categorization has been developed. The information was divided into three tiers to take full advantage of the security of cloud computing. The study recommended using hybrid encryption approaches like RSA Digital Signature and Blowfish algorithm for encoding and decoding with Feistel structure and Exclusive-OR operations [10]. Furthermore, to overcome the security issues of Cloud Storage, [10] combined two independent encryption techniques. It conducted a survey of past studies devoted to cloud data security and then proposed a combined form of defense encryption system based on MD5 and Blowfish to increase cloud server security.

Furthermore, [11] investigated the combination of several cryptographic methods to protect cloud data using attribute-based, homomorphic, searchable contemporary cryptography. A hybrid encryption model was constructed to maximize each system's ability to safeguard cloud information and an assessment of data encoding and decoding of AES-256 SHA-512, IDAs, and 3DES was implemented by [11]. For large and small data files, the technique delivers a substantially greater level of safety and performance. These efforts have not examined the issue in relation to healthcare system.

A great number of studies on cryptography techniques that are symmetric were developed for various applications using lightweight encryptions such RECTANGLE, TWINE, CLEFIA and others [12]. A lightweight encryption system is a n-bit block size having m-bit key generated cryptosystem and repeated in a number of rounds, using mainly XOR combined with left or right rotations LR or RR operation. It serves the purpose of provide protection for ubiquitous resource constrained devices like RFID tags and wireless sensor nodes.

CLEFIA-128 is a symmetric block cipher developed by Sony for both hardware and software [13]. The 128-bit block size encryption uses 128-bit key size, with 28 rounds of Feistel structure. TWINE cryptosystem is a lightweight block encryption for multiple stage using wide range Feistel structure cipher. It has a 64-bit block size and 36 cycles of either 80-bit or 128-bit key size. Every round contains a layer with a 4-bit S-box combined with a 4-bit block permutation in addition to layer of nonlinear substitution for diffusion and confusion purposes.

Another lightweight encryption is RECTANGLE cryptosystem. It is a 64 bits block cipher with a key size of either 80-bit or 128-bit, but runs 25 rounds only for its encryption and decryption processes to achieve optimized protection [14]. Stable IoT (SIT) is a 64-bit block lightweight encryption algorithm using a 64-bit address to encrypt data.

The combination of Feistel structure and a substitution-permutation architecture enable the network algorithm protects the data. The lightweight encryption algorithm encrypts data in IoT devices utilizing a block cipher with a 64-bit size and an 80-bit key length. As an alternative, Data Encryption Lightweight Systems (DESXL), rather than multiple S-Boxes without any initial and final permutations, a single S-Box is used to improve security with a 184-bit key. No successful case of attack has been reported against DESXL [15].

The benchmarking framework [16] compares the execution times, RAM footprints, and binary code size of 19 existing lightweight cryptographic algorithms such as AES, Chaskey, and Speck. [16] indicates that the best cipher is Chaskey, with Speck second, when using C and assembly languages to create block encryption on 8-bit AVR, 16-bit MSP430, and 32-bit ARM microprocessor platforms. This benchmarking tool is used to assess lightweight cryptographic methods for IoT-enabled devices. The lightweight cryptography standardization for Internet of Things devices are being given consideration [17]. RSA, Attribute Based Encryption (ABE), Identity-Based Encryption (IBE), Elliptic Curve Cryptography (ECC), and other cryptographic methods and protocols have been employed in IoT. RSA is a cryptographic algorithm that encrypts and transports encrypted symmetric keys for use with cryptography techniques that are symmetric such as AES [18]. To ensure the safety of communication between IoT devices, ECC is utilized for key agreement and authentication [19].

IBE [20] streamlines certificate administration in IoT systems, allowing, without accessing the public key certificate, a sender (smart node) to encrypt a message for an entity. Another technique that has been offered is ABE, which has been adjusted to be suited for IoT devices by utilizing a proxy [21]. These fundamental cryptography techniques are either changed or used in conjunction with protocols or other algorithms to achieve confidentiality in IoT, as detailed below. Maintaining data privacy at restricted devices and the Internet of Things (IoT) connectivity is a critical responsibility. The IoT's resource restrictions and asymmetric capabilities make it difficult to achieve this using the conventional cryptographic methods outlined above. [21] look at various specific algorithms that have been recommended for use in IoT security, including their architecture, benefits, and limitations. Communication devices over the integrated network suffer confidentiality difficulties in a smart home context.

Salami et al. (2016) [22] propose a lightweight cryptographic encryption strategy with two type of algorithms. The algorithm for key encryption, which only encrypts the session key currently in use once, and the algorithm for data encryption, which encrypts many messages using the encrypted session key. Without any additional processing or communication overheads, this system delivers secrecy services to devices and users. To provide flexible public key management, the stateful IBE approach is used. [22] does not, however, enable prior verification of the sender or receiver. This approach also necessitates more Diffie-Hellman (DH) key operations algorithm, particularly on the node that sends the message [23], and this is subject to the chosen plaintext attack [24]. Usman et al. [25] present a Secure IoT (SIT) algorithm with 64-bit size (2017) based on the combination of a Feistel and a uniform substitution permutation network (SPN). The SIT uses round of five encryptions on an 8-bit microcontroller, resulting in simpler code, lower memory usage, and shorter encryption and decryption cycles.

The literature reviewed of related works show the need for a secure system in healthcare sector for the proper monitoring of patient data and information. This will also give a sense of belonging to patients that are using the IoT-based system for their wellbeing.

The speck encryption can greatly increase the security and privacy of healthcare data in IoT-based system.

## 3 Methodology

Speck belongs to a family of lightweight symmetric block ciphers having a pair of varying block and key sizes. A block cipher is an encryption algorithm that enables users to have a common key secretly and securely encrypt/decrypt blocks of data. Lightweight encryption is built for efficient implementation of extremely constrained platforms, such as RFID tags, smart sensors, microcontrollers, and other resource-limited devices.

The term "lightweight" does not mean how secured the algorithm is, but rather refers to its suitability for use on highly constrained devices. Thus, a secure lightweight block cipher having a given block and key size pairs offers the equivalent level of security as any other secure block cipher with that same block and key size.

Lightweight encryption addresses the request to secure resource constraint devices which the conventional encryptions could not adequately attend to, using algorithms and protocols designed to perform well on platforms for devices with resources constrained.

Speck has support for a block of various sizes such as 32, 48, 64, 96, and 128 bits, and up to three different key sizes to go along with each size of a block. Speck family has ten different algorithms with various block and key sizes, as shown in the Table 1. All values are in bits with n as the word size and M as the number of words.

**Table 1.** The speck parameters

| Block size | Key sizes | n | M |
|---|---|---|---|
| 32 | 64 | 16 | 4 |
| 48 | 72 | 24 | 3 |
| 48 | 96 | 24 | 4 |
| 64 | 96 | 32 | 3 |
| **64** | **128** | **32** | **4** |
| 96 | 96 | 48 | 2 |
| 96 | 144 | 48 | 3 |
| 128 | 128 | 64 | 2 |
| 128 | 192 | 64 | 3 |
| 128 | 256 | 64 | 4 |

Speck has operations that are highly efficient for software platforms of varying requirements. There are a quite few numbers of the operation such as NOT, OR, AND, XOR, rotations, modular addition, and subtraction use to achieve nonlinearity as against S-Boxes used in conventional encryptions. Since Speck is not using S-Boxes, then it is not Substitution-Permutation Networks (SPNs). Rather than being SPNs, Speck is an

Add–Rotate–XOR (ARX) cipher with Feistel Structure round functions, which provides an adequate equilibrium between operations of nonlinear confusion and linear diffusion. Using the bit permutation rotation operation may not achieve sufficient diffusion, but with additional functional rounds, an adequate level of security is achieved like in the S-Boxes permutation of SPN. Speck nonlinearity is achieved with modular addition. Thus, its functions are secured cryptographically and effectively suitable for IoT constraint devices software implementations.

### 3.1 The Proposed Model

The notations used in the definition of the proposed model is given in the table below.

| | |
|---|---|
| n | A word size |
| 2n | A block size |
| M | Number of words |
| R | Number of rounds |
| PT | 2n-bit input plaintext |
| CT | 2n-bit output ciphertext |
| Ki | n-bit round subkey for round i |
| K | mn-bit Master key from which round subkeys are generated |
| ⊕ | Bitwise exclusive OR operation |
| Sj | Left cyclic shift by n bits |
| S-j | Right cyclic shift by n bits |
| + | Mod 2n addition operation |

Speck block cipher denoted as Speck2n has an n-bit word, where n represents either 16, 24, 32, 48, or 64. Speck2n having an m-word (mnbit) key is denoted as Speck2n/mn. For instance, Speck64/96 denotes the Speck version having plaintext blocks of 64-bit and with a key having 96-bit (n = 32 and m = 3).

The round function of Speck2n encryption is denoted by the mapping.

$R_k(x, y) = \left(\left(S^{-\alpha}x + y\right) \oplus k, S^{\beta}y \oplus \left(S^{-\alpha}x + y\right) \oplus k\right)$, as shown in Fig. 1.

where x and y are n-bit halves of 2n-bit plaintext, and k is round key.

| | |
|---|---|
| x = RCS(x , α) | Right shift x by α and assign the result to x |
| x = x + y | modulo 2n addition of x and y and assign the result to x |
| x = x ⊕ k | XOR x and round key k and assign the result to x |
| y = LCS(y, β ) | Left shifty by β and assign the result to y |
| y = y ⊕ x | XOR y and x and assign the result to y |

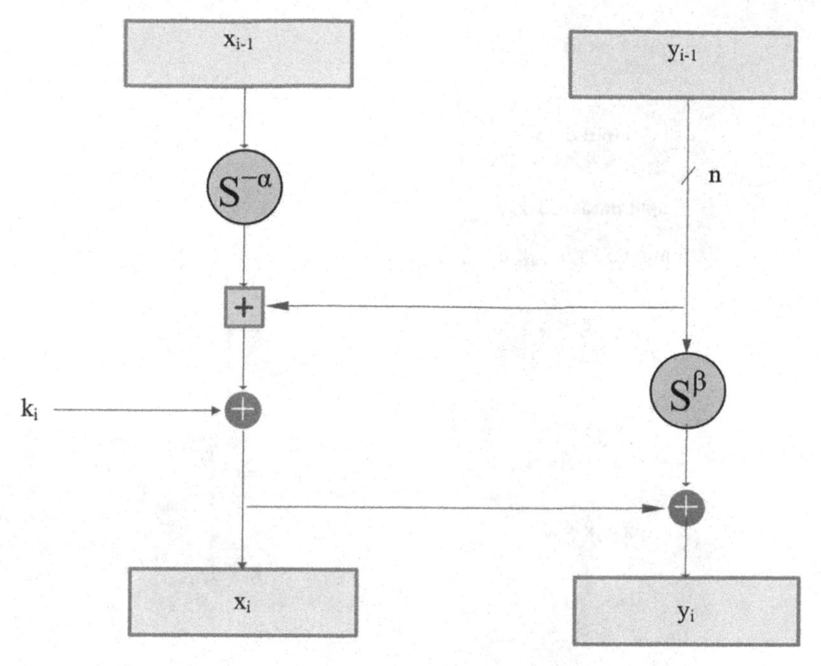


**Fig. 1.** Speck round function

For decryption, the inverse of the round function is used with modular subtraction instead of modular addition and is given by

$$R_k^{-1}(x, y) = \left(S^{\alpha}\left((x \oplus k) - S^{-\beta}(x \oplus y)\right), S^{-\beta}(x \oplus y)\right)$$

The parameters α and β are 7 and 2 respectively, for Speck32/64 and they are 8 and 3 for other variance of speck.

Figure 2 presents the speck encryption flow of the proposed model.

Key Schedule

Speck uses key schedules of 2-, 3-, and 4-word depending on Speck variance. Key schedules for Specks use around function, as given below.

Suppose m is the number of words for a key, and the key K can be written as $(l_{m-2},\ldots,l_0,k_0)$.

Then, to generate sequences of two words $k_i$ and $l_i$ by

$$l_{i+m-1} = (k_i + S^{-\alpha} l_i) \oplus \mathrm{i}$$

and

$$k_{i+1} = S^{\beta} k_i \oplus l_{i+m-1}$$

The value $k_i$ is the ith round key, for $\mathrm{i} \geq 0$

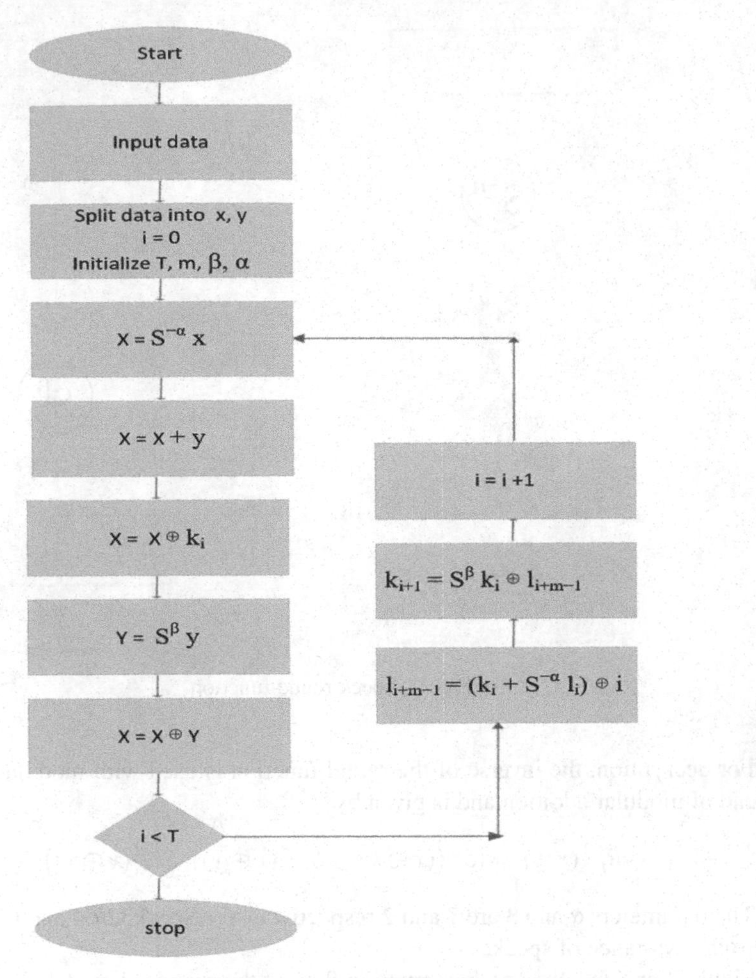


**Fig. 2.** The speck encryption flow

The Table 2 below shows the Speck rounds and the required parameters

The Speck key schedules receive an input key K and produce from it a sequence of T key words$k_0,\ldots,k_{T-1}$, where T is the number of rounds. Encryption is the composition $R_{k_{T-1}} \circ \cdots \circ R_{k_1} \circ R_{k_0}$, read from right to left.

Algorithm 1 below presents the pseudo-code of the SPECK encryption procedure. The key scheduling function forms the encryption key K $=l_{m-2},\ldots,l_0$, $k_0$ and generates the T –rounds keys sequence$k_0,\ldots$, $k_{T-1}$ using the procedures presented by Algorithm 2.

**Table 2.** The speck rounds

| Block size 2n | n | M | key size mn | Speck rounds | $\alpha$ | B |
|---|---|---|---|---|---|---|
| 32 | 16 | 4 | 64 | 22 | 7 | 2 |
| 48 | 24 | 3 | 72 | 22 | 8 | 3 |
| 48 | 24 | 4 | 96 | 23 | 8 | 3 |
| 64 | 32 | 3 | 96 | 26 | 8 | 3 |
| 64 | 32 | 4 | 128 | 27 | 8 | 3 |
| 96 | 48 | 2 | 96 | 28 | 8 | 3 |
| 96 | 48 | 3 | 144 | 29 | 8 | 3 |
| 128 | 64 | 2 | 128 | 32 | 8 | 3 |
| 128 | 64 | 3 | 192 | 33 | 8 | 3 |
| 128 | 64 | 4 | 256 | 34 | 8 | 3 |

```
Algorithm 1: Speck Encryption
input: plaintext, K encryption key
output: ciphertext
split plaintext into n-bit x, y
    initialize T, β, α
    generate round keys k_0, . . . , k_{T-1}
    for i = 0 to T-1
        x = (S^{-α} x + y) ⊕ k_i
        y = S^β y ⊕ x
    end for loop
```

```
Algorithm 2: Round Key generation
Input: Encryption key
Output: sequence of keys
initialize m;
generate l_{m-2}, . . . , l_0, k_0
for I = 0 to T-2
    l_{i+m-1} = (k_i + S^{-α} l_i) ⊕ i
    k_{i+1} = S^β k_i ⊕ l_{i+m-1}
end for loop
```

## 4 Results and Discussion

The proposed encryption method was developed on the microcontroller (MCU) AVR 8-bit RISC (Reduced Instruction Set Computing) architecture that has programmable on-chip flash memory, SRAM, IO storage space, and EEPROM, as shown in Table 3. Different block size data was processed for encryption in KB and our experimental findings are seen in Table 2. The execution time tests the amount of encryption and decryption block cycles used. This is the contrast between the beginning and the end of the process.

**Table 3.** Feature of the device

| Name | Value |
|---|---|
| MCU | ATmega328PB |
| Program memory type | Flash |
| Program memory size (KB) | 32 |
| CPU speed (MIPS/DMIPS) | 20 |
| SRAM (B) | 2,048 |
| Data EEPROM/HEF (bytes) | 1024 |
| Digital communication peripherals | 2-UART, 2-SPI, 2-I2C |
| Capture/compare/PWM peripherals | 3 Input Capture, 3 CCP, 10PWM |
| Timers | 2 x 8-bit |
| Number of comparators | 1 |
| Operating voltage range (V) | 1.8 to 5.5 |
| Pin count | 32 |
| Low power | Yes |

**Table 4.** Performance analysis of the proposed model

| File size (kilobytes) | Encryption time (mm) | Decryption time (mm) |
|---|---|---|
| 0.82 | 0.121 | 0.120 |
| 1.65 | 0.216 | 0.215 |
| 12.32 | 0.893 | 0.891 |
| 36.50 | 2.014 | 2.012 |
| 50.2 | 3.142 | 3.138 |
| 100.7 | 5.461 | 5.459 |

Other block cipher techniques such as AES, DES, SIMON were compared to evaluate the analysis of memory use and speed. Table 4 presents the findings of the performance of the proposed system using encryption time and decryption time, from table 4 the results show that the high the file size (KB) the high the encryption and decryption time. The findings as shown in Table 5 indicate that the memory use of AES, DES, and SIMON is greater than the proposed memory usage of the proposed system. In comparison, the time to encrypt and decrypt is less in the proposed system than in most of the lightweight ciphers.

**Table 5.** Comparison results of our proposed model with another model

| Algorithm | Speed (Clockcycle /bytes) | Memory usage (bytes) | Encryption time (ms) | Decryption Time (ms) |
|---|---|---|---|---|
| AES | 24695 | 1709 | 650 | 124 |
| DES | 28401 | 1608 | 797 | 281 |
| SIMON | 39313 | 755 | 700 | 300 |
| SPECK | 30957 | 1364 | 619 | 236 |

Table 5 uses various algorithms to display the different encryption times of the same input data. This illustrates that the suggested encryption solution takes the same amount of time as other regular block ciphers.

The proposed system was still tested using the EEG datasets taken from Bonn database a widely used dataset [30]. The dataset contains five various datasets labels as A, B, C, D and E. The dataset A, D, and E was analysed to measure the latency in both cloud and cloud computing using computational time and transmission delay. Dataset A was taken from a stable individual, while dataset D was taken during an interracial situation. During seizure activity, dataset E was collected from the epileptegenic region (ictal state). Each dataset contains 100 EEG epochs, each of which contains 4097 samples. The EEG data was collected at a sampling rate of 173.61 Hz. Tables 6 and 7 demonstrate the statistical parameters that were derived from the sub-bands [31]. The values of all statistical parameters are clearly higher for dataset E. The derived attribute estimates for datasets A and D are extremely similar. Dataset A, D, and E versus Dataset E were used to assess identification in this study.

On an Intel Celeron processor, for example, each cycle requires about 270 mA on average. For example, an encryption of 20,000 cycles would burn 7.7 J on a 700 MHz CPU operating at 1.35 V. As a result, program P's energy consumption to achieve its aim (encryption or decryption) is provided by:

$$E = Vcc * I * N * \tau \quad (1)$$

Both Vcc and are fixed for a given piece of hardware,

$$E \alpha I \times N. \quad (2)$$

However, at the application level, talking about T rather than N is more useful, thus we write energy as $EalxT$. The duration time for encryption or decryption is replaced by the total number of clock cycles divided by the clock frequency. The amount of energy used by program P to achieve its purpose (encryption) is then calculated as follows:

$$\therefore\ E_{cost} = V_{cc} \times I \times TJoules \quad (3)$$

Where $V_{cc}$ is the system's supply voltage and I is the average current consumed from the power source in amperes. The number of clock cycles is denoted by the letter $N.\tau T$ the clock period. T – N/processor's speed (seconds).

**Table 6.** Coefficients of extracted features for dataset A

| Coefficient | Variance | Standard deviation | Energy |
|---|---|---|---|
| D1 | 25.2164 | 5.0216 | 2.8564e + 04 |
| D2 | 587.553 | 24.2395 | 3.0435e + 05 |
| D3 | 5.3957e + 03 | 73.4555 | 1.4426e + 06 |
| D4 | 9.9058e + 03 | 99.5279 | 1.9874e + 06 |
| A4 | 1.5439e + 04 | 124.2539 | 4.0502e + 06 |

**Table 7.** Coefficients of extracted features for dataset E

| Coefficient | Variance | Standard deviation | Energy |
|---|---|---|---|
| D1 | 1.4426e + 03 | 37.9819 | 1.8934e + 06 |
| D2 | 6.4382e + 04 | 253.736 | 4.8707e + 07 |
| D3 | 7.0151e + 05 | 837.560 | 3.0676e + 08 |
| D4 | 6.9684e + 05 | 834.769 | 1.887e + 08 |
| A4 | 1.7177e + 06 | 1.310e + 03 | 4.0854e + 08 |

## 5 Conclusion and Future Research Directions

The IoT-based system is an upcoming information and communication science technology. A large number of IoT technologies have been developed in the healthcare field to remotely detect, track, predict, and manage chronic diseases. The data generated by these devices is huge and the processing and transmissions need typical infrastructure to cope with them. Cloud computing can be applied to overcome the challenges of IoT-based systems. The cloud computing technology with lower latency, reduced computational time, and scalability will help IoT-based system devices in real-time data collection and processing. The deployment of cloud data is far away from the network; this causes the response time delay in real-time. Moreover, the cloud may cause significant overhead on the backbone network to the user application due to the huge amount of big data sent to the cloud. Hence, the application of cloud computing will bring the storage resources and computational closer to the end-user devices, thus reducing the burden on the cloud. The information transmitted by the smart sensor in smart healthcare systems (SHS) is personal, requiring secrecy, trustworthiness, and usability for healthcare practitioners to take timely and accurate decisions, so the data required to be transmitted and stored safely. For the IoT based Smart Healthcare system, a lightweight ciphering technique was implemented. The proposed approach is based on basic operations of ARX Addition, Rotation XORing, swapping, slicing, among others. The outcome of the implementation reveals that the memory occupation is limited while the speed for producing a successful cipher is important. The results reveal that the longer the encryption and decryption time, the larger the file size (KB). The results show that the memory usage of AES, DES, and SIMON is higher than the proposed system's memory usage. In different constrained devices, the proposed technique can be applied and tested. It is also possible to perform differential and linear crypto-analysis of this algorithm in the future to ensure the cipher's robustness.

## References


1. Awotunde, J.B., Adeniyi, A.E., Ogundokun, R.O., Ajamu, G.J., Adebayo, P.O.: MIoT-based big data analytics architecture, opportunities and challenges for enhanced telemedicine systems. Stud. Fuzziness Soft Comput. **2021**(410), 199–220 (2021)
2. Maskeliūnas, R., Damaševičius, R., Segal, S.: A review of internet of things technologies for ambient assisted living environments. Future Internet **11**(12), 259 (2019)
3. Abiodun, M.K., Awotunde, J.B., Ogundokun, R.O., Adeniyi, E.A., Arowolo, M.O.: Security and information assurance for IoT-based big data. Stud. Computat. Intell. **2021**(972), 189–211 (2021)
4. Azeez, N.A., Van der Vyver, C.: Security and privacy issues in e-health cloud-based system: a comprehensive content analysis. Egypt. Inf. J. **20**(2), 97–108 (2019)
5. Abikoye, O.C., Ojo, U.A., Awotunde, J.B., Ogundokun, R.O.: A safe and secured iris template using steganography and cryptography. Multimedia Tools Appl. **79(31–32)**, 23483–23506 (2020)
6. Ogundokun, R.O., Awotunde, J.B., Adeniyi, E.A., Ayo, F.E.: Crypto-Stegno based model for securing medical information on IOMT platform. Multimedia Tools Appl. 1–23 (2021)
7. Tabrizchi, H., Rafsanjani, M.K.: A survey on security challenges in cloud computing: issues, threats, and solutions. J. Supercomput. **76**(12), 9493–9532 (2020). https://doi.org/10.1007/s11227-020-03213-1
8. Awotunde, J.B., Chakraborty, C., Adeniyi, E.A., Abiodun, K.M.: Intrusion detection in industrial internet of things network based on deep learning model with rule-based feature selection. Wirel. Commun. Mob. Comput. **2021**, 1–17 (2021)
9. Thabit, F., Alhomdy, S., Jagtap, S.: A new data security algorithm for the cloud computing based on genetics techniques and logical-mathematical functions. Int. J. Intell. Netw. **2**, 18–33 (2021)
10. Abdulraheem, M., Awotunde, J.B., Jimoh, R.G., Oladipo, I.D.: An efficient lightweight cryptographic algorithm for IoT security. Commun. Comput. Inf. Sci. **2021**(1350), 444–456 (2021)
11. Mohammed, K.M.A.K.: Confidentiality of data in public cloud storage using hybrid encryption algorithms. Doctoral Dissertation, Sudan University of Science and Technology
12. Singh, P., Acharya, B., Chaurasiya, R.K.: Lightweight cryptographic algorithms for resource-constrained IoT devices and sensor networks. In: Security and Privacy Issues in IoT Devices and Sensor Networks, pp. 153–185. Academic Press
13. Makarenko, I., Semushin, S., Suhai, S., Kazmi, S.A., Oracevic, A., Hussain, R.: A comparative analysis of cryptographic algorithms in the internet of things. In: 2020 International Scientific and Technical Conference Modern Computer Network Technologies (MoNeTeC), pp. 1–8. IEEE, Oct 2020

14. Nayancy, Dutta, S., Chakraborty, S.: A survey on implementation of lightweight block ciphers for resource constraints devices. J. Discrete Math. Sci. Cryptogr. 1–22 (2020)
15. Saddam, M.J., Ibrahim, A.A., Mohammed, A.H.: A lightweight image encryption and blowfish decryption for the secure internet of things. In: 2020 4th International Symposium on Multidisciplinary Studies and Innovative Technologies (ISMSIT), pp. 1–5. IEEE, Oct 2020
16. Dinu, D., Le Corre, Y., Khovratovich, D., Perrin, L., Großschädl, J., Biryukov, A.: Triathlon of lightweight block ciphers for the internet of things. J. Cryptogr. Eng. **9**(3), 283–302 (2019)
17. Turan, M.S., McKay, K.A., Çalik, Ç., Chang, D., Bassham, L.: Status report on the first round of the NIST lightweight cryptography standardization process. National Institute of Standards and Technology, Gaithersburg, MD, NIST Interagency/Internal Rep. (NISTIR) (2019)
18. Kraft, J.S., Washington, L.C.: An Introduction to Number Theory with Cryptography. Chapman and Hall/CRC (2018)
19. Das, A.K., Wazid, M., Yannam, A.R., Rodrigues, J.J., Park, Y.: Provably secure ECC-based device access control and key agreement protocol for IoT environment. IEEE Access **7**, 55382–55397 (2019)
20. Shamir, A.: Identity-based cryptosystems and signature schemes. In: Blakley, G.R., Chaum, D. (eds.) CRYPTO 1984. LNCS, vol. 196, pp. 47–53. Springer, Heidelberg (1985). https://doi.org/10.1007/3-540-39568-7_5
21. Fischer, M., Scheerhorn, A., Tönjes, R.: Using attribute-based encryption on IoT devices with instant key revocation. In: 2019 IEEE International Conference on Pervasive Computing and Communications Workshops (PerCom Workshops), pp. 126–131. IEEE, Mar 2019
22. Al Salami, S., Baek, J., Salah, K., Damiani, E.: Lightweight encryption for smart home. In: 2016 11th International Conference on Availability, Reliability and Security (ARES), pp. 382–388. IEEE, Aug 2016
23. Naoui, S., Elhdhili, M.E., Saidane, L.A.: Lightweight enhanced collaborative key management scheme for smart home application. In: 2017 International Conference on High Performance Computing Simulation (HPCS), pp. 777–784. IEEE, July 2017
24. Syal, R.: A comparative analysis of lightweight cryptographic protocols for smart home. In: Shetty, N.R., Patnaik, L.M., Nagaraj, H.C., Hamsavath, P.N., Nalini, N. (eds.) Emerging Research in Computing, Information, Communication and Applications. AISC, vol. 882, pp. 663–669. Springer, Singapore (2019). https://doi.org/10.1007/978-981-13-5953-8_54
25. Awotunde, J.B., Jimoh, R.G., Folorunso, S.O., Adeniyi, E.A., Abiodun, K.M., Banjo, O.O.: Privacy and security concerns in IoT-based healthcare systems. Internet Things, 105–134 (2021)

# Hybrid Algorithm for Symmetric Based Fully Homomorphic Encryption

Kamaldeen Jimoh Muhammed[1(✉)], Rafiu Mope Isiaka[2], Ayisat Wuraola Asaju-Gbolagade[3], Kayode Sakariyah Adewole[3], and Kazeem Alagbe Gbolagade[2]

[1] Department of Educational Technology, University of Ilorin, Ilorin, Nigeria
jimoh.km@unilorin.edu.ng

[2] Department of Computer Science, Kwara State University, Malete, Nigeria
{abdulrafiu.isiaka,kazeem.gbolagade}@kwasu.edu.ng

[3] Department of Computer Science, University of Ilorin, Ilorin, Nigeria
adewole.ks@unilorin.edu.ng

**Abstract.** Fully Homomorphic Encryption (FHE) supports realistic computations on encrypted data and hence it is widely proposed to be used in cloud computing to protect the integrity and privacy of data stored in the cloud. The existing symmetric-based FHE schemes suffer from insecurity against known plaintext/ciphertext attacks and generate a large ciphertext size that required a large bandwidth to transfer the ciphertext over the network. To ameliorate these weaknesses is the aim of this paper. A hybrid algorithm for symmetric-based Fully Homomorphic Encryption that combines N-prime model and Matrix Operation for Randomization and Encryption with Secret Information Moduli Set (MORES-IMS) is proposed. The results show that the proposed hybrid symmetric-based FHE framework satisfied homomorphism properties with robust inbuilt security that resists known plaintext, ciphertext and statistical attacks. The ciphertext size produced by the proposed hybrid framework is less than 4 times of its equivalent plaintext size with a considerable encryption execution time and fast decryption time. Thus, guaranteed to provide optimum performance and reliable solutions for securing integrity and privacy of user's data in the cloud.

**Keywords:** Homomorphic Encryption · Fully Homomorphic Encryption · Residue Number System · Symmetric encryption · Hybrid encryption · Cloud computing

## 1 Introduction

Cloud technology allows paradigm of data storage to be shifted from local storage to cloud storage which has brought tremendous improvement over the years. Despite numerous advantages offered by cloud technology in term of flexibility and cost-effectiveness, security and privacy have grown to be of significant concern in cloud technology and is discouraging factor for potential adopters [1]. For instance, unencrypted sensitive data on the cloud are vulnerable to different attacks such as unauthorized access


H. Florez and M. F. Pollo-Cattaneo (Eds.): ICAI 2021, CCIS 1455, pp. 377–390, 2021.
https://doi.org/10.1007/978-3-030-89654-6_27

to the server by the adversary, spy on secluded data by legitimate administrator at hosting providers and stealing or destruction of data by attacker with physical access to the server [1].

Traditional encryption schemes such as Data Encryption Standard (DES) and Advanced Encryption Standard (AES) have been used to address the issue of security and privacy facing the data stored in the cloud. However, cloud technology has advanced to the level of cloud computing where the stored data on the cloud are used for computation. In 2008, Cloud computing has evolved as a new revolution in information technology as it provides a variety of services and applications to be run from anywhere in the world to its users through the Internet, traditional encryption schemes do not fit enough to protect the integrity and privacy of data in cloud computing server as it requires the release of secret key prior the computation on the cloud server which threatens the privacy of such computation and the data [2].

Consequently, Homomorphic Encryption (HE) was introduced as an encryption scheme that allows computation to be performed on encrypted data without the use of secret key before the computation. This scheme was proposed by Rivest, et al. [3] aliased as Partial Homomorphic Encryption (PHE) due to its support in either additive or multiplicative homomorphic operation not until Gentry [4] introduced Somewhat Homomorphic Encryption (SHE) scheme that allows both additive and multiplicative operations with limited number of addition and multiplication. He further proposed Fully Homomorphic Encryption (FHE) scheme that allows unlimited number of arbitrary functions that can be evaluated on an encrypted data [5].

FHE schemes were categorized as asymmetric and symmetric cryptosystem as shown in Fig. 1. Asymmetric approach as in [5–15] uses public and private keys for encryption and decryption respectively. This has been proved to be secured [16], however, the scheme is far from being practicable because of its large computational cost and large ciphertexts generation. Since then, though efforts have been made to devise more efficient schemes. Most of the asymmetric based FHE schemes still have large ciphertexts (millions of bits for a single ciphertext). This presents a considerable bottleneck in the practical deployments [17].

Symmetric approach on the other hand, as in [18–26] uses a single private or secret key for both encryption and decryption. It has its strength in low computational overhead and practical consideration for real-world deployments. However, some of these symmetric based FHE schemes such as [24, 26–29] still suffer from either large ciphertext expansion or low immunity against security attack as argued in [25, 30].

To ameliorate the drawbacks, such as low immunity against security attack, ciphertext expansion, encryption and decryption time; hybrid symmetric-based FHE is proposed which comprises of two layers. The top layer enhances N-prime model proposed in Mohammed and Abed [24] with Residue Number System (RNS) to address ciphertext expansion problem. The bottom layer is based on Enhanced Matrix Operation for Randomization and Encryption (EMORE) as presented in Muhammed and Gbolagade [31] to further strengthen the security of the proposed hybrid framework.

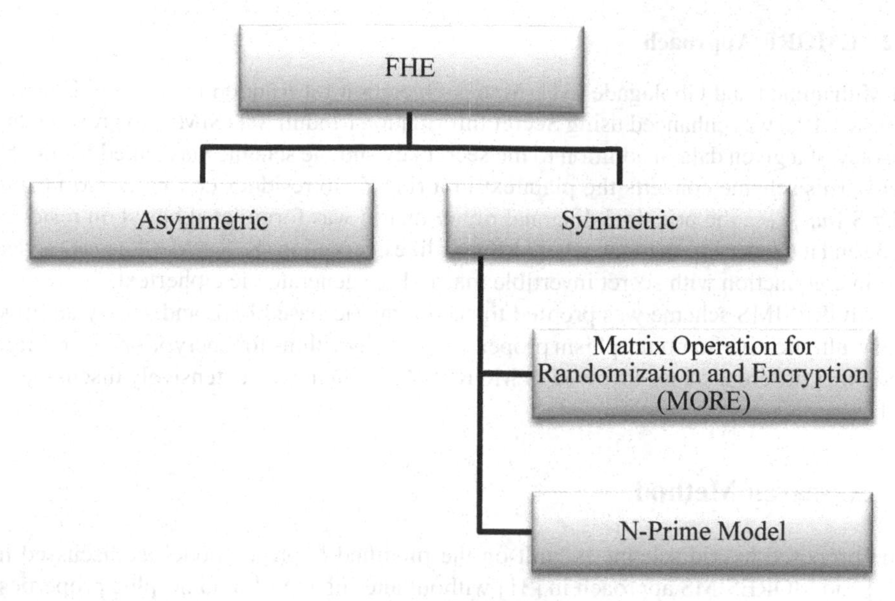


**Fig. 1.** Taxonomy of FHE

The rest of this paper is organized as follows. Section 2 Discusses N-prime Model and EMORE Approaches. Section 3 Describes the Research Method in line with Modified N-prime Model hybridize with EMORE. Results Findings and Discussion are given in Sect. 4. Conclusions to the study are drawn in Sect. 5.

## 2 N-Prime Model and EMORE Approach

The two major approaches involve in this study are discussed as follow:

### 2.1 N-Prime Model

In Mohammed and Abed [24], a new mathematical model was developed and designed to achieve symmetric-based FHE scheme that supports additive and multiplicative homomorphism called N-prime model. The security of this model depends on the problem of factorization of integers to their primary numbers. It operates in a way that, instead of dealing with two prime numbers, it is expanded to deal with n prime numbers, this forms the bases for the security of the presumptive algorithm to be more efficient in the face of the security challenges facing cloud computing. The N-prime model operates in three stages which are Key Generation, Encryption and Decryption.

The proof for both additive and multiplicative homomorphism properties for this model was extensively discussed in [24]. However, the encryption stage of this model generates excessively big ciphertext which serves as the main drawback of this model due to the multiplication of big prime number *Sk*, *n* (product of numerous prime numbers) and random number *r* that also serve as noise to the ciphertext. However, the drawback of N-prime model has been addressed in [32].

### 2.2 EMORE Approach

In Muhammed and Gbolagade [31], Matrix Operation for Randomization and Encryption(MORE) was enhanced using Secret Information Moduli Set (SIMS) to preserve the privacy of a given data in addition to the secret key and the scheme was called MORESIMS. This scheme converts the plaintext in a ring $Z_n$ to residues $(x_1, x_2, ......x_k)$ using SIMS $\{m_i\}_{i=1,k}$, the principal diagonal of the matrix was formulated based on residues and send it to encryption block where MORE-like encryption ($K^{-1} * M * K$) was carried out in conjunction with secret invertible matric K to generate the ciphertext.

MORESIMS scheme was proofed to be symmetric-based FHE and satisfy additive and multiplicative homomorphism properties. The algorithms for encryption, decryption and other different stages involved in MORESIMS scheme are extensively discussed in [31].

## 3 Research Method

The proposed hybrid scheme is built on the modified N-prime model as discussed in [32] and MORESIMS approach in [31] without altering their homomorphic properties. The steps are illustrated in Fig. 2 and explained as follow:

### 3.1 Encryption Process

Using SIMS approach in Stage 2 of Fig. 2, parameters such as plaintext ($x$), random number ($r$), big integer ($l$) and N-prime ($np$) at Stage 1 of Fig. 2 will be converted to residues using Residue Number System (RNS) forward conversion module which addresses ciphertext expansion problem in N-prime model and performs the first encryption based on RNS arithmetic operation with respect to SIMS as shown in Stage 1 of Fig. 2. Thus, the result of stage 1 produces ciphertext ($ct$).

$$\begin{aligned} Ciphertext\ ct = \{x_1, x_2, x_3, ...x_k\}_{mi} + \{r_1, r_2, r_3, ...r_k\}_{mi} \\ * \{l_1, l_2, l_3, ...l_k\}_{mi} * \{np_1, np_2, np_3, ...np_k\}_{mi} \end{aligned} \tag{1}$$

$$ct = \{ct_1, ct_2, ct_3, ...ct_k\}_{mi} \tag{2}$$

where $ct_1$, $ct_2$, $ct_3, \ldots ct_k$ are ciphertext residues based on modified N-Prime model and $mi\ \forall i = 1, k$ is the moduli set.

Stage 2 of the framework in Fig. 2 receives the output of Stage 1 which is Eq. (2) as an input to form the principal diagonal of matrix $M$ ($M = diag(ct)$) as shown in Eq. (3) and pass it to MORE-like encryption block.

$$M = \begin{bmatrix} m \cdots 0 \\ \vdots \ddots \vdots \\ 0 \cdots r \end{bmatrix} = \text{diag(ct)} = \begin{bmatrix} ct_1 \cdots 0 \\ \vdots \ddots \vdots \\ 0 \cdots ct_k \end{bmatrix} \tag{3}$$

The second encryption at this stage requires the generation of Secret Invertible Matrix ($K$) and find $K^{-1}$ to perform encryption operation as shown in Eq. (4). Thus, $E(m, k)$

as an output with ciphertext C, will be sent to cloud server for storage and to perform homomorphic computation on the ciphertext as presented in Stage 3 of Fig. 2.

$$E(m,k) = K^{-1}\begin{bmatrix} ct_1 & \cdots & 0 \\ \vdots & \ddots & \vdots \\ 0 & \cdots & ct_k \end{bmatrix} K = C \tag{4}$$

The operations in Stage 3 of Fig. 2, shows that homomorphic computation can be done in two ways in the cloud. The first one is through Normal Arithmetic Based Operation Cloud Server (NABOCS) shown at the right-hand side where the cloud server receives the output of stage 2 together with evaluation function ($f(x, C)$) provided by the user and perform homomorphic operation (addition and multiplication) based on the normal arithmetic operation. The second one is through RNS Based Arithmetic Operation Cloud Server (RNS-BAOCS) at the left-hand side of Stage 3 where further optimization is provided for homomorphic bit operation. This optimization stage requires the use of Public Moduli Set (PMS) that comprises of Information Moduli Set (IMS) and Redundant Moduli Set (RMS) to convert the ciphertext C received from stage 2 to residues using RNS Forward Conversion Module with respect to PMS and perform homomorphic operation (evaluation function ($f(x, C)$)) on the residues with respect to PMS in parallel with no carry propagations using RNS based arithmetic operations.

### 3.2 Decryption Process

The inverse of encryption process is simply the decryption process. User with the following secret parameters can decipher the result of homomorphic operation as shown in Fig. 3:

i. Invertible Matrix ($k$)
ii. Secret Key ($l$)
iii. SIMS and
iv. Value of $n$

Depending on which side of the server (either NABOCS or RNS-BAOCS) is ciphertext coming from, the decryption model directs the ciphertext ($ct$) into CRT module if it contains PMS before sending it to MORE-like decryption block otherwise it goes straight down to MORE-like decryption block where it requires secret invertible matrix ($k$) from the user to decrypt $c \leftarrow ct$ and pass $c$ together with a secret key ($l$), *SIMS* and $n$ from the user's end to RNS based N-prime decryption block where $diag(c)$ form the residues which need to be converted using CRT with respect to *SIMSandn* to get its equivalent weighted number ($wn$) and finally compute $|wn|_{SecretKey(l)}$ as the original message or result of an operation as summarized in Table 1.

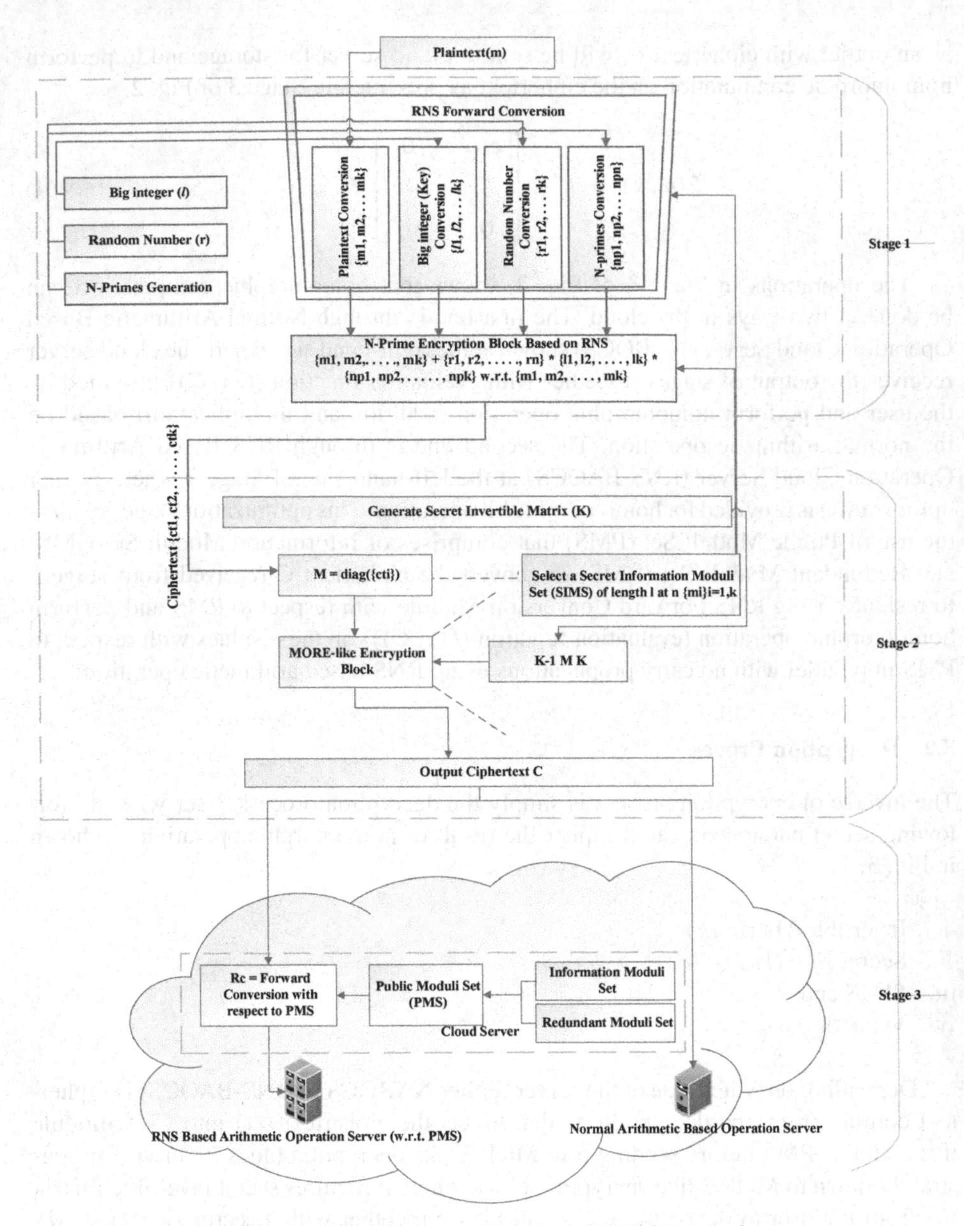


**Fig. 2.** Hybridized symmetric based FHE framework

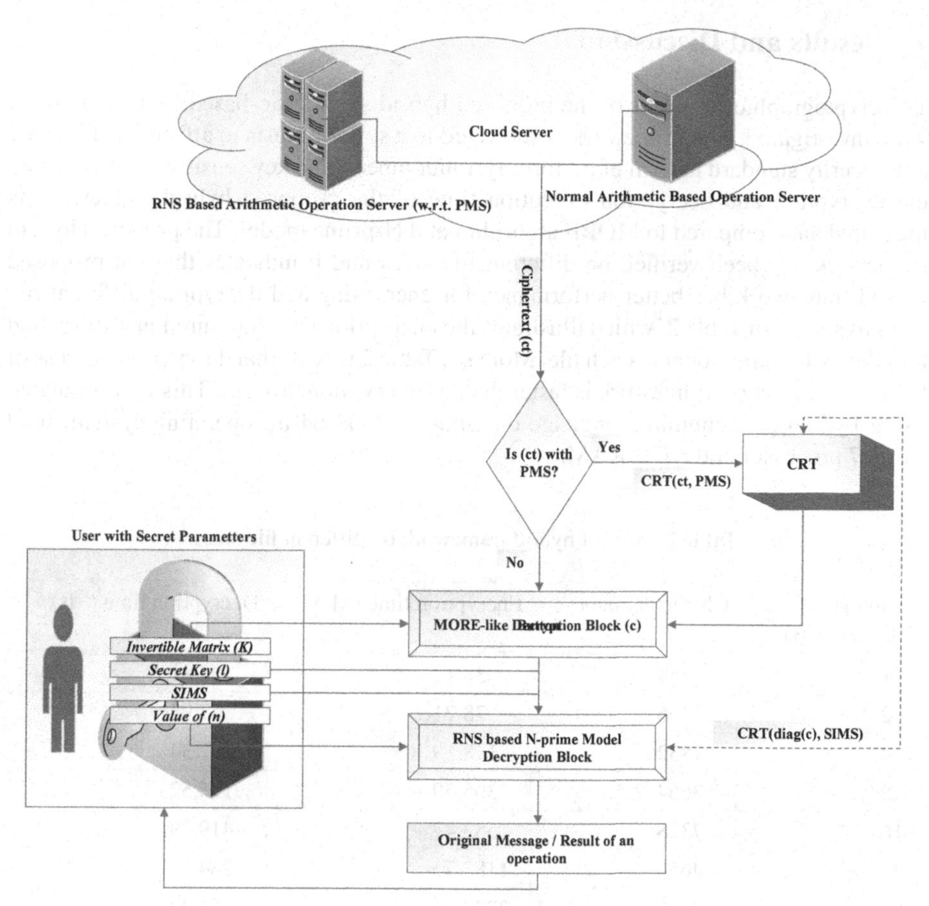


**Fig. 3.** Decryption model

**Table 1.** Decryption algorithm

| Stages in Decryption Process |
|---|
| 1: $Input \rightarrow Dec_{ct}(ct, k, l, sims, n, pms = "")$ |
| 2: $Check\ if\ (pms\ != "")\ then$ |
| $Store\ t\ e\ output\ of\ CRT(ct, pms) \rightarrow ct$ |
| 3: $Compute\ MORE - Like\ Decryption\ (c \leftarrow k \times ct \times k^{-1})$ |
| 4: $Compute\ RNS - Based\ N - Prime\ model\ Decryption$ |
| $(wn = CRT(diag(c), sims_n)$ |
| 5: $Compute\ \lvert wn \rvert_l\ and\ return\ the\ result\ as$ |
| $Original\ Message\ or\ the\ Result\ of\ the\ Operation$ |

## 4 Results and Discussion

The cryptographic strengths of the proposed hybrid symmetric-based FHE framework were investigated. The framework is subjected to a series of tests to affirm its efficiency and security standard in term of uniformity, randomness, and key sensitivity. In addition, the encryption and decryption execution time of the proposed hybrid framework is measured and compared to MORE algorithm and N-prime model. The proposed hybrid framework has been verified on different file sizes and it indicates that the proposed hybrid framework has better performance for encrypting and decrypting different file sizes as shown in Table 2, which illustrates the encryption time measured in millisecond together with items count in each file. More so, Table 2 reveals that decryption process of the proposed hybrid framework is faster than its encryption process. This was simulated using Python programming language on Ubuntu 18.04 64-bit operating system, Intel Core i7 processor and 8 GB RAM.

**Table 2.** Test of hybrid framework on different file sizes

| Plaintext File size (KB) | Character count | Encryption time (Ms) | Decryption time (Ms) |
|---|---|---|---|
| 1 | 458 | 38.12 | 23.81 |
| 2 | 916 | 78.71 | 47.98 |
| 4 | 1832 | 148.19 | 97.34 |
| 8 | 3664 | 305.59 | 195.52 |
| 16 | 7328 | 583.47 | 419.79 |
| 32 | 14656 | 1183.75 | 784.53 |
| 64 | 29312 | 2333.04 | 1554.43 |
| 128 | 58624 | 4630.17 | 3104.95 |
| 256 | 117248 | 9273.98 | 6212.47 |
| 512 | 234496 | 18666.11 | 12755.58 |

Furthermore, Table 3 shows ciphertext expansion ratio for each plaintext files that was investigated and findings show that averagely for any given plaintext size, proposed hybrid framework generates ciphertext size that is less than 4 times of its original plaintext size. In comparison with existing studies, Table 4 shows that the proposed hybrid symmetric based FHE framework outperforms the existing state-of-art schemes in term of reduction in ciphertext size.

In addition, the proposed hybrid framework was compared with MORESIMS model and RNS based N-Prime model. The results show that approximately, there is no significant difference in encryption execution time between the proposed hybrid framework and MORESIMS model as shown in Fig. 4. However, Fig. 5 shows that there is a slightly difference in encryption execution time between the proposed hybrid framework and RNS based N-prime model while Fig. 6 reveals that RNS based N-prime model took

**Table 3.** Size ratio (ciphertext size/plaintext size)

| Plaintext file Size (KB) | Ciphertext Size (KB) | Size ratio (Plaintext: Ciphertext) |
|---|---|---|
| 1 | 3.68 | 1: 3.68 |
| 2 | 7.8 | 1: 3.90 |
| 4 | 16.18 | 1: 4.05 |
| 8 | 29.42 | 1: 3.68 |
| 16 | 60.0 | 1: 3.75 |
| 32 | 122.0 | 1: 3.81 |
| 64 | 247.7 | 1: 3.87 |
| 128 | 502.52 | 1: 3.93 |
| 256 | 1019.11 | 1: 3.98 |
| 512 | 1836.73 | 1: 3.59 |
| **Average size ratio** | | **1: 3.82** |

**Table 4.** Comparing the proposed hybrid framework with related studies.

| Authors (year) | Size ratio (Plaintext: Ciphertext) |
|---|---|
| Chillotti et al. [10] | 1: 400,000 |
| Mohammed and Abed [24] | 1: 10.21 |
| Proposed Hybrid framework | 1: 3.82 |

the least encryption execution time in compared with proposed hybrid framework and MORESIMS model because it involved single stage of encryption and RNS arithmetic operations were used in the encryption process of RNS based N-prime model with carry free propagation and parallelism.

For the security of the proposed hybrid framework, correlation coefficient, uniform distribution and key sensitivity was used to measure its security strength against attacks.

**Correlation Coefficient:** For any strong encryption scheme to be achieved, among the important requirement is that the ciphertext should be greatly different from its plaintext form and the correlation coefficient between the plaintext and ciphertext should be kept so low. Thus, the correlation coefficient between the plaintext and ciphertext is computed as follows:

$$\rho_{x,y} = \frac{cov(x, y)}{\sqrt{D(x) \times D(y)}} \tag{5}$$

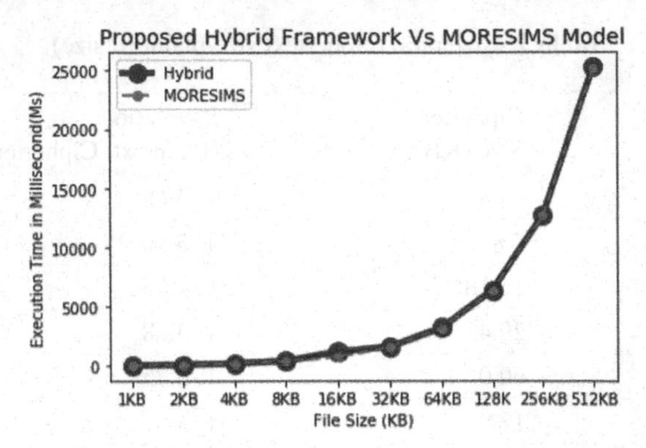


**Fig. 4.** Encryption execution time between proposed Hybrid Framework and MORESIM Model

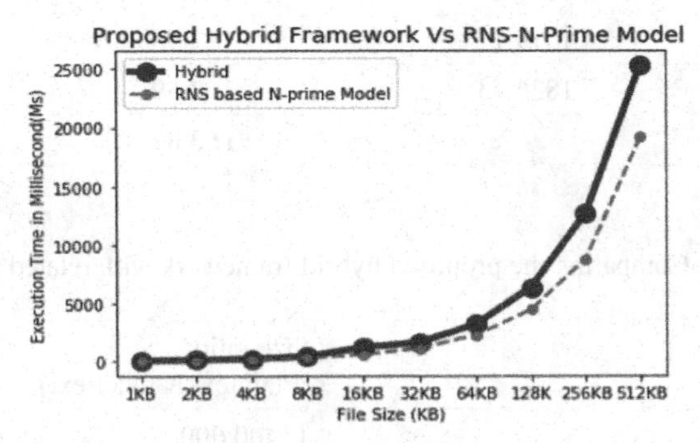


**Fig. 5.** Encryption execution time between proposed Hybrid Framework and RNS based N-Prime Model

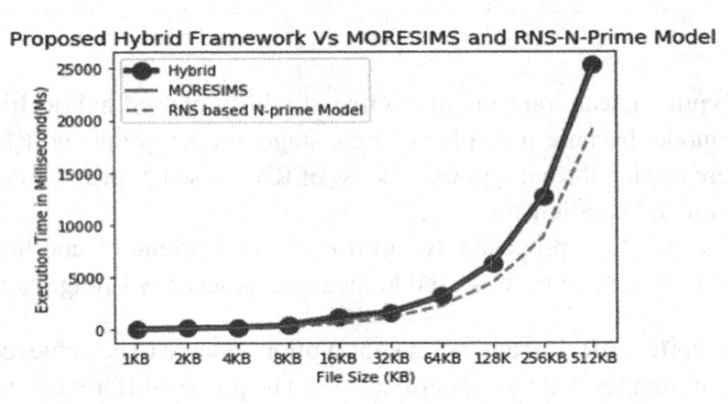


**Fig. 6.** Comparing the proposed Hybrid Framework with MORESIMS Model and RNS based N-Prime Model

*where* $cov(x, y) = E[\{x - E(x)\}\{y - E(y)\}]$,

$$E(x) = \frac{1}{n} \times \sum\nolimits_{k=1}^{n} x_i,\ E(y) = \frac{1}{n} \times \sum\nolimits_{k=1}^{n} y_i,\ D(x) = \frac{1}{n} \times \sum\nolimits_{k=1}^{n} \{x_i - E[x]\} \text{ and } D(y) = \frac{1}{n} \times \sum\nolimits_{k=1}^{n} \{y_i - E[y]\}$$

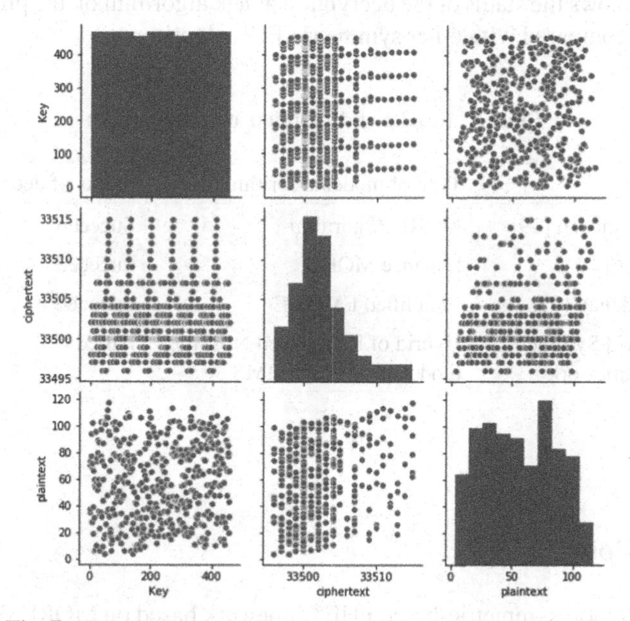


**Fig. 7.** Correlation matrix between plaintext, ciphertext and key

Figure 7 shows the correlation coefficient between the plaintext and ciphertext with different keys. This result shows that there is a very weak/low correlation that is nearly undetectable between the plaintext and its corresponding ciphertext.

**Uniform Distribution:** Looking at the distribution of plaintext and its corresponding ciphertext in Fig. 7, the result clearly shows that the contents of the ciphertexts have a good level of mixing and it is uniformly distributed. If the frequency counts of the encrypted generation are close to a uniform distribution, then it is possible to categorize that the concerned cipher under test has a good level of mixing.

**Key Sensitivity:** The sensitivity of random keys K and plaintext in bit-level are computed and analyzed using hamming distance with the following equations.

$$K = \frac{\sum_{k=1}^{T} E_k(P) \oplus E_{k'}(P)}{T} \times 100\% \tag{6}$$

Where T is the length of the plaintext and ciphertext message (in bits), All the elements of $k$ are equal to those of $k'$ and $P$ is the plaintext or ciphertext.

The correlation sensitivity of the plaintext and ciphertext with respect to random keys are shown in Fig. 7. This shows the vast change in the ciphertext with respect to a small change in the secret keys or the plaintext.

The proposed hybrid framework was subjected to decryption attack algorithm on symmetric FHE proposed by Vizár and Vaudenay [25] and compared with the existing state-of-art schemes. The results show that the decryption attack algorithm failed to recover the original plaintext message after several tests using different series of ciphertexts, Table 5 shows the status of the decryption attack algorithm on the proposed hybrid framework as compared with other symmetric FHE algorithms.

**Table 5.** Results of decryption attack algorithm

| Authors | Type of model/algorithms | Status of decryption attack |
|---|---|---|
| Kipnis and Hibshoosh [29] | MORE Algorithm | Succeed |
| Hariss, et al. [26] | Enhance MORE | Succeed |
| Vizár and Vaudenay [25] | Simplified EMORE | Succeed |
| Proposed Hybrid Symmetric based FHE Framework | Hybrid of RNS based N-prime model and MORESIMS | Failed |

## 5 Conclusion

In this paper, hybrid symmetric-based FHE framework based on MORESIMS and modified N-prime model was proposed as an improvement over existing symmetric FHE schemes for real-world applications. The proposed hybrid framework takes in integer-based plaintext together with generated secret keys plus secret information moduli set (SIMS), passes it to RNS forward conversion module and RNS arithmetic operation to produce the first level of ciphertext $ct_i$ as residues. MORESIMS channel received $ct_i$ and secret invertible matric $k$ to produce MORRE-like encryption as the second level of encryption. The ciphertexts produced at the end of the proposed hybrid framework were tested and analyzed, the results show that the ciphertext size is less than 4 times of its equivalent plaintext size, this is a promising result as compared with the existing state-of-art schemes that produce exceptionally large ciphertext size of like 10 times of its plaintext size.

Despite double encryption layers involved in proposed hybrid symmetric based FHE framework to resist plaintext/ciphertext attacks or any other forms of statistical attacks, we found out that the framework still has a very considerable encryption execution time as compared with existing state-of-art schemes. Further research can be investigated to determine the effect of Public Moduli Set (PMS) on the proposed hybrid symmetric based FHE framework on the RNS-BAOCS. Also, optimization of RNS based matrices arithmetic operations with PMS can as well be investigated on the cloud server.

## References


1. Song, X., Wang, Y.: Homomorphic cloud computing scheme based on hybrid homomorphic encryption. In: 2017 3rd IEEE International Conference on Computer and Communications (ICCC), pp. 2450–2453. IEEE (2017)
2. Seth, B., Dalal, S., Kumar, R.: Hybrid homomorphic encryption scheme for secure cloud data storage. In: Kumar, R., Wiil, U.K. (eds.) Recent Advances in Computational Intelligence. SCI, vol. 823, pp. 71–92. Springer, Cham (2019). https://doi.org/10.1007/978-3-030-12500-4_5
3. Rivest, R.L., Adleman, L., Dertouzos, M.L.: On data banks and privacy homomorphisms. Found. Secur. Computat. **4**(11), 169–180 (1978)
4. Gentry, C.: A fully homomorphic encryption scheme (no. 09). Stanford University Stanford (2009)
5. Gentry, C.: Fully homomorphic encryption using ideal lattices. In: Stoc, vol. 9, no. 2009, pp. 169–178 (2009)
6. Gentry, C., Halevi, S.: Fully homomorphic encryption without squashing using depth-3 arithmetic circuits. In: 2011 IEEE 52nd Annual Symposium on Foundations of Computer Science, pp. 107–109. IEEE (2011)
7. Brakerski, Z., Vaikuntanathan, V.: Fully homomorphic encryption from ring-LWE and security for key dependent messages. In: Rogaway, P. (ed.) CRYPTO 2011. LNCS, vol. 6841, pp. 505–524. Springer, Heidelberg (2011). https://doi.org/10.1007/978-3-642-22792-9_29
8. Brakerski, Z., Vaikuntanathan, V.: Efficient fully homomorphic encryption from (standard) LWE. SIAM J. Comput. **43**(2), 831–871 (2014)
9. Chen, Y., Gu, B., Zhang, C., Shu, H.: Reliable fully homomorphic disguising matrix computation outsourcing scheme. In: 2018 14th International Wireless Communications and Mobile Computing Conference (IWCMC), pp. 482–487. IEEE (2018)
10. Chillotti, I., Gama, N., Georgieva, M., Izabachène, M.: Faster fully homomorphic encryption: Bootstrapping in less than 0.1 seconds. In: Cheon, J.H., Takagi, T. (eds.) ASIACRYPT 2016. LNCS, vol. 10031, pp. 3–33. Springer, Heidelberg (2016). https://doi.org/10.1007/978-3-662-53887-6_1
11. Ducas, L., Micciancio, D.: FHEW: bootstrapping homomorphic encryption in less than a second. In: Oswald, E., Fischlin, M. (eds.) EUROCRYPT 2015. LNCS, vol. 9056, pp. 617–640. Springer, Heidelberg (2015). https://doi.org/10.1007/978-3-662-46800-5_24
12. Gentry, C., Sahai, A., Waters, B.: Homomorphic encryption from learning with errors: conceptually-simpler, asymptotically-faster, attribute-based. In: Canetti, R., Garay, J.A. (eds.) CRYPTO 2013. LNCS, vol. 8042, pp. 75–92. Springer, Heidelberg (2013). https://doi.org/10.1007/978-3-642-40041-4_5
13. Hu, Y.: Improving the efficiency of homomorphic encryption schemes (Doctoral dissertation, Worcester Polytechnic Institute) (2013)
14. van Dijk, M., Gentry, C., Halevi, S., Vaikuntanathan, V.: Fully homomorphic encryption over the integers. In: Gilbert, H. (ed.) EUROCRYPT 2010. LNCS, vol. 6110, pp. 24–43. Springer, Heidelberg (2010). https://doi.org/10.1007/978-3-642-13190-5_2
15. Brakerski, Z.: Fundamentals of fully homomorphic encryption. In: Providing Sound Foundations for Cryptography, pp. 543–563. ACM (2019)
16. Alaya, B., Laouamer, L., Msilini, N.: Homomorphic encryption systems statement: trends and challenges. Comput. Sci. Rev. **36**, 100235 (2020)
17. Cheon, J.H., Kim, J.: A hybrid scheme of public-key encryption and somewhat homomorphic encryption. IEEE Trans. Inf. Forensics Secur. **10**(5), 1052–1063 (2015)
18. Burtyka, P., Makarevich, O.: Symmetric fully homomorphic encryption using decidable matrix equations. In: Proceedings of the 7th International Conference on Security of Information and Networks, p. 186. ACM (2014)

19. Gupta, C., Sharma, I.: A fully homomorphic encryption scheme with symmetric keys with application to private data processing in clouds.: In: 2013 Fourth International Conference on the Network of the Future (NoF), pp. 1–4. IEEE (2013)
20. Li, J., Wang, L.: Noise-free symmetric fully homomorphic encryption based on noncommutative rings. IACR Cryptol. ePrint Archive **2015**, 641 (2015)
21. Liang, M.: Symmetric quantum fully homomorphic encryption with perfect security. Quantum Inf. Process. **12**(12), 3675–3687 (2013). https://doi.org/10.1007/s11128-013-0626-5
22. Umadevi, C., Gopalan, N.: Outsourcing private cloud using symmetric fully homomorphic encryption using $Q^n p$ matrices with enhanced access control. In: 2018 International Conference on Inventive Research in Computing Applications (ICIRCA), pp. 328–332. IEEE (2018)
23. Qu, Q., Wang, B., Ping, Y., Zhang, Z.: Improved cryptanalysis of a fully homomorphic symmetric encryption scheme. Secur. Commun. Netw. **2019**, 1–6 (2019)
24. Mohammed, M.A., Abed, F.S.: An improved fully homomorphic encryption model based on N-primes. Kurdistan J. Appl. Res. **4**(2), 40–49 (2019)
25. Vaudenay, D.V.S.: Cryptanalysis of enhanced MORE. Tatra Mt. Math. Publ. **73**, 163–178 (2019)
26. Hariss, K., Noura, H., Samhat, A.E.: Fully enhanced homomorphic encryption algorithm of MORE approach for real world applications. J. Inf. Secur. Appl. **34**, 233–242 (2017)
27. Hariss, K., Noura, H., Samhat, A.E., Chamoun, M.: An efficient solution towards secure homomorphic symmetric encryption algorithms. In: ITM Web of Conferences, vol. 27, p. 05002. EDP Sciences (2019)
28. Hariss, K., Noura, H., Samhat, A.E., Chamoun, M.: Design and realization of a fully homomorphic encryption algorithm for cloud applications. In: Cuppens, N., Cuppens, F., Lanet, J.-L., Legay, A., Garcia-Alfaro, J. (eds.) CRiSIS 2017. LNCS, vol. 10694, pp. 127–139. Springer, Cham (2018). https://doi.org/10.1007/978-3-319-76687-4_9
29. Kipnis, A., Hibshoosh, E.: Efficient methods for practical fully homomorphic symmetric-key encrypton, randomization and verification. IACR Cryptol. ePrint Archive **2012**, 637 (2012)
30. Vizár, D., Vaudenay, S.: Cryptanalysis of chosen symmetric homomorphic schemes. Stud. Sci. Math. Hung. **52**(2), 288–306 (2015)
31. Muhammed, K.J., Gbolagade, K.: Enhanced MORE algorithm for fully homomorphic encryption based on secret information moduli set. In: 2019 IEEE International Conference on Industrial Engineering and Engineering Management (IEEM), pp. 469–473. IEEE (2019)
32. Muhammed, K.J., Isiaka, R.M., Asaju-Gbolagade, A.W., Adewole, K.S., Gbolagade, K.A.: Improved cloud-based N-primes model for symmetric-based fully homomorphic encryption using residue number system. In: Chiroma, H., Abdulhamid, S.M., Fournier-Viger, P., Garcia, N.M. (eds.) Machine Learning and Data Mining for Emerging Trend in Cyber Dynamics, pp. 197–216. Springer, Cham (2021). https://doi.org/10.1007/978-3-030-66288-2_8

# Information Encryption and Decryption Analysis, Vulnerabilities and Reliability Implementing the RSA Algorithm in Python

Rocío Rodriguez G.[✉], Gerardo Castang M., and Carlos A. Vanegas

Universidad Distrital Francisco José de Caldas, Facultad Tecnológica, Bogotá, Colombia
{rrodriguezg,gacastangm,cavanegas}@udistrital.edu.co

**Abstract.** The processing and transmission of information has increased its effectiveness in recent decades. From mathematical models the security and integrity of the data are guaranteed. In spite of that, interceptions in the signal, attacks and information theft can happen in the transmission process. This paper presents a RSA algorithm analysis, using 4, 8 and 10 bits prime numbers with short messages. The encryption and decryption process implemented in python allowed the computational resources use. Processing time and data security are evaluated with a typical computational infrastructure required for its operation; in order to identify vulner-abilities and their reliability level when ideal conditions are available to perform a cryptanalysis.

**Keywords:** Encryption · Decryption · Security · Cryptography · RSA · Cryptanalysis

## 1 Introduction

Information theory is based on mathematical and probabilistic concepts that allow to create a communications system. This theory aims to achieve efficient and reliable communication in the transmission of information. To be able to achieve the adequate communication. It is necessary to take into account the variables that influence the process of transmission and communication of information from the source to the destination through the communication medium where the channel is generally affected by noise. In 1928 Hartley formulated the first mathematical laws that regulate a communication system. These ideas were considered by Shannon allowing to develop the fundamental principles of the theory.

The physical model exposed by Shannon is shown in Fig. 1. It indicates how the communication can be expressed in a process of information transfer. In which a transmitter, issues a codified signal (message) that travels along a channel, the signal transits through the channel able or not able to be affected by


H. Florez and M. F. Pollo-Cattaneo (Eds.): ICAI 2021, CCIS 1455, pp. 391–404, 2021.
https://doi.org/10.1007/978-3-030-89654-6_28

noise, it finally arrives to the receiver who codifies the signal, for Shannon the quantity of contained data in a message is tantamount to a math data well defined and measurable, this quantity refers to the possibility that the message within a set will be received and not to the quantity of sent data articles [1].

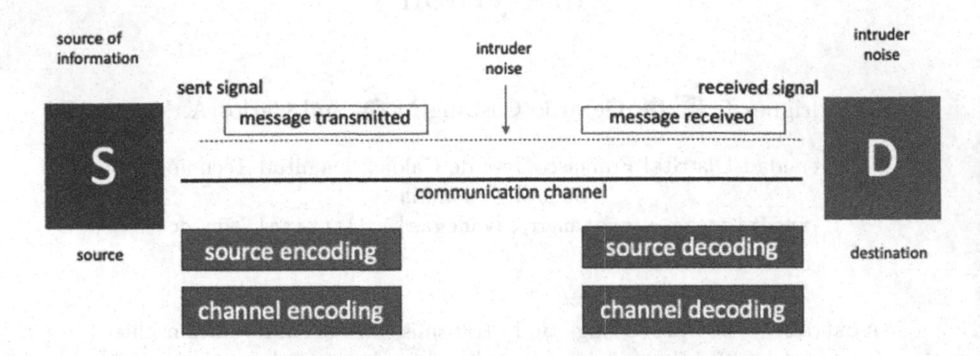


**Fig. 1.** Transmission and communication process of information.

The codification message (text, image, voice) makes reference to its transformation in symbols. This subsequently changes to electrical signals that provide security to the information that travels from a place to another as we can see at Fig. 1. In this case, the noise represents the effect of an attacker when intercepting and possibly decodifying the coded message (intruder).

The development of secure applications in mobile environments, or limited computational requirements, makes it necessary to create information security solutions in this area, especially for the processing and transmission of short messages.

Generally, the RSA algorithm is used for lengths of keys in high bits. The factoring of prime numbers is a complex process and requires a lot of computational time, this is the principal strength of the algorithm providing security to the information.

Since the RSA works efficiently for this length of keys in high bits, we intend to validate the security of the algorithm, with short key lengths. Once the investigation concluded if needed improvements will be made to the algorithm. All of this to provide protection and security of the information that is transmitted and that can be intercepted by an attacker.

## 2 Network Security Aspects

The topic of network security and the problems associated with it can be divided in four big interrelated areas. These areas are: confidentiality, authentication, non-repudiation and message integrity check [2].

Confidentiality consists in exchanging, delivering, sending and sharing the information, with the corresponding or appropriate user.

Authentication consists in validating, identifying or authenticating the user with whom you interact before starting the transfer or exchange process of the information.

Non-repudiation handles the verification and validation of the established terms among the users, to perform a process or a transaction.

Message integrity check validates that the received message is the same as the one that was sent, and that the sent message was not altered, replaced or modified by an intruder.

In each one of the TCP/IP model layers the concept of the information security and integrity, can be involved on each layer [3]. These qualities can be implemented logically and physically. Logically through the models and algorithms and physically, through hardware devices that perform processes such as calculations and validations, with the purpose of maintaining the information security and integrity qualities in the network.

The cryptography allows to implement math processes with a certain level of complexity to the information prior to be transmitted, processed or stored. The cryptographic algorithms and methods play a very significant role in the communication systems [4].

Encrypted messages that have been transmitted or stored can be intersected, copied and stolen with the aim of decrypt encrypted messages, it is called cryptanalysis.

The group of methods, algorithms and procedures that perform the cryptography and cryptanalysis processes is known as cryptology. Cryptology is one of the most important branches in security systems of communications networks. It allows to identify flaws in the used methods, algorithms and procedures for the information protection, storing and transmission [5].

**Encryption and Decryption Processes of a Message:** Cryptography is based on two fundamental processes known as encryption (E) and decryption (D). These are mathematical functions that are involved through the parameter (K), which corresponds to the set of keys used to the implementation of the procedure. The arguments of the functions are the message or plain text (P), and the cipher text (C).

The **methods** of encryption and decryption of a message can be expressed generally, through the following expressions:

$$C = E_K[P] \tag{1}$$

$$P = D_K[C] \tag{2}$$

Replacing the value of the parameter C, of the expression (1) in (2), we get:

$$P = D_K\{E_K[P]\} \tag{3}$$

Because the functions of encryption and decryption have an inverse relation, we get:

$$P = 1 \cdot P = P \tag{4}$$

This allows to obtain the original message or plain text.

**Kerckhoff's Principle:** One of the rules of the cryptanalysis is that the encryption and decryption methods are known, this means that the cryptanalyst knows its functioning [6]. The cryptanalyst job is to find the parameter (K) to decrypt the message. The user job is to generate or change the keys continuously each time that the information encryption process is performed. The Kerckhoff''s principle is set out as follows: all algorithms must be public just keys should be secret.

## 3 The Cryptology

The cryptology has as the main objective to encrypt the information, with purpose of protecting it. For that end, it employs algorithms that tend to use one or more keys, in such a way the communication between sender and receiver is made with security and privacy.

Cryptography is located at the cryptology branch along with cryptanalysis [7]. Cryptanalysis is the study of everything involved with deciphering data technics. Some of the features of cryptanalysis applied to a cryptographic system are:

- To ensure its robustness and resistance.
- To discover weaknesses to avoid possible attacks.
- To strength it in order to increase its security.

The cryptographic methods used by the cryptosystems to perform the encryption are classified in asymmetric and symmetric. The classifications are described in Table 1 [8].

**Table 1.** Symmetric vs asymmetric cryptosystems.

| | Public key | Private key |
|---|---|---|
| key management | It is only necessary to memorize the sender's private key and the receiver's public key | you have to memorize a high keys number |
| length and key space | key is in the order of thousands of bits | the order key is of hundreds of bits |
| key's life | key duration is usually long | the key duration is short, it usually ends when the session is over. |
| Authentication | By having a public and a private key, the sender and the mes-sage can be authenticated | It is possible authenticate the message only |
| Speed cipher | Speed encryption is slow | Speed encryption is high |
| Use | They are used for key exchanges and digital signatures | They are algorithms used for encryption |

## 4 Cryptosystem Analysis RSA

Rivest, Shamir and Adleman developed an asymmetric cryptosystem [9][10], from the product of two prime numbers (p and q) previously selected. These numbers multiplied allow to get an n number.

$$n = p \cdot q \tag{5}$$

Using the *Euler function(φ)* the size of the integers multiplicative group is generated.

$$\varphi(n) = (p-1)(q-1) \tag{6}$$

Subsequently it is proceeded to calculate two parameters. One is used as public key($e$) and the other as private ($d$). They are used for encryption and decryption respectively [11]. To make this calculation, one of the parameters is chosen and the other one is calculated.

$$e \cdot d = 1 + k \cdot \varphi(n) \tag{7}$$

For an integer k is relevant that k is less than $\varphi(n)$ and coprime. The following equation must be verified:

$$e \cdot d = 1\,(mod\varphi(n)) \tag{8}$$

Therefore

$$d = e - 1\,(mod\varphi(n)) \tag{9}$$

$$e = d - 1\,(mod\varphi(n)) \tag{10}$$

From this algorithm the public key is obtained:

$$K_{pb} = (e, n) \tag{11}$$

From this algorithm the private key is obtained:

$$K_{pr} = (d, n) \tag{12}$$

The resulting key is obtained from the two components:

$$< K_{pb}, K_{pr} > \tag{13}$$

The RSA algorithm works with the premise of calculating $n$ by multiplying $p$ and $q$. It is difficult to factorize the product since it prevents to obtain the private key $K_{pr}$ from public key $K_{pb}$.

The Python code of the implemented RSA algorithm is described as follows:

```
Begin
import time
import random
def getD(e, z):
     for x in range(1, z):
          if (e * x) % z == 1:
               return x
def values E(a):
     list_obj = []
     for x in range(2, a):
          if mcd(a, x) == 1 and encrypt_function(x,z) !=
None:
               list_obj.append(x)
     for x in list_obj:
          if x == encrypt_function(x,z):
               list_obj.remove(x)
     return list_obj
def mcd(a, b):
     while b != 0:
          c = a % b
          a = b
          b = c
     return a
def encrypt_function(e, z):
    return e%z
def encrypt_block(x):
     encryption = encrypt_function (x**e, n)
     return encryption
def decrypt_block(encryption):
     decrypted_message =encrypt_function(encryption**d, n)
     return decrypted_message
def encrypt_message(message):
     return ''.join([chr(encrypt_block(ord(x))) for x in
list(message)])
def decrypt_message(message):
     return ''.join([chr(decrypt_block(ord(x))) for x in
list(message)])
p=int(input('Enter prime number P: '))
q=int(input('Enter prime number Q: '))
print("\nSelected prime numbers: p=" + str(p) + ", q=" +
str(q) + "\n")
n=p*q
print("Value of (n = p * q) = " + str(n) + "\n")
z=(p-1)*(q-1)
print("Euler's function [Z(n)]: " + str(z) + "\n")
print("Possible values for number 'e':\n")
print(str(valuesE(z)) + "\n")
e = random.choice(valuesE(z))
print ("Random selected value:",e)
d=getD(e,z)
```

```
print("\nPublic key e,n (e=" + str(e) + ", n=" + str(n)
+ ").")
print("Private key  d,n (d=" + str(d) + ", n=" + str(n)
+ ").")
message = input("Plain text message: ")
begin=time.time()
encryption = encrypt_message (message)
print("\nEncrypt message: " + encryption + "\n")
decrypted = decrypt_message (encryption)
print("Decrypt message: " + decrypted + "\n")
final=time.time()
print("Processing time in seconds:",round(final - begin,
10))
end
```

## 4.1 RSA Algorithm Source Code's Analysis

At first, the random and time libraries are imported. Time is a Python library that provides a set of functions to work with dates and/or time, and Random contains different functions related with random values.

Subsequently, the following functions are implemented:

- ***getD(e,z)***: This function receives two parameters (e, z). A *for* cycle is executed from an $x$ value equals *1* to the $z$ parameter value. The if structure verifies if the $e * x$ modulus (%) product of $z$ is equal to *1.* If the condition is met, the $x$ value is re-turned.
- ***valuesE(a)***: This function receives a parameter (a). First, an empty list (*list_obj*) is declared. The *for* cycle iterates from $x$ equal two, to *a value.* The *if* structure verifies that the *mcd(a, x)* function is equal to *1* and the *encrypt_function(x, z)* function is different to *empty.* If the condition is met, it is added to the list (*list_obj*) the value of the x variable. Furthermore, in a second *for* cycle, the function iterates from one of *x's* values equal to *0* to the quantity of the list (*list_obj*) elements. With the *if* structure it is verified that the value of the $x$ variable is equal to the value that the *encrypt_function(x, z)* function returns. If the condition is met, the $x$ variable value is deleted from *list_obj.* Finally, the list *(list_obj)* elements are sent back.
- ***mcd(a,b)***: The function receives two parameters (a, b). The *While* cycle iterates while the value $b$ variable is different from *zero*, the modulus of $a$ and $b$ are assigned to $a$ variable called $c$, the $b$ value is assigned to the $a$ variable and the $c$ value to the $b$ variable. When the cycle is done, the value of $a$ variable is returned.
- ***encrypt_function(e, z)***: The function receives two parameters (e, z). It returns the modulus result among the *e, z* variables.
- ***encrypt_block(x)***: This function receives one parameter (x), the *encryption* variable is assigned the value that returns the *encrypt_function(x**e, n)* function. This function returns the value of the *encryption* variable.

- ***decrypt_block(encryption)*:** This function receives one parameter (*encryption*), the *decrypted_message* variable is assigned the value that returns the *en-crypt_function(encryption**d, n)* function. This function returns the value of the *decrypted_message* variable.
- ***encrypt_message(message)*:** The function receives one parameter (*message*). The *encrypt_block()* function returns the ascii value from x variable that belongs to the list (*list_obj*). The function receives one parameter when performing the loop.
- ***decrypt_message(message)*:** The function receives one parameter (*message*). The *decrypt_block()* function returns the ascii value from x variable that belongs to the list (*list_obj*). The function receives one parameter when performing the loop.

In addition, the program carries out the process to obtain the input and output values of the information. First, the $p$ and $q$ variables are declared, and a prime number is assigned to them. The prime number is captured from the keyboard with the *input()* function and is converted to an integer with the *int function*, in order to be printed with *print()* function. Then, a variable called $n$ is declared, and the product of $p * q$ is assigned to it, that also is printed. The $z$ variable is declared as well, and the product of $(p - 1) * (q - 1)$ is assigned to it and it is printed.

Eventually, an obtained random value with the random library's choice method is assigned and printed to a variable called e. Such value is returned by the *valuesE(z)* function.

Subsequently, a value returned by the *getD(e, z)* function is assigned to a variable called *d*, and the variable *e* value is printed as public key and the variable *d* value as private key. In a variable called *message*, the text to be encrypted is saved, and the initial processing *time* (time.time()) is assigned to a variable called *begin*. Other variables are also declared: *encrypted*, to this variable is assigned the value that returns the *encrypt_message(message)* function. The *decrypted* variable is assigned the value that returns the *decrypt_message (message)*. These variables are also printed.

The final processing time is assigned to a variable called *final*, and the result of the *final-begin* subtraction operation is printed. The *roud()* method is used to show the result with ten decimals maximum.

The algorithm execution can be seen in Fig. 2.

```
Enter prime number P: 11

Enter prime number Q: 13

Selected prime numbers: p=11, q=13

Value of (n = p * q) = 143

Euler´s function [Z(n)]: 120

Possible values for number 'e':

[11, 17, 23, 31, 41, 47, 53, 61, 71, 77, 83, 91, 101, 107, 113]
Random selected value: 107

Public key e,n (e=107, n=143).
Private key  d,n (d=83, n=143).

Plain text message: prueba del algoritmo RSA para el congreso 2021

Encrypt message: ⊡R'_6%L开_rL%r&CRv^\⊡CLAL⊡%R%L_rL,C⊡&R_:CL⊡⊡⊡E

Decrypt message: prueba del algoritmo RSA para el congreso 2021

Processing time in seconds: 0.0019950867
```

**Fig. 2.** Transmission and communication process of information.

In order to perform tests on the implemented algorithm we used: Intel (R) Core (TM) i5-10210U Processor CPU @ 1.60 GHz 2.11 GHz, 8.00 GB RAM and Windows 10 × 64 b Operating System.

## 4.2 Length of Encrypted Message Test

A series of tests carried out with values for p and q, using prime numbers of 4, 8 and 10 bits. Using short message lengths (Typically 2 KB). This allowed establishing the values for e and the processing time shown on Table 2.

**Table 2.** Values and data of tests results on the RSA algorithm

| Number of bits to generate the prime numbers | Values for P | Values for Q | Values of E | Processing time (In seconds) |
|---|---|---|---|---|
| 4 | 11 | 13 | 23 | 0.003 |
| 8 | 191 | 227 | 42529 | 20.467 |
| 10 | 937 | 1019 | 952787 | 1697.762 |

The tests results allowed establish the following analysis:

- To the extent that small keys in number of bits are used, the processing time is very small.

- To go from four to eight bits, the time spent in processing is multiplied by an approximated factor of 1000.
- To go from eight to ten bits, the time spent in processing is multiplied by an approximated factor of 10.
- To the extent that increases the number of bits used to generate the prime numbers p and q, the processing time to code the message increases exponentially and presents a relation by ten times as shown above.
- We attempted to perform tests using prime numbers of 12 bits, where the e generation was about 3 h and 30 min and the message could not be encrypted because a memory overflow was presented.

### 4.3 Cryptanalysis Test

This section proposes to carry out the cryptanalysis to validate the ability to decrypt the information, in order to find the user's private key with an ideal environment for the attacker [13]. That is, knowing part of the transmitted message, the length in bits of the prime numbers used to encrypt and decrypting the message and the user's public key [14,15].

A decryption algorithm developed by the authors was designed. First described with a natural language and then implemented with python's programming language [16]. In purpose to verify the processing time and the reliability of the decrypted in-formation.

The aim goes to validate the strength of the encryption algorithm implemented and the speed or time required to decrypt an encrypted message, using typical computational capabilities of a user.

We performed a test with the same available hardware infrastructure, taking into account the following ideal conditions for the attacker:

- Intercepts and knows the encrypted message.
- Knows part of the original message.
- Knows the bit number of $P$ and $Q$.
- Knows the public key of the sender $(e, n)$.
- Lacks to determinate the private key $(d, n)$ for deciphering the message.

The following is the algorithm in natural language proposed to obtain the private key and decipher the message by an attacker.

1. Knows the public key $(e, n)$.
2. Knows the number of bits to generate the prime numbers.
3. Generates the set of primes for that given number of bits.
4. Performs operations in order to find both prime numbers that multiplied give the value of $n$, to get $p$ and $q$ values subsequently.
5. Finds $(p - 1)$ and $(q - 1)$ and multiply them to get the Euler function *(called Z)*
6. Deduces a $d$ number, that is within range from *1* to *z*, when the remainder of the operation is equal to *1*, $d$ will be the result.

7. With the $d$ number generates the private key $(d, n)$, thus proceeds to decipher the message.

The previously algorithm was implemented in python:

```
Begin
import time
encrypted_message="#"
bit=4
public_key=[23,143]
z=0
d=0
def decrypt_message(encrypted_message):
  return ''.join([chr(decrypt_block(ord(x))) for x in
list(encrypted_message)])
def encrypt_function(e, n):
  return e%n
def decrypt_block(x):
    decrypted_message = encrypt_function(x**d,
public_key[1])
    return decrypted_message
def prime(i):
  x=2
  flag=0
  while (x<i):
    if (i%x==0):
      flag=1
      return flag
    else:
       x=x+1
  return flag
nro_primes=[]
for i in range(2,15,1):
    if(prime(i)==0):
     nro_primes.append(i)
for j in range(len(nro_primoes)):
   for y in range(len(nro_primes)):
     values=nro_primes[j]*nro_primes[y]
     if(values==public_key[1]):
       z=(nro_primes[j]-1) *(nro_primes[y]-1)
       print ("z",z)
for account in range(z):
    value_d=(account*public_key[0])%z
    if(value_d==1):
      d=account
      print ("private key:",d)
begin=time.time()
decrypted = decrypt_message(encrypted_message)
print ("The original message is:",decrypted)
final=time.time()
```

```
print("processing time in seconds:",round(final - begin,
10))

end
```

With the implementation of cryptanalysis, the values for $p$, $q$, $e$ and the *time of processing* were established as shown on Table 3.

The tests results allowed to establish the following analysis:

- For 4 bits the decryption time is: 0.0019946098 milliseconds. To generate the e parameter, the time processing was less than a millisecond.
- For 8 bits the decryption time is: 6.6627941132 s. To generate the $e$ parameter, the approximate time processing was 0.4679889679 millisecond.
- For 10 bits the decryption time is greater than 11 min. To generate the $e$ parameter the approximate time processing was 173.8833482265 min.
- For 12 bits with values of $P = 2059, Q = 4067, E = 8366779$ spent about 3 h to get the possible values of $e$.
- To go from 4 to 8 bits the time spent in processing is multiplied by an approximated factor of 1000.
- To go from 8 to 10 bits the time spent in processing is multiplied by an approximated factor of 100.
- To go from 10 to 12 bits the estimated time for processing is multiplied by an approximated factor of 10.

**Table 3.** Values and data of tests results on the RSA algorithm

| No of bits to generate the prime numbers | Values for P | Values for Q | Values of E | Time to generate E | Processing time (in seconds) to decipher the message |
|---|---|---|---|---|---|
| 4 | 11 | 13 | 17 | millisecond | 0.0019946098 |
| 8 | 191 | 227 | 47737 | millisecond | 6.6627941132 |
| 10 | 937 | 1019 | 205157 | 10 min | 689.0595903397 |

## 5 Conclusions

When the number of bits used to generate the primes $p$ and $q$ increases, the time processing to decipher the message gets exponentially higher whit a deduced relation of the ten times factor.

When working with larger key sizes, the processing time increases causing the computational requirements to be stronger.

For internal messaging applications with functional levels of security the RSA algorithm for low-bit lengths is useful.

After reviewing the specialized literature, we can conclude that one way to improve the security of the RSA algorithm against possible cyberattacks is obtaining a suitable bit length for the encryption and decryption keys. Unfortunately, there are not procedures that can accurately determine the length of the keys.

When the attacker has the perfect conditions to perform the cryptanalysis, there is a greater risk in the interception of the transmitted information. This is due to a short-er processing time to discover the message, according to the computational capabilities used by the attacker.

## 6 Future Work

We propose to modify the operations performed in the RSA algorithm to use it with short key lengths that can be mathematically proven by including more elements that increase the security level.

We propose to experiment with key lengths for public and private keys with twelve, fourteen, sixteen, twenty and thirty-two bits. To validate the security and strength of the encryption and decryption process information. The challenge is to use any hardware available for the majority of the community, without the processing time being quite high and expensive, slowing down the processing of the keys.

One challenge is to increase the security levels in the message processing when the bit sizes of the primes p and q are small to generate the public and private keys.

## References

1. Shannon, C.E.: A mathematical theory of communication. Bell Syst. Tech. J. **27**(3), 379–423 (1948). https://doi.org/10.1002/j.1538-7305.1948.tb01338.x
2. Redes de Computadoras, Andrew S. Tanenbaum; cuarta edición, editorial: Pearson-Prentice Hall
3. Mateti, P.: Chapter 1 Security Issues in the TCP/IP Suite (2007). https://doi.org/10.1142/9789812770103_0001
4. Satish, G., Raghavendran, dr. Ch., Varma, dr.: Secret key cryptographic algorithm. Researchgate.net (2012). https://www.researchgate.net/publication/266389826_secret_key_cryptographic_algorithm
5. Awad Al-Hazaimeh, O.: A new approach for complex encrypting and decrypting data. Int. J. Comput. Networks Commun. **5**(2), 95–103 (2013)
6. Petitcolas, F.: Kerckhoffs Principle (2011). https://doi.org/10.1007/978-1-4419-5906-5
7. Shamir, A., Rivest, R.L., Adleman, L.: A method for obtaining digital signatures and public-key cryptosystems Mag. Commun. ACM **21**, 120–126 (1978). https://doi.org/10.1145/359340.359342
8. Gabriel, E.M.: Clúster de alto rendimiento en un cloud: ejemplo de aplica-ción en criptoanálisis de funciones hash. Universidad de Almería, p. 60 (2011). http://repositorio.ual.es/bitstream/handle/10835/1202/PFC.pdf?sequence=1
9. Asjad, S.: The RSA Algorithm. Researchgate.net, pp. 5–15 (2019). https://www.researchgate.net/publication/338623532_The_RSA_Algorithm

10. Fonseca-Herrera, O.A., Rojas, A.E., Florez, H.: A model of an information security management system based on NTC-ISO/IEC 27001 standard. IAENG Int. J. Comput. Sci. **48**(2) (2021)
11. Goldreich, O.: Modern Cryptography, Probabilistic Proofs and Pseudorandomness (Second Edition - author's copy), pp. 1–2. Springer (2000). http://www.wisdom.weizmann.ac.il/~oded/PDF/mcppp-v2.pdf
12. Castro Lechtaler, A., Cipriano, M., García, E., Liporace, J., Maiorano, A., Malvacio, E. and Tapia, N.: Estudio de técnicas de criptoanálisis.XXI Workshop de Investigadores en Ciencias de la Computación.Sedici.unlp.edu.ar (2021). http://sedici.unlp.edu.ar/handle/10915/77269
13. Al-hazaimeh, O.: A new approach for complex encrypting and decrypting data. Int. J. Comput. Networks Commun. **5**, 95–103 (2013). https://doi.org/10.5121/ijcnc.2013.5208
14. Tiwari, G., Nandi, D., Mishra, M.: Cryptography and cryptanalysis: a review. Int. J. Eng. Res. Technol. **2**, 1898–1902 (2013)
15. Bokhari, M., Alam, S., Masoodi, F.: Cryptanalysis tools and techniques (2014)
16. Rodríguez, G.R., Vanegas, C., Castang, G.: Python a su alcance. EditorialUD, 2020. Páginas, pp. 100–120 (2020). ISBN 978-958-787-181-4

# Simulation and Emulation

# Optimization of Multi-level HEED Protocol in Wireless Sensor Networks

Walid Boudhiafi[1(✉)] and Tahar Ezzedine[2]

[1] Faculty of Economics and Management of Tunis, University of Tunis - El Manar, Tunis, Tunisia
walid.boudhiafi@fsegt.utm.tn

[2] National Engineers School of Tunis, University of Tunis - El Manar, Tunis, Tunisia

**Abstract.** Increasing the lifetime of the wireless sensor network after deployment is a challenge. Thus, the one-hop routing protocols associated with agglomeration techniques are not scalable resistant for the large wireless sensor network. In this paper, we try to remedy these problems, by proposing a new approach based on the hierarchical routing technique (multi-hop), this approach presents a solution to maximize the lifetime of the large wireless sensor network. Our approach consists of the design of a new ML-HEED (Multi level Hybrid, Energy-Efficient, Distributed approach) protocol. ML-HEED is a multi-hop routing protocol, this protocol is based on the HEED protocol to organize the wireless sensor network into clusters. A cluster is formed by a set of sensor nodes and elected sensor node. Multi-hop communication takes place between the elected nodes and the base station. The simulation results show the effectiveness of this protocol in maximizing the lifetime of wireless sensor networks. In this approach, we have noticed, the further the base station is, the more the elected nodes consume more energy, in order to route their data to the latter. Thus, in this paper, we will propose an optimization of the Multi-Level HEED protocol in terms of energy consumption. This optimization method takes into account two criteria for the selection of the elected nodes of the higher levels: the distance between the elected nodes and the base station, and the residual energy of each elected node of higher-level. this method becomes effective when the elected nodes are far from the base station.

This approach has been evaluated by simulation according to existing metrics and new proposed metrics and they have proven their performance.

**Keywords:** WSN · Energy · Distance · HEED protocol · MHEED protocol · HEED protocol with f(Er/d) hierarchical routing · MATLAB

## 1 Introduction

The wireless sensor network defines as a set of sensor nodes. It monitors an area and participates in a collective way to send data to the base station directly or through intermediate sensor nodes. Wireless sensor networks have many application prospects in a wide variety of fields. We cite a few examples of applications: the case of human safety, monitoring of building structures presented by monitoring of the structure of a bridge. There is another economic objective monitoring application to improve the


H. Florez and M. F. Pollo-Cattaneo (Eds.): ICAI 2021, CCIS 1455, pp. 407–418, 2021.
https://doi.org/10.1007/978-3-030-89654-6_29

information tracking process in the case of monitoring free areas in a parking lot, or ecological objective to have information on migratory phenomena. Thus, we can find other type of monitoring application such as the case of understanding natural phenomena such as desertification [1, 2].

The emergence of this technology paves the way for the deployment of new large-system monitoring applications, especially those that span large geographic areas and require large-scale instrumentation. These applications bring new scientific and technological challenges that have captured the attention of a large number of researchers in recent years [3, 4].

The diversification of wireless sensor network monitoring application returns to their feature as a self-contained sensor node in power resource, low cost, it can easily and large-scale deployment, so the sensor node size is small allows to operate easily in industrial or environmental areas, without influencing the operation [5].

On the other hand, the sensor node is characterized by an energy limitation. The requirement to optimize power consumption is necessarily through the microprocessor it uses is low power and short range for wireless communication, moreover, the small size of sensor node requires that the storage capacity information is weak. Therefore, all of these features make the implementation and design of protocols for the wireless sensor network very complex. Throughout the article, we will attempt to account for these characteristics. Thus, routing in this type of network is one of the main tools to minimize the power consumption in each sensor node, which results in extending the life of the wireless sensor network. The routing protocols proposed for sensor networks can be classified into two classes [6–8]: data-centric routing and hierarchical routing:

- (Data-centric) routing: This type of routing is focused on minimizing data redundancy in WSNs. This given redundancy shows a source of wasted Energy of the sensor, we find in the literature several types of protocols: Minimum Cost-Forwarding [7, 9], Direct Diffusion [10, 11], Energy Aware-Routing [5]. These different protocols are based on finding the best path for sending data.
- Hierarchical routing: The protocols use different techniques of clustering and data aggregation to minimize of messages transmitted to the BS. The formation of clusters and the choice of their elected nodes are based on the energy reserve of the sensors and on the proximity of the latter. Among these protocols, there is LEACH [10] which builds the clusters in a completely distributed way. HEED [3, 6] is also another protocol of this family. It uses a principle similar to LEACH while guaranteeing a good distribution of the clusters. There are also other protocols such as: PEGASIS [11], TEEN [12] or even DirQ [13].

In the following sections, we present HEED protocol, the Multi-level HEED protocol optimization approach.

## 2 Description of the Optimization Approach of the Multi-level HEED Protocol

In the topic of large wireless sensors network, there are several hierarchical routing approaches, efforts have been directed to defining a hierarchical topology based on clusters. In this paper, the Multi-level HEED protocol which is based on the cluster technique and the multi-level hierarchy technique. This approach is based on the HEED protocol to organize the network in the form of a cluster and the notion of elected node representatives appears for each cluster.

### 2.1 HEED Protocol

The HEED protocol chooses the elected according to a hybrid criterion grouping the residual energy and the degree of the nodes. It achieves a uniform distribution of elected representatives in the network and to generate clusters balanced in size. A node is selected elected according to a probability $P_{ch}$ equal to (1)

$$P_{ch} = C_{prob}\frac{E_n}{E_{Total}} \tag{1}$$

where $E_n$ is the remaining energy of node n, $E_{Total}$ is the overall energy in the network and $C_{prob}$ is the optimal number of clusters. However, the evaluation of $E_{Tota}$ presents a certain difficulty, due to the absence of any central control. Another challenge is determining the optimal number of clusters[14]. On the other hand, the cluster topology does not achieve minimum power consumption between elected nodes of the intra-clusters, thus, the generated clusters are not balanced in number of nodes. These critical points in the HEED protocol, we propose the Multi-Levels HEED protocol which supports the communication between the elected nodes and these nodes in the cluster and the elected nodes between them[14].

### 2.2 Multi-levels HEED Protocol

The first phase of the HEED protocol is to organize the wireless sensor network into clusters. Each cluster is characterized by an elected node, the role of each elected node is to collect information from these members and send it to BS (sink).

The HEED protocol problem: The elected nodes farthest from the BS will suffer in terms of power consumption, as a result, those elected nodes will die quickly. Following this critical point, we try to apply the Multi-Levels HEED protocol, in fact, the elected nodes furthest from SB do not send data directly to the BS but to neighboring elected nodes. Multi-Level HEED (Multi-Levels Hybrid, Energy-Efficient, Distributed) is a dedicated hierarchical protocol for large wireless sensor networks.

After the execution of the HEED protocol to organize the network in the form of a cluster and after the selection of the elected nodes of level1, we still apply the HEED protocol on the elected nodes which we selected at the start, then we repeat this process which depends on network size; the larger the network size the more levels there are in the network organization.

In this case, the election of the level 2 super-elected nodes is done by reapplying the HEED protocol on the level 1 elected nodes and using a larger cluster radius.

After running the HEED protocol, a second time, the network will be organized into clusters of two categories. The first category of clusters consists of a set of regular nodes located in the same radius *Tr1* and an elected node of level 1. The elected role is to collect data from the different members of its cluster (regular nodes) and to send them to his super-elected. The second category of clusters consists of a set of elected level 1 nodes which are located within the same radius of the *Tr2* cluster and an elected level 2 node, which has the role of collecting data from the different members of its cluster (elected nodes of level one) and routing them to the base station, this cluster category is generated following the second execution of the HEED protocol on the elected nodes of level one. If we want to create another level of hierarchy, we only have to reapply the HEED protocol a third time on the nodes elected from level 2 and thus elect nodes from level three. The previous figure shows the principle of the HEED multi-level protocol:

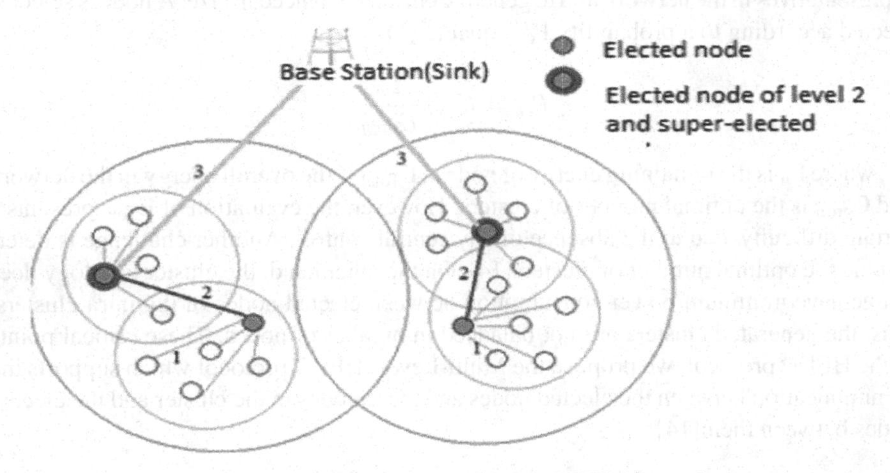


**Fig. 1.** Principle of the multi-levels HEED protocol

First, the HEED protocol will be executed and a number of clusters will be formed. These clusters are represented in the figure using green circles. In each cluster formed, there is only one elected node; the latter are denoted elected nodes of level 1 and are represented in the figure by the color green.

Secondly, we will rerun the HEED protocol on the elected nodes of level 1, this time with a larger coverage radius represented in the figure by the circles in red.

At the end of the execution of the HEED protocol on the elected nodes of level 1, we will have in each cluster in red color an elected node of level 2, it plays the role of a gateway between the elected nodes of level 1 and the base station. The goal is to reapply each time the HEED protocol on the elected nodes of the last level. It routes the data step by step towards the base station passing through the elected nodes of the lowest level to the elected nodes of the highest level, instead of transmitting the data of the elected nodes directly to the base station. In this way, we reduce the number of messages sent to the base station, consequently, we save the total energy consumed by all the sensor nodes of the network.

It is necessary to ensure that the radius of coverage (radius of the cluster) increases from one level to another because according to Lemma 5 [14] proposed by the owner of the HEED protocol, the possibility of finding two nodes elected in the same radius is almost impossible. the HEED protocol is reapplied on the elected nodes of the last level, it is necessary to increase the coverage radius, because otherwise no super-elected node of the last level created will be within the scope of another.

In the next section, we present the performance of the multi-level HEED protocol.

### 2.3 Performance Evaluation of Multi-levels HEED Protocol

We evaluated the performance of the Multi-Levels HEED protocol on the MATLAB simulator. We present some of the results obtained in this next section. We start first by specifying the hypothesis and metrics we used for the simulation.

### 2.4 Hypothesis

- The links between the sensor nodes are symmetrical
- Nodes have no information about their geographic locations
- All nodes have the same capabilities (processing/communication as well as battery capacity).
- Each node has a fixed number of transmit power levels.
- It is assumed that node failures are caused solely by the depletion of their batteries.
- The data delivery model that will be used during the simulation is the periodic model, the nodes send their data periodically to the base station.

### 2.5 Metrics

we quote in Table 1, the different parameters of the simulation:

**Table 1.** Simulation parameters

| Parameters | Value |
|---|---|
| Size of data packet | 800bits |
| Size of broadcaste packet | 200bits |
| $T_{no}$ | 5 TDMA |
| $E_{elec}$ | 50nj/bit |
| $E_{fs}$ | 10pj/bit/m^2 |
| $E_{amp}$ | 0.0013pj/bit/m^4 |
| $E_{fusion}$ | 5nj/bit/signal |
| $C_{prob}$ | 5% |
| E | 1 J |
| $d_0$ | 87 m |

The radio model used $E_{elec}$ is the electrical energy used for data detection, data transmission and reception as well as actuation when the node is asleep and wakes up; in Indeed, the node elected are the amplifying energies [1, 3, 10]. To calculate the energy consumed when sending n bits, proceed as follows:

- If the distance d which separates the transmitter from the receiver is less than or equal to $d_0$, where $d_0$ is a constant which depends on the environment, it is assumed that the radio model is a free space model where the energy consumed is calculated as a function of the square of the distance d and of $E_{fs}$.

$$E_c = n \times \left(E_{elec} + E_{fs} \times d^2\right)J \tag{2}$$

- If the distance d between the transmitter and the receiver is greater than $d_0$, it is assumed that the radio model is a multiple path model where the energy consumed is calculated as a function of the distance d to the power of four and of $E_{amp}$.

$$E_c = n \times \left(E_{elec} + E_{amp} \times d^4\right) \tag{3}$$

## 2.6 Validation and Results

In this section, we validate our approach. We simulate the life of Multi-Level HEED The parameters of my simulation are in Table 2 and the rest of parameters are declared in Table 1.

**Table 2.** Simulation parameters

| Parameters | Figure 1 | Figure 2 | Figure 3 |
|---|---|---|---|
| Area(m) | 200 * 200 | | 1000 * 1000 |
| Number of nodes | 200 | | 1000 |
| Er(J) | 1 | 2 | 3 |
| Radius of the level 1(m) | 25 | | 25 |
| Radius of the level 2(m) | 50 | 50 | 50 |
| Radius of the level 3(m) | 75 | 75 | 75 |
| SB(x,y)(m) | (100,250) | (250,550) | (500,1050) |

From the previous figures, we see that the Multi-level HEED 3 curvel offers better results than the Multi-level HEED 2 curve. The Multi-level HEED 2 curve gives better results than the Multi-level HEED 1 curve.

All this is true when the number of desired sensor nodes in the network is large, but the last turns are reached and the number of desired nodes in the network becomes too small, the Multi-level HEED 2 and Multi-level HEED 3 curves no longer provide any improvement compared to the Multi-level HEED 1 curve and the three curves meet.

This would be very logical because for twenty or even ten sensor nodes; there is not too much interest in creating a hierarchy.

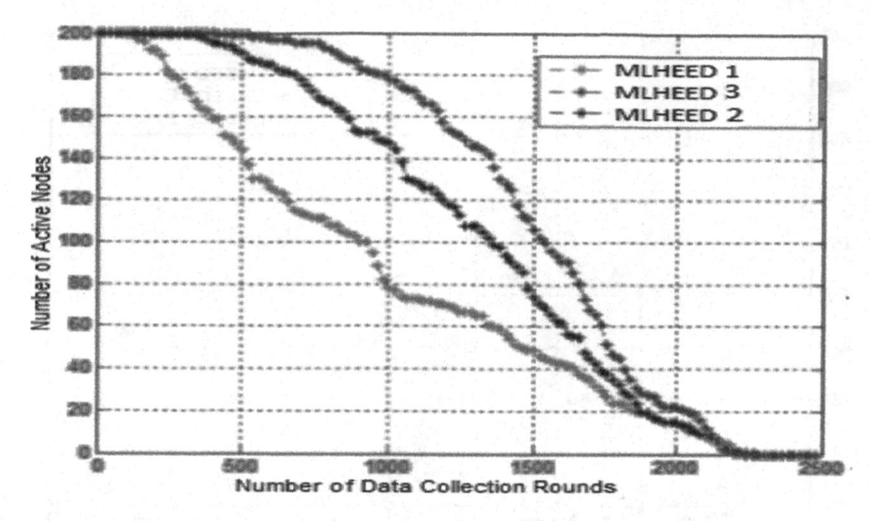


**Fig. 2.** Lifetime of WSN between the different levels in area 200 * 200

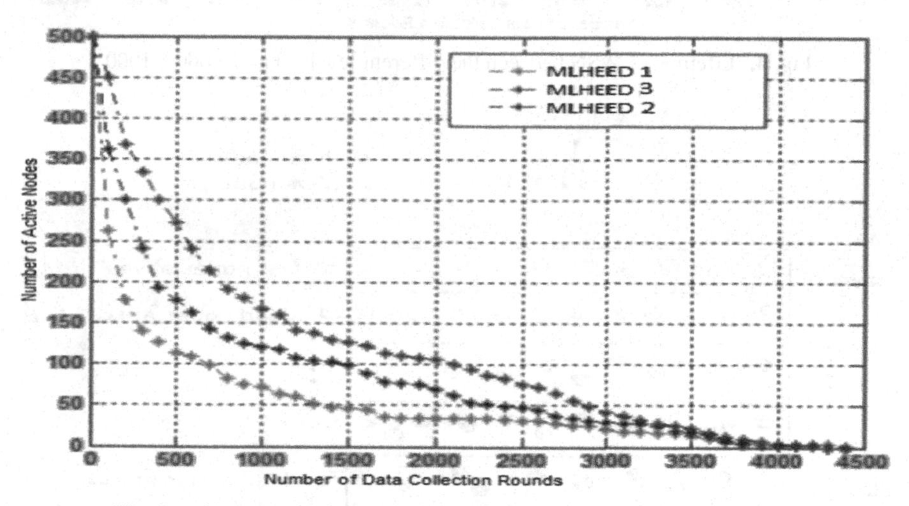


**Fig. 3.** Lifetime of WSN between the different levels in area 500 * 500

Moreover, if we examine the shape of the curves carefully, we notice a considerable drop in the number of desired sensor nodes when we enlarge the surface of the simulation. When, we move away the base station which is outside this surface. The further away the base station is, the more power the nodes consume in order to carry their data. This rapid drop shows that the sensor nodes which are furthest from the base station die first and so on until there are only the nodes that are closest to the base station left. “Sink”. This phenomenon can be described by the following Fig. 5:

In this context, we will cite an approach for the improvement of Multi-Level HEED protocol. This optimisation is the objective of the Following section.

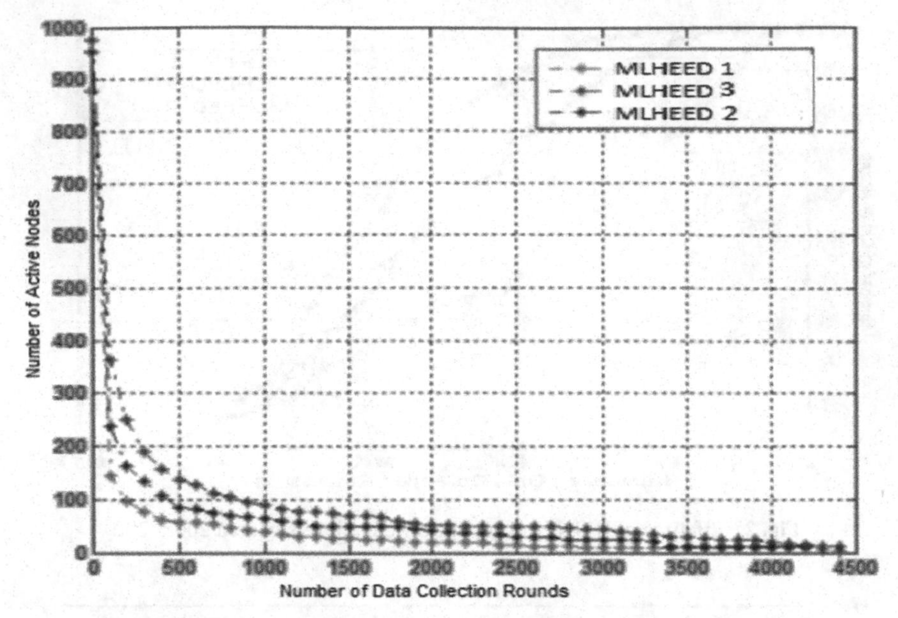


**Fig. 4.** Lifetime of WSN between the different levels in area 1000 * 1000

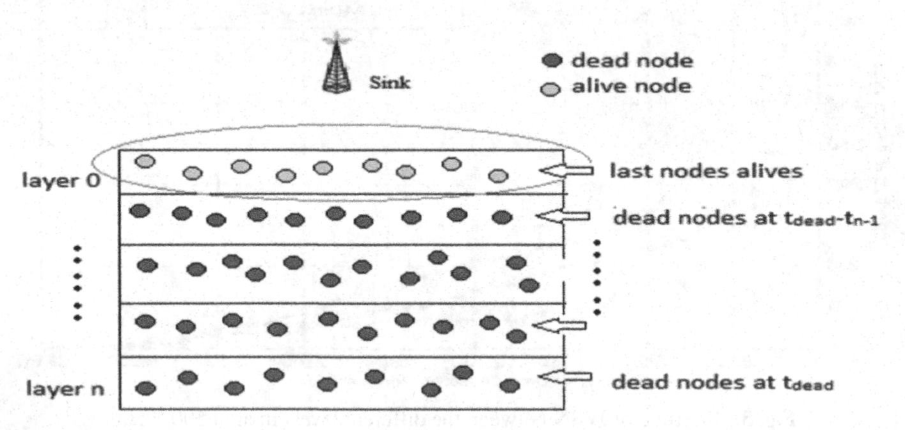


**Fig. 5.** The last nodes that remain alive in the network using the Multi-level LHEED protocol

# 3 Amelioration of Multi-level HEED Protocol

## 3.1 Description

In the previous section, we applied the HEED protocol for each step from one level to another, therefore, the performance of the Multi-level HEED protocol deteriorates and it results in a loss of energy of each elected node.

In this section, we propose a solution to remedy this problem, this solution consists of applying the HEED protocol to have the level 1, After the election an elected node of each cluster (level 1), we run our algorithm on them to have the super-elected nodes.

### 3.2 Algorithm

- Each node elected from level 1 will calculate a factor(4):

$$\mathrm{f} = \mathrm{E_r}/\mathrm{d} \tag{4}$$

- $E_r$ is the residual energy: d is the distance between the chosen one and the base station (sink).
- Each elected node will diffuse its ratio $f$ by a scope that is twice the radius of the lower level.
- Each elected node will receive the messages containing the values of the factors of the neighboring elected nodes,
- An elected node with the greatest ratio will be the super-elected node.
- If an elected node of level one is within range of two super-elected nodes of level two, it will choose to join the nearest super-elected node of level two.
- The data transfer phase:
  - Each sensor sends the data to its elected level 1 node. Each elected node collects all the data from the members of its cluster, then it merges (data aggregation) the data and it will transmit the merged data to the super-elected node.
  - If a super-elected node of level two, it will wait for the other elected nodes of level one to send it their data. Then merge them and transmit them to the base station (sink).

### 3.3 Validation and Results

In this section, we validate our approach for the improvement Multi-Level HEED Protocol. We simulate the life of Multi-Level HEED (Level 1) and the ratio f = (E/d) at two levels. The parameters of my simulation are in Table 2 and the rest of parameters are declared in Table 1 (Figs. 6, 7, 8 and 9) (Table 3).

**Table 3.** Simulation parameters

| Parameters | Figure 1 | Figure 2 | Figure 3 | Figure 4 |
|---|---|---|---|---|
| Area(m) | 100 * 100 | 200 * 200 | 500 * 500 | 1000 * 1000 |
| Number of nodes | 100 | 200 | 500 | 1000 |
| Er(J) | 0.5 | 1 | 3 | 3 |
| Radius of the level 1(m) | 25 | 25 | 25 | 25 |
| Radius of the level 2(m) | 50 | 50 | 50 | 50 |
| SB(x,y)(m) | (50,150) | (100.250) | (250,550) | (500,1050) |

This idea constitutes an improvement of the Multi-level HEED 2 protocol: we achieved two hierarchy levels without re-executing the HEED protocol on the super-elected nodes of level 1, located close to each other over a radius of 50m.

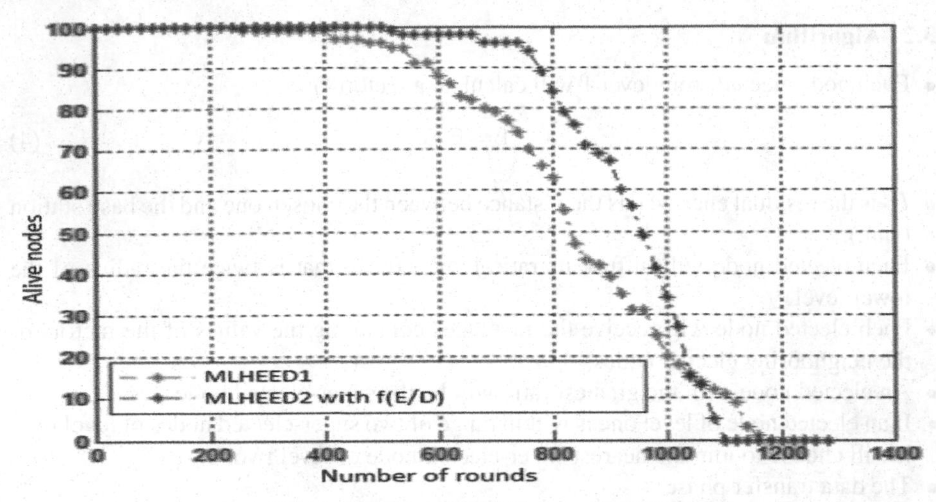


**Fig. 6.** Lifetime of WSN in area (100/100)

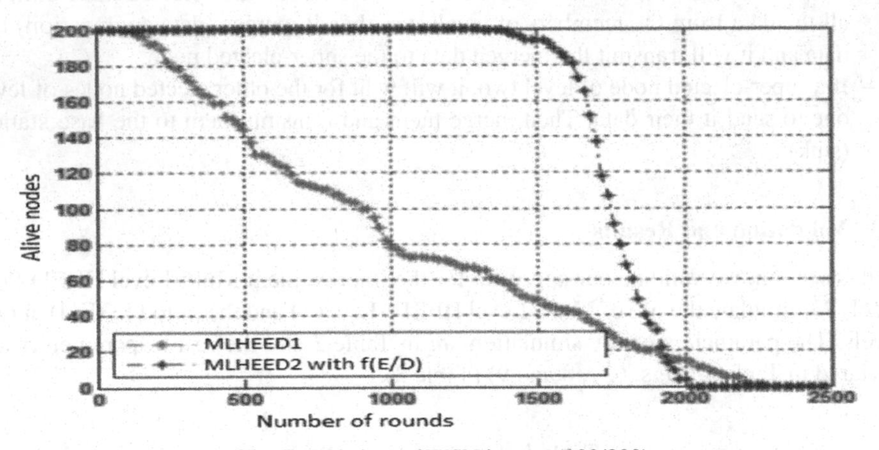


**Fig. 7.** Lifetime of WSN in area (200/200)

But each node elected from level 1 calculates a ratio f = Er / d and it broadcasts to all of its neighboring nodes. The elected node of level 1 which will be elected super-elected node of level 2 on this staff, in this way, we have reduced the number of messages exchanged during the cluster creation phase when we run the HEED protocol on the level 2 super-elected nodes as in the case of approach 1.

Also, we have guaranteed that the node elected from level 2 will have the greatest energy value, either it will be the closest to the base station, or it will have both conditions met. in the previous figures, we notice that there is a more efficient energy consumption because the number of nodes alive in the network drops less than for the other curves.

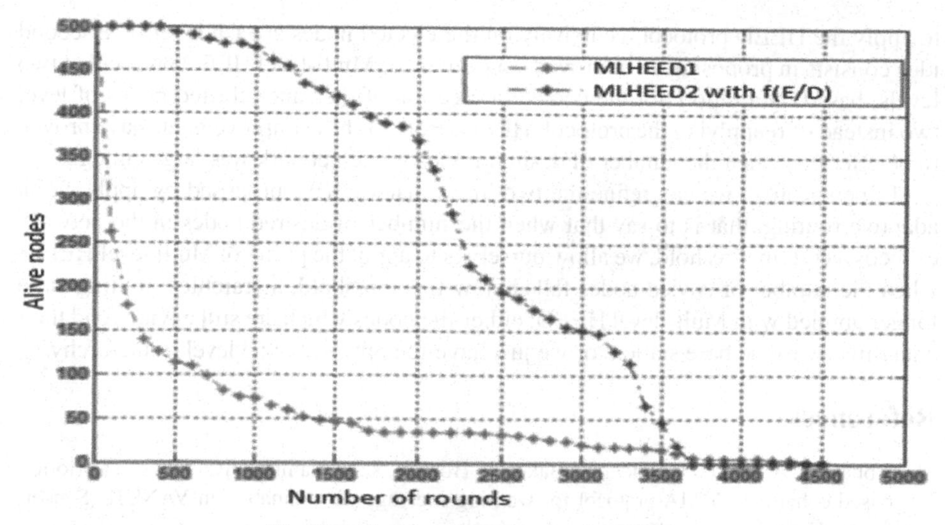


**Fig. 8.** Lifetime of WSN in area (500/500)

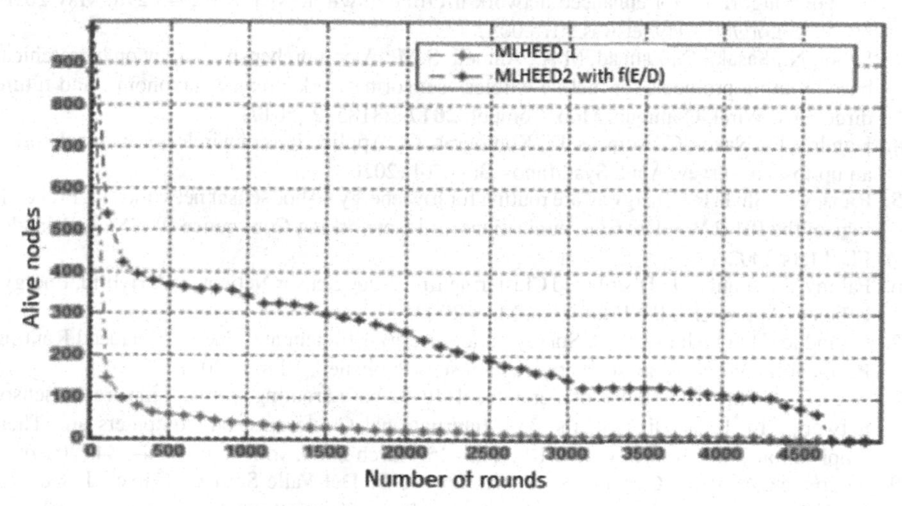


**Fig. 9.** Lifetime of WSN in area (1000/1000)

However as soon as the number of envy nodes in the network drastically decreases the life curve of wireless sensor network using the Mulit-level HEED with the retio f becomes below that of the Multi-level HEED 1 protocol, also, we notice that the realization of a hierarchy on a small number of nodes is not profitable.

## 4 Conclusion

in this paper, two ideas for improving the HEED protocol have been proposed. The first idea is to implement the Multi-level HEED protocol at two and three levels, in order to apply the HEED protocol recursively on the elected nodes at each level. The second idea consists in proposing an improved version of the Multi-level HEED protocol at two levels, based on a ratio *f* for the choice of election of the super-elected nodes of level two instead of reapplying the protocol. HEED. Both of these improvements have proven to be effective when the number of desired nodes in the network was large enough.

In perspective, we can refine the two ideas that we have proposed by applying an adaptive routing, that is to say that when the number of desired nodes in the network exceeds a certain threshold, we allow ourselves to apply the protocol Multi level HEED. when the number of envied nodes falls below this threshold, hierarchical routing is no longer applied with Multi-level HEED: either the nodes which are still envied send their data directly to the base station, or we just have to apply only one level of hierarchy.

## References


1. Abbas, G., Abbas, Z.H., Haider, S., Baker, T., Boudjit, S., Muhammad, F.: PDMAC: a priority-based enhanced TDMA protocol for warning message dissemination in VANETs. Sensors **20**(1), 45 (2020)
2. Ray, A., De, D.: Energy efficient clustering protocol based on K-means (EECPK-means)-midpoint algorithm for enhanced network lifetime in wireless "ISSN 2043 27th May 2016. https://doi.org/10.1049/iet-wss.2015.0087
3. Sabor, N., Sasaki, S., Zahhad, M.A., Ahmed, S.M.: A comprehensive survey on hierarchical based routing protocols for mobile wireless sensor networks: review, taxonomy, and future directions. Wirel. Commun. Mob. Comput. **2017**, 2818542 (2017)
4. Kandris, D., Nakas, C., Vomvas, D., Koulouras, G.: Applications of wireless sensor networks: an up-to-date survey. Appl. Syst. Innov. **3**(1), 14 (2020)
5. Rabaey, J., Shah, R.: Energy aware routing for low energy ad hoc sensor networks. In: Proceedings of the IEEE Wireless Communications and Networking Conference (WCNC), Orlando, FL, Mars 2002
6. Fahmy, S., Younis, O.: Distributed Clustering in Ad-hoc Sensor Networks: A Hybrid, Energy-Efficient Approach", IEEE Infocom, Mars 2004
7. Ogundile, O.O., Alfa, A.S.: A Survey on an Energy-Efficient and Energy-Balanced Routing Protocol for Wireless Sensor Networks Sensors Published, 10 May 2017
8. Nam, D.: Comparison studies of hierarchical cluster based routing protocols in wireless sensor networks. In: Proceedings of the 35th International Conference on Computers and Their Applications, San Francisco, CA, USA, 23–25 March 2020, vol. 69, pp. 334–344 (2020)
9. Rodríguez, A., Pérez-Cisneros, M., Rosas-Caro, J.C., Del-Valle-Soto, C., Gálvez, J., Cuevas, E.: Robust clustering routing method for wireless sensor networks considering the locust search scheme. Energies **14**, 3019 (2021). https://doi.org/10.3390/en14113019
10. Supriyo, C., De, L.S., Havinga, P.J.M.: A directed query dissemination scheme for wireless sensor networks In Abraham
11. Lindsey, S., Raghavendra, C.S.: PEGASIS power E_cient GAthering in sensor information systems. In: Proceedings of IEEE Aerospace Conference (2002)
12. Heinzelman, W., Chandrakasan, A., Balakrishnan, H.: Energy-efficient communication protocol for wireless sensor networks. IEEE 06 August 2002, USA (2002)
13. Arati, M., Agrawal, D.P.: TEEN a routing protocol for enhanced efficiency in wireless sensor networks. In: Proceedings of the 15th International Parallel & Distributed Processing Symposium, IPDPS 2001. IEEE Computer Society (2001)
14. Arifa, A., Wail, M., Abuein, Q.Q., Mazen, K.: A hybrid and efficient algorithm for routing in wireless f networks. In: 2020 11th International Conference on Information and Communication Systems (ICICS) Date Added to IEEE Xplore, 27 April 2020

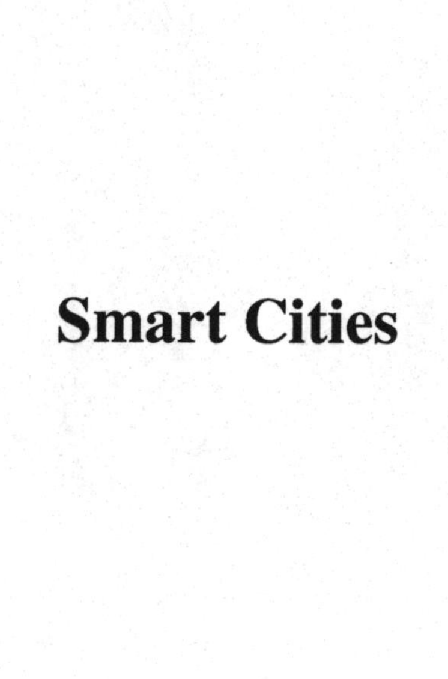

# LiDAR and Camera Data for Smart Urban Traffic Monitoring: Challenges of Automated Data Capturing and Synchronization

Gatis Vitols[1], Nikolajs Bumanis[2](✉), Irina Arhipova[1], and Inga Meirane[1]

[1] WeAreDots Ltd., Elizabetes Str. 75, Riga, Latvia
inga.meirane@wearedots.com
[2] Faculty of Information Technologies, University of Life Sciences and Technologies, 2 Liela Str., Jelgava, Latvia
nikolajs.bumanis@llu.lv

**Abstract.** Availability and range of sensors and other type of smart hardware are growing. Implementation of such solutions is becoming a crucial part of development strategies for towns and cities. Usually, hardware and software are proposed by various vendors and proper implementation can raise various tasks. One of the tasks is to establish communication between sensors and synchronization of, for example, data capturing. The aim of this article is to analyze available solutions for data capturing and synchronization and propose solutions for real world applications. Available solutions are analyzed and a general process model is proposed involving synchronization between cameras and LiDAR sensors. The proposal is installed and tested to analyses traffic flow on a 4-lane street in Latvia and capture vehicles that are not allowed to use a particular road such as heavy machinery and trucks. Synchronization proposal includes application of timestamps of capturing devices as a most feasible solution for real-world application.

**Keywords:** LiDAR · Data synchronization · Data capturing · Smart cities

## 1 Introduction

With the development of smart city solutions, including autonomous driving in cities, monitoring, and other tasks, more sensors and cameras are installed and connected to the network. Cameras and sensors are used for a broad set of tasks focused mainly on observation, planning, and security [1–3]. Available research often focuses on people and vehicle flow monitoring and prediction in cities [4, 5], violation of rules in cities [6–8], and detection of anomalies [9–11].

For the identification of objects, various machine learning and computer vision solutions are used. Common solutions include localization and classification of objects in images, use of object tracking algorithms for estimation of motion trajectories, and creation of an association between frames that involves synchronization of data between capturing devices [4, 12–14].


H. Florez and M. F. Pollo-Cattaneo (Eds.): ICAI 2021, CCIS 1455, pp. 421–432, 2021.
https://doi.org/10.1007/978-3-030-89654-6_30

Such development raises a set of challenges that needs to be addressed. One of the challenges is the synchronization of captured data from cameras and sensors. Data synchronization can be a built-in solution if installed cameras and sensors come from one manufacturer or have been installed as one solution. In many cases it is not applicable and real word constraints prohibit scenarios when all equipment in particular setting is from one manufacturer. Types of devices and sensors used for capturing can raise additional challenges where multiple environmental and technological factors are involved such as light intensity, object type and count in the frame, type of cameras and sensors, the intensity of traffic, weather, etc.

Typical solutions for capturing data involve such steps as data capture by camera or sensor, pre-processing of data, detection of objects in captured data, frame and data post-processing to improve detection and feature extraction for classification tasks. For video and image processing Convolutional neural networks and its derivatives are widely used [15, 16].

In a multi-sensor and camera environment, proper synchronization between devices are important for the precision of solutions.

The aim of this article is to analyze available solutions for data capturing and synchronization and propose a solution for real world application, which differs from existing solutions with integration and use of multiple video data sources, functionally complementing each other Article is divided into two main parts: the first section (numbered 2) describes general data capturing and synchronization problems that occur, where multiple sensors are used to capture video data in an urban setting; the second section (numbered 3) describes the concept of proposed the solution and implementation results.

## 2 Data Capturing and Synchronization Between Sensors and Cameras in Urban Setting

Prior to practical implementation two steps of data capturing and synchronization must be applied. While data capturing plays an important role in determining the quantity and quality of pre-processed data by ensuring that all objects are captured and the background is correctly subtracted, data synchronization determines if the final data set will be usable for decision making, i.e., offense validation processing.

### 2.1 Data Capturing in Urban Setting

Successful video surveillance in an urban environment requires an understanding of multiple factors and a set of actions which improve, mitigate and adapt to the influences of those factors. These factors are road condition, including the type of pavement including the number and width of lanes, traffic flow, including pedestrian crossings, weather conditions, possibility to predict, detect and prevent anomalous traffic events. The first challenge of urban surveillance is to develop and implement a complex set of task specific algorithms. According to Pramanika et al. [17] multiple algorithms are needed to address these factors. Pramanika et al. proposed five task specific algorithms just for anomalous traffic pre-event detection and prevention, for example speeding detection and potentially dangerous overtaking detection.

With regards to the vehicle and pedestrian object detection, modern object detection and tracking algorithms [18–20] can offer a high accuracy and precision. However, most of these state-of-the-art algorithms have been tested in good weather conditions. Urban environment, especially in static video surveillance scenarios, can ensure that background stays almost constant, thus background subtraction techniques can be taken advantage of [21, 22]. However, although the road pavement and road lanes are static, the colour, illumination and reflectivity can change according to the weather conditions, such as, sunrise, rainfall, and snowfall. Such precipitations as rain (Fig. 1) and snow affect both video camera precision [23] and LiDAR intensity [24].

**Fig. 1.** Example of rainfall during video surveillance, where on the left are shown raindrops on camera lens and on the right - reflections of vehicle's body and road pavement [23].

Heavy rain reduces visibility and depth of field, while raindrops on the camera lens creates blur and diffuse scattering of light which, in turn, leads to missed detections in the area of a raindrop. In addition, the previously static background now becomes non-constant which consequently creates more false-detections. A similar situation can be observed during heavy snow. Filguera et al. [24] proved that the rain affects the count of 3D point cloud points LiDAR can create, where having a stronger rain decreases the total count of points due to the increased reflection of the vehicle's body and road pavement caused by puddles. One of the solutions is to remove rain when post-processing each frame [25] which can be done by using rain removal algorithms. This, however, can only be performed for videos with low to moderate rainfall intensity [23]. According to Yi et al. [26] rain removal algorithms focus on nearly consistent rain with invariant intensity, whereas practical implementation requires processing dynamic rains. Yi et al. proposed a method based on Gaussian distribution to model individual rain streaks, thus addressing dynamic nature of rain over longer video surveillance periods. The proposed methods showed great results with regards to leaving low amount of evident rain texture. It can be concluded that although the rainfall and snowfall are the major constraints and challenges for urban surveillance, the existing and newly proposed methods are capable of solving and eliminating these factors.

## 2.2 Data Synchronization in Urban Setting

The synchronization between various sensors and cameras is explored in the research of autonomous driving and autonomous mooring. For example, heavy vehicle autonomous systems in off-road environments use cameras, LiDAR and radar sensors providing synchronization of data [27] or in a regular city setting using hybrid LiDAR-camera data processed with Faster R-CNN networks for improved object in real time [28]. The majority of autonomous driving solutions include multiple sensor synchronization, typically cameras, whereas LiDAR's and radars - with an addition of deep neural networks. Its specific setup and typical solutions involve configuration of intrinsic and extrinsic parameters, as well as use of specific algorithms for marching point clouds captured by the sensors. In a situation where it is not essential to perform synchronization in the real time other methods can be applied [29].

Outside of autonomous driving tasks synchronization between cameras and sensors is still in many cases performed with manual interference. There are solutions when synchronization between cameras is automated with the help of video and audio signal recorded by cameras [17]. Such solutions can be applied if cameras support audio signal recording. Also, possibility of noise makes synchronization based on the audio challenging.

Timestamps are one of the most prevalent method available for synchronization between sensors and cameras in various settings, even personal setting to synchronize wearable devices [30] or sensors in robotics [31] and sensors on autonomous driving cars [32]. The accuracy of time synchronization is important in these cases as there could be such constraints as incorrectly configured devices, delays in operating systems, signal delivery delays, etc. As a solution IEEE 1588 Precision Time Protocol can be used [32].

## 3 Proposed Synchronization Solution

The proposed synchronization solution involves the application of multiple road surveillance that comes with LiDAR sensor for various tasks – fixation of traffic intensity, identification of illegal parking, measurement of vehicle dimensions, etc.

The idea of using a LiDAR alongside multiple cameras is the possibility to receive increasingly accurate data and to prevent cases of missing objects entirely. Firstly, each camera is set up with a particular Field of View (FoV) in order to be limited to a couple of road lines. It is assumed that multiple cameras with overlapping FoV will film in a different spectre – infrared and visible light. In addition, increasing the number of cameras may lead to solving the occlusion problem that occurs when larger size vehicle, especially average to large cargo vehicles, like cargo vans, hide the vehicles closely driving behind. A similar approach was found to be effective in multicamera setup for pedestrian tracking [33], where two and three overlapping FoV camera setups were tested and proven to be superior to the monocular system.

Secondly, LiDAR is set up to monitor all road lines by taking advantage of its dynamic Area-of-effect FoV.

Objects of interest may be captured by either one or multiple cameras and either may or may not be captured by LiDAR. The synchronization is required between LiDAR's XML data and multiple images provided by video cameras (Fig. 2).

Our iteration of the data capturing and synchronization proposal can be seen in Fig. 3.

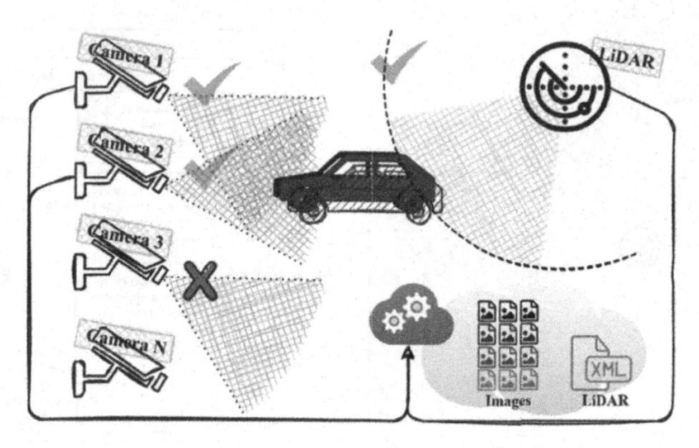


**Fig. 2.** The concept of multicamera setup

An event is defined as a database record with particular datetime, identification numbers of associated images and XML data. In order to register an event at least one iteration of data capturing and synchronization must be performed. Event registration is initialized as soon as at least one video camera is able to provide any data. The LiDAR data are always added to an existing event record; however, the LiDAR sensor itself always provides data. In particular, each vehicle, prior to being detected by video camera, has already been detected by LiDAR sensor and has appropriate data about it. One of the problems is that the data itself is not linked to a car. LiDAR XML contains data about all vehicles, therefore, the objective of processing is to annotate and synchronize data gathered by LiDAR sensor with those, gathered by video cameras.

Firstly, each video camera takes a shot when any contrast-based changes are detected, i.e., foreground segmentation using contrast is possible. A similar approach was developed into a method proposed by Fuentes and Velastin [34] where it was stated that such approach can reduce the processing power needs as only one background image is required and no additional pixel-wise information is needed. Video camera continues to take shots until there are no more changes persistent. This, in turn, results in having multiple shots for a particular vehicle. Additional shots may be taken if the vehicle changes lanes, as it introduces heavier contrast changes.

Secondly, LiDAR creates a point cloud for multiple vehicles simultaneously and organizes this data based on timestamp values. This data is then encapsulated into XML file for further processing. As LiDAR includes data about all vehicles, there are no additional "instances" of LiDAR XML being created with time or when the same vehicle changes lines.

Lastly, whenever this data is available for exporting, an event registration is started and data is prepared for processing.

There are various processes and steps regarding a complete event registration. However, as many steps may be software depended and very specific especially for larger organization the more finite details are not being discussed. Instead, the general outline and main processes are described below.

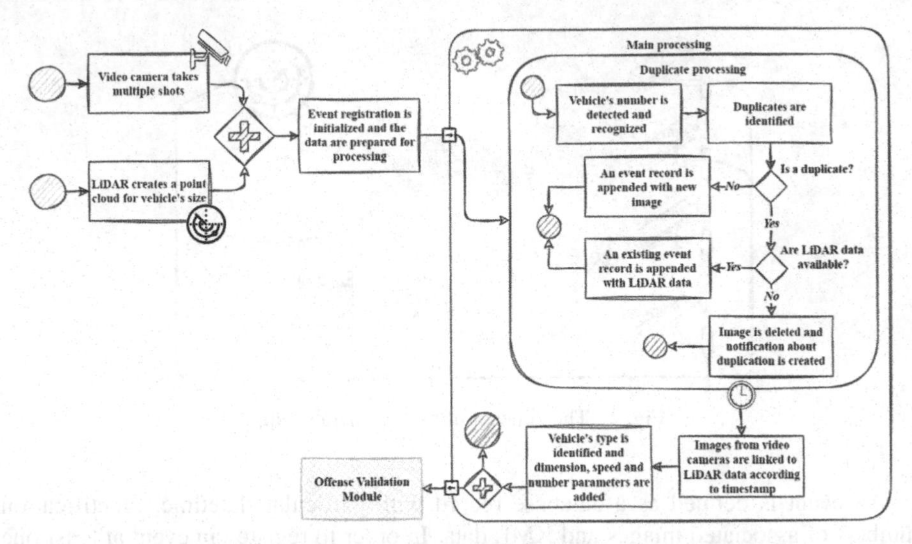


**Fig. 3.** The general iteration of data capturing and synchronization

In order to register a single event a lot of images must be processed, while many of them may be corresponding to the same event. We, therefore, suggest processing each image individually and LiDAR data in consequently. In order to fully register an event a pair of LiDAR data and image must be linked. This can be done in multiple steps and iterations. A fully registered event has the following information: type of vehicle, numberplate and dimension; moving direction and speed; offense validation result.

The main processing starts with determining duplicate events that are being registered. Firstly, an imagine of a newly registered event record taken in order to recognize vehicle's numberplate. This can be performed by separate numberplate recognition process using a recognition algorithm. Secondly, another recognition algorithm determines vehicles movement direction. The following parameters are used to determine if an event is a duplication: vehicle's movement direction, date and time of an image being exported (should be configurable to optimize performance/accuracy), the vehicle's number. In case there is an event that matches these parameters, meaning there is already an image associated with this event, a check for LiDAR data availability is performed. Available LiDAR data are appended to an existing record.

The result of multiple runs of duplicate processing is a single event with video and LiDAR data that are synchronized according to the timestamps and previously mentioned parameters. This data is used to get the following information about a vehicle: its type, dimensions, speed, and numberplate. The data is added to an existing event after processing.

Full information about the vehicle is forwarded to the offense validation module. This module performs validation in compliance with the road traffic regulations. In case an offense is identified, the appropriate notification is sent to a local enforcement department. This information is also added to event data. Offensive validation module can have other types of events, for example parking restriction event or event that process city transportation such as public bus.

To verify proposed solution 4 cameras and 2 LiDAR sensors were installed in Jelgava city, Latvia on central road (Fig. 4). Installed hardware includes:

- Mobotix M16B body 4 units
- Day lens Mx-O-SMA-S-6D23 4 units
- Night lens Mx-O-SMA-S-6L237 4 units
- Emit light Infrared illuminator M-series 860 nm, 15°, Pure White, DC 12–30 V, 38 W max (24 V DC)
- LiDAR sensor type TIC501 manufactured by SICK AG 2 units

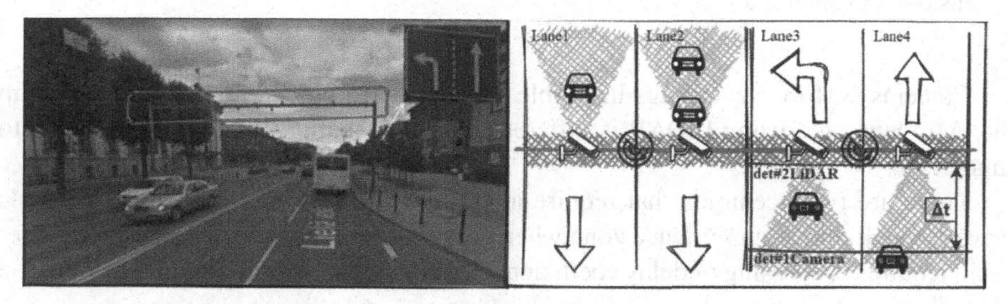


**Fig. 4.** Google Maps view from selected location of installed sensors and cameras (marked with red rectangle) and surveillance system setup in Jelgava City (Color figure online)

The concept of solution is to capture the vehicles that pass through this crossroad and measure dimensions of the vehicle. The measurement of dimensions is performed using LiDAR while the cameras capture registration numberplates of cars. At the moment of this research a prototype platform was being developed using Azure services to store captured data and perform synchronization between devices. LiDAR generates an XML file about each vehicle passing through the area of interest (Fig. 5). There are 4 cameras capturing images of vehicles. In certain conditions we did not receive an image of a vehicle, at the same time in the intense traffic we received up to 100 images about a particular vehicle. Camera takes an image of a vehicle whenever the contrast changes in the area of interest. That also means that the cameras sometimes generate images without any vehicles in case of bad weather conditions or other disturbing objects, such as multiple birds crossing the area of interest.

Due to the way LiDAR is set up to scan the area below it, the set of camera-LiDAR is providing the data for the same detected car at different timestamps. There are multiple cases of the car being detected, i.e., detected by both camera and LiDAR or detected by the camera or LiDAR only. Assuming that the weather conditions and the traffic flow allows to detect a car by both camera and LiDAR, the camera will always detect the car first (see det#1Camera, Fig. 4), prior to LiDAR (see det#2LiDAR, Fig. 4). Thus, the delay $\Delta t$ between the registration timestamps exists. In our practical assessment due to the traffic conditions, the delay was as long as 30 s. The Lane 3 and Lane 4 were used for a practical assessment.

An example of a generated LiDAR XML file can be seen in Fig. 4.

```
  </TimePositionSpeedDiagram>
- <Triggers>
  - <VehicleStartTrigger>
      <TimeStamp DateTime="2021-03-15T06:42:11.688656Z">63751387331688656</TimeStamp>
      <TriggerTimeStamp DateTime="2021-03-15T06:42:11.678641Z">63751387331678641</TriggerTimeStamp>
      <UniqueObjectId>595135</UniqueObjectId>
      <MaxX Unit="m">4.481</MaxX>
      <MinX Unit="m">3.395</MinX>
    </VehicleStartTrigger>
  - <VehicleStopTrigger>
      <TimeStamp DateTime="2021-03-15T06:42:12.069268Z">63751387332069268</TimeStamp>
      <TriggerTimeStamp DateTime="2021-03-15T06:42:12.029204Z">63751387332029204</TriggerTimeStamp>
      <UniqueObjectId>595135</UniqueObjectId>
      <MaxX Unit="m">5.139</MaxX>
      <MaxY Unit="m">1.443</MaxY>
      <MinX Unit="m">3.177</MinX>
      <StartTimeStamp DateTime="2021-03-15T06:42:11.678641Z">63751387331678641</StartTimeStamp>
      <StopReason>Normal</StopReason>
    </VehicleStopTrigger>
  </Triggers>
</Vehicle>
```


**Fig. 5.** Example from installed LiDAR generated XML file structure where particular vehicles dimensions are captured

Cameras perform recordings in visible and infrared light. In real world timestamp in XML generated using LiDAR is different than timestamp generated or assigned to images.

There are two scenarios that require attention. The first one (Fig. 6) is a simple drive-through under surveillance zone when cars stay on their selected lanes. This scenario requires processing of delay coefficient Δt, calculated prior to running recognition algorithm.

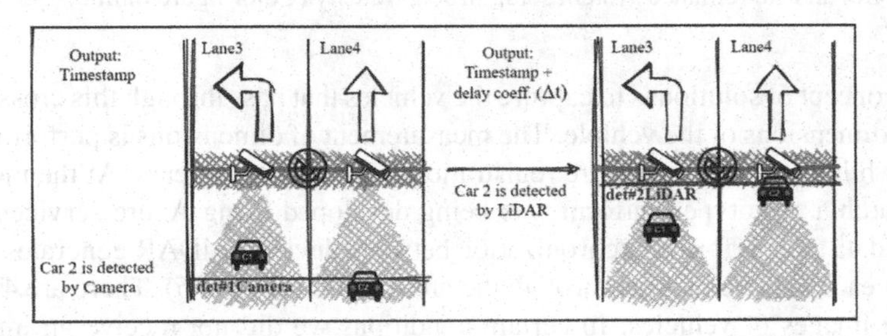


**Fig. 6.** Single car drive-through scenario

The second scenario (Fig. 7) considers the lane switching performed by a detected car. In this scenario, Car 2 is firstly detected by Camera on Lane 4. This results in a recorded detection event with the timestamp. Next, the Car 2 decides to turn left which requires for it to change lanes. After changing the lanes, the Car 2 is detected by Camera on Lane 3. This results in another detection event with another timestamp. In the end, the Car 2 is detected by LiDAR that is set up over both lanes.

It is worth mentioning that detection by Camera on Lane 3 does not require processing of delay coefficient because a vehicle's numberplate is being recognized, thus allowing to sync these two events. However, the delay between the 2nd and 3rd detection may vary heavily, thus increasing the chance of misaligning these events.

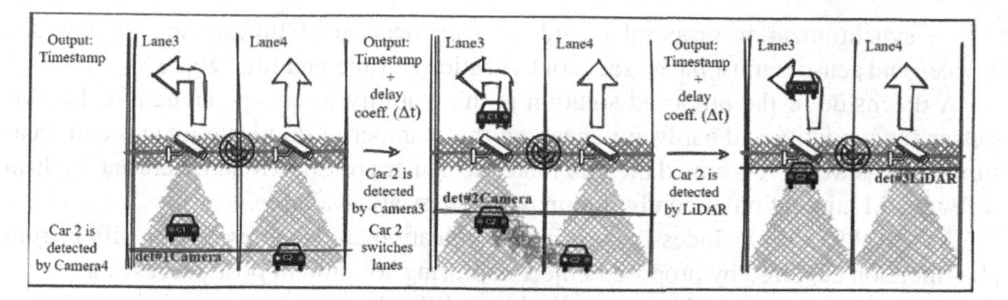


**Fig. 7.** Single car multi lane detection scenario

The proposed procedure for synchronization (Fig. 3 step "Images from video cameras are linked to LIDAR data according to timestamp") include the following steps:

- From configuration parameters LiDAR and camera max delta parameter is retrieved. This parameter stores maximum delay for particular location defined, for example in seconds.
- Information about the installed hardware which captures events is retrieved such as camera identifier.
- Search is performed to find LiDAR measurement that fits following criteria:

– measurement start time and image creation time (timestamp + max delta);
– confirmation that LiDAR and camera data is from right direction on street.

- If a measurement is absent the search is repeated for up to 5 times (in case LiDAR measurement is still processed).
- In the case of a particular timestamp there are multiple measurements, measurement timestamp which is the closest to the image timestamp is further processed.
- If a camera image and appropriate measurement is found, further image processing is performed, such as, an identification of the registration numberplate in case the measurements of the vehicle exceed allowance, for example, when a truck is passing through the city centre when it is not allowed.

The following solution is in operation in Jelgava city since April, 2021 and has been performing with satisfactory outcomes. Further improvements include an expansion of information system, such as definition of exception vehicles, for example military transport. There is a plan to scale the solution by adding multiple locations and install solutions in other cities.

## 4 Conclusions

The precision of proposed synchronization solution is directly related to precision of LiDAR measurement and camera precision. In the situations where an object is being missed in the camera recording or where LiDAR does not perform a measurement can disturb synchronization.

As synchronization proposal includes an application of timestamps all included devices and sensors must have exact configuration of time and time zones.

A downside of the proposed solution is the inability to change camera or LiDAR sensor functionality and hardware, meaning that number of duplicate images and measurements as well as corrupted files are linked to camera or LiDAR hardware and built-in software. Manipulation of configuration parameters are available.

As LiDAR also includes a built-in classification of vehicles it can differ from classification captured by proposed object capturing solution in post-processing.

Proposed solution could potentially have difficulties with accumulation of extra duplicates in case of extended traffic jams.

The proposed capturing and synchronization solution use vehicle registration numberplates for identification, therefore, any visibility disadvantages of the registration plate (dirt, reflection, shade, etc.) can impact identification of a particular vehicle.

**Acknowledgements.** The research leading to these results has received funding from the project "Competence Centre of Information and Communication Technologies" of EU Structural funds, contract No. 1.2.1.1/18/A/003 signed between IT Competence Centre and Central Finance and Contracting Agency, Research No. 1.12 "Multi-object detection and tracking in vehicle traffic surveillance: fusing 3D-LiDAR and camera data".

## References

1. Jodoin, J.P., Bilodeau, G.A., Saunier, N.: Urban Tracker: Multiple object tracking in urban mixed traffic. In: 2014 IEEE Winter Conference on Applications of Computer Vision, WACV 2014. pp. 885–892. IEEE Computer Society (2014). https://doi.org/10.1109/WACV.2014.6836010.
2. Ooi, H.-L., Bilodeau, G.-A., Saunier, N., Beaupré, D.-A.: Multiple object tracking in urban traffic scenes with a multiclass object detector. In: Bebis, G., et al. (eds.) ISVC 2018. LNCS, vol. 11241, pp. 727–736. Springer, Cham (2018). https://doi.org/10.1007/978-3-030-03801-4_63
3. Feng, J., Wang, F., Feng, S., Peng, Y.: A multibranch object detection method for traffic scenes. Comput. Intell. Neurosci. **2019**, 1–16 (2019). https://doi.org/10.1155/2019/3679203
4. Luo, W., et al.: Multiple object tracking: a literature review. **5**, 1–18 (2017)
5. Barthélemy, J., Verstaevel, N., Forehead, H., Perez, P.: Edge-computing video analytics for real-time traffic monitoring in a smart city. Sensors **19**, 2048 (2019). https://doi.org/10.3390/s19092048
6. Tsakanikas, V., Dagiuklas, T.: Video surveillance systems-current status and future trends. Comput. Electr. Eng. **70**, 736–753 (2018). https://doi.org/10.1016/j.compeleceng.2017.11.011
7. Olszewska, J.I.: Tracking the invisible man: hidden-object detection for complex visual scene understanding. ICAART 2016, In: Proceeding of 8th International Conference Agents Artificial Intelligence, vol. 2, pp. 223–229 (2016). https://doi.org/10.5220/0005852302230229
8. Li, G., Song, H., Liao, Z., Deng, K.: An effective algorithm for video-based parking and drop event detection. Complexity **2019**, (2019). https://doi.org/10.1155/2019/2950287
9. Lee, J.T., Ryoo, M.S., Riley, M., Aggarwal, J.K.: Real-time illegal parking detection in outdoor environments using 1-D transformation. IEEE Trans. Circuits Syst. Video Technol. **19**, 1014–1024 (2009). https://doi.org/10.1109/TCSVT.2009.2020249

10. Tang, H., Peng, A., Zhang, D., Liu, T., Ouyang, J.: SSD real-time illegal parking detection based on contextual information transmission. Comput. Mater. Contin. **62**, 293–307 (2020). https://doi.org/10.32604/cmc.2020.06427
11. Ibadov, S.R., Kalmykov, B.Y., Ibadov, R.R., Sizyakin, R.A.: Method of automated detection of traffic violation with a convolutional neural network. EPJ Web Conf. **224**, 04004 (2019). https://doi.org/10.1051/epjconf/201922404004
12. Yilmaz, A., Javed, O., Shah, M.: Object tracking: a survey (2006). https://doi.org/10.1145/1177352.1177355
13. Ciaparrone, G., Luque Sánchez, F., Tabik, S., Troiano, L., Tagliaferri, R., Herrera, F.: Deep learning in video multi-object tracking: a survey. Neurocomputing **381**, 61–88 (2020). https://doi.org/10.1016/j.neucom.2019.11.023
14. Verma, R.: A review of object detection and tracking methods. Int. J. Adv. Eng. Res. Dev. **4**, (2017)
15. Girshick, R.: Fast R-CNN. In: Proceedings of IEEE International Conference Computer Vision 2015 International, pp. 1440–1448 (2015). https://doi.org/10.1109/ICCV.2015.169
16. Ren, S., He, K., Girshick, R., Sun, J.: Faster R-CNN: towards real-time object detection with region proposal networks. IEEE Trans. Pattern Anal. Mach. Intell. **39**, 1137–1149 (2017). https://doi.org/10.1109/TPAMI.2016.2577031
17. Pramanik, A., Sarkar, S., Maiti, J.: A real-time video surveillance system for traffic pre-events detection. Accid. Anal. Prev. **154**, 106019 (2021). https://doi.org/10.1016/j.aap.2021.106019
18. Bochkovskiy, A., Wang, C.-Y., Liao, H.-Y.M.: YOLOv4: optimal speed and accuracy of object detection (2020)
19. Bumanis, N., Vitols, G., Arhipova, I., Solmanis, E.: Multi-object Tracking for Urban and Multilane Traffic: Building Blocks for Real-World Application. In: Proceedings of the 23rd International Conference on Enterprise Information Systems, pp. 729–736. SCITEPRESS - Science and Technology Publications (2021). https://doi.org/10.5220/0010467807290736
20. Wang, C.-Y., Yeh, I.-H., Liao, H.-Y.M.: You Only Learn One Representation: Unified Network for Multiple Tasks (2021)
21. Lo Bianco, L.C., Beltrán, J., López, G.F., García, F., Al-Kaff, A.: Joint semantic segmentation of road objects and lanes using convolutional neural networks. Rob. Auton. Syst. **133**, 103623 (2020). https://doi.org/10.1016/j.robot.2020.103623
22. Cheung, S.S., Kamath, C.: Robust techniques for background subtraction in urban traffic video. In: Panchanathan, S. and Vasudev, B. (eds.) Visual Communications and Image Processing 2004, p. 881. SPIE (2004). https://doi.org/10.1117/12.526886
23. Bahnsen, C.H., Moeslund, T.B.: Rain removal in traffic surveillance: does it matter? IEEE Trans. Intell. Transp. Syst. **20**, 2802–2819 (2018). https://doi.org/10.1109/TITS.2018.2872502
24. Filgueira, A., González-Jorge, H., Lagüela, S., Díaz-Vilariño, L., Arias, P.: Quantifying the influence of rain in LiDAR performance. Meas. J. Int. Meas. Confed. **95**, 143–148 (2017). https://doi.org/10.1016/j.measurement.2016.10.009
25. Wang, J., Gai, S., Huang, X., Zhang, H.: From coarse to fine: a two stage conditional generative adversarial network for single image rain removal. Digit. Signal Process. A Rev. J. **111**, 102985 (2021). https://doi.org/10.1016/j.dsp.2021.102985
26. Yi, L., Zhao, Q., Wei, W., Xu, Z.: Robust online rain removal for surveillance videos with dynamic rains. Knowl.-Based Syst. **222**, 107006 (2021). https://doi.org/10.1016/j.knosys.2021.107006
27. Yeong, D.J., Barry, J., Walsh, J.: A Review of Multi-Sensor Fusion System for Large Heavy Vehicles off Road in Industrial Environments. In: 2020 31st Irish Signals and Systems Conference, ISSC 2020. Institute of Electrical and Electronics Engineers Inc. (2020). https://doi.org/10.1109/ISSC49989.2020.9180186

28. Banerjee, K., Notz, D., Windelen, J., Gavarraju, S., He, M.: Online camera LiDAR fusion and object detection on hybrid data for autonomous driving. In: IEEE Intelligent Vehicles Symposium, Proceedings, pp. 1632–1638. Institute of Electrical and Electronics Engineers Inc. (2018). https://doi.org/10.1109/IVS.2018.8500699
29. Ravindran, R., Santora, M.J., Jamali, M.M.: Multi-Object Detection and Tracking, Based on DNN, for Autonomous Vehicles: A Review (2021). https://doi.org/10.1109/JSEN.2020.3041615.
30. Zhang, Y.C., et al.: SyncWISE: Window induced shift estimation for synchronization of video and Accelerometry from wearable sensors. In: Proceedings of ACM Interactive, Mobile, Wearable Ubiquitous Technol. **4**, (2020). https://doi.org/10.1145/3411824
31. Tschopp, F., et al.: VersaVIS: An Open Versatile Multi-Camera Visual-Inertial Sensor Suite. Sensors (Switzerland). **20**, (2019). https://doi.org/10.3390/s20051439
32. Rangesh, A., Yuen, K., Satzoda, R.K., Rajaram, R.N., Gunaratne, P., Trivedi, M.M.: A Multimodal, Full-Surround Vehicular Testbed for Naturalistic Studies and Benchmarking: Design, Calibration and Deployment (2017)
33. Yan, Y., Xu, M., Smith, J.S., Shen, M., Xi, J.: Multicamera pedestrian detection using logic minimization. Pattern Recognit. **112**, 107703 (2021). https://doi.org/10.1016/j.patcog.2020.107703
34. Fuentes, L.M., Velastin, S.A.: Foreground segmentation using luminance contrast. In: WSES International Conference on Speech, Signal and Image Processing, Malta (2001)

# Mobility in Smart Cities: Spatiality of the Travel Time Indicator According to Uses and Modes of Transportation

Carlos Alberto Diaz Riveros[1(✉)], Karen Astrid Beltran Rodriguez[1], Cesar O. Diaz[2], and Alejandra Juliette Baena Vasquez[3]

[1] Corporación Universitaria del Meta, Villavicencio, Colombia
{carlos.diaz,karen.beltran}@unimeta.edu.co
[2] Universidad ECCI, Bogota D.C., Colombia
cdiazb@ecci.edu.co
[3] Universidad Antonio Nariño, Bogota D.C., Colombia
alejandra.baena@uan.edu.co

**Abstract.** The objective of this document is to present the analysis of the level of vehicular congestion that occurs in the surrounding sectors, the downtown neighborhood, specifically on Alfonso López Avenue, to analyze and propose, what is the best solution, and thus improving the mobility of the road network, and minimizing congestion in the present and the future. The results of the analysis of the measurement of the speeds that were obtained taking the different modes of transportation: private, public: taxis and buses, and plate method, speeds were taken at peak and off-peak hours, in the morning and afternoon days, for 15 continuous days; through cycles that incorporate series of fluid and congested traffic or combined; taking into account the relationship variables between flow, velocity, density, interval, and spacing; Finally, the results were spatialized, to expand the crossovers of variables. The previous analysis seeks to contribute to the discussion about the impact of the different modes of transportation on traffic congestion and how to diagnose and propose solutions to improve mobility.

**Keywords:** Spatialization · Transport · Travel time · Modes of transport · Villavicencio

## 1 Introduction

Information and communication technologies ICT are the foundations of modern life in many aspects, and they are determining changes in how cities work. The use of ICT to integrate transport, energy, health, resource management, and other systems in the cities is the concept of smart cities that seek to improve citizens' quality of life, disposing of all services accessible to them [1,2]. The smart

Villavicencio, Meta, Colombia. Supported by organization RUMBO-Red Universitaria Metropolitana de Bogotá.


H. Florez and M. F. Pollo-Cattaneo (Eds.): ICAI 2021, CCIS 1455, pp. 433–448, 2021.
https://doi.org/10.1007/978-3-030-89654-6_31

cities are inspired by the sustainable development goals adopted by world leaders on September 25, 2015, where were planned the masters lines to bring about the world's global development. In this context, the improvement of mobility impacts 13 of the 17 SDGs [3].

Considering that 85% of people will live in cities, it is crucial to rethink mobility within a smart city [4]. The aim is to use the new generation of information technologies to obtain and manage the city's data to mitigate risks, plan, build, and implement a services citizen city [5]. Within smart mobility, strategies have been proposed that range from urban planning, means and modes of transport and intelligent transport systems that, based on the capture of information, allow the generation of adaptable projects for the improvement of mobility and road safety in a city [6,7].

Regarding mobility, one of the biggest challenges in cities, as in LATAM countries, is the transport infrastructure. It has been built a disorderly way and planned for a small population over the years; it is summing to the increasing number of vehicles generating affections such as traffic congestion, longer traveling time for the road users, and even affecting road safety. In the light of all this, it is crucial to begin rigorous research to study the cities context. The way that citizens move, habits, and traffic flow by the hour, the traffic lights conditions, among the others, are essential variables to know the core of the mobility problem and propose viable solutions towards more sustainable mobility and smarter. This article presents a diagnostic study of vehicular mobility, focused on the central area of the city of Villavicencio, based on field work and the uses of ICT to propose solutions that can improve the quality of life of the community.

Villavicencio City, located in the east of Colombia, is focusing on becoming a smart city. In 2020 the city participated in the process led by the Ministry of Information Technologies and Telecommunications Min TIC of self-diagnosis in the measurement of the level of maturity of Smart Cities, within the opportunities for improvement and sub-dimensions with lower indicators are: - Climate Change (score 1) - Environment Dimension. - Digital Government (score 1) - Governance Dimension. - Multilevel Governance (score 1) - Governance Dimension. - Intelligent Mobility (score 1.8) - Habitat Dimension. - Risk Management (score 2.2) - Environment Dimension. In this sense, Villavicencio's city focuses on improving the planning, organization, and direction, formulation and application of policies on mobility, road configuration, signage, and public transport, which consult the community's needs articulate the different modes.

The study of vehicular mobility was developed with graduates of the faculty of civil engineering of the University Corporation of Meta. Firstly, the speed of the vehicles was acquired in Alfonso López Avenue by plate method analyzing two paths: from the GRAMA to the 33rd street with 32rd street and from the 33rd street with the 32nd street to the GRAMA. Secondly, it was made a study of the intersections of Carrera 33 with Carrera 29 and Carrera 33 with Calle 34 based on the methodology of the six steps recommended by CAL Y MAYOR (2007). It consists of establishing the approach to the problem and a logical and adequate solution through: observation of the problem, formulation of hypotheses, data collection, data analysis, and results to propose solutions.

This diagnostic study of vehicular mobility in Villavicencio is directly related to the projections of the city focusing on the general objectives to ensure social development, political, economic, physical, and environmental of the Municipality; the general well-being and the continuous improvement of the quality of life of its population.

## 2 Methodology

This project is carried out with a quantitative and qualitative approach, experimental, descriptive, explanatory with the simulation model; based on data collection, on the observation of means of transport and speeds; The analysis seeks to establish the vehicular flow, of the means of transport, deliberating in favor of the improvement or increase of the mobility of the existing traffic, and future 15 years, in this area; Alfonso López Avenue was taken as the object of study, from the roundabout of the grass, including the flag park; the length of the sections to be intervened was measured, and travel times and speeds were studied by the plate method; on the other hand, numerical methods referring to counts were used to evaluate the traffic volume, vehicle composition and projection of the results.

### 2.1 Study Zone

Avenida Alfonso López from the roundabout of the grass, between two paths, path 1 Glorieta de la GRAMA to Carrera 33 with Calle 32, and path 2 return in the opposite direction, from Carrera 33 and Carrera 29, to the roundabout from the grass; These in turn were subdivided into 13 sub-sections (See Fig. 1).

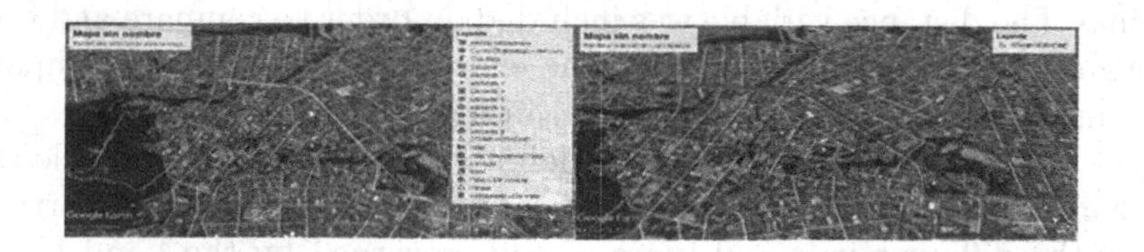

**Fig. 1.** Left Section 2, Right. Section 1. Google Earth, 2018.

### 2.2 Experimental Design

In Alfonzo López Avenue, through the plate method, the measurement detail was sought, the total was subdivided, into Sections 1 and 2 north - south, and south - north, and these in turn, into thirteen sub-sections. The three practitioners, "Paola Andrea Velásquez Parrado", "Juan Camilo Cubidez Ordoñez", "Jefferson Clavijo Morales", carried out the data collection; the two graduates Ruby Lorena

Álvarez Vanegas and Camila Andrea Carvajal Pardo, carried out the modeling of the civil engineering program of the Meta University Corporation - UNIMETA. They sought to know the distance, through the odometer measurement, in the total of the thirteen sub-sections, of the two sections and thus totalize, the tabulation of the information, by hand in a format, see image 1, where the last one was written plate letter, with its respective number; With the help of the cell phone timer, the exact time in which the vehicle passed in front of the practitioner was recorded and so on until 40 data were obtained; The measurement was proposed to be carried out at the following times: off-peak hour (8:30 am–11:30 am; 2:30 pm–5:00 pm), and peak hour (7:30 am–8:15 am; 11:45 am–12:30 pm; 5:30 pm–6:45 pm). The travel time method and registration plates annotation were combined, the results of these two methods being obtained, then the data was tabulated. It is to clarify, the concepts to be used, in the field, data on travel time and travel speed are obtained, in this way: Travel time, Time of march or circulation, Movement rate, Travel speed, and Walking speed or circulation.

The plate method, allowed to show in the following table, the travel times, when comparing the data obtained in the sample, the time in seconds that each vehicle took to move from the initial point to the final point was determined, taking into account the waiting time at traffic lights, and at what time of day the shift is made to determine peak and off-peak hours. The analysis of the tables of the plates showed coincidences, in both formats: the travel time from one point to another of each vehicle, the difference in the time it took the vehicle to move from one corner to another, the speed of the section, the mean, the standard deviation, the standard error, the minimum and maximum value of the seconds, the range, the sum of the seconds and the number of tabulated data; then data collection.

It was continued, with this technique each sub-section, of the two sections, object of study was systematized; and the next step, to carry out an analysis compared by days, the speeds of the daytime off-peak, versus the afternoon off-peak time. The distance variable was included, in order to compare and broaden the horizon of the outcome scenarios. The elapsed time was taken, compared to the distance, by bus, private car and taxi in each section.

In the last phase, a step prior to modeling, the behavior of the cycle time of the traffic lights was made known, a physical review of the place was carried out, denoting the different uses and types of soil, governed by the Land Use Plan; as well as its spatial conditions of high densification; Also, an intervention of the municipal administration was found with a view to improving the fluidity of vehicles at intersections, with studies of vehicular traffic, and the management of green light times at traffic lights, which did not work, even though the problem persists. These intersections have a high flow due to the connectivity that it establishes between the city center with the different communes, sidewalks and neighboring municipalities, from Carreras 33, Calle 34 and Carrera 33, they conclude at Carrera 33, see Fig. 3 (Fig. 2).

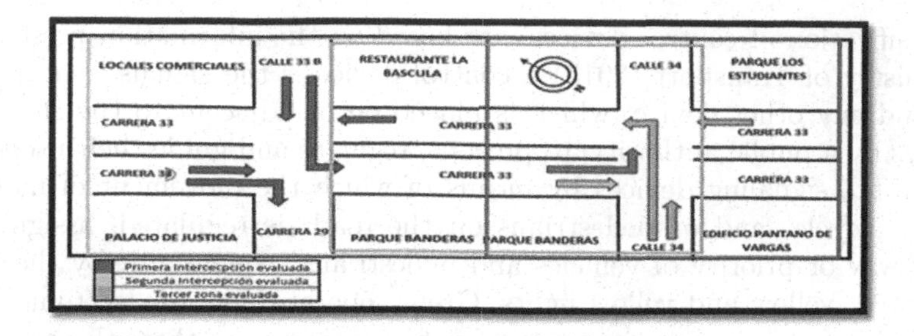


**Fig. 2.** Scheme of areas to be evaluated. Own source

The data collected were the turns given by the vehicles, in the 3 most representative times, from 9am–10am, 4pm–5pm and 8 pm–8:30pm. They are represented with bars followed by colors: red (time stopped) and green (time in motion), in order to show graphically which are the longest times at each of the two intersections. Now, with the observation of the speed study, we will measure "The speed that vehicles can develop on a road, its inverse and the time it takes to travel a given distance. It is usually measured at a point or short section of a road, to determine the speed with which vehicles pass through there, while the travel time is observed in sections of track of a certain length to know the quality of the overall service they provide or its variations throughout them". It was necessary to use chronometers, explicitly in Carrera 33 in both directions; 60 motorized vehicles were evaluated, from these figures obtained, it was possible to determine the descriptive statistics, and in this way to know the variation data such as the standard deviation, the sample variation, the typical error, the range, among others. relevant data.

It allowed to recognize the distances, and the speeds, that were presented in the results, supported by the norms that are presented in the following section: Data acquisition. Then, it will be visualized in a geographical framework, where it was approached from the approach of human geography, according to [10], a "Cultural turn and spatialization of social science: historical geography, cultural geography and environment" quotes Carl Saúer, who exercises human geography.

Also, he is the one who studies the territory and the human aspects, inherent to the analysis of transport networks and flows, applying spatialization, a technique that seeks to lead to planimetry, data analysis, allowing to cross technical variables, with what are the means of transport: vehicle flow, speed, density, interval and spacing; ending, thus visualizing the deductions of this forecast, taking them to software modeling, [8] the types of intersections and the allowed turns, [11].

In controlled intersections such as intersections, vehicles traveling through different accesses can meet simultaneously; some of them traffic signs, the traffic light [11]. On the other hand, the Intersections without control are those that do not use a type of control, there the rule is the priority for the first vehicle that arrives.

Classification of control devices, are based on the information presented by the Ministry of Transport (2015) a control device is the signals, marks, traffic lights and any other device, which is placed on or adjacent to the streets and highways by a public authority, to prevent, regulate and guide their users. Traffic lights are signaling devices by means of which the movement of motorized vehicles, bicycles and/or pedestrians on the roads is regulated, assigning the right of way or priority of vehicles and pedestrians sequentially, by the indications of red, yellow and yellow lights. Green, operated by an electronic control unit. Vehicle counts are the counts of vehicles that pass through certain network arches. Manual Gauges are generally used to record turnover volumes and classified volumes. The duration of the capacity varies with the purpose of the capacity. Some classified gauges can last up to 24 h.

### 2.3 Data Acquisition

The data collection was carried out by means of procedures: the method of travel time, and annotation of license plates, tabulation of information; having as a guide the text of Cal y Mayor, R., and Cárdenas, J. Ingeniería de Tránsito. Fundamentals and applications, from 2007. "seeks to establish the approach to the problem and a logical and adequate solution, as follows: 1. Observation of the problem, 2. Formulation of hypotheses of the problem and its solution, 3. Data collection, 4. Data analysis, 5. Specific and detailed proposal, 6. Study of the results obtained".

To achieve each of these steps, tools such as: Taking of vehicular and pedest-The data collection was carried out by means of procedures: the method of travel time, and annotation of license plates, tabulation of information; having as a guide the text of Cal y Mayor, R., and Cárdenas, J. Ingeniería de Tránsito. Fundamentals and applications, from 2007. "seeks to establish the approach to the problem and a logical and adequate solution, as follows: a. Observation of the problem, b. Formulation of hypotheses of the problem and its solution, c. Data collection, d. Data analysis, e. Specific and detailed proposal, f. Study of the results obtained". To achieve each of these steps, tools such as: Taking of vehicular and pedestrian counts in the study interceptions. Analysis of the data collected at the traffic gates and the information collected from the traffic congestion through the Google Maps application based on the information and observation collected. Taking traffic light times and study of vehicle flow speeds in the study area Collection and processing of data and information on the road section studied in different software. The travel time methods, in which the travel time between two established points is measured, go hand in hand with the Colombian Traffic Code, in TITLE III "BEHAVIOR RULES", explicitly in CHAPTER XI "Limits of Speed"-"ARTICLE 106. SPEED LIMITS IN URBAN AREAS PUBLIC. Modified by art. 1, Law 1239 of 2008. Amended by art. 1, National Decree 15 of 2011NOTE: National Decree 15 of 2011 was declared Inexpensive by means of Judgment of the Constitutional Court C-219 of 2011. On urban roads the maximum speeds will be sixty (60) kilometers per hour except when the competent authorities for signs indicate different speeds."

It is important to bear in mind that the speed of this road should not exceed 60 km, as indicated above. On the other hand, it is important to mention (Cal, Mayor and Cárdenas, 2007, p. 15).

As an element of general knowledge it is important to mention that for Cal y Mayor - Cárdenas (2007). Who in his text relates that the efficiency of the mobility of a road depends on the volume of traffic that circulates on it. It also expresses the concept of vehicle demand is the number of vehicles that require moving through a certain road system or road supply. It is understood that within the vehicular demand are those vehicles that are circulating on the road system, those that are in a queue waiting to circulate. The vehicle counts will be carried out in order to study the traffic volumes which "are carried out with the purpose of obtaining real data related to the movement of vehicles and/or people, on specific points or sections within a road system of roads or streets. These data are expressed in relation to time, and their knowledge makes it possible to develop methodologies that allow a reasonable estimate of the quality of the service that the system provides to users" When it comes to analyzing how the vehicle operation works, which must occur in conditions of stable or saturated flow, then the road supply, which represents the physical space, that is, streets and highways, in terms of their cross section or capacity. In this way, the road supply or capacity represents the maximum number of vehicles that can finally move or circulate in said physical space. (Cal, Mayor and Cárdenas, 2007, p. 15) As an element of general knowledge it is important to mention that for Cal y Mayor - Cárdenas (2007). Vehicle demand and road supply is the capacity of a road system that encompasses quantitative and qualitative characteristics, allowing to determine the sufficiency (quantitative) and quality (qualitative) of the service offered by the system, that is, the supply and users who are the demand.

## 2.4 Modelling

It is the Synchro software which has begun to be implemented around the world for the modeling of the traffic operation of road corridors, In Saudi Arabia in 2009, by Ratrout and Maen; In Lima, Peru in 2012 by VERA. In Colombia has not been alien to the application in the cities of Medellín and Tunja. In order to carry out the mobility modeling at the intersections of Carrera 33 with Carrera 29 and Carrera 33 with Calle 34, the Synchro 8 software was used, the location of existing traffic lights, travel speeds All this through the use of the results of the analysis of the gauges and physical characterization.

One of the analyzes used is the Webster Method, "it is based on the study of 100 intersections in the city of London whose analysis allowed to conclude on the influences exerted on the traffic or the saturation flow, the distribution of green: by having the cycle time must be distributed for each phase as green time, this distribution is carried out proportionally to the load factor of each of the phases and of the intersection (Highway Capacity Manual, 2010).

$g_i = y_i/Y(C_o - L)(1)$

Where
$g_i$ = Duration of effective green of phase i.
$y_i$ = load factor of phase i.
$Y$ = load factor for the entire intersection.
$C_o$ = Optimal cycle.
$L$ = total lost time of the intersection.

It also analyzes, average delay per vehicle, is given for any access of a traffic light intersection by the following expression:

$d = 9/10(c(1-\lambda)^2/2(1-\lambda x) + x^2/2q(1-x))(2)$
Where:
$C$ = duration of the traffic light cycle (s).
$\lambda = g/c$ = effective green ratio.
$g$ = duration of effective green (s).
$q$ = access flow $(veh/h)$.
$s$ = saturation flow (veh.lig / h.v) = 525w.
$x$ = degree of saturation = $q/\lambda s$.
$D$ = total delay for each access = $dq$

Traffic Modeling software that offers traffic analysis, optimization and simulation applications. It combines the modeling capabilities of Synchro and the micro-simulation and animation capabilities of Sim Traffic with the 3D authoring viewer. (Methods 2000 and 2010) for signposted intersections and roundabouts. It also implements the intersection capacity utilization method to determine the intersection capacity. Signal optimization routine allows the user to weight specific phases, thus providing users with more options in developing signal frequency plans. In the processes of analysis, evaluation and optimization of road networks, specialized computer programs are currently being used, such as Synchro, which applies the HCM 2010 method. The following advantages will be established:

- Optimization of cycle lengths and distribution of green times per phase, eliminating the need to carry out multiple tests of plans and times in search of the optimal solution.
- Generation of optimal time plans in less time than any other existing program today.
- Interaction, in such a way that when changes are made to the input data, the results are automatically updated, and the operation plans are shown in easy-to-interpret time-space diagrams.
- Application in networks of up to 300 intersections with enough success, being able to disaggregate larger networks and then join them.

The data previously obtained, in the interest of the elaboration of the model to be simulated, such as: the vehicular flows at peak hours for turns, and taking into account the type of vehicle, the taking of gauges carried out on 02/27/2018, 28/02/2018, 01/03/2018 and 04/03/2018. Then, the summary tables of the volumes obtained are presented, with all the vehicles that pass through this intersection, including bicycles in rush hour each day; In addition, its equivalent composition in percentage and its daily FHP are disclosed; obtaining the highest vehicle flows in one hour; with the two intersections, conceiving the results.

# 3 Results

## 3.1 Travel Time Estimation

The fieldwork described in Sect. 2 allowed collecting data related to the traffic flow using the license plate method. Data from each segment for both paths of the Alfonzo López avenue gave the mean velocity values for valley and peak times, as shown in the bar graphs presented in Fig. 1. For path 1, the velocities at valley times were similar for all of the segments. In addition, the same behavior was presented for peak times. However, there are differences between the peak and valley moments. The speeds at the valley moment are more significant than peak one, which is expected due to the peak time is related to more traffic congestion. Valley times, there was an exception roundabout of the Grama up to Calle 41 and Calle 32, which showed a higher velocity in respect with the other segments; It could be explained that the buses go down fast because this segment has no stops at this location. On the other hand, path 2 presented similar velocities for all segments except for Palacio de Justicia - Calle 35 and Calle 41 - Grama, which had higher speeds. Likewise, Calle 32 with Calle 33 exhibits an anomaly speed with lower velocities; It can be explained due to traffic jams generated by the cars from Av. 40 (Unicentro). It is crucial to consider that when traffic is slower along the peak hours during the day, people go to their jobs and studies. On the other hand, at night the peak hours is due to the people usually return to their homes. Figure 4 shown the type of vehicle versus velocity at path 1. The data related to the bus, taxi, and private car speeds show that the buses always present lower velocity. Because buses are parking in different areas to wait for passengers, mobility is difficult and obstructs the road. On the other side, the speed obtained on the taxi route on occasions was more significant than the individual. In general, taxi drivers are more impatient and without enough caution to drive. In contrast, the driver is more careful and probably did not hurry to get the destiny sometimes. The behavior of velocities as a function of the mean of transport was similar to path 2, not presented here.

## 3.2 Traffic Flow Indicators at Intersections

**Volume of Vehicles.** The evaluation of vehicular traffic at the intersections between Carrera 33 between Carrera 29 and Calle 34 allowed obtaining the total volume of each type of vehicle, including all movements for periods of 15 min, per hour, the hourly and daily variation, the maximum and minimum total volume for all the vehicles per hour, among the others crucial indicators. As discussed later, these indicators were crucial for modeling the traffic flow conditions at both intersections.

Figure 5 shows the number of vehicles at each intersection where the motorcycle is predominant in both intersections under study. The fieldwork data gives

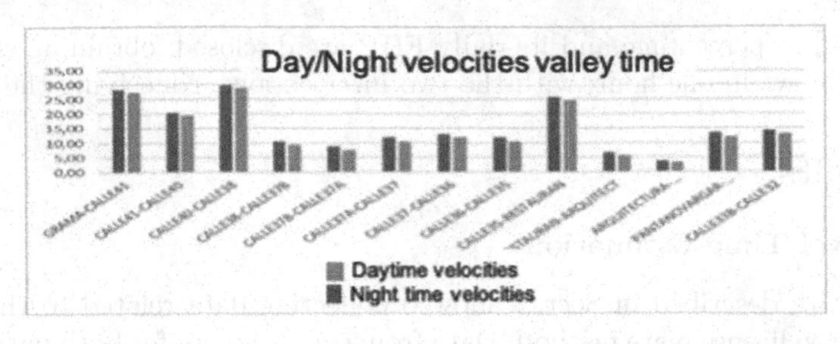


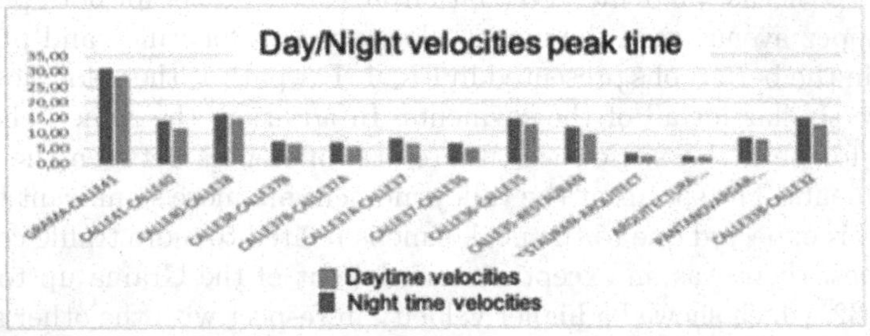


**Fig. 3.** Day/night velocities for Section 1 of the Avenue under study. The mean velocity is around 15 Km/h.

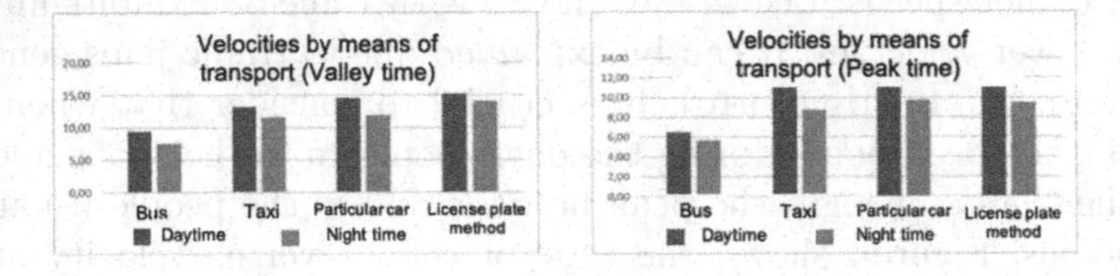


**Fig. 4.** Velocities according to the type of vehicles (means of transport) at Section 1 of the Avenue under study

to concluded that it is possible to identify the difficulty of mobility between the crossroads of the Carrera 33 when it crosses Carrera 29 (namely, Intersection 1) and Calle 33 with Calle 34 (namely, Intersection 2). Firstly, on the sidewalk with two lanes, which generates an impact on the other advancing movements and causes vehicular bottling. Additionally, movements 1 and 8 (described in Sect. 2) from the intersection between Carrera 33 and Calle 34 show significant congestion on the traffic flow.

**Traffic Lights Cycles.** To study mobility conditions, a evaluation of the semaphore times was also necessary for each intersection by phases. For example, intersection 1, located on Carrera 33 with Carrera 29, has four stages, called 1,2,3 and 4, as shown in the Fig. 6 Left. In contrast, intersection 2, which is located on Carrera 33 with Calle 34, has three stages, called 5,6 and 7, as presented in Fig. 6 Right.

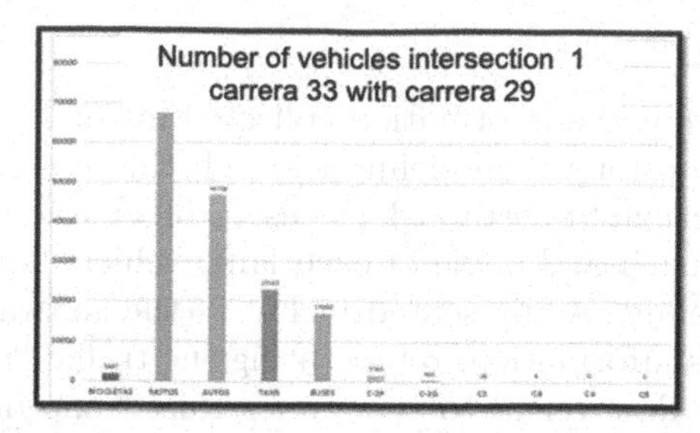


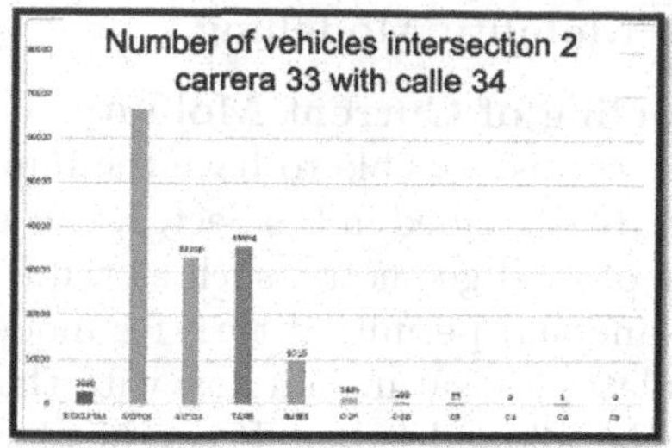


**Fig. 5.** Number of vehicles per week. Each color is associated with a specific means of transport.

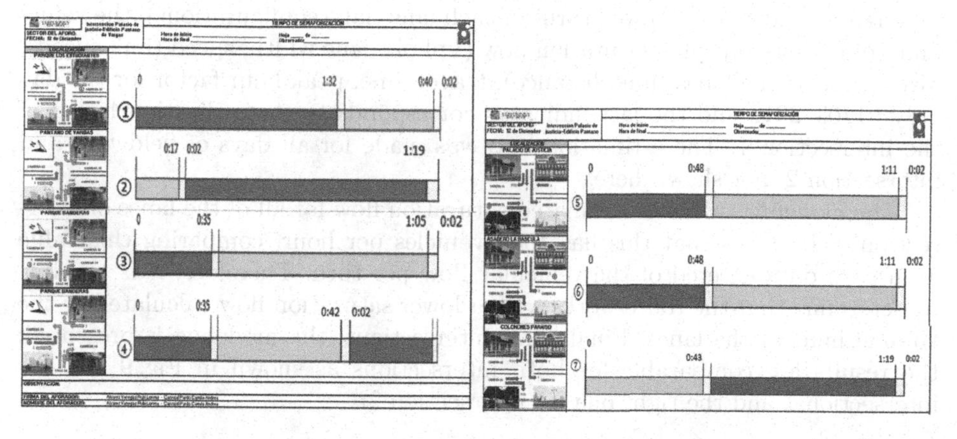

**Fig. 6.** Traffic lights phases for the intersections. Left, for intersection 1 and Right for intersection 2.

Figure 6, Left, shows the scheme of phases and time belonging to intersection 1. The most prolonged phase was on the third and fourth phases at Carrera 33 with Centro (Downtown) and Barrio Barzal. Likewise, it is possible to see that the second phase, located in front of the Pantano de Vargas building, has a section with less time in green than in red. On the other hand, in the four phases, it is possible to appreciate a similarity of seconds and the average totality of the cycle yellow is 106-second at intersection 1. For intersection 2, phases 5 and 6, as shown in Fig. 6 Left, which are located in Carrera 33 in double direction, are synchronized and more prolonged in 120 s. In the same way, it was possible to observe that the seventh phase, located in Calle 33b corresponds to the section with less time and has a duration in the green of 71 s. Furthermore, in the three phases, it is possible to appreciate a similarity of seconds in yellow light and the average totality of the cycle for the intersection in 120 s.

### 3.3 Mobility Modelling

**Modeling of Current Mobility.** From the information collected in the fieldwork, it was possible to have the inputs for the modeling through the Synchro software described in the Sect. 2. It is crucial to establish the essential characteristics of road geometry, such as the length and angle of each lane, vehicle flows per lane and permitted turn for modeling. As presented in Fig. 7, the software calculates the saturation flow with this information, representing the traffic flow and the allowed turns. The vehicular flow refers to the values raised one day corresponding in this study to the data collected at fieldwork for July 1, 2018, a day representing a maximum vehicular volume of 4,018, between the hours of 11:30 am to 12:30 pm.

Figure 8 shown the allowed turns of each lane (lane configurations), the vehicular volumes used (vph), saturation flow (vphpl), lane width, speed (link speed), distance from the lanes (link distance), travel time, peak hour factor for the day July 01/03/2018 and the last indicator corresponds to the utilization factor of the intersection 1. The estimation also was made for all days of fieldwork and intersection 2; not shown here.

The modelings give the estimated saturation flow for all of the lanes of intersection 1. It shows that this has 1900 vehicles per hour, comparing this value with each data entered of the vehicular flow per turn. Therefore, it is possible to determine that the route works with a lower saturation flow calculated as the volume limit of the lanes. Finally, for intersections, the modeling is presented; the results are comparable for both intersections as shown in Fig. 9 left, for intersection 1 and the right one for intersection 2.

**Level of Service.** To model mobility, it is crucial to consider the service level of a road. This category explains how an intersection works and how much additional capacity is available to handle the traffic fluctuations. This classification is estimated using protocols (Highway Capacity Manual, 2010 [9]) and permits the analysis of the conditions expected at the intersection.

The analysis showed that the intersection Carrera 33 with Carrera 29 (Intersection 1) at "La Bascula-Juzgados Restaurant" presents a general level of service

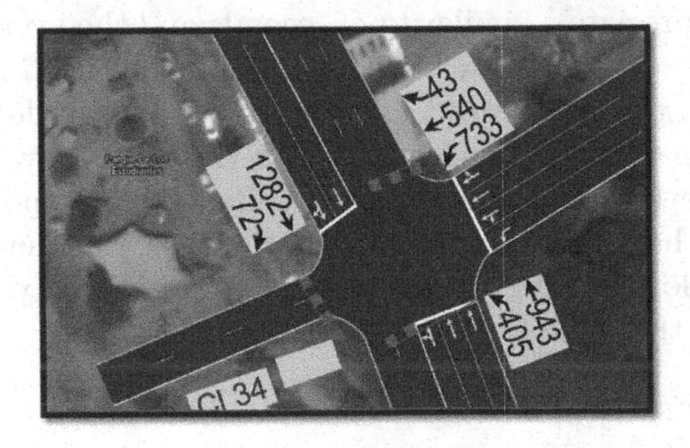


**Fig. 7.** Traffic flow and allowed turns simulation for Intersection 1.

| Lane Group | EBL | EBT | EBR | WBL | WBT | WBR | NBL | NBT | NBR | SBL | SBT | SBR |
|---|---|---|---|---|---|---|---|---|---|---|---|---|
| Lane Configurations | | | | | | | | | | | | |
| Volume (vph) | 0 | 0 | 0 | 733 | 540 | 43 | 405 | 943 | 0 | 0 | 1282 | 72 |
| Ideal Flow (vphpl) | 1900 | 1900 | 1900 | 1900 | 1900 | 1900 | 1900 | 1900 | 1900 | 1900 | 1900 | 1900 |
| Lane Width (m) | 3.6 | 4.6 | 3.6 | 3.6 | 3.3 | 3.6 | 3.6 | 3.6 | 3.6 | 3.6 | 3.6 | 3.6 |
| Link Speed (k/h) | | 20 | | | 20 | | | 20 | | | 20 | |
| Link Distance (m) | | 50.2 | | | 103.3 | | | 114.6 | | | 135.0 | |
| Travel Time (s) | | 9.0 | | | 18.6 | | | 20.6 | | | 24.3 | |
| Peak Hour Factor | 0.81 | 0.81 | 0.81 | 0.81 | 0.81 | 0.81 | 0.80 | 0.81 | 0.81 | 0.81 | 0.81 | 0.81 |

**Fig. 8.** Traffic flow variables and lane configuration abstract estimated by Synchro for Intersection 1.

**Fig. 9.** Vehicle flows per lane and allowed turns of the current state. Left image, for intersection 1. Right image, for intersection 2.

D. The level of service D represents a circulation of high density, only stable. Speed and freedom from maneuvers are severely restricted, and the driver experiences a low general level of comfort and convenience. Small increments in the flow usually cause problems with functioning. Additionally, the intersection of Carrera 33 with Calle 34 (2) is on a service level "F" (low). It means the intersection is not working optimally, and it represents adverse effects on mobility and, therefore, on user's daily lives.

**Mobility Solutions Analysis.** Based on the previous results supported by the software Synchro 8, considering the experiences of other projects, three models of a solution have been proposed and analyzed to improve mobility in intersections 1 and 2.

- Solution 1-Optimize the Semaforization Cycles. The modeling shows that this solution improves from "F" to "E" in some lanes in the entire study zone. Therefore, it was determined that optimizing only the signalization cycles is not an efficient solution. This solution is represented in Top Fig. 10.
- Solution 2-Exchange Type of Intersection. This proposal would also negatively impact citizens' daily lives because it will be necessary to closc the intersections to construct the roundabouts. In addition to a significant purchase of buildings, which would be a great inconvenience, this route is of great

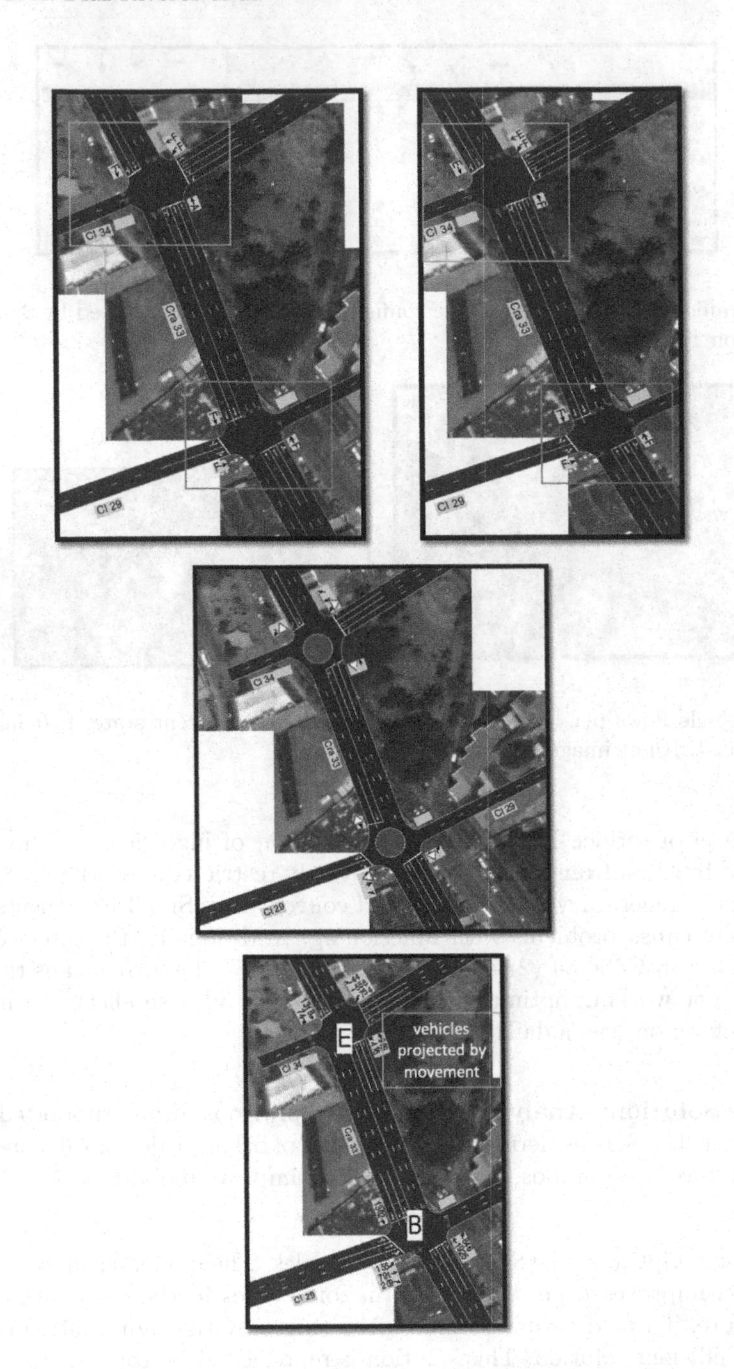


**Fig. 10.** Models of Solution. Top pictures represents optimizing the semaforization cycles. Middle picture shown the modeling for the exchanging type of intersection. Bottom picture is related to the implementation of new lanes solution

importance for the connectivity that the center represents with a large part of Villavicencio. The proposal simulation is shown in middle Fig. 10.
- Solution 3-Implementation of new lanes. This proposal seeks to increase the road capacity and thus improve the level of service, reducing travel times and improving users' convenience.
  Thanks to the modeling and simulation, it could visualize that the proposed solution is efficient. It improves the level of service from the intersections 1 in E and a service level B to the other intersections 2. Service level B, corresponds to a stable flow range, even though you can take pleasure in observing other vehicles that are part of the circulation. On the other hand, the level of service E corresponds to a level of operation close to the limit of the capacity. Therefore, this is considered a viable alternative. The simulation of solution 3 is represented in Bottom Fig. 10. The results support the planning projects focusing on smart mobility in Villavicencio. The purpose is to continue supporting the local government from the academy in the next step that considers traffic lights infrastructure for smart mobility.

## 4 Conclusions

- In path 1 in, the sub-section with the lowest speed is in Arquitectura - Pantano de Vargas where most of the time, especially at nighttime rush hour, there is greater damming of cars, either because the buses are parked to wait for the Unimeta students who use this transport, or because of the gauges that are in charge of giving information to public transport drivers on how long it takes a bus to another or through the traffic light at the corner of the Vargas swamp.
- When making the tables with the speeds of the bus, taxi and private car, it can be observed that in all the samples the buses always presented the lowest speed since they have the habit of parking in different areas to wait for passengers, which makes mobility difficult in the sector by obstructing the road.
- The speed obtained in the taxi route was sometimes higher than that of the individual, since in general taxi drivers are more impatient and not very cautious to walk while in the car the driver was more cautious and did not have a desire to get there to his destiny.
- A comparison was also made with the Google Maps application and the route in the private car in the two sections studied, where the application times were always shorter than the real ones, always giving a 2-min difference. It can be inferred that the application is not very accurate because it does not have complete and exact information on the traffic in the area. Even so, it is a very useful tool that allows us to approximate the time it may take to move from one point to another, especially in very long stretches.

## References

1. Smart Cities y Acústica Sostenible. http://ww.editorialbonaventuriana.usb.edu.co. Accessed 16 June 21

2. Ochoa, N., et al.: Revista Facultad de Ingeniería Universidad de Antioquia **93**, pp. 41–56 (2016)
3. Transforming our world: the 2030 Agenda for Sustainable Development. https://sdgs.un.org/2030agenda. Accessed 16 June 21
4. World Urbanization Prospects. https://population.un.org. Accessed 16 June 21
5. Benevolo, C., Dameri, R.P., D'Auria, B.: Smart Mobility in Smart City. Lecture Notes in Information Systems and Organisation 11 (2016)
6. Stolfi, D.H., Alba, E.: Sustainable Transportation and Smart Logistics: Decision-Making Models and Solutions: Sustainable Road Traffic Using Evolutionary Algorithms. Elsevier, Amsterdam (2018)
7. Sánchez-Vanegas, M.C., et al.: Towards the Construction of a Smart City Model in Bogotá. In: ICAI Workshops, pp. 176–193 (2020)
8. Miralles-Guasch, C.: City and transportation. the imperfect binomial. Ariel Geografía, Barcelona, p. 256 (2002)
9. MLA, HCM 2010: Highway Capacity Manual. Washington, D.C., Transportation Research Board (2010)
10. Merodio, G.G.G.: Geografía histórica y medio ambiente, Temas Selectos de Geografía de México. (I.1.9), Instituto de Geografía, UNAM, México, p. 111 (2012)
11. Pinzón, A., Mauricio, C., Parrado, H., Alejandro, Ó.: Estudio de intersección semaforizada entre calle 15 y carrera 45, tesis pregrado ing. civil Universidad Cooperativa (2018)
12. Wanumen, L., Moreno, J., Florez, H.: Mobile based approach for accident reporting. In: International Conference on Technology Trends, pp. 302–311 (2018)
13. Velosa, F., Florez, H.: Edge solution with machine learning and open data to interpret signs for people with visual disability. In: CEUR Workshop Proceedings, pp. 15–26 (2020)

# Preliminary Studies of the Security of the Cyber-Physical Smart Grids

Luis Rabelo[1](✉), Andrés Ballestas[2], Bibi Ibrahim[1], and Javier Valdez[1]

[1] University of Central Florida, Orlando, FL 32816, USA
luis.rabelo@ucf.edu
[2] Universidad de La Sabana, Chia, Cundinamarca, Colombia

**Abstract.** The electric power network is a cyber-physical system (CPS) that is important for modern society. Smart Grids (SG) can improve the electric power system's profitability and reliability by integrating renewable energy and advanced communication technologies. The communication network that connects numerous generators, devices, and controllers distributed remotely plays a vital role in controlling electrical grids, and the trends favor the implementation of IoT devices. However, it is vulnerable to cyber-attacks. This paper presents our current studies of potential malware that can attack these power systems and investigate future strategies to improve that security and risk management. The paper explains as the first step the explanation of a Delphi with experts from the top security companies in order to understand the trends and malware beyond an extensive literature survey. Also, the paper studies the propagation of the selected malware using System Dynamics. Finally, conclusions are provided, allowing future work to be accomplished using multiple resolution modeling (MRM) (i.e., a form of distributed simulation).

**Keywords:** Smart grids · IoT · Cybersecurity · Delphi · System dynamics

## 1 Introduction

As we know them today, cities consume more than 75% of the world's energy production. Some studies also indicate that by 2050 it will be residing in these cities, 85% of the population [1]. Electricity is the most versatile and known form of energy. Infrastructure is necessary to generate, transmit, and distribute electricity. This infrastructure was conceived and designed over 100 years ago. This infrastructure has served us well and has been an essential contribution to industrialization and economic growth. There is almost no process in the industry or application in personal life that does not use electricity. Electricity demand is growing faster than any other form of energy, especially in rapid industrialization areas, such as China and India. Simultaneously, economies' growing digitalization imposes greater demands on the electricity supply's reliability; even minor interruptions cause substantial economic losses.

A smart grid (SG) is a system that allows bidirectional communication between the final consumer of energy, whether a private user or industry and electricity companies.


H. Florez and M. F. Pollo-Cattaneo (Eds.): ICAI 2021, CCIS 1455, pp. 449–461, 2021.
https://doi.org/10.1007/978-3-030-89654-6_32

The information obtained in this communication process allows electricity companies to perform a more efficient electricity grid operation [2].

The communication network that connects numerous generators, devices, and controllers distributed remotely plays a vital role in controlling electrical networks. However, it is vulnerable to cyber-attacks. In particular, due to its frequency and lasting impact, the distributed denial of service (DDoS) attack represents a significant threat to SGs [3].

Another related concept is the one of the Internet of Things (IoT). The IoT is a heterogeneous network involving different devices, such as electronic devices, mobile devices, industrial equipment, etc. Different devices may have different communication platforms, networks, data processing, storage capabilities, and transmission power. These devices must be connected using communication and network protocols to communicate and cooperate to share their data (Fig. 1).

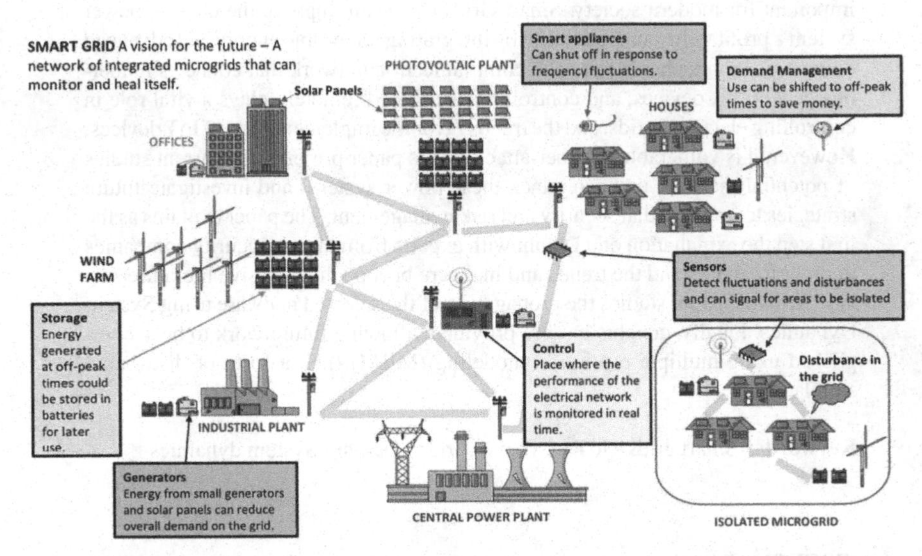


**Fig. 1.** A comprehensive view of a smart grid (SG) with the different components

Cybersecurity in electrical systems (and IoT) is vital since modern society's functioning depends on them. For example, cutting off the electrical energy supply could cause severe consequences in hospitals, the industrial sector, airports, telecommunications, etc. Furthermore, because new energy systems incorporate data transmitted and access to a network, some vulnerabilities are possible that hackers can attack and impact a government or a private company. Therefore, cybersecurity in electrical systems is an important area of research. An excellent example of this type of research is Yin [4], which develops network intrusion detection systems (NIDS) for IoT-based systems. Yin [4] describes the concept of an IoT-based system for Smart Grids using the SCADA/DNP3 communication protocol. Therefore, the NIDS supports the development of a security layer to protect the SCADA system. First, a new method to assess the DNP3 protocol's vulnerability is discussed, providing an idea of where to carry out the attack ("white hacker"). Then, based on the vulnerabilities discovered, they developed an attack model aimed at the data link layer, the transport link layer, and the application link layer of the DNP3 protocol. Finally, it presents experimental results, showing that the proposed method could detect intrusions in the SCADA system and classify them with detailed information on the compromised fields of the DNP3 package [5].

## 2 Delphi Methodology

After an extensive literature survey conducted between 2017 and 2019, we decided to validate the literature review with a Delphi method due to the rapid developments in this field. It is well recognized that a literature review usually has two to five years behind current developments. The conclusions of this literature survey that involved more than 50 documents were that given the increasing penetration of IoT devices across the spectrum of daily activities and critical infrastructures (e.g., smart grids), cybersecurity incidents are increasing. Although threats continue to evolve and target devices with new techniques, by 2021, it is possible to identify these threats in three main categories: Exploits or security holes, User-related practices, and Malware.

The Delphi method is included within the prospective methods, which study the future evolution of the factors of the techno-socioeconomic environment and their interactions. Delphi's predictive capacity is based on the systematic use of an intuitive judgment issued by a group of experts. The objective of the successive questionnaires is "to reduce the interquartile space. It is important to observe how much the expert's opinion deviates from the whole's opinion, specifying the median" [6] of the answers obtained. It allows obtaining information from points of view on very broad or very specific topics. The Delphi Exercises are considered "holistic," covering a wide variety of fields. Three characteristics make the sector vulnerable to cyber threats. The first focuses on public sector companies' many threats, such as cybercriminals, political factors, or external terrorism, to provoke security and economic problems in society and profit. Second, the expansion of public sector companies brings this geographic, and organizational complexity, including that cybersecurity leadership has a decentralized nature. The third is the electrical system's interdependence with the existing infrastructure and cybernetics elements [6].

Cybercriminals prefer to attack the public sector and other critical factors to obtain an economic profit from this criminal activity. For instance, a company in the electricity sector in Puerto Rico was subjected to cyberattacks on its IoT devices, which cost the company 400 million dollars [7]. However, as electronic devices are deployed in the electrical system and connected to the Internet, intelligent meters will contain and eliminate threats. At present, the attacks have occurred in different sectors of government agencies. For example, recently, ransomware affected a gas company causing loss of income [8].

### 2.1 Interviews

We decided to study the potential malware trends affecting IoT implementations, such as smart grids and microgrids. Therefore, experts from several top cybersecurity companies such as Mcafee (https://mcafee.com), Norton (https://us.norton.com/), and other cybersecurity consulting organizations were interviewed. After the experts' panel was selected, a list of the most common types of cybersecurity threats was identified. Then, the panelists' field of experience was compared against the classifications found, and the scope was narrowed to focus on the taxonomy of malicious activities.

**Round 1**

In the first round, emails were sent to nine experts to identify three potential malware attacks that commonly attack IoT devices. These results are demonstrated in Table 1.

**Table 1.** Results of first round

| Participants | Expert 1 | Expert 2 | Expert 3 | Expert 4 | Expert 5 | Expert 6 | Expert 7 | Expert 8 | Expert 9 |
|---|---|---|---|---|---|---|---|---|---|
| Option 1 | Mirai Botnet | Intel Spoiler | Emotet | BrickerBot | Mirai Botnet | Mirai Botnet | Stuxnet | TrickBot | Moose/Elan |
| Option 2 | Stuxnet | Meltdown | muBot | Mirai Botnet | Stuxnet | Stuxnet | Mirai Botnet | Psyb0t | Meltdown |
| Option 3 | Bashlite | Mirai Botnet | Hydra | IoT Troop / Reaper | IoT Troop / Reaper | BrickerBot | BrickerBot | Moose/Elan | Mirai Botnet |

**Round 2**

In the second round, a list of the most common malware found within the IoT environment was given to the same group of experts and was asked to rank the top five most harmful malware. This round aimed to identify if the experts' opinions significantly varied from the first round. The results are presented in Table 2.

**Table 2.** Results of second round

| Participants | Expert 1 | Expert 2 | Expert 3 | Expert 4 | Expert 5 | Expert 6 | Expert 7 | Expert 8 | Expert 9 |
|---|---|---|---|---|---|---|---|---|---|
| Option 1 | Meltdown | Psyb0t | Emotet | Moose/Elan | Moose/Elan | Stuxnet | Bashlite | Mirai Botnet | Moose/Elan |
| Option 2 | BrickerBot | Intel Spoiler | muBot | Mirai Botnet | Stuxnet | Hydra | Mirai Botnet | Psyb0t | Stuxnet |
| Option 3 | Mirai Botnet | Meltdown | Hydra | BrickerBot | IoT Troop / Reaper | Mirai Botnet | Stuxnet | Moose/Elan | Mirai Botnet |
| Option 4 | Stuxnet | Mirai Botnet | Stuxnet | Iot Troop /Reaper | Mirai Botnet | Iot Troop /Reaper | BrickerBot | TrickBot | Emotet |
| Option 5 | Bashlite | Stuxnet | Mirai Botnet | Intel Spoiler | BrickerBot | BrickerBot | muBot | Intel Spoiler | Meltdown |

**Round 3**

In the last round, the experts were asked to identify the most potential impact malware. Again, the majority of the experts agreed that the malware Mirai Botnet was the most harmful. The results are presented in Table 3.

**Table 3.** Results of the third (final) round

| Participants | Expert 1 | Expert 2 | Expert 3 | Expert 4 | Expert 5 | Expert 6 | Expert 7 | Expert 8 | Expert 9 |
|---|---|---|---|---|---|---|---|---|---|
| Final selection | Mirai Botnet | Mirai Botnet | Stuxnet | Iot Troop /Reaper | Mirai Botnet | Mirai Botnet | Stuxnet | Mirai Botnet | Mirai Botnet |

## 3 Results

With the Delphi methodology carried out in this study, it was obtained that the malware that would be one of the main future threats to intelligent electrical systems will be the malware of the type Mirai Botnet. A description of this malware is given below.

**Description of the Mirai Botnet**

The Mirai Botnet was built and deployed in several campaigns in 2017 [9]. As a result, it has infected hundreds of thousands of IoT devices.

The attacker (Botmaster) initiates the process by scanning the IoT devices' Internet (Fig. 2). The code is then initiated to execute the module scanner, c, which starts to scan the Internet, searching for vulnerable IoT devices with open Telnet ports 23 or 2332. Once a device is detected, the malware tries to forcefully log in to the device using a list of 62 default usernames and passwords. The login session credentials and the device's information are sent back to the report server if successful. The scanner server will use this information later to initiate the session. An infection command is sent from the control server to the scanner server, containing all the necessary IP direction and hardware architecture. Mirai accepts multiple hardware architectures, which include ARM, MIPS, Sparc, and PowerPC. The scanning server uses this information to initiate the session and indicates the vulnerable device to establish a communication connection and discharge and execute the binary file. Once executed, the first IoT device infected becomes part of the Mirai botnet and can communicate with the control server. The binary file is eliminated and is only executed inside the memory to avoid detection. The botmaster can now issue attack commands, specifying the attack duration and objective parameters. The malware includes ten types of DDoS attacks, including overloading the UDP, DNS, and the SYN and ACK packages that can be used to attack another user on the Internet. The first bot tries to repeat the infection process and spread the bots' network by scanning the Internet for additional vulnerable IoT devices with open Telnet and ports. The new information of the device is sent back to the control server. A new infection command is emitted to the scanning server. The binary file is loaded on the new IoT device discovered. The search for additional IoT devices is repeated to expand the botnet.

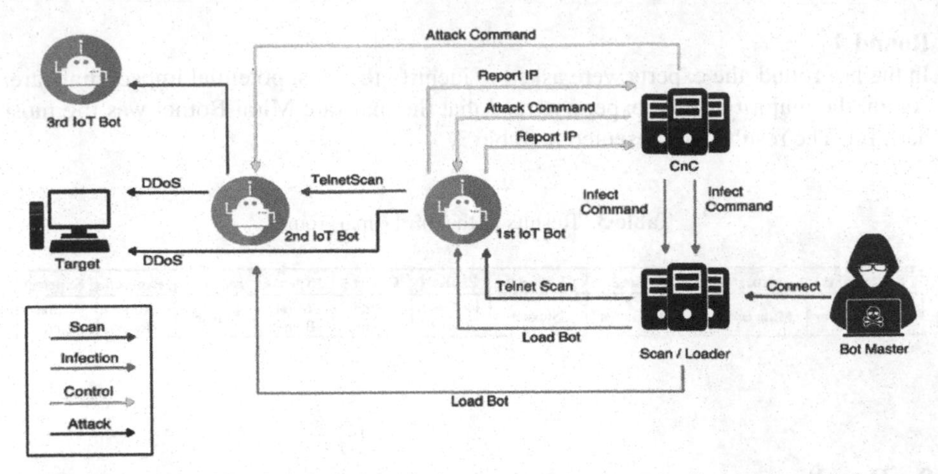


**Fig. 2.** Description of Mirai Botnet with the respective sequence of activities and flows of information/commands

With the continuous integration of the Internet of Things (IoT) and increasing device connectivity, threat actors continue to focus their efforts against information technology (IT) networks. Therefore, identifying one of the future threats could give us the advantage of studying it and being forewarned to protect the elements used by the electrical systems connected to the Internet.

**Using System Dynamics, the SEIR Model, and the Satori Botnet**
The Mirai Botnet was studied. This botnet has generated a family of malware. We decided to investigate one of the latest botnets based on the Mirai Botnet: the Satori Botnet [10]. The Satori Botnet is an evolution of the Mirai Botnet, and it was used in several campaigns starting in June 2018. We modeled one of these campaigns to get some of the parameters of propagation. The differential equations of Fig. 3 (as recommended by model [11] the Mirai Botnet) were utilized to model this propagation. However, we have to modify and reduce them to three differential equations to model using system dynamics.

The structure of equations in Fig. 3 represents behavior. Behavioral models in epidemiology simplify the mathematical modeling of infectious diseases. The population is assigned to compartments with labels, for example, S, I, or R (Susceptible, Infectious, or Recovered). People can progress between different states. The order of the labels generally shows the flow patterns between the states; for example, S stands for Susceptible, Exposed (E), Infectious (I), Recovered (R), then Susceptible (S) again. Therefore, this model is denominated SEIR (see Fig. 4). The infectious rate, $\beta$ beta, controls the rate of spread that represents the probability of transmitting disease between a susceptible and infectious individual. The incubation rate, $\sigma$ sigma, is the rate of latent individuals that become infectious (the average duration of incubation is $1/\sigma$). The recovery rate, $\gamma$ gamma, is determined by the infection's average duration, D. For the SEIRS model, $\xi$ (Xi) is the rate that recovered individuals return to the susceptible state due to loss of immunity. The models are run more frequently with ordinary differential equations

(1)
$$\frac{dS}{dt} = \lambda(1-\gamma) + \mu R(t) - \frac{\alpha I(t)S(t)}{N(t)} - \Phi_3 S(t)$$
(2)
$$\frac{dE}{dt} = \frac{\alpha I(t)S(t)}{N(t)} - (\beta + \Phi_2)\, E(t)$$
(3)
$$\frac{dI}{dt} = \beta E(t) - \Phi_1 I(t)$$
(4)
$$\frac{dR}{dt} = \lambda\gamma + \Phi_1 I(t) + \Phi_2 E(t) + \Phi_3 S(t) - \mu R(t)$$
where:

(5)
$$N(t) = I(t) + E(t) + S(t) + R(t)$$

**Fig. 3.** Differential equations of the IoT-BAI as presented by [11] where S is the susceptible population of IoT devices, E is the exposed population of IoT devices, I is the number of infected IoT devices, R represents the recovered devices, and N is the entire population of IoT devices of that particular ecosystem.

but can also be used with a stochastic framework, which is more realistic but more complicated to analyze. The models try to predict things like how a disease spreads, the total number of infected, or the duration of an epidemic, and to estimate various epidemiological parameters such as the reproductive number.

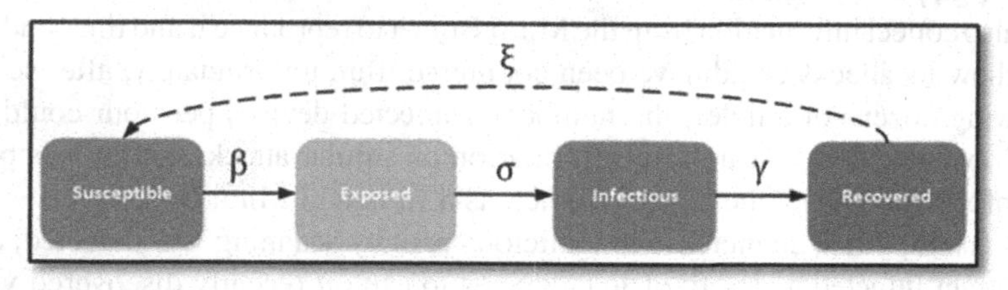


**Fig. 4.** The SEIR Model with the Nodes Susceptible, Exposed, Infectious, and Recovered and the respective transfer rates

The action of malware through a network can be studied through epidemiological models for the spread of diseases. Based on the SEIR model, it is possible to establish dynamic models for the propagation of malicious objects, providing estimates of the temporal evolutions of the infected nodes depending on the parameters and topological aspects of the network. The SEIR model was created to analyze the spread of the virus in networks, particularly social ones. Each node in the figure represents a user. The edge connects node i, and node j represents that user i and user j are related. This model defines the number of users in the contact list as the grade of this node. According to the virus propagation rules, the nodes are divided into four categories: susceptible (S), exposed (E), infectious (I), and recovered (R). Susceptible nodes represent those capable of contracting viruses; exposed nodes represent those infected but not yet infectious; the infectious points represent those infected capable of transmitting the disease. Finally, the recovered nodes represent those who are permanently immune. The SEIR model defines the virus spread rules as follows:

- If a susceptible node contacts an infectious node, then the probability that the susceptible node will be transmitted to an exposed node is P.
- An exposed node will transmit to an infectious node with speed without contact with any other node.
- An infectious node will not spread viruses incessantly. Instead, an infectious node will quickly transmit to a recovered node without contact with any other node.

**Using System Dynamics, the Satori Botnet, and Optimization**
A widely used simulation technique is system dynamics. It takes a slightly different approach to discrete event simulation, focusing more on flows around networks than on individual entity behavior. System Dynamics considers three main objects: stocks, flows, and delays. Stocks are primary warehouses for objects. Flows define the movement of items between different stocks in the system. Finally, the delays are the delay times between the system that measures something and the action on that measure. System Dynamics works with differential equations. Therefore, we implemented the SEIR model in System Dynamics to capture the structure of the selected malware (Satori Botnet) and its infectious pattern. There is one problem with this. You need to capture enough data of the selected malware to optimize the model and obtain the different transfer rates from the data (e.g., $\beta$, $\sigma$, $\gamma$, and $\xi$).

We also collect information from the Mirai Botnet to reproduce it and thus study it and analyze how its attacks could have been countered. But, unfortunately, after searching and viewing dozens of articles, the number of infected devices per hour could not be found. So we decided to search for information on similar attacks, and it was possible to find information about the Sartori Botnet, as a variant of Mirai Botnet.

On June 15, 2018, an increase in malicious activity scanning was detected, and the Satori Botnet infected a variety of IoT devices to exploit recently discovered vulnerabilities (Fig. 18). The payload, never seen before, is delivered by the infamous Satori Botnet, taking advantage of a worm-like form of propagation. As a result, an exponential increase was observed in the number of attack sources spread throughout the world and reaching a maximum of more than 2500 attackers in a single 24-h period. Interestingly, the Sartori Botnet only attacked the D-Link DSL-2750B router (https://support.dlink.com/list.aspx?type=1&model=DCS-93) and also liked the XiongMai uc-httpd 1.0.0 cameras (https://www.forbes.com/sites/thomasbrewster/2016/10/07/chinese-firm-xm-blamed-for-epic-ddos-attacks/?sh=22a3365242d7) that had to be withdrawn from the market by the Chinese XiongMai Company, about 10,000 of these cameras.

We fit the Satori Botnet's original campaign (we could get hourly information of its propagation) using Markovian Chain Monte Carlo Simulation [12] to get the transfer rates. Markovian Chain Monte Carlo Simulation is stochastic optimization and was used to adjust the proposed ranges of the investigation and reproduce it. Figure 5 and Fig. 6 show the model and the reproduction of the Satori Botnet attack with excellent accuracy.

The model uses a differential equation for the susceptible, another differential equation for the infected, and the third for the recovered devices. The parameter to be obtained from the stochastic optimization was the Factor from Susceptible to Recovered Before, the inverse denoting a time constant. Based on the literature, the range of this factor ranged from 0.01/hour to less than 0.2 /hour. Therefore, the optimization found for this factor the value of 0.033 / hours. Also, the optimization found the value of the Infected Factor at Recobrados ("Factor from Infected to Recovered Before") that resulted in 0.5 / hours (and the range was from 0.02 / hours to 0.2 / hours according to the literature.

Another critical value was finding the contact rate, and the optimization found it to be 0.45 (which is relatively high due to the vulnerabilities found in the attacked devices). For its part, (Fig. 6) shows the result after stochastic optimization. Again, the red line is the actual curve of the Satori Botnet propagation, and the green line of the graph shows those reproduced by the System Dynamics model developed.

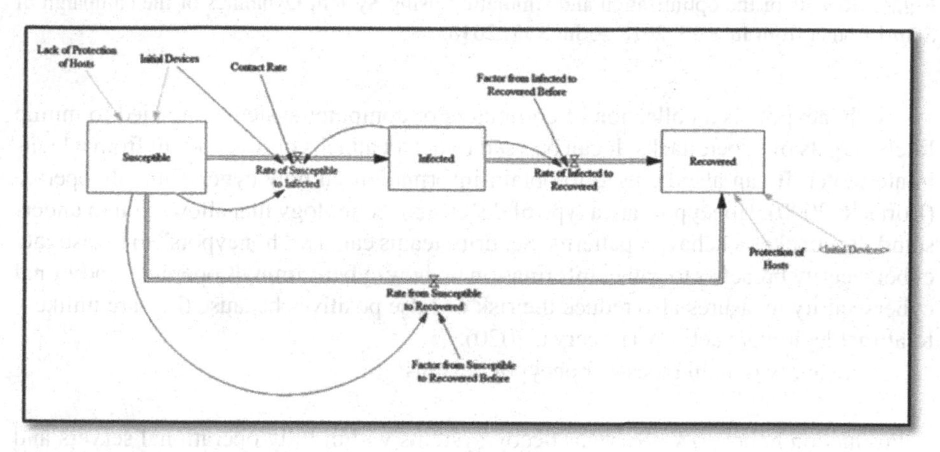


**Fig. 5.** The model was developed in System Dynamics using three differential equations represented by Susceptible, Infected, and Recovered to simulate the behavior of the Satori Botnet. The solving method to execute the three differential equations in parallel was Euler.

This model implemented in System Dynamics can provide, using sensitivity analysis, more details of the malware diffusion. Furthermore, this model can support the analysis of the different policies to avoid exploitation by hackers.

**Sensitivity Analysis**

Now, with the model built, many scenarios can be conducted. One of them is to assume an attack threshold (limit) to detect attack activities to try to infect devices. This "Attack Threshold" is the number of unique IP addresses within a four-hour interval that could indicate that there is an attack in progress. So a sudden increase in unique IP addresses within periods of four hours could indicate a DDoS attack. This implementation can be done with "honeypot" systems to keep statistics using dashboards with the respective metrics in real-time.

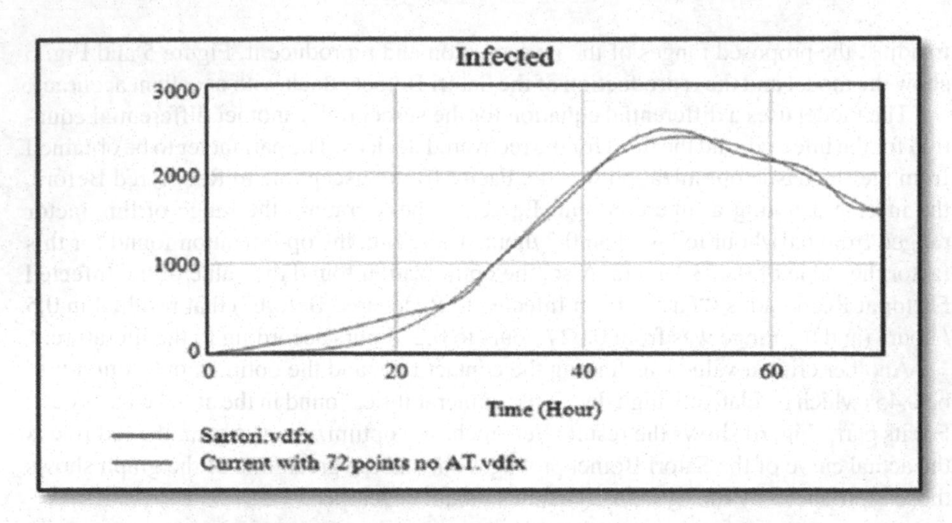


**Fig. 6.** Results of the optimization and simulation using System Dynamics of the campaign of Satori Botnet from June 15, 2018, to June 17, 2018.

A "honeypot" is a collection of computers or computer systems intended to mimic likely targets of cyberattacks. It can be used to detect attacks or divert them from a legitimate target. It can also be used to obtain information on how cybercriminals operate (Loras R, 2000). Honeypots are a type of deception technology that allows you to understand the attacker's behavior patterns. Security teams can use "honeypots" to investigate cybersecurity breaches to gather information on how cybercriminals operate. Traditional cybersecurity measures also reduce the risk of false positives because they are unlikely to attract legitimate activity (Imperva, 2020).

There are two main types of honeypot designs:

- Production honeypots - Serve as decoy systems within fully operational servers and networks, often as part of an intrusion detection system (IDS). They divert criminal attention from the entire system while analyzing malicious activity to help mitigate vulnerabilities.
- Research honeypots, used for educational purposes and to improve security. They contain traceable data that you can track when stolen to analyze the attack (Loras R, 2000).

Figure 7 shows the results for the attack thresholds of 1000 and 1500 infected IPs. Figure 7 shows the graphs, and the blue line corresponds to a response to the attack when it reaches 1000 infected IPs and the red line to a response to the attack when the attack threshold reaches 1500 infected IPs. These attack thresholds significantly limit the potential damage from the malware.

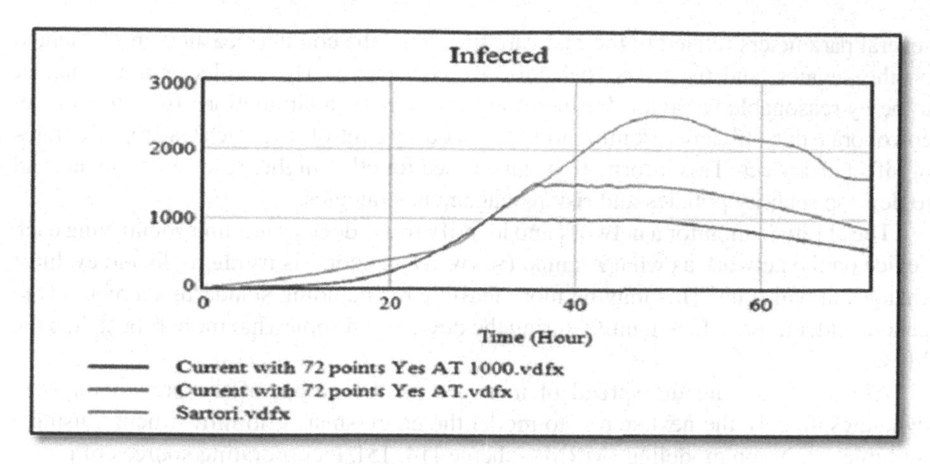


**Fig. 7.** Results using Attach Thresholds of 1000 (blue line) and 1500 (red line). The green line is the original Satori Botnet. (Color figure online)

## 4 Conclusions

Both the internet and technological advances have provided significant progress in society in recent years. Still, they have also been exposed to new dangers that did not exist before, such as cyberattacks. Cyberattacks are increasingly numerous and sophisticated. Moreover, they continue to evolve so that traditional analysis tools no longer work, and the only solution is to improve security, perfecting prevention, analysis, and mitigation strategies.

Malignant programs and the respective vulnerabilities they can exploit, either at the hardware or software levels, remain a weakness in implementing modern systems in any economic sector. Since the continuous evolution and generation of malware as intended by attackers cannot be eliminated, management or technology decision-makers must invest in advancements related to secure IT environments to minimize threats and take corrective measures regarding potential malicious software attacks.

There are two primary sources of information in the literature found: the commercial one supplied mainly by cybersecurity companies that research to inform and generate needs regarding threats on the web in readers and interested parties and, on the other hand, the academic, in which it is intended from the research to understand and model the behaviors of harmful malware to be able to carry out different actions of mitigation proposals. This work involved consulting both sources to generate a consolidated knowledge input about the impact of malware on the new IoT ecosystem, a fact recorded during the documentation of the state of the art and theoretical framework.

Based on the opinion of experts from the top cybersecurity companies mentioned above, it was possible to identify that the malware Mirai would be one of the main future threats to IoT devices in Smart grids. One area to explore more is ransomware [13].

In this work, an SEIR mathematical model is introduced to simulate the propagation of the Mirai Botnet through a computer network. Also, the SEIR model was reduced to a SIR model to simulate the Satori Botnet attack (2018) accurately. In this model, several parameters related to the malware life cycle, the countermeasures implemented on the devices, and the users' behavior are considered. The results obtained appear to be by reasonable behavior. Furthermore, the numerical simulations from this model corroborate that efficient risk mitigation and security control strategies lead to lower rates of infected devices. This information can be used for other higher resolution models and to develop security policies and risk management strategies.

The ability to monitor a network and identify rogue devices requires identifying each device on the network as what Akamai (www.akamai.com) is trying to do and evaluate changes in real-time. This may be more feasible by including standardized methods to generate identifiers when manufacturing the device (and somewhat more robust than the MAC).

Already following the spread of malware (as the output of the previous system dynamics model), the next step is to model the entire smart grid infrastructure using a multiple resolution modeling (MRM) scheme [14, 15], incorporating sources of power generation, distribution, and consumers. MRM can then model the different components (with agent-based simulations), including cyberattacks. After that, we can improve risk management strategies. In terms of development and MRM for modeling smart grids, high-resolution models that specifically describe the fundamental phenomenon are desirable. "However, building high-resolution models with its insatiable demand for data is expensive, time-consuming, and often, is not available or is not reliable" [14]. But we can make MRM in a hierarchical form that contains different levels of abstraction. These different abstraction models will allow a certain level to focus on the critical information required by the parts. Stakeholders and relegate irrelevant data to other levels. MRM enables designers to identify the conceptual and analytical model and entity breakdowns. The importance of MRM is based on the need for different levels of resolution to understand the problem from various time dimensions, space, detail, actors, and policies (e.g., blockchain).

Finally, it is necessary to analyze, record, and notify each of the incidents detected, promote the culture of cybersecurity, share the lessons learned to implement a more effective defense, and develop policies focused not only on the recovery of systems but also on the economy and of the society after an attack.

## References

1. Steinbrink, C.S.: Future perspectives of co-simulation in the smart grid domain. In: IEEE International Energy Conference (2018)
2. El-Hawary, M.E.: The smart grid state of the art and future trends. Electric Power Components Syst. **42**, 239–250 (2014)
3. Yi, X.: Simulation-based stochastic programming to guide real-time scheduling for smart power grids under cyber-attacks. In: Proceedings of the 2018 Winter Simulation Conference, pp. 1180–1191 (2018)
4. Yin, X.C.: Toward an applied cyber security solution in IoT-based smart grids: an intrusion detection system approach. J. Sens. **19**, 2–22 (2019)
5. Abdul Wahid Mir, R.K.: Security gaps assessment of smart grid-based SCADA systems. J. Inf. Comput. Secur. 434–452 (2019)

6. Lange, T., et al.: Comparison of different rating scales for the use in Delphi studies: different scales lead to different consensus and show different test-retest reliability. BMC Med. Res. Methodol. **20**(1) (2020). https://doi.org/10.1186/s12874-020-0912-8
7. McKinsey & Company: Mckinsey (2020). Source from https://www.mckinsey.com/business-functions/risk/our-insights/the-energy-sector-threat-how-to-address-cybersecurity-vulnerabilities
8. Krebs, B.: FBI: Smart meter hacks likely to spread, Krebs on Security, krebsonsecurity.com, 12 April 2012
9. Alert (AA20–049A): Ransomware impacting pipeline operations," Cybersecurity and Infrastructure Security Agency, 18 February 2020
10. De Donno, M., Dragoni, N., Giaretta, A., Spognardi, A.: DDoS-Capable IoT Malwares: Comparative Analysis and Mirai Investigation. Secur. Commun. Networks (2018). https://doi.org/10.1155/2018/7178164
11. Radware: Satori IoT Botnet Variant. https://security.radware.com/ddos-threats-attacks/threat-advisories-attack-reports/satori-iot-botnet/
12. Gardner, M.T., Beard, C., Medhi, D.: Using SEIRS epidemic models for IoT botnets attacks. In: DRCN 2017 - 13th International Conference on Design of Reliable Communication Networks (2017)
13. Chib, S., Greenberg, E.: The American Statistician, vol. 49, No. 4, pp. 327–335, Nov. 1995
14. Arghire, I.: Ransomware operators demand $14 million from the power company, SecurityWeek, 2 July 2020, securityweek.com.
15. Lee, K., Lee, G., Rabelo, L.: A systematic review of the multi-resolution modeling (MRM) for integration of live, virtual, and constructive systems. Information **11**(10), 480 (2020). https://doi.org/10.3390/info11100480
16. Cortes, E., Rabelo, L., Sarmiento, A.T., Gutierrez, E.: Design of Distributed Discrete-Event Simulation Systems Using Deep Belief Networks. Information, **11**(10), 2078–2489 (2020)

# Software and Systems Modeling

# Enterprise Modeling: A Multi-perspective Tool-Supported Approach

Paola Lara Machado[1,2(✉)], Mario Sánchez[1,2], and Jorge Villalobos[1,2]

[1] Technical University of Eindhoven, Eindhoven, The Netherlands
p.lara.machado@tue.nl
[2] Universidad de los Andes, Bogotá, Colombia
{mar-san1,jvillalo}@uniandes.edu.co

**Abstract.** Enterprises are inherently complex systems, which involve a multitude of components working dynamically and in coordination to obtain a desired end goal. The intricacy of an enterprise and how the different components can be modeled is the area of research of Enterprise Modeling (EM). When holistically modeling an enterprise, it is required to use more than one modeling language, including general and domain specific modeling languages (DSML), to boost the comprehension of different enterprise areas. However, the use of multiple languages can lead to a lack of detail and understanding of the relationships between enterprise components that are modeled in different domains. Analyzing the relationships between various enterprise domains can generate valuable insight into understanding the uncertainty, complexity, and operations of an enterprise. In this paper, we present an approach to enable EM from multiple perspectives allowing enterprises to use different languages to describe various domains. This approach supports the creation of relationships and dependencies between domains, to analyze from different depths and viewpoints desired enterprise perspectives. Considering the needed requirements for multi-perspective modeling, we developed a workbench that allows the composition and merger of different EM languages and creates ad-hoc EM tools based on the necessities of the enterprise.

**Keywords:** Enterprise Modeling · Language workbench · Domain specific modeling languages

## 1 Introduction

A model is an abstraction of a system or artifact that is being studied [1]. Therefore, Enterprise Modeling (EM) is the process of creating a unified enterprise model that represents different aspects of the enterprise in its current or future state [2]. This model facilitates sharing a common vision among the different enterprise stakeholders and allows more efficient communication among them. Nevertheless, completely modeling an enterprise entails a complex process where smaller and specific models are used to represent different domains. These smaller models are viewpoints of particular enterprise components such as processes, actors, applications, infrastructure, etc. To create such enterprise models, there are two crucial elements: EM languages and modeling tools. EM


H. Florez and M. F. Pollo-Cattaneo (Eds.): ICAI 2021, CCIS 1455, pp. 465–480, 2021.
https://doi.org/10.1007/978-3-030-89654-6_33

tools facilitate the creation, support, and modification of the models, while the languages play a key role as they are used to create enterprise models. Generally, graphical modeling languages which use visual notation, are preferred in EM over textual languages.

EM tools have become fundamental in the field of EM as they must be able to support capabilities such as creating graphical diagrams based on a modeling language [3]. Some commonly used EM editors have been developed for both general modeling languages and enterprise DSML. Likewise, besides actually allowing models to be created and edited, the functionality of any modeling tool can also include functional components such as modeling viewpoints, tool customization, and extensibility.

Accordingly, enterprise modelers need to diagram a specific problem or context in one or more domains or perspectives to model the enterprise as a whole. Thus, when creating a complete enterprise model more than one language is needed to represent and properly describe the complexity of the enterprise and create different viewpoints. Experts favor these domain languages as they allow them to communicate and validate assignments in their domain [4]. Domain languages also promote the convenience and productivity of modeling and contribute to model quality and integrity with their special graphical notations [4].

Multi-perspective modeling, like multi-level modeling, offers a series of benefits over traditional modeling. It allows additional abstractions, that cannot be done with a single language approach. Additionally, multi-perspective modeling enables using concepts and terminology that correspond directly to the domains being modeled [5]. However, modeling different domains through multiple languages can create communication barriers in the understanding of different perspectives or viewpoints and the relationships between overlapping elements in different languages, although, possibly representing akin contexts. Analyzing the relationships between different viewpoints can generate valuable insight in understanding the complexity and operations of an enterprise. Creating composite viewpoints, which can jointly model different domains using a merger between languages, can heighten the understanding and communication of enterprise aspects among stakeholders. In a recent survey [6], 84% of practitioners call for metamodeling tools to enable defining languages using composition, for instance, by reusing semantical definitions of existing languages.

With the advances of new technological solutions such as intelligent information systems, IoT, industry 4.0, blockchain, new models of each of the emerging solutions appear. Enterprises adopt these new technologies, and they must integrate and change accordingly the enterprise's architecture. Through multi-perspective modeling, it is possible to analyze different integration viewpoints between different technologies and analyze how they interrelate. In this paper, we examine the possibility of multi-perspective EM through a tool-supported approach. The goal of our work is to develop a workbench that creates a modeling tool to facilitate and provide proper means to model different EM needs and perspectives. The created tool, named ModelFusion, allows merging different modeling languages and generating an ad-hoc modeling tool with the desired languages. The ModelFusion workbench contains two main components and a language repository. The first component, Greta, is an engine that takes the language definition and generates the modeling tool. The second component, Gromp, allows merging different EM languages to create different modeling perspectives.

The rest of the paper is structured as follows. Section 2 presents an overview of modeling in enterprises. Then, Sect. 3 presents the requirements for multi-perspective EM. Section 4 introduces our approach towards creating an EM language workbench. Section 5 illustrates a multi-perspective industry 4.0 scenario using our EM workbench. Lastly, Sect. 6 presents the main conclusions of the paper.

## 2 Modeling in Enterprises

EM is the process of creating holistic models which can describe multiple aspects of an organization such as its structure, goals, actors, processes, information systems, among others. The different perspectives that can be modeled are viewpoints of the enterprise that focus on a particular domain. Enterprise models are fundamental in the design and operations of an enterprise as they help deal with routine business challenges such as changes in an organization, understanding organizational dependencies, improving business processes, and analyzing and developing IT and its strategies [2]. The analysis methods and techniques applied to enterprise models are remarkably beneficial for understanding the current and future state of an organization even with ongoing enterprise shifts caused by different factors such as innovation, alliances, new market incursions, etc.

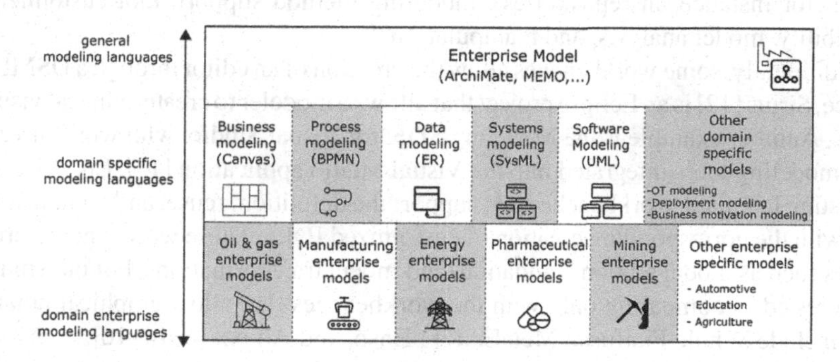


**Fig. 1.** Enterprise modeling: from general languages to domain specific languages

To build enterprise models, it is necessary to have EM languages and the corresponding tools that facilitate the model design. EM languages can have a general scope, or they can be domain specific as shown in Fig. 1. In EM, general-purpose languages (GPML) are modeling languages that cover a wide extension of domains, such as ArchiMate or MEMO, which allows modeling business, information systems, and technology elements. On the contrary, domain specific modeling languages (DSML) are used to model specific enterprise concerns. Respectively, it is possible to have specific DSMLs for particular industry needs such as in oil and gas, manufacturing, energy, and others, which can be an extension of other languages or stand-alone. GPMLs lack detail into specific domains, in which case, it is necessary to use more specific languages such as BPMN or SysML. These DSML cover the necessities of a particular domain with a higher level of abstraction allowing greater expressiveness. For instance, with the rise of trends such as industrial IoT and cloud manufacturing, that seek convergence of IT technologies in OT (operational technology) networks, the need to model OT is rising in asset intensive industries. OT modeling requires a DSML as it varies greatly throughout different enterprises such as in the oil and gas industry, in which OT encompasses oil pumps, vessels, reactors, valves, meters [7].

As mentioned previously, EM tools have become fundamental in the field of EM, as they support the capabilities of creating diagrams based on a modeling language [3]. Some commonly used EM editors have been developed for both GPML and DSML. For example, Enterprise Architect for modeling business and IT systems with UML [8], Archi for general EM with ArchiMate [9], Bizagi for modeling business processes with BPMN [10]. The most important functional aspect for any modeling tool is the model creation and editing. Additionally, tools should also ensure that the rules and guidelines provided by the language (abstract syntax) are adhered to. Most available EM tools focus on a single modeling language, which means that only models in the specific supported modeling language are possible [11]. Some EM tools, such as Archi, offer the functionality of creating viewpoints to facilitate the understanding of intricate large models. This also facilitates the use of the models and makes it easier to navigate through them. Moreover, some modeling tools can be extended and modified, for instance, by adding new viewpoints, languages, or internal functions, such as EnterpriseArchitect through third-party extension. Other EM tool functions include model storage (for instance, in repositories), modeling method support, tool customization, extensibility, model analysis, and manipulation.

Additionally, some workbenches allow the creation of an editor through a DSML. For instance, Sirius [12] is an Eclipse project that allows a modeler to create, edit and visualize models. Another example is the Modeling SDK for Visual Studio, with which users can create modeling tools integrated into the Visual Studio application [13]. Likewise, there are existing language workbenches that support the definition, reuse, and composition of DSLs with the corresponding creation of customized IDEs. These workbenches provide features such as model edition, validation, and model transformations. For our study, we are interested in comparing only with the workbenches that allow graphical notations. These include Whole Platform, MetaEdit +, Ensō, and Adoxx [6, 14–16].

## 3 Requirements of Multi-perspective Enterprise Modeling

To build enterprise models that can be documented and communicate different enterprise perspectives, we propose the following set of minimum requirements: multiplicity of modeling languages; composition between modeling languages; model validation; viewpoint creation; automated analysis methods; modeling tool support.

As previously mentioned, when creating enterprise models different modeling languages are used to model, understand, and accurately diagram the enterprise as a whole. The use of multiple languages allows an enterprise to express different features and components. Thus, the fundamental requirement to create a multi-perspective EM is the ability to analyze the enterprise from different depths and domains. Additionally, models must be validated and verified syntactically, to ensure integrity and compliance of the elements within. In conjunction with the multiplicity of modeling languages, the ability to relate one enterprise domain to another is a fundamental requirement. It should be possible to relate a higher-level or a lower-level enterprise domain to express different depths. Modeling different domains through multiple languages can create communication barriers in the understanding of different perspectives and the relationships between overlapping elements in different languages, although, possibly representing akin aspects. In broad terms, it is possible to map concepts from one EM language to another to create a bridge between modeling domains using overlapping concepts.

Many EM languages can be combined with other general or specific modeling languages [17–19]. ArchiMate, for example, can be used in unison with the business motivation model, balance scorecard, canvas, value maps, service blueprint, UML, SysML, ITIL, and many more. With all of these aforementioned languages, it is possible to create a mapping scheme in which elements from one language are represented by the elements of the other language (e.g., [20]). For instance, the business model canvas has key partners and customer segment elements which can both be represented through ArchiMate's business actor element.

Another important requirement is the creation of viewpoints. Viewpoints are used to show the different perspectives of an enterprise. Said viewpoints can help reduce complexity by including only the necessary elements instead of a complete and intricate model and enhance the comprehension of the modeled perspective. For example, assume an enterprise model for an organization contains elements such as the IT applications, infrastructure, services, process, and actors, a process viewpoint would only display the model's process and the relevant information for the processes.

To develop multi-perspective enterprise models, viewpoints must allow the use of one, two, or more EM languages that can express specific EM requirements or domains. The use of multiple modeling languages generally rules out the possibility of creating viewpoints that reference cross sections between different domains, which in turn might cause points of failure in the perception of the enterprise. Creating composite viewpoints, which can jointly model different domains using a merger between languages, can heighten the understanding and communication of enterprise aspects among stakeholders. Analyzing the relationships between different viewpoints can generate valuable insight in understanding the complexity and operations of an enterprise.

To analyze different models automated analysis methods can be used. Automated analysis methods are algorithms that extract information from the model and perform calculations with the available information. Automating these procedures for extracting and calculating information from complete models or viewpoints allows using large amounts of elements and relations. Model analysis has become a critical task that results can be used to support decision making processes in an enterprise and evaluate the current state of the organization. Multiple analysis methods span over different EM languages that can be reused in the context of multi-perspective modeling [21]. Lastly, A modeling tool that supports the creation of multi-perspective enterprise models is essential to maintain the models and facilitating their use. Very few modeling tools allow the composition of different modeling languages and their graphical notation causing a scarcity of tools to efficiently generate the models. Considering the previously mentioned requirements, a tool for multi-perspective EM must include the ability to create viewpoints, use different analysis methods, and validate the model, among other additional functions. Nevertheless, this is an immense challenge as different enterprises have different modeling needs, which in turn leads to near endless combinations of modeling languages. Thus, an ad-hoc modeling tool must be created for the specific needs of each enterprise. To create such a tool for an enterprise it is necessary to further specify this requirement [6, 22]:

**RQ1:** ***Implement a modeling language***. To create the modeling tool, it is necessary to introduce and specify different modeling languages. A graphical modeling language has two parts. Firstly, it is necessary to describe its abstract syntax, typically through a metamodel [23] that defines the elements of the language and their relationships. Secondly, it is necessary to define the concrete syntax for the language. This includes the graphical notation for each of the elements and relationships, as well as other grammar rules typically based on layouts, color, size, and other graphical qualities [24]. Figure 2 shows the elements of a graphical modeling language. On one hand, it is composed of one metamodel which contains an array of elements that describe the language. On the other hand, it contains a graphical notation that is composed of multiple graphical elements which are used to represent one element of the metamodel. The metamodel elements can have one or more graphical notations.

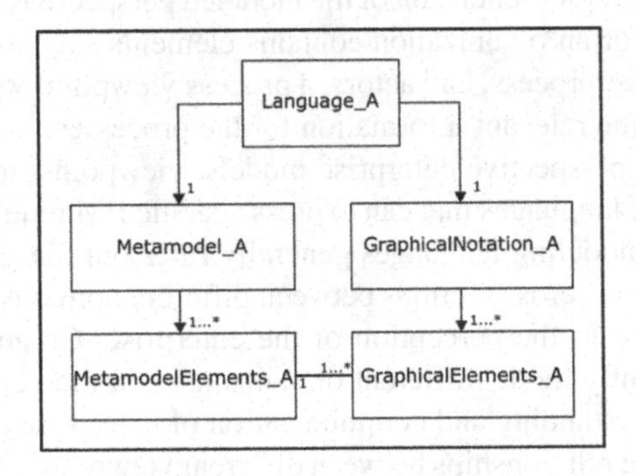


**Fig. 2.** Elements of a modeling language

**RQ2:** ***Create language compositions***. Assume a language A and a language B. Both languages are conformed by metamodels A and B, and graphical notations A and B. The metamodels and graphical notations consist of multiple elements or concepts that conform the language. With multi-perspective modeling, we can apply composition functions to both the metamodel elements, and the graphical notations of these languages as shown in Fig. 3. There are a set of composition functions for the metamodel elements [25, 26] and a set of composition functions for the graphical elements. The metamodel composition functions include: (i) deleting a metamodel element from language A or language B, (ii) linking a metamodel element from language A with an element from language B, and (iii) merging a metamodel element from language A with an element from language B. The new entity is a mixture of the two previous entities and groups all attributes and links that existed before. This set of functions result in a newly composed metamodel AB. The set of composition functions on the graphical elements include:

(i) adding a new graphical element that corresponds to any element of the language, and (ii) merging two graphical elements of languages A and B (iii) choosing one of the pre-established notations. This graphical set of functions result in a newly composed graphical notation AB.

**RQ3:** ***Repository with reusable languages.*** The repository should allow reusing modeling languages that were already defined to create the ad-hoc modeling tool. Not only should singular modeling languages be reusable but also the created composite modeling languages. This avoids having to recreate a modeling tool from scratch. Additionally, the repository can also store viewpoint and analysis method definitions for specific modeling languages.

**RQ4:** ***Multi-perspective viewpoints, analysis methods, and validations***. The tool must allow multi-perspective viewpoints and analysis methods that take into account singular or multiple modeling languages.

**RQ5:** ***Generate a modeling tool with the ad-hoc specifications***. The modeling tool must consider all of the requirements **RQ1** through **RQ4**. To do so we propose a flexible workbench that supports the definition of all of these requirements to create multi-perspective enterprise models. This workbench will be detailed in the following chapter.

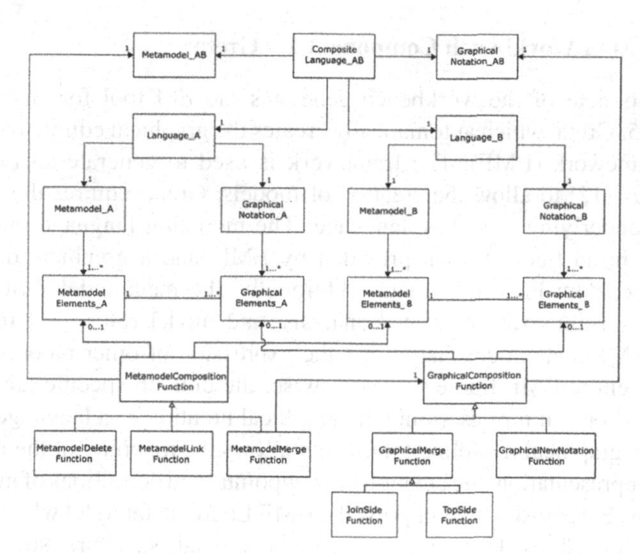


**Fig. 3.** Language composition through metamodel and graphical functions

## 4 A Flexible Workbench for Supporting Multi-perspective Enterprise Modeling

Our proposal consists of a workbench, named ModelFusion. This workbench has two main components and a repository, which together meet the needs of the requirements specified throughout Sect. 3. Figure 4 shows the elements of ModelFusion and specifies which requirements are met by each element.

Broadly speaking, on one hand, Greta (component 1 – **RQ1, RQ4, RQ5**) is the engine in charge of generating the ad-hoc EM tool that is used by the enterprise modeler. On the other hand, Gromp (component 2 – **RQ2, RQ4**) is the engine in charge of composing the languages for the multi-perspective modeling tool. The repository plays the role of storing the elements of the modeling language.

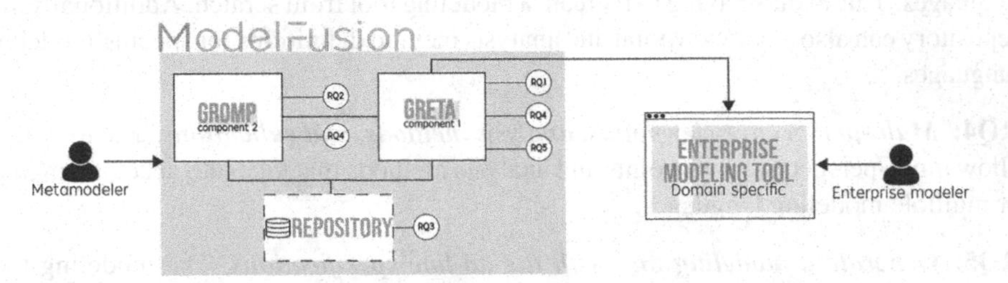


**Fig. 4.** ModelFusion workbench elements

## 4.1 ModelFusion Workbench Component 1 – Greta

The first component of the workbench generates the EM tool for any language, as shown in Fig. 5. Greta, which automatically creates the graphical editor, uses the Eclipse Modeling Framework (EMF). The framework is used to generate an Eclipse plugin that uses Sirius [12] to allow the creation of models. Greta requires the metamodeler to define the underlying modeling language. The modeling language must consist of a metamodel, in an Ecore format provided by EMF, and a graphical description, in our proposed custom Picture format. Additionally, the metamodeler can also define viewpoints, algorithms for automated analysis, and model rules used for validation. Ecore is the EMF metamodel that allows the expression of other models and enables the use of the entire EMF framework. Likewise, the domain specific language (DSL) Picture, is our proposal for describing the graphical notation of a language.

To create a graphical modeling tool, Greta allows the mapping of the metamodel to the graphical representation, the creation of viewpoints, and definitions of modeling rules and algorithms. Foremost, Greta imports the EMF Ecore metamodel which contains the entities and relationships. These are then mapped onto data structures such as hashmaps and the root element of the metamodel is appointed. Greta also imports the graphical definition from the corresponding Picture file. The engine can then correlate the elements in the metamodel with the graphical definitions for each of the elements.

To create viewpoints, Greta uses the definition specified in the Picture file. To do so, it is necessary to define and map from the imported graphical definition file which elements are used in the viewpoint onto the Ecore metamodel. This is done to have, only the elements that belong to a specific viewpoint. Additionally, the created modeling tool exposes an API that lets third parties use its services. It allows that third party to read the models created and get information from the model which enables automated analysis.

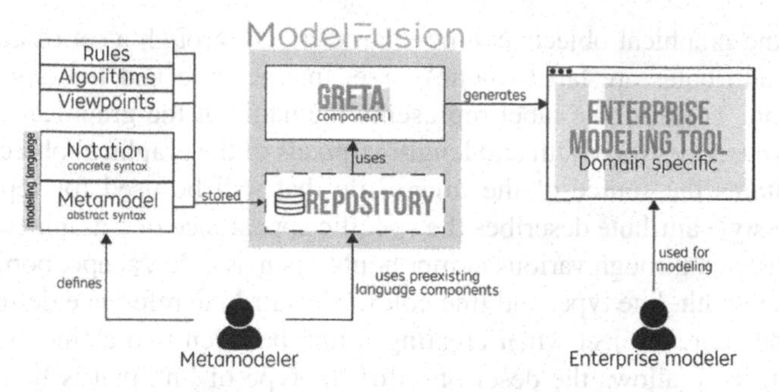


**Fig. 5.** Use of Greta (component 1) of the ModelFusion workbench

Figure 6 shows an example of a BPMN graphical modeling tool created from Greta, a BPMN Ecore metamodel (using the BPMN specification [27]), and the Picture graphical representation for BPMN. The editor contains a project explorer zone, a view zone where the diagram is shown in the center of the editor, and the palette zone which is on the right hand of the editor and shows all the possible elements and relationships.

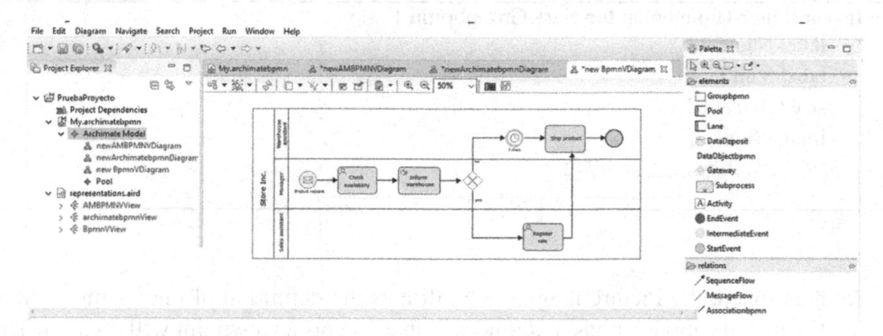


**Fig. 6.** BPMN graphical modeling tool example

**Expressing the Graphical Representation through Picture.** Picture is a graphical description language that allows the definition of the concrete syntax of the modeling language. With Picture, it is possible to generate a Picture file (. picture) that can subsequently be used as an input for creating the modeling tool. With this language, all the graphical representations of the metamodel elements can be specified. The graphical description contained in a Picture file must match the elements of the metamodel that it corresponds to. The Picture language can be used to describe the specification of a graphical description as well as the definition of viewpoints.

The first aspect of the Picture language is the definition of the graphical representation of a modeling language. There are three types of graphical objects that the Picture language can describe, which are containers, nodes, and links. A node represents a unique element of the language; a container is an element that contains subsets of other elements; and a link is used to represent relationships between elements.

Each of the graphical objects can be characterized through a set of attributes. The most crucial attributes are label (name), size, image, style and line style, reference, and decoration. The attribute label represents the name of the graphical element. The attribute size describes the width and length in pixels of the graphical object. The image attribute contains the source of the image file that will be used for representing an element. The style attribute describes the specific appearance of a graphical object. The style is constructed through various components, such as color, shape, border definition line style, line width, line type, and line color. The attribute reference defines the target object and the source object when creating a link between two elements. Finally, the decoration attribute allows the description of the type of link that is being used. It is possible to describe the source decoration and the target decoration. Decorations include output/input arrow, output/input closed arrow, output/input filled closed arrow, diamond, fill diamond, input arrow with diamond, input arrow with fill diamond. The attributes for line style, decoration, and reference can only be used when describing links. Table 1 shows a simple example of the Picture language.

**Table 1.** Graphical description in Picture language for containers, nodes and links

```
Node_container Groupbpmn for class Groupbpmn {
    label Name
    label icon false
    size ( 300 , 302 )
    image figure {
        "/images/Group.gif" }
}
```

As mentioned, the Picture language facilitates the definition of one or more viewpoints. For this, the modeler has to define which elements a viewpoint will contain. This is simply a list of the elements and the relationships that the palette viewpoint contains which have to have been previously defined both in the graphical description and in the metamodel. Table 2 shows an example of a business viewpoint definition based on the ArchiMate language.

## 4.2 ModelFusion Workbench Component 2 – Gromp

The strategy for the Gromp component was to create an engine that could generate a composite language. For this, Gromp only requires the specifications of which languages will be merged and how they will be merged. This specification is described using the Gromp language. The Gromp language takes two pre-existing languages and merges the metamodel and the graphical notation as specified by the metamodeler. The coupled languages can be used in Greta to generate a multi-perspective ad-hoc modeling tool with the coupled (Fig. 7).

The Gromp language can merge the metamodel as specified in **RQ2.** One of the most interesting functions of the language is merging two pre-existing graphical notations.

**Table 2.** Viewpoint definition

```
ViewsDefinition {
  ViewsDefinition businessView{
    paletteName "elements" elements {
      Group, Location, BusinessActor, BusinessRole, BusinessCollaboration, BusinessPro-
      cess}
    paletteName "relationships" elements {
      Specialization, Composition, Aggregation, Assignment, Realization, Triggering, Flow,
      Serving, Access}
  }
```

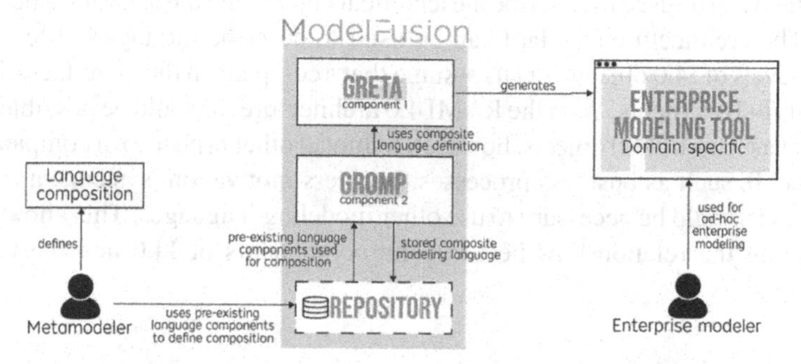


**Fig. 7.** Use of Gromp (component 2) of the ModelFusion workbench

One of the merge functions is "Join Side", which aligns the graphical representations of the elements horizontally (as shown in Table 3). Another merge function is "Join Top", which places the graphical representation of one element on top of the other graphical representation (aligned vertically). When merging two graphical notations it is also possible to define certain characteristics of the merger. For instance, it is possible to select whether to place the graphical representations internally or externally and the size of the graphical elements once they are merged, among others.

**Table 3.** Graphical representation of language notation mergers

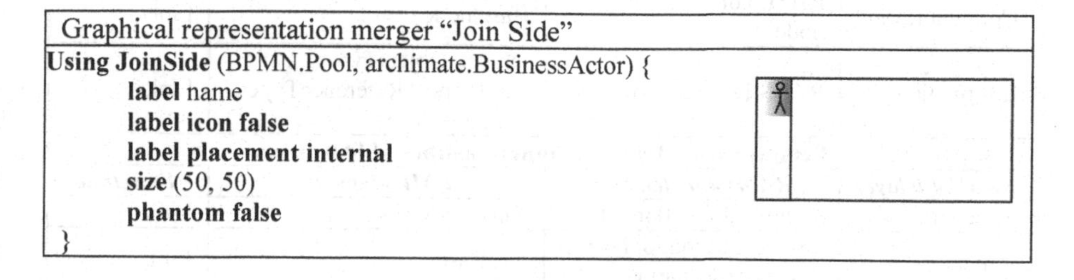

Graphical representation merger "Join Side"

```
Using JoinSide (BPMN.Pool, archimate.BusinessActor) {
    label name
    label icon false
    label placement internal
    size (50, 50)
    phantom false
}
```

The Gromp language also allows the definition of new viewpoints. The required syntax to create a viewpoint is the same as previously specified in Table 2. However, with Gromp, metamodelers can use metamodel elements from the preceding languages or can use elements that are being merged to create the composite language.

## 5 Multi-perspective Enterprise Modeling in Industry 4.0

To illustrate the use of multi-perspective EM and the use of the ModelFusion workbench for composing EM languages and generating the respective modeling tool, we will use industry 4.0 (I4.0) as an example. I4.0's purpose is to enable and facilitate cooperation and collaboration between technological assets. RAMI 4.0 is a reference architecture model for I4.0 which can be used to describe the technical objects and the relevant aspects related to them. The architecture standard can be used to describe the layers, life cycle, and hierarchy levels of I4.0 objects. Let us assume that a company in the manufacturing sector implements I4.0 elements. With the RAMI 4.0 architecture, it would be possible to model the relevant aspects of I4.0 objects, however, to model other aspects of a company that are related to I4.0, such as business processes, business motivation, systems, applications, and services it would be necessary to use other modeling languages. This allows viewing and analyzing the relationships between the components of I4.0 and the rest of the organization.

**Table 4.** Different perspectives related to RAMI 4.0

| **Perspective 1: I4.0 business layer (process & motivation)** | | | |
|---|---|---|---|
| ***RAMI 4.0 layer*** | ***RAMI 4.0 elements*** | ***ArchiMate elements*** | ***Function*** |
| Business | Business actor | Business actor, Business role | Merge |
| Business | Business goal | Goal | Merge |
| Functional | Technical functionality | Service | Merge |
| Asset | I4.0 asset, non I4.0 asset | Device | Merge |
| Business | Business process | Process | Merge |
| ***RAMI 4.0 layer*** | ***RAMI 4.0 elements*** | ***BPMN elements*** | ***Function*** |
| Business | Business process | Process | Link |
| Asset | I4.0 asset, non I4.0 asset | Activity | Link |
| **Perspective 3: Information and communication between I4.0 assets** | | | |
| ***RAMI 4.0 layer*** | ***RAMI 4.0 elements*** | ***OPC-UA/UML elements*** | ***Function*** |
| Communication | RAMI communication node | Base node | Link |
| Information | RAMI information model | Object, ObjectTypes, DataTypes, ReferenceTypes, VariableTypes | Link |
| **Perspective 4: Technical functionalities of I4.0 assets** | | | |
| ***RAMI 4.0 layer*** | ***RAMI 4.0 elements*** | ***UML elements*** | ***Function*** |
| Functional | Technical functionality | Class <<Service>> | Link |
| Functional | Use case scenario, business case scenario | UseCase | Link |
| Business | Business actor | Actor | Merge |

To model different perspectives of an I4.0 enterprise, we will use multiples languages and map the relationships between them. This allows us to create an ad-hoc EM tool to facilitate the creation of models considering different domains. We analyzed different modeling languages and their possible relations to RAMI 4.0. For this, we considered the following perspectives: business motivation, business process, information and communication, and technical functions. To model these perspectives, we chose ArchiMate, BPMN, OPC-UA, and UML languages respectively. ArchiMate is a powerful EM language that can model both the business and the IT domain. ArchiMate contains a metamodel that can be used to model the business' motivation. BPMN is the business process modeling notation for specifying processes. OPC-UA can be used to model interconnectivity between I4.0 assets. Finally, UML can be used to express I4.0 use cases and technical functionalities. These other domain languages and general-purpose language are needed because modelers might require more than RAMI 4.0 can offer, hence they can use these modeling languages to produce a more complete enterprise archetype. However, the tradeoff behind using these domain languages is that they lack the ability to relate smoothly to other domains. Table 4 shows different perspectives that can be related to RAMI 4.0. These are merely examples of how multi-perspective modeling can be mapped using our approach. To create the EM tool for RAMI 4.0 we used the ModelFusion workbench. The workbench allowed us to merge and link each of the elements specified in Table 4.

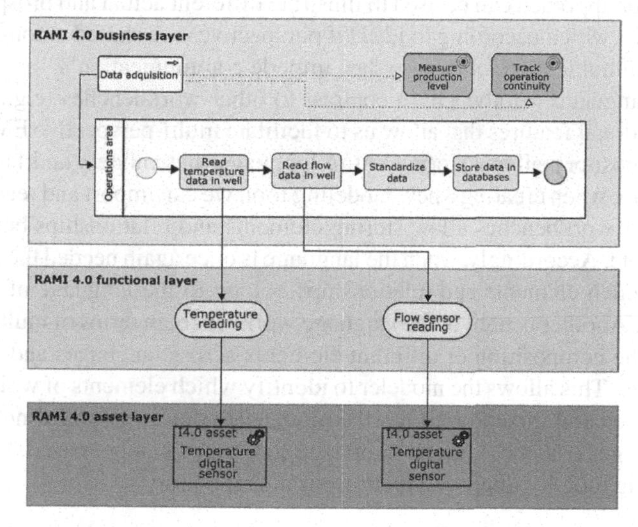


**Fig. 8.** I4.0 Business layer perspective: process & motivation elements

To model the I4.0 perspectives in the workbench we created a RAMI 4.0 metamodel based on each of the layer's functions. We will use perspective 1 in Table 4 as an example. As previously mentioned, the first step was to map the RAMI4.0 metamodel to the BPMN metamodel and the ArchiMate metamodel. For the creation of each modeling language, we used the Greta component of our ModelFusion workbench. We created the metamodel and each graphical notation using the Picture language. Afterward, when the languages were created, we firstly merged RAMI 4.0 with ArchiMate. Then we took the composite language and merged it with BPMN using the Gromp component. Figure 8 shows a model created with the ad-hoc EM tool representing the data acquisition process in a well for the oil and gas industry.

## 6 Conclusions

Modelers require ways to express the enterprise, its architecture, and changes in the architecture as clearly as possible: for their understanding and for communication with other stakeholders, such as system developers, end-users, and managers. It is common, for modelers in different domains, even within the same enterprise, to use their own modeling, techniques, conventions, languages which can hinder communication [11]. Modelers prefer these languages when modeling in a specific domain, as they provide more detailed descriptions than what general EM languages can offer. Those languages, such as UML, BPMN, and others, have a narrower scope, but it can be difficult to relate one domain language to another.

In this paper, we present an approach for multi-perspective EM. Our ModelFusion workbench tool contributes on one hand, to the creation of a custom graphical modeling tool for any EM language by specifying a metamodel and a graphical notation. On the other hand, modelers can merge two or mode languages and can create composite viewpoints. Our approach can be used to illustrate different actual and prospective views of an enterprise, which according to [22] "if perspectives are shared among individuals, they foster communication, otherwise they impede communication".

Our EM language workbench, in contrast to other workbenches (e.g., [9, 14–16]), has some additional features that allow us to facilitate multi-perspective EM. Firstly, the workbench repository allows storing entire language metamodels and language notations. Therefore, when creating a new modeling tool, we can import and reuse a language instantly. Other workbenches allow storing elements and relationships but not the language as a whole. Accordingly, when the language is once again needed the metamodeler must define which elements and relationships belong to the language of the tool they want to create. Another benefit of the language workbench, in terms of multi-perspective modeling, is the composition of different elements across languages and the graphical notation merger. This allows the modeler to identify which elements of which languages have been reused and mixed or allows the modeler to detect graphical notations in the original languages. Likewise, these composite languages can be reused, when creating a new modeling tool, by importing them from the repository.

## References

1. Da Silva, A.R.: Model-driven engineering: a survey supported by the unified conceptual model. Comput. Lang. **43**, 139–155 (2015). https://doi.org/10.1016/j.cl.2015.06.001
2. Sandkuhl, K., Stirna, J., Persson, A., Wißotzki, M.: Enterprise Modeling. Springer Berlin Heidelberg, Berlin, Heidelberg (2014). https://doi.org/10.1007/978-3-662-43725-4
3. ter Doest, H.W.L., et al.: Tool support. Enterprise Eng. Ser. 277–299 (2017). https://doi.org/10.1007/978-3-662-53933-0_11

4. Frank, U.: Domain-specific modeling languages: requirements analysis and design guidelines. In: Reinhartz-Berger, I., Sturm, A., Clark, T., Cohen, S., and Bettin, J. (eds.) Domain Engineering, pp. 133–157. Springer Berlin Heidelberg, Berlin, Heidelberg (2013). https://doi.org/10.1007/978-3-642-36654-3_6
5. Frank, U.: Designing models and systems to support IT management: a case for multilevel modeling. In: 3rd International Workshop on Multi-Level Modelling, vol. 1722, pp. 3–24 (2016)
6. Ozkaya, M., Akdur, D.: What do practitioners expect from the meta-modeling tools? A survey. J. Comput. Lang. **63**, 101030 (2021). https://doi.org/10.1016/J.COLA.2021.101030
7. Lara, P., Sánchez, M., Villalobos, J.: Enterprise modeling and operational technologies (OT) application in the oil and gas industry. J. Ind. Inf. Integ. **19**, 100160 (2020). https://doi.org/10.1016/j.jii.2020.100160
8. Achillas, C., Aidonis, D., Vlachokostas, C., Moussiopoulos, N., Banias, G., Triantafillou, D.: A multi-objective decision-making model to select waste electrical and electronic equipment transportation media. Resour. Conservation Recycling. **66**, 76–84 (2012). https://doi.org/10.1016/J.RESCONREC.2012.01.004
9. Archi – Open Source ArchiMate Modelling, https://www.archimatetool.com/. Accessed 6 Sept 2021
10. Bizagi - Low-Code Automation Leader, https://www.bizagi.com/en. Accessed 6 Sept 2021
11. Lankhorst, M.: Enterprise architecture at work : modelling, communication and analysis. Springer (2009). https://doi.org/10.1007/978-3-642-29651-2
12. Sirius, E.: https://www.eclipse.org/sirius/. Accessed 6 Sept 2021
13. Microsoft: Modeling SDK for Visual Studio - Domain-Specific Languages|Microsoft Docs, https://docs.microsoft.com/en-us/visualstudio/modeling/modeling-sdk-for-visual-studio-domain-specific-languages?view=vs-2017. Accessed 25 Feb 2019
14. Negm, E., Makady, S., Salah, A.: Survey on domain specific languages implementation aspects. Int. J. Adv. Comput. Sci. Appl. **10**, 624–633 (2019). https://doi.org/10.14569/IJACSA.2019.0101183.
15. Erdweg, S., et al.: Evaluating and comparing language workbenches: existing results and benchmarks for the future. In: Computer Languages, Systems and Structures, pp. 24–47. Elsevier Ltd (2015). https://doi.org/10.1016/j.cl.2015.08.007
16. Fill, H.-G., Redmond, T., Karagiannis, D.: FDMM: A formalism for describing ADOxx meta models and models. In: ICEIS 2012 - Proceedings of the 14th International Conference on Enterprise Information Systems, pp. 133–144 (2012)
17. Lankhorst, M.M., Aldea, A., Niehof, J.: Combining ArchiMate with Other Standards and Approaches. Presented at the (2017). https://doi.org/10.1007/978-3-662-53933-0_6
18. Vicente, M., Gama, N., Mira da Silva, M.: Modeling ITIL business motivation model in ArchiMate. In: Falcão e Cunha, J., Snene, M., Nóvoa, H. (eds.) Exploring Services Science. IESS 2013. Lecture Notes in Business Information Processing, vol. 143. Springer, Berlin, Heidelberg (2013). https://doi.org/10.1007/978-3-642-36356-6_7
19. de Kinderen, S., Gaaloul, K., Proper, H.A.: On transforming DEMO models to ArchiMate. In: Bider, I., et al. (eds.) Enterprise, Business-Process and Information Systems Modeling. BPMDS 2012, EMMSAD 2012. Lecture Notes in Business Information Processing, vol. 113. Springer, Berlin, Heidelberg (2012). https://doi.org/10.1007/978-3-642-31072-0_19
20. Ceri, S., Valle, E.D., Pedreschi, D., Trasarti, R.: Mega-modeling for big data analytics. In: Atzeni, P., Cheung, D., Ram, S. (eds.) Conceptual Modeling. ER 2012. Lecture Notes in Computer Science, vol. 7532. Springer, Berlin, Heidelberg (2012). https://doi.org/10.1007/978-3-642-34002-4_1
21. Florez, H., Sánchez, M., Villalobos, J.: A catalog of automated analysis methods for enterprise models. SpringerPlus. **5**, 406 (2016). https://doi.org/10.1186/s40064-016-2032-9

22. Frank, U.: Multi-perspective enterprise modelling: Background and terminological foundation. Universität Duisburg-Essen, Institut für Informatik und Wirtschaftsinformatik (ICB), Essen (2011)
23. Cengarle, M.V., Grönniger, H., Rumpe, B.: Variability within modeling language definitions. Model Driven Eng. Lang. Syst. **5795**, 670–684 (2009). https://doi.org/10.1007/978-3-642-04425-0_54
24. Moody, D.: The "Physics" of notations: toward a scientific basis for constructing visual notations in software engineering. IEEE Trans. Software Eng. **35**, 756–779 (2009). https://doi.org/10.1109/TSE.2009.67
25. Anwar, A., Benelallam, A., Nassar, M., Coulette, B.: A graphical specification of model composition with triple graph grammars. In: Machado, R.J., Maciel, R.S.P., Rubin, J., Botterweck, G. (eds.) Model-Based Methodologies for Pervasive and Embedded Software. MOMPES 2012. Lecture Notes in Computer Science, vol. 7706. Springer, Berlin, Heidelberg (2013). https://doi.org/10.1007/978-3-642-38209-3_1
26. Fleurey, F., Baudry, B., France, R., Ghosh, S.: A generic approach for automatic model composition. In: Giese, H. (eds) Models in Software Engineering. MODELS 2007. Lecture Notes in Computer Science, vol. 5002. Springer, Berlin, Heidelberg (2008).https://doi.org/10.1007/978-3-540-69073-3_2
27. Object Management Group: The Business Process Model and Notation Specification Version 2.0, https://www.omg.org/spec/BPMN/2.0/. Accessed 25 Feb 2019

# Software Design Engineering

# Architectural Approach for Google Services Integration

Hamilton Ricaurte and Hector Florez(✉)

Universidad Distrital Francisco Jose de Caldas, Bogota, Colombia
hricaurtec@correo.udistrital.edu.co, haflorezf@udistrital.edu.co

**Abstract.** Every day, new software applications appear and, commonly, these applications need to implement an authorization system. Usually, some applications are focused on specific communities such as universities. In this case, managing the credentials of every application could be tedious for the members of the community. In addition, applications need different hardware requirements for the right performance of some features, which might be very expensive in some cases. Fortunately, big companies such as Google and Microsoft offer suites to some organizations like universities; then, universities can seize said suites to improve their business processes. In this paper, we present an approach for building applications using a microservices architecture and taking advantage of Google Workspace.

**Keywords:** Microservices architecture · Google services · Google OAuth

## 1 Introduction

Currently, several organizations such as universities have acquired suites from Google or Microsoft improving their technology and business functions. In particular, The *University xxx* has implemented the suite from Google called *Google Workspace* taking advantage particularly of the following services: Gmail, Google Meet, Google Calendar, and google classroom. Nevertheless, this university (like others) needs other kinds of business functionalities; then, they constantly need to build a lot of software projects to tackle their needs. Those projects solve services related to academic information, human resources, payroll, research information management, library information management, among others.

Unfortunately, although in the university we have Google services at our disposal, some software developers prefer to build systems without using the available Google services increasing costs as well as difficulting the assurance of quality attributes such as maintainability, concurrency, security, or scalability. When a new project is deployed, the most important issue might be the login process because users (e.g., students and professors) will have different login (i.e., user and password) information for each project. This becomes a usability issue because users will need to keep in mind a lot of information for the


H. Florez and M. F. Pollo-Cattaneo (Eds.): ICAI 2021, CCIS 1455, pp. 483–496, 2021.
https://doi.org/10.1007/978-3-030-89654-6_34

same organization. Another important issue is the lack of integration because some information might be replicated in different systems and when updating is required, users must do it in every project.

Then, in this paper, we present a microservices architectural approach to build software applications, specifically web applications and mobile applications, based on Google services integration.

The paper is structured as follows. Section 2 presents the background in which we describe the features of Google Workspace and its mains services. Section 3 presents the proposed approach to integrate Google services. Section 4 illustrates a case study in which we use the proposed architecture to build a software application to manage academic events using several Google services. Finally, Sect. 5 concludes the paper.

## 2 Background

### 2.1 Google Workspace

Google Workspace, previously known as GSuite, is a set of online web tools focused on business environments, which allows companies to save infrastructure resources as well as to keep safe access information [8]. Among the most outstanding tools of Google Workspace we can find:

- Gmail: The email management tool whose domain and users can be managed directly by the company that acquires the service.
- Docs: The office automation tool for creating and editing documents.
- Forms: A tool for creating dynamic forms.
- Spreadsheets: A tool for creating and editing spreadsheets.

### 2.2 Google Account

It is a user account that allows the use of different Google Applications like Gmail, Youtube, Blogger, etc. through an email address and password. This account can be or cannot be part of Google Workspace.

### 2.3 Service Account

It is one kind of Google Account that can be used for an application and its authentication is not in the traditional way (i.e., email and password) [9]. These accounts could be useful when we need to manage sensitive data or files; for instance, files that the user opens with our application but can be later modified by the user using the official Google Applications. Then, to avoid this situation we can do a copy of the file in our Service Account or changing the privileges of the files.

## 2.4 Cloud Services

They are a set of subscription services for infrastructure that aims to be an option for the implementation of different software solutions [14]. These services can be classified in:

- Infrastructure-as-a-Service (IaaS): provides processing, networking, and storage resources. These services usually have fixed costs in base to resources acquired and need constant maintenance [12].
- Platforms-as-a-Service (PaaS): provides a platform where the users can implement, monitor, and manage their applications [1].
- Software-as-a-Service (SaaS): provides to the user an application implemented in its respective platform and infrastructure [3].
- Function-as-a-Service (FaaS): provides a set of services that allow developers to implement their apps without maintaining the infrastructure [10].

The most popular platforms that offer these kinds of services are Amazon Web Services, Microsoft Azure, Google Cloud Platform, IBM Cloud, and Alibaba Cloud.

## 2.5 Microservices

A microservice is an atomic application with a single responsibility that can be independently deployed, scaled, and tested [15]. The use of microservices is an alternative to the classic monolithic architecture that is used in many webs platforms. Multiple services are implementing and migrating to microservices patterns because, in this way, there are several benefits such as [5,6]:

- Extensibility: It is possible to add new features to one microservice without affecting other microservices.
- Maintainability: Developers can focus on specific systems allowing them a better understanding of how the system works.
- Diversity: It is possible to develop each microservice in the most convenient language/framework without affecting the other microservices.
- Low cost: In the cloud services, there are flexible options that allow configuring the resources for each service dedicating the needed hardware resources. It might be done through containers like Kubernetes.

Figure 1 presents a graphical comparison between the classic monolithic architecture and the microservices architecture, where the main difference is regarding how the different parts of the system interact with each other. So, it is possible to appreciate in the monolithic case, how the different parts of the system are all embedded in one single application, which is not bad, but when a system starts to grow, the resources management, maintainability, integration of new services, deployment, etc. could be very complicated and thus, one little error could crash the entire system [13]. To avoid these situations, many companies have migrated to the microservices architecture, which is one implementation of the client-server pattern, where we use multiple servers in at least

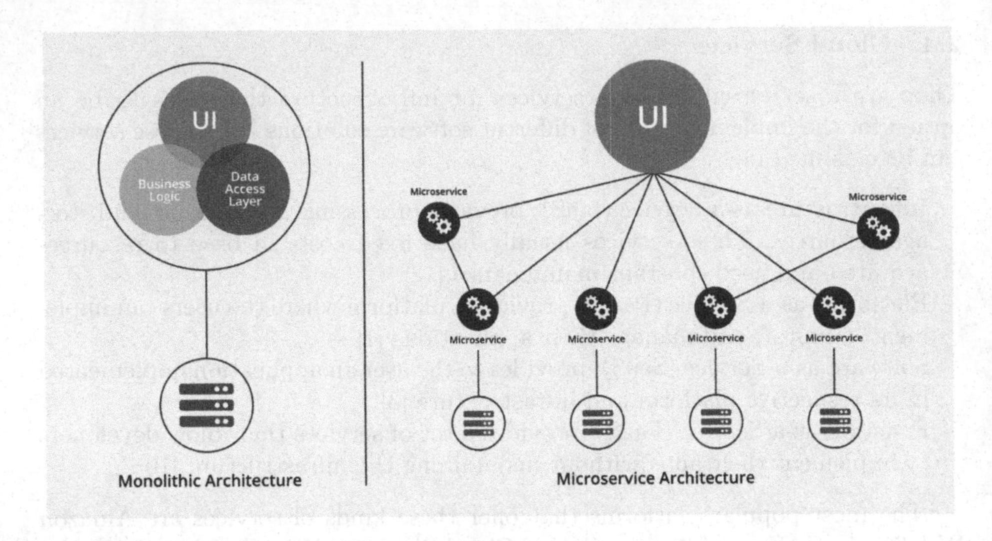


**Fig. 1.** Architectures comparison [11]

one client application obtaining better management of resources, distributed development flows (for maintenance, escalation, etc.) enabling the possibility to implement new client applications and to integrate new services. These architectures also include database connections; however, in the microservices architecture, the connection to one database for each service is optional.

### 2.6 Google OAuth

It is the authentication strategy by Google that allows developers easy access to the user information and services. In addition, the users have the possibility of quick sign up and sign in with just one password and taking advantage of the Google security systems [2]. Figure 2 presents the basic authentications workflow that allows the authentication of the user in the system. The second step `User login & consent` is made by the client in a Google service. This process could be made client-side or server-side using special libraries provided by Google or using another authentication service (these services use the Google libraries but offer a simplified workflow mainly focused on authentication within the system and not on the consumption of services). The recommendation is to use the client-side mode (emergent window) when the client application is a single page application (made using AJAX or if the re-directions are not supported); otherwise, it is necessary to use the server-side mode (re-direction).

When using Google OAuth, it is necessary to set up the service scopes before using it. Then, in the step 5 `Token response`, the application has access to 3 different tokens: `identity token` (used if only implementing an authentication system is needed), `access token`, and `refresh token`. The recommendation is using the `identity token` for validating the existence of the user in the

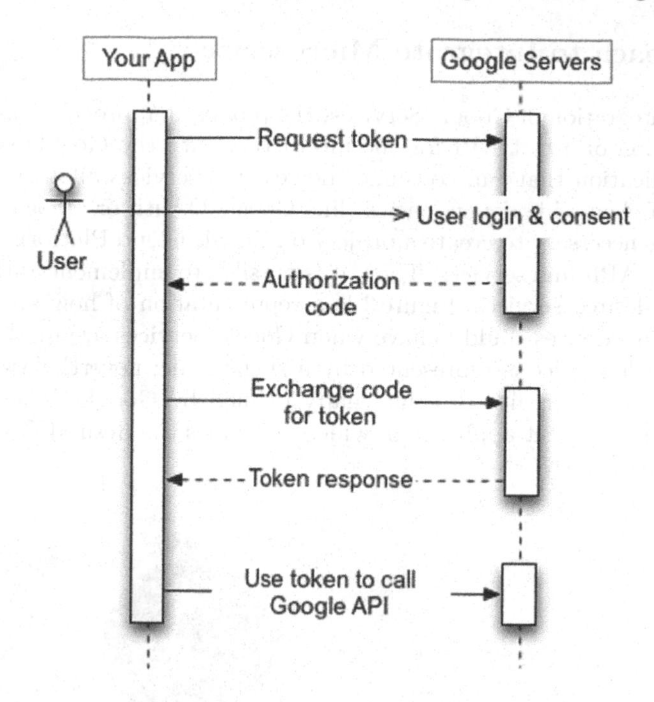


**Fig. 2.** Google OAuth workflow [4]

application. The `access token` can be used to check in every transaction if the user has granted permission for access to their apps (this token is needed in every Google API request so it is necessary to save this token on the client-side). Finally, the `refresh token` will be used for refreshing the user session in the application because when the user grants permission to their apps this permission can last up to an hour, so when the system detects that the permissions are expired, it is necessary to use the `refresh token` for using the services one more hour. The application can ask the user for permissions again; however, it might not be comfortable for the client.

### 2.7 Google APIs

They are a set of services exposed by the Google Cloud Platform that allows developers to integrate their applications with the features of the Google Apps. For the correct implementation of these services, it is necessary the implementation of Google OAuth, in the way that the user can permit the services that the application needs (the user can give the required permissions, but it is possible to obtain a better experience if the application asks for permission only once). The better way to integrate these services is using the Google official libraries[1] because, in this way, it is possible to build applications preventing them from becoming very robust and preventing security issues [7,16].

[1] https://developers.google.com/api-client-library.

## 3 Approach to Integrate Microservices

To get the integration of Google Services, the proposed approach is based on the implementation of a rest API for the interaction with the Google services and a client application that can consume the resulted services allowing the implementation of client-side dependencies (like Google OAuth or Google Picker).

First, it is necessary to create a project on Google Cloud Platform and enable the necessary APIs and services. Then, it is possible to implement authentication and use the desired services. Figure 3 is a representation of how an application based on microservices could behave when Google Services are implemented, in which a Google service is represented by a single node; nevertheless, said node has a microservice architecture internally defined by Google. The big orange node represents a client application, which consumes the desired Google service through an API Rest.

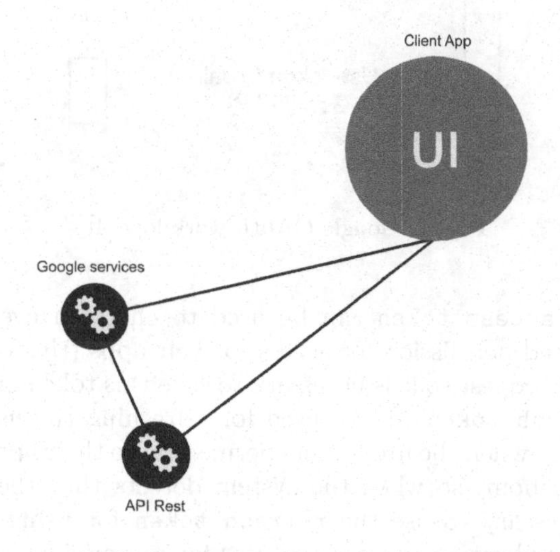


**Fig. 3.** Microservices structure

### 3.1 Authentication

When the client application just requires user authentication without using any other service, the recommendation is to use a social authentication tool because this offers a simple way to developers and can be used not only with Google but also with other social platforms. However, if the client application requires a service implementation, the best solution is to use the Google official tools. With this in mind, the first recommendation is creating in the client application a class and/ or database table with at least the following attributes: `googleId`, `email`, `name`, `lastname`. Thus, the password field is optional; developers can implement an email/password strategy too, but the application will need to ask for permissions when the user needs a Google function. So, when the user signs in to the platform using the Google OAuth.

Later on, it is necessary to decode the Google token (identity or access) and compare the fields email and id to the application database. If the user does not exist, the application can ask if it wants to sign up and fill the fields with the information of the identity token. This step is also important when implementing services to set the *scopes* of the application. Basically, there is an array of strings that defines which Google Services the client application wants to implement. Since there are a variety of scopes for every Google Service, it is necessary to check the official documentation of the required service to decide which scopes are the right choice.

In addition, it is desirable to use the field `googleId` as the primary key (if the database allows it), although the client application can use its primary key. However, the best option is that the user can authenticate in the application comparing the fields `googleId` and `email` with the information provided by the tokens. Also, in most cases, it is better to configure the Google Auth Service `Force prompt` with the `Concert screen` option because in this way, the privileges of the client application upon the Google Services are clear and concrete. Finally, to test the client application, the Google OAuth Playground[2] might be used.

### 3.2 Service Implementation

This approach is focused on a microservice architecture for Google services integration. Figure 4 illustrates a deployment diagram, in which the Google APIs are integrated allowing the inclusion of the nodes that represent new Google services.

The main node is the `Rest API Server` because it is managing the Google Services Integration, which is made up of the following components:

- `Routes`. They expose the available service endpoints.
- `Middlewares`: It is a layer of validations, which is in charge to validate that the user is logged in and has granted privileges to the client application.
- `Controllers`: If the request passes the validations, the client application can call the needed functions that could have relations with a Google Service.
- `Models`: In this component, the business logic is defined.
- `Entities`: This component represents the persistence layer of the client application, which can be built with an ORM (Object-Relational-Mapping) if it is needed.
- `Services`: This component layer is in charge of doing requests to external services including the Google Services and is made using any of the official Google APIs libraries. In Fig. 5, the recommended structure for the Google Services class is presented (one for each service)

[2] https://developers.google.com/OAuthplayground.

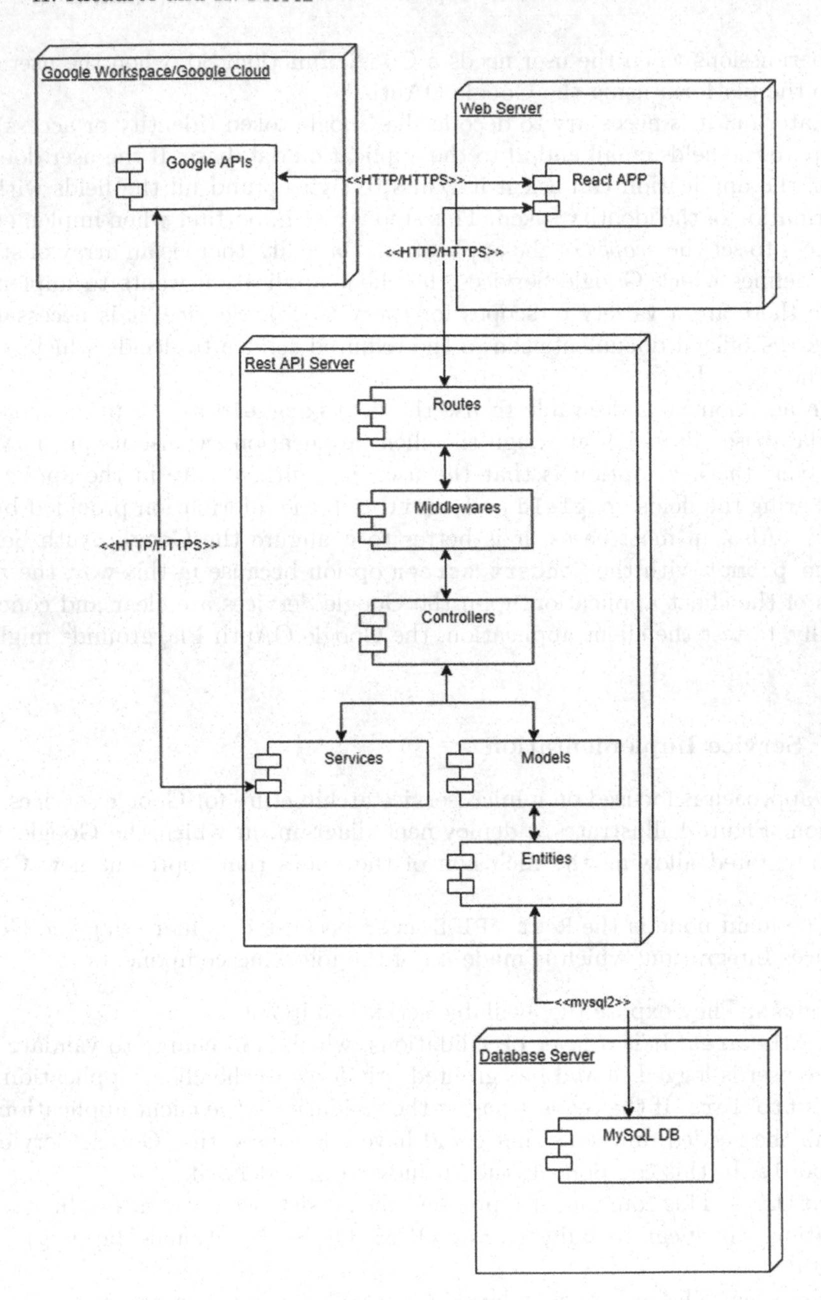


**Fig. 4.** Deploy diagram

Based on the service architecture described, Fig. 6 presents a components diagram that illustrates how this architecture can implement some Google Ser-

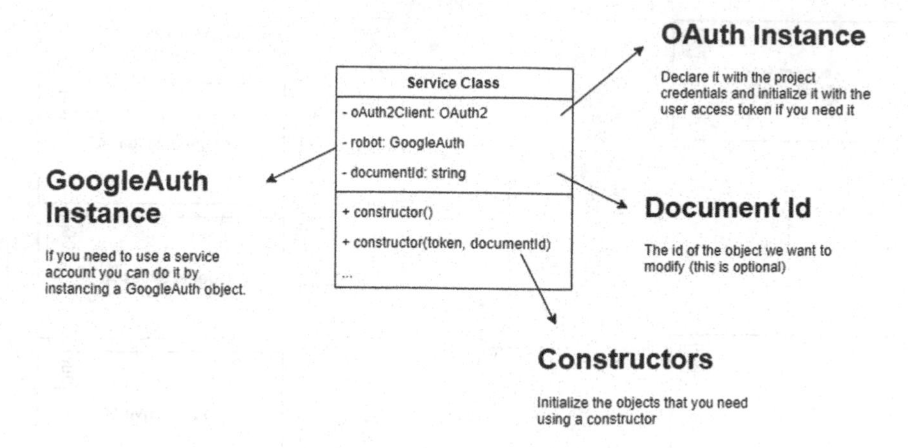


**Fig. 5.** Suggested service class structure

vices such as Google Calendar, Google Slides, Google Drive, Google Picker, and Google OAuth; nevertheless, it can integrate further services.

## 4 Case Study

Considering that the *University xxx* has acquired the Google Workspace, we created a web application called *GudDay*, which is a progressive application developed using the proposed microservices architecture. This application has been designed for events management. Thus, with this application, the members of the university can organize and attend different kinds of academic events such as seminars, conferences, meetings, contests, among others.

To achieve this, it was necessary the implementation of the Google OAuth to get easy and secure authentication. In addition, during the analysis phase, it was evidenced that the implementation of Google Services was a good way to save computing resources.

The result was an application that behaved like an official Google App with tools to make easy the creation, administration, and search of academic events of the university taking advantage of the following Google services:

- Google Calendar: To save the events in the personal Google Calendar of the users and using the Google Calendar notification system to allow users to receive the events notifications in a mobile device or email.
- Google Slides: To enable massive generation of attendance certificates using predefined templates. In this way, the event organizers can customize the certificate templates of the events. Google Slides was selected because the edition is easy and unlike other office tools, new elements do not damage the document. Since Google Drive is optimized for the management of the Google Office Tools, so the size of these files is 0 bytes in the user Google Drive account.

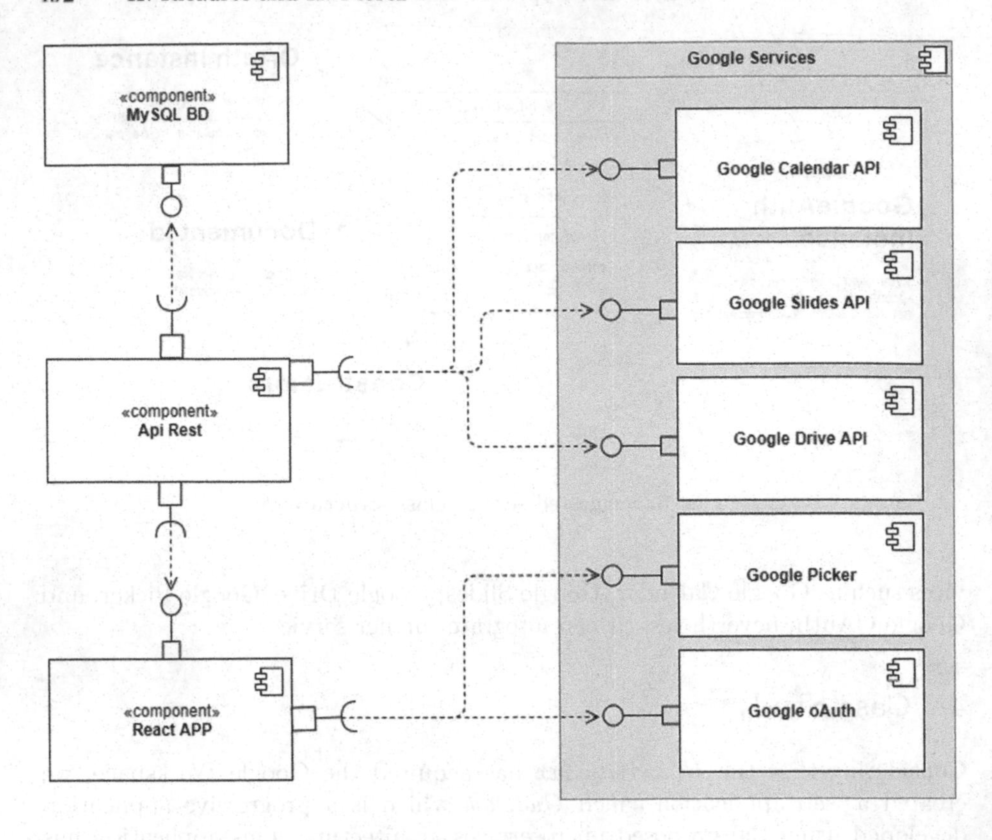


**Fig. 6.** Components diagram

- Google Drive: For file storing and file exporting.
- Google Picker: To allow users to select template files.
- Google Service Account: To prevent the files modification by the user through the Google Official Applications.

This system considers three different kinds of users: `Organizers`, `Attendees`, and `Exhibitors`. In addition, the system considers three different kinds of event status: `Created`, `In progress`, and `Ended`. Figure 7 describes the main process workflow of the platform that is initiated by the user `Organizer`, when he decides to organize an event through the activity `Organize event`. In this stage, the user must set the event duration and details through the activities `Set duration` and`Add details` to give more information using text, images, embedded videos, embedded pages, etc.; and optionally adding one certificate template. Also, the user optionally can `Enable enrollments`, `Set enrollment time limit`, as well as `Set Attendee Limit`. Finally, when the user posts an event using the activity `Post`, he can access some Organizer tools like `Invite registered users to exhibit`, where the organizer can invite people using registered emails. Then, the invited people can `Accept invitation` or `Reject`

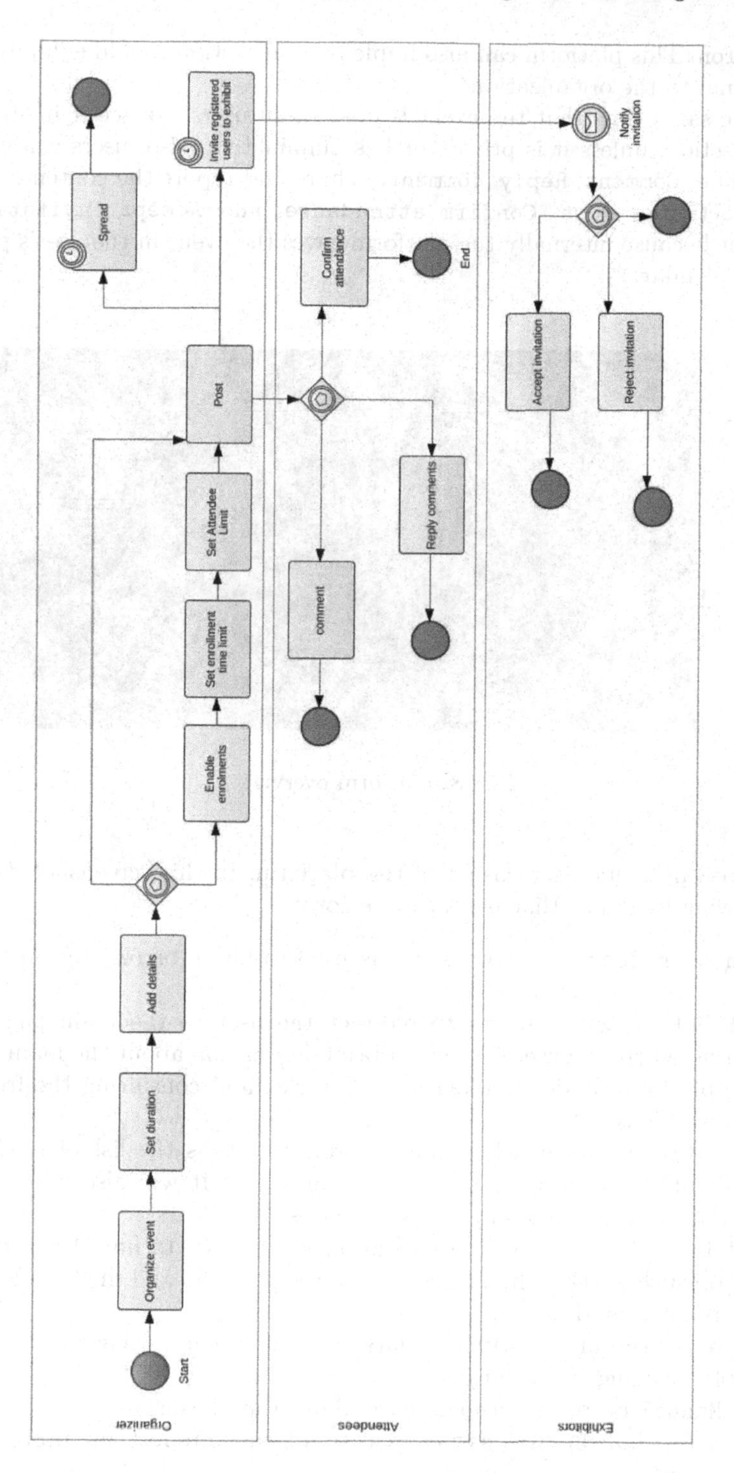


**Fig. 7.** Process diagram

invitation. This platform can also implement an option to add exhibitors that are external to the organization.

At the same time that the event is posted, all users can see it in the publications section, unless it is private or has culminated. Also, users can `Confirm attendance`, `Comment`, `Reply comments`, share and report the content.

The activities `Post`, `Confirm attendance`, and `Accept invitation` are important because internally the platform saves the event in the user's personal Google Calendar.

**Fig. 8.** Platform overview

Figure 8 presents a screenshot of the platform. In this screenshot, there are the following functions that users can perform:

1. `Login with Google` button, which is used to log in or register on the platform.
2. `Start` button, which serves to redirect the user to the main page of the platform, where the user can read a brief description about the main features of the platform, understanding how it works, and consulting the frequently asked questions.
3. `Publications` button, which allows users to access the list of forthcoming events published by other users of the platform. It will also allow users to filter the content based on their preferences.
4. `About` button that allows users to know some more technical details of the platform such as the technologies used in the platform and the Google services that are integrated.
5. `Change theme` button switch to dark or light theme to navigate more comfortably avoiding visual fatigue.
6. `User Manual` button to deepen more about the platform.
7. `Report Bug` button gives users access to a form to report any problems with the application.

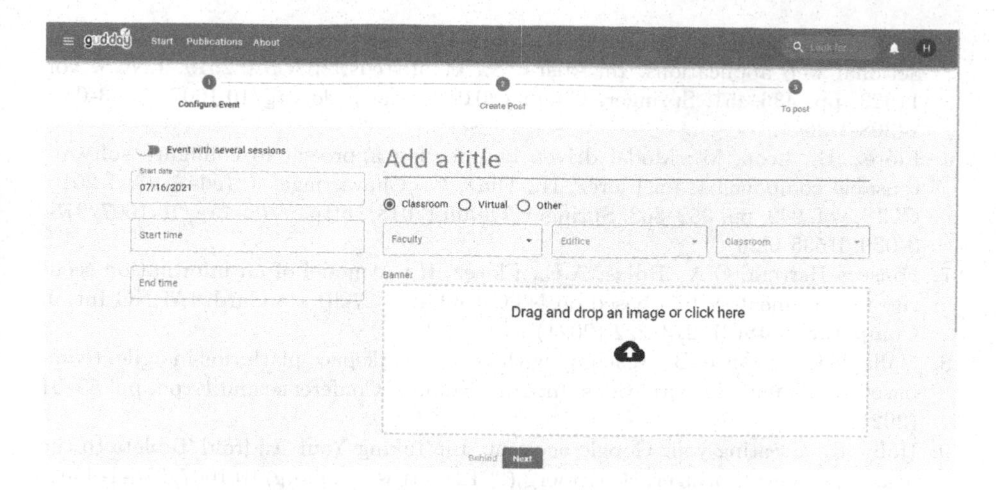


**Fig. 9.** Event form 1

Figure 9 shows the *Organize Event* form, which is composed by three steps. In this figure, the first step is presented, where users can fill in the basic information. After filling in the information, the system uses the API Google Picker, which is launched through a button called `Select file from Google Drive` to select *Presentation Files.*

## 5 Conclusions

The proposed architectural approach to integrate Google microservices allows systems to be easily maintainable and extensible giving the possibility of implementing new modules or functionalities, new third-party services, new client-side applications, etc.

The developed application based on the proposed offers to the members of the university community a pleasant space, easy to use, and capable of being installed on different devices, allowing them to easily manage their events.

The developed application takes advantage of some of Google's free services obtaining a software product that does not require many hardware resources for its correct operation.

## References

1. Beimborn, D., Miletzki, T., Wenzel, S.: Platform as a service (PaaS). Bus. Inf. Syst. Eng. **3**(6), 381–384 (2011)
2. Boyd, R.: Getting Started with OAuth 2.0. O'Reilly Media, Inc., Sebastopol (2012)
3. Cusumano, M.: Cloud computing and SaaS as new computing platforms. Commun. ACM **53**(4), 27–29 (2010)
4. Google Developers. Using OAuth 2.0 to access Google APIs. https://developers.google.com/identity/protocols/oauth2

5. Florez, H., Garcia, E., Muñoz, D.: Automatic code generation system for transactional web applications. In: Misra, S., et al. (eds.) ICCSA 2019. LNCS, vol. 11623, pp. 436–451. Springer, Cham (2019). https://doi.org/10.1007/978-3-030-24308-1_36
6. Florez, H., Leon, M.: Model driven engineering approach to configure software reusable components. In: Florez, H., Diaz, C., Chavarriaga, J. (eds.) ICAI 2018. CCIS, vol. 942, pp. 352–363. Springer, Cham (2018). https://doi.org/10.1007/978-3-030-01535-0_26
7. Fonseca-Herrera, O.A., Rojas, A.E., Florez, H.: A model of an information security management system based on NTC-ISO/IEC 27001 standard. IAENG Int. J. Comput. Sci. **48**(2), 213–222 (2021)
8. Gallagher, P., Vance, B.: Teaching with google workspace platforms in agile, team-based communication situations. In: 2021 Summit Conference and Expo, pp. 55–61 (2021)
9. Holly, R.: Creating your Google account. In: Taking Your Android Tablets to the Max, pp. 37–48. Springer, Heidelberg (2012). https://doi.org/10.1007/978-1-4302-3690-0_3
10. Lynn, T., Rosati, P., Lejeune, A., Emeakaroha, V.: A preliminary review of enterprise serverless cloud computing (function-as-a-service) platforms. In: 2017 IEEE International Conference on Cloud Computing Technology and Science (CloudCom), pp. 162–169. IEEE (2017)
11. Malav, B.: Microservices vs monolithic architecture. https://medium.com/startlovingyourself/microservices-vs-monolithic-architecture-c8df91f16bb4
12. Manvi, S.S., Shyam, G.K.: Resource management for infrastructure as a service (IaaS) in cloud computing: a survey. J. Netw. Comput. Appl. **41**, 424–440 (2014)
13. Sanchez, D., Mendez, O., Florez, H.: An approach of a framework to create web applications. In: Gervasi, O., et al. (eds.) ICCSA 2018. LNCS, vol. 10963, pp. 341–352. Springer, Cham (2018). https://doi.org/10.1007/978-3-319-95171-3_27
14. Sunyaev, A., Schneider, S.: Cloud services certification. Commun. ACM **56**(2), 33–36 (2013)
15. Thönes, J.: Microservices. IEEE Softw. **32**(1), 116 (2015)
16. Wanumen, L., Florez, H.: Architectural approaches for phonemes recognition systems. In: Florez, H., Diaz, C., Chavarriaga, J. (eds.) ICAI 2018. CCIS, vol. 942, pp. 267–279. Springer, Cham (2018). https://doi.org/10.1007/978-3-030-01535-0_20

# Proposal to Improve Software Testing in Small and Medium Enterprises

Melisa Argüello, Carlos Antonio Casanova Pietroboni(✉), and Karina Elizabeth Cedaro

Facultad Regional Concepción del Uruguay, Universidad Tecnológica Nacional, 3260 Concepción del Uruguay, Argentina
{casanovac,cedarok}@frcu.utn.edu.ar

**Abstract.** Faced with their massive demand for software products, their quality has been increasingly questioned and, consequently, the quality of the processes by which they are produced. Different institutions have developed different models and standards for the continuous improvement of these development processes, mainly focused on the testing stage. Even in this proliferation of improvement models, there are none applicable to software developing SMEs in our region, industries that count with many opportunities but face challenges such as lack of resources, skills and experience in their pursuit to create quality software and survive in the market. This article presents a management proposal that allows to improve the testing process to obtain higher quality software products in SMEs in the region, based on a set of successful methods for software testing in large companies.

**Keywords:** Software testing · Small and medium enterprises · Software quality · Software Process Improvement

## 1 The Quality of the Software Development Process Today

One of the software development process' main activities, regardless of the methodology used, is software testing. Testing is about providing information on an item's quality and any residual risk in relation to how much the item has been tested; to find defects before it is released for use; and to mitigate the risks of poor product quality for stakeholders [1]. Throughout the evolution of testing, its application in software development is essential when it comes to guaranteeing the quality of the software product, through verification and validation of the specifications, behavior, design and coding. As customers have become increasingly selective and rejected products that are unreliable or did not really meet their needs, software quality has been gaining importance in software engineering and with it, the role of testing in quality assurance.

Traditionally, efforts to improve software quality focused on the end of the product development cycle, during the testing phase, with an emphasis on detecting and correcting defects. Instead, the new approach to quality improvement must encompass all phases in the product development process, from the elicitation of requirements to the


H. Florez and M. F. Pollo-Cattaneo (Eds.): ICAI 2021, CCIS 1455, pp. 497–512, 2021.
https://doi.org/10.1007/978-3-030-89654-6_35

final delivery of the product to the customer. Every step in the development process must be performed to the highest possible standard [2].

Throughout history, there have been several institutions that have been motivated by quality improvement through the refinement of the software process. Some results of these entities are recapitulations and catalogs of best practices that have given rise to models for the improvement of software processes [3].

SPI (Software Process Improvement) models provide a framework for developing software according to established planning, while simultaneously improving the developer's ability to create better products. A software process model can be used by an organization to ensure its maturity, identifying and prioritizing the areas of greatest importance. In this context, the term maturity refers to the ability of an organization to reach a defined, continuous and optimized state. There are numerous SPI models in use today. The most important includes: the CMM and TMM family [4–6], TPI, TMAP [7–9], IT Mark [10] and COMPETISOFT [11]. In addition, the ISO / IEC 29110 [12] and ISO / IEC 29119 [1] standards were included in the analysis since, although they do not include specific activities, they show paths for process improvement.

## 1.1 Software Development SMEs

Small and medium-sized enterprises (SMEs), both in Argentina and in the rest of the world, are an important part of the economy as they promote contributions and the distribution of goods and services. In addition to generating wealth, SMEs are major employment generators and, therefore, locally rooted. They have flexibility that allows them to adapt to technological and economic changes and in many cases detect new processes, products and markets. Above all, they have a dynamic capacity and great growth potential.

Recently in Argentina a series of policies and institutions have been promoted for the development and revaluation of software developers SMEs. Among them we can name the Knowledge Economy Law 27.506, which aims to promote new technologies, generate added value, foster quality employment, facilitate the development of SMEs and increase exports of companies that engage in knowledge-based services. However, a necessary requirement to achieve these benefits is the accreditation of continuous improvement and quality certification [13, 14].

## 1.2 Quality Assessment in Software Development SMEs

According to the "*Reporte anual sobre el Sector de Software y Servicios Informáticos de la República Argentina*" (Annual Report on the Software and Computer Services Sector of the Argentine Republic) for the year 2018 carried out as part of the *Observatorio Permanente de la Industria de Software y Servicios Informáticos* (Permanent Observatory of the Software and Information Services Industry, OPSSI) program, an initiative of the CESSI (*Cámara de Empresas de Software y Servicios Informáticos*, Chamber of Software and Computer Services) [15], 64% of the companies said they had some type of certification as of December 2018 (60% had certified at least ISO 9001). This 64% is composed as follows: 4% has some CMMI level, 16% ISO 9001 guide ISO 90003 and the remaining 44% ISO 9001 only. This high proportion is not surprising since quality certifications are one of the requirements to access the regime of promotion of the Software Law (most of the companies surveyed receive benefits from the scheme or are in the process of registration). However, ISO 9001 does not define a specific process improvement model for the software industry, even less a specific one for quality assurance.

It is likely that the software development SME sector has the same characteristics as the entire software development industry in Argentina. In fact, in [16] a survey of software developing SMEs in the region including Buenos Aires, Córdoba, Entre Ríos and Santa Fe provinces, the following conclusions were reached:

- Most companies use agile lifecycle models, characterized by short development cycles and highly adaptable to changes in context.
- Many lifecycle models are being used for software development. In addition, a considerable percentage of organizations do not follow any model. Therefore, it is concluded that a good methodology must be independent of the lifecycle model.
- Most of the SMEs surveyed carry out their tests informally, characterized by the lack of execution of tests at different levels: Unit Tests, Functional Tests and Acceptance Tests. Which places them at the lowest level of methodologies such as CMMI or TPI.
- In general, development projects invest scant resources in testing activities. This was detected by the time they spend planning and executing tests.
- The surveyed companies demonstrated a good predisposition for the use of registration and tracking tools for requirements and detected bugs, though a large number of them are not using the most appropriate tools.
- With regard to certifications, it is concluded that having any of them generates activities and tasks more focused on the continuous improvement of processes. Consequently, more focused on the quality of the final product. However, in the context of SMEs, these certifications are arduous to accomplish. The survey results carried out emphasize the need for an achievable improvement model for SMEs.
- Based on the above, it can be said that most of the surveyed organizations fall into the category of immature organizations, defined by CMM.

These findings, in conjunction with a consistent list of articles on SMEs [17–20] in various countries similar to Argentina (in general, developing countries), show a lacking in applicable SPI models for SMEs. Nevertheless, it is reasonable to adapt the methodologies conceived for large companies to fit the SMEs contexts.

Although other models with minimal practices have been proposed, such as [21], which includes risk management practices, this proposal is an approach that aims to implement quality assurance in software development processes characterized by lighter practices and less documentation.

The rest of the paper is structured as follows. In Sect. 2 the proposed methodology is introduced. Section 2.1 presents the axes and levels structure of the proposal. Section 2.2 shows the matrixes that facilitate the diagnosis and evolution of the testing process. Section 2.3 shows the application process of the proposal. Sections 2.4 and 2.5 present the testing roles and the role of communication respectively. Section 2.6 describes the full implementation result. Finally, the Sect. 3 presents the expert evaluation results, the conclusions and future work.

## 2 Proposed Methodology

In order to establish a path for the improvement of the testing process, a series of axes were defined. These axes have a double objective: diagnosis and direction of improvements. They have a diagnostic objective because the structure of levels allows evaluating the degree of evolution in which they are with respect to certain characteristics of the axis. On the other hand, they have a directional objective which indicate where the improvements have to be made, establishing the necessary conditions to position themselves on the next level.

The axes are divided into three levels: No implementation, Partial Implementation and Full Implementation (See Fig. 1). The fact that there are only three levels permits a quick diagnosis, focusing on specific issues. Also allowing, to carry out short cycles, easier to manage.

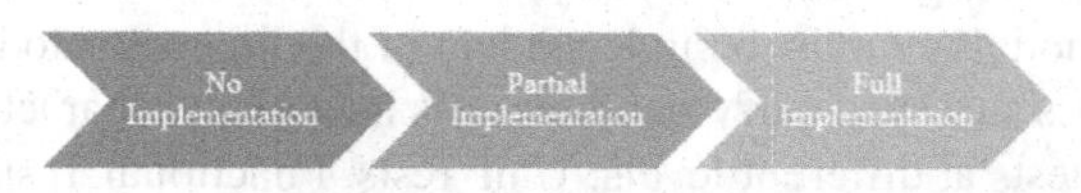


**Fig. 1.** Axes evolution levels.

The tiered structure is one of the characteristics that has allowed many of the improvement models to have wide acceptance globally, because it allows changes to be made in a progressive and organized way. It should be noted that, in the case of the improvement models analyzed, their structure is usually five levels or more, which is very complex to implement when they are carried out in informal settings such as the organizations surveyed. That is why this three-level approach is superior. Additionally, it enables laying the necessary foundations to obtain some internationally recognized certification, which can be done in a faster way and with the previous experience provided by this methodology.

### 2.1 Axes and Levels

The axes proposed below are based on the most popular SPI models in use today [1, 4–12]. They were selected by its impact on the final product and for its ability to be divided into levels achievable for SMEs in the sector.

**Plan Definition.** Planning involves activities that define both the test objectives and the approach to achieving those objectives within the constraints imposed by the context. Planning depends on the organization's test policy and strategy, development lifecycles and methods used, test scope, objectives, risks, limitations, criticality, testability and resource availability [22]. However, as a first approach to planning and setting a schedule within budget, a test plan should be developed. This plan will then be executed, its progress will be tracked, and corrective and preventive measures will be taken if it is necessary. For that reason, the plan should include information about the test basis, as well as the criteria to be used during monitoring and control.

The level of detail and the way in which this plan is communicated is what will determine the level of evolution on this specific axis. If planning activities are not performed, there is No Plan. An Informal Plan is one in which one or more activities and planning decisions are carried out or assumed, but are not fully formalized. It is understood by formalized that it is documented and distributed to all stakeholders. Finally, in a Formal Plan all the previous formalization conditions are met. That is, there is a plan written on paper or some monitoring tool that was distributed and is known to all those involved.

Ideally, as proposed by widely accepted test models [4–11], test planning should begin simultaneously with obtaining requirements. Faster and closer interactions between testers and other team members lead to a better understanding of testing needs in terms of resources and challenges. In turn, as the project and test planning evolve, more information becomes available and more details can be included in the plan. Test planning is an activity that must be carried out continuously throughout the entire product lifecycle. Feedback from testing activities should be used to detect changes in risks so that planning can be fine-tuned. Finally, it should be noted that in the way that the plans are functional to the objectives of the organization, they institutionalize certain issues that can be reused and adapted to new projects.

**Affected Test Levels.** Testing processes cannot focus only on system level testing. Before a complete system is subjected to the rigors of testing at this level, unit-level and integration-level testing must be performed. The efficiency of low-level testing allows defects to be detected and corrected much earlier in the development cycle. The same is true of higher levels of proof. Ideally, after system testing is complete, acceptance testing should be performed to establish confidence in the product quality, validate that it is complete, and that it will perform as expected.

To determine the levels affected, the tests that should be performed have to be determined. For that, it is recommended to analyze the available test basis, in relation to the different test levels, to identify the features to be tested. Then, define and prioritize test conditions for each level.

Probably, in most cases, SMEs do not have their complete test basis. The important thing is to spend the testing time wisely, covering as many levels as possible, so that testing can detect deviations early. Consequently, it is important to keep the traceability between the requirements, test objects and test conditions.

During this analysis, having to detect what is to be tested allows a review process. This review is very useful when there are no other activities of this type in the process. At this point, the level of evolution of the axis analyzed is given by the test levels that are covered.

**Definition of Test Cases.** Just as the affected test levels must determine "what to test", the definition of test cases answers the question "how to test". This activity involves:

- Designing the test cases, identifying the possible input data, the execution conditions and the expected results.
- Prioritizing them.
- Specifying and recording them for follow-up.

Applying a test specification technique allows to properly perform those activities. There are multiple test specification techniques that allow deriving test cases from a test object. Ideally, it is intended that each test case can be traced bidirectionally to the test condition(s) it covers.

Generally, it is good practice to design high-level test cases, without concrete values for the input data and the expected results (also known as checklists). These high-level test cases are reusable over multiple test cycles. It should be noted that the generation of test cases also leads to the identification of other needs early, such as the necessary infrastructure, data and tools.

This axis is at the lowest level when no test cases are generated. In the partial implementation level, high-level test cases are identified. Finally, it is at the highest level when the test cases have a complete specification of their inputs, execution conditions, and expected results.

**Test Activities.** In a broad sense, there are two types of test activities: static and dynamic. Dynamic tests involve the actual code execution, while static tests evaluate the code or other work product that is being tested without actually executing it. Both complement each other by finding different types of defects.

In the case of dynamic tests, the use of test execution tools (test automation) is recommended. The potential benefits of using them include: reduction of repetitive manual work, time savings, greater consistency, objectivity and repeatability.

Static testing is based on manual evaluation of work products (reviews) or on tool-based code evaluation or other work products (static analysis). Almost any work product can be examined by a static test, for either or both options, reviews and static analysis, e.g.: specifications of functional, security, architecture and design requirements; user stories and acceptance criteria; Code and others [23].

Static, type-independent, and dynamic testing can have the same objectives, such as providing a quality assessment of work products and identifying defects as early as possible. The two forms of testing have their own advantages and limitations. However, what is clear is the need to perform both types of tests for the overall result to be effective and economical. That is why, in this axis, the level of evolution is evaluated based on the ability to perform both types of tests.

**Traceability.** In any process with a certain degree of complexity for which a plan was developed, monitoring and control activities must be included. Test monitoring involves the continuous comparison of actual progress against the test plan. Test control involves taking actions necessary to meet the objectives of the test plan (which can be updated over time). The monitoring and control is supported by the evaluation of the exit criteria [22].

To implement effective test monitoring and control, it is important to maintain traceability throughout the testing process. Good traceability makes the analysis of tests coverage possible. Furthermore, it allows to analyze the impact of the changes, make the test auditable and provide information to evaluate the quality of the products, the capacity of the processes and the progress of the projects in relation to the business objectives.

There are many tools to solve these issues. The level of traceability between the test basis and the work products is what will determine the evolution of this axis.

**Test Environment.** Ideally, system-level testing should be conducted in the system's operating environment. However, in most cases this is not possible because it is very expensive, impractical or, if the system is highly available and in operation, almost impossible. Therefore, test environments should be as close to reality as possible.

Keep in mind that different test environments are needed to perform varied test categories. A performance test is not the same as an installation test. However, even if a single dedicated environment for testing is available, the most important thing is its similarity with the production environment. If these differ significantly from the operating environments, it is meaningless to detect failures under those conditions. Due to that, the evolution of this axis is evaluated. It is considered the lowest level when there is no specific environment for testing. The intermediate implementation level is when a managed and controlled dedicated testing environment exists but is not complete, either because integrations with other services or modules are missing, they are not configured correctly or the database is not updated with productive data. Finally, the highest level of implementation occurs when the environment is adequate, that is, when it fully resembles the production environment.

**Testing Tools.** The tools can be used in multiple ways to support one or more activities in the testing process. There are tools that are used directly in testing, such as test execution tools. There are tools to help manage requirements, test cases, test procedures, automated test scripts, test results, test data, and defects. These also allow reporting and monitoring test execution. Either way, any tool that assists in testing is considered a testing tool [22].

Some benefits of using testing tools are: they improve the efficiency of testing activities by automating the manual, repetitive, or resource-intensive tasks (for example, running regression tests); support manual testing activities throughout the testing process; increase the reliability of tests; improve quality by allowing more consistent testing and a higher level of defect reproducibility. In general, the use of tools makes the activity less prone to errors. However, simply acquiring a tool does not guarantee success. Every new tool introduced into an organization requires an effort to achieve real and lasting benefits.

This particular axis aims to assess the level at which the different types of tools are integrated into the overall process. That is why it is determined it is at the lowest level if it incorporates non-specific tools such as spreadsheets or emails for both the test management and test monitoring. It is considered medium level if it uses specific tools for the test execution and monitoring, and it is at the top level if it has extended automation of the process. We refer to extended automation, when they have a set of tools that allow end-to-end management, i.e., when there are tools that support all the proposed axes.

**Feedback.** While the project is in progress, there are a number of milestones that are important to collect data from the testing activities performed. Some of them can be: a software release, a completed (or canceled) project, when an iteration of an agile project ends, when a test level is completed, or when a maintenance release is completed. This collection allows the consolidation of experience, test products, and any other relevant information. In addition, it is the first step to be able to establish metrics.

There are different types of feedback or reports involved in the testing process, including internal and external reports. By providing accurate and timely reports, both internally and externally, it is possible to detect deviations early, as well as failure patterns that lead to the correction of defects. Also, providing feedback on the process and activities contributes to building good relationships since they create a space for dialogue that allows visibility of progress and leads to continuous improvement. Finally, it is essential to carry out the management of improvement actions for the entire process.

## 2.2 Diagnosis Matrix and Evolution Matrices

Based on what was previously analyzed, the following matrix was derived to facilitate the diagnosis (See Table 1). It allows to establish the level of evolution for each of the defined axes.

**Table 1.** Axis diagnostic matrix, own elaboration.

| | No Implementation | Partial implementation | Full Implementation |
|---|---|---|---|
| Plan definition | No plan | Informal plan | Formal Plan |
| Affected test levels | No testing | Unit and system testing | Unit tests, Integration tests and higher levels |
| Definition of test cases | No test cases | High-level test cases | Fully specified test cases |
| Test activities | No testing | Static or dynamic tests | Static and dynamic tests combined |
| Traceability | No traceability | Partial traceability | Full traceability |
| Test environment | No specific testing environment | Managed and controlled environment | Testing in a suitable environment |
| Testing tools | Non-specific tools | Specific tools for monitoring and executing tests | Extensive process automation |
| Feedback | No reports | Internal defect reports | Progress reports, activities and priority defects |

In a complementary way, a matrix was prepared for each of the levels with the necessary improvement activities to evolve to the next level of each axis. Although the levels of evolution correspond to situations surveyed in SMEs, the axes work independently. The table corresponds to the evolution from the level without implementation, while Table 2 corresponds to the evolution from the "No implementation" level, the Table 3 corresponds to the evolution from the "Partial implementation" level. It should be noted that it is common for an organization to have some axes more developed than others, so both matrices must be used to work on their improvement plan.

**Table 2.** Evolution matrix from No Implementation level

| | Improvement activities |
|---|---|
| Plan definition | Answer the questions: What do we test? What did we not test? And why do we do it?<br>Define the general approach to testing<br>Generate a high-level test schedule. When starts each test, when it ends and the most important milestones |
| Affected test levels | Analyze the available objects from the unit test level and system test level test<br>Identify the features to be tested from the previously analyzed test objects, as well as the conditions under which it will be tested |
| Definition of test cases | Create high-level test cases. A checklist is often recommended as a first approximation |
| Test activities | Perform static review tests, e.g., code reviews or inspections, walkthroughs, algorithm analysis, and functional documentation reviews<br>Perform exploratory dynamic tests manually |
| Traceability | Maintain traceability between test objects and test cases |
| Test environment | Build a test environment outside of the development and production environment<br>Prepare it with test data, where possible, that is similar to production data |
| Testing tools | Incorporate requirements and incident monitoring tools. As a first approximation, they can be carried in a spreadsheet, but it is advisable to use specific tools for this type of task<br>Incorporate test execution tools that allow them to be automated |
| Feedback | Generate reports of internal defects. The idea is to communicate the quality level of the product under test. It is suggested to keep a count of the test cases executed, the result obtained and the defects reported once the derived adjustments have been made. It is also important to be able to estimate the resources / hours used to carry out this task |

## 2.3 Application Process

The improvement process consists mainly of making small yet significant improvements in short cycles. These “baby steps”, based on what has been analyzed, is the way in which the best results are observed in SMEs. This approach allows both the development team and stakeholders to focus on specific activities, obtaining results faster and with less resistance to change.

**Table 3.** Evolution matrix from Partial Implementation level

| | Improvement activities |
|---|---|
| Plan definition | Document the plan, either in a specific tool or spreadsheet, and distribute it to the development team and stakeholders<br>Integrate and coordinate testing activities with development lifecycle activities<br>Estimate people and resources to carry out the planned activities. If possible, make a tentative assignment<br>Generate a calendar with specific dates or in the context of each iteration<br>Make an estimated time budget required to carry out the test activities |
| Affected test levels | Analyze the available integration test level objects and higher level tests such as acceptance tests<br>Study the need for tests at specialized levels such as Installation tests<br>Identify the features to be tested in the different levels<br>Define and prioritize the conditions for each level while maintaining, as far as possible, traceability with the previous objects |
| Definition of test cases | Generate test cases identifying: the possible input data, the execution conditions and the expected results<br>Prioritize them based on the features to be tested<br>Specify and record them for follow-up |
| Test activities | Perform static analysis tests guided by tools such as source code analysis, model checking and business process simulation<br>Automate dynamic tests |
| Traceability | Maintain traceability between requirements, test objects and test cases, so that it is possible to determine the coverage of the tests |
| Test environment | Include in the test environment: test harness, service virtualization, simulators and other necessary infrastructure elements<br>Prepare test data from production data and keep it up to date. The result should be an environment as similar to the productive environment within the possibilities and restrictions of the business |
| Testing tools | Incorporate tools for test management and test products<br>Incorporate test design and implementation tools to create maintainable work products, including test cases, test procedures, and test data<br>Incorporate tools for the execution and registration of tests<br>Incorporate tools for performance measurement and dynamic analysis<br>Incorporate tools for specialized testing needs in the event that higher priority tests require it. For example: security tests, portability, and accessibility and others |

(*continued*)

**Table 3.** (*continued*)

| | Improvement activities |
|---|---|
| Feedback | Generate progress reports to evaluate the evolution of the tests and track associated activities<br>Prepare defect reports to communicate the testing process and product quality. It should allow interested parties to provide feedback about the quality achieved and priorities for future releases |

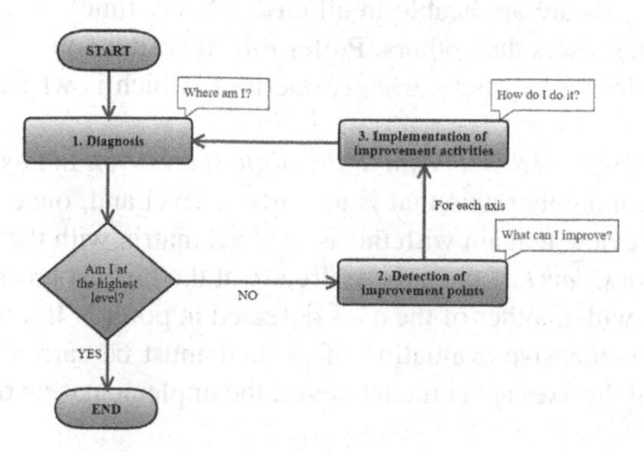


**Fig. 2.** Application process

The application process, summarized in Fig. 2, is implemented as follows:

1) **Diagnosis.** Before implementing any improvement, the starting point must be established, that is, the current state of the organization's testing process. For this, a level must be established for each of the axes described in the diagnosis matrix. The objective at this point is to answer the question "Where am I?" for each of the axes. It is recommended that everyone involved in the project participate in this diagnosis.
2) **Detection of improvement points.** Once the diagnosis has been made, axes which have the potential to be improved must be detected, that is, those that are not at the highest level. The purpose of this step is to determine what can be improved. The axes act independently, i.e., the level of one does not depend on or condition the degree of evolution of the rest. That is why it is recommended to start with those that are less developed and in the order in which they are found in the table. This order is not random, as they are ordered by the stage of the development process in which these activities are involved. As mentioned above, the earlier the testing activities are integrated into the process, the better their impact on the quality of the final product.
3) **Implementation of improvement activities.** Once the improvement points have been detected and the order in which they will be treated, the iterations begin incrementally. It begins by taking an axis, seeing at what level it is and begin implementing the proposed activities detailed in the specific matrix of the level. It is recommended

to work one axis at a time, incorporating one activity per iteration. It is advisable to use support tools for the tests as they have shown their positive effects in the process.

4) **Verify results.** Once the improvement activities have been carried out, the degree of evolution of the axis that was worked must be re-measured, using the diagnosis matrix again. It is recommended to verify the changes after achieving a project milestone. During this verification, the following situations may occur:

a) *The level could not be improved.* The axis remains at the same level of evolution as when it was first diagnosed. In this case, you can take another of the activities to achieve implementation or a new tool to deepen the change. Not all improvement activities and tools are applicable in all cases. Many times certain tools are better accepted by their users than others. Preferably, this situation should be detected as soon as possible to carry out corrective actions, which is why progress control is necessary.

b) *Level has been improved, but is not at the highest level yet.* In this case, it is possible to continue with another axis that is at a lower level and, once the rest have been processed, to review it again with the associated matrix with the next level.

c) *It is at the highest level.* If the axis is already at the highest level, the improvement must continue with another of the axes detected in point 3. If all have been treated, the same comprehensive evaluation of point 1 must be carried out again. In this diagnosis, if all the axes are at the top level, the implementation of the methodology is Complete.

## 2.4 Test Roles

One of the first considerations to take into account when applying this methodology is that testing tasks can be performed by people who play a specific test role or by people who play another role (for example, customers or product owners). As many improvement models have shown, a certain degree of independence makes the evaluator more effective at finding defects. However, independence is not a substitute for familiarity, and developers can efficiently find many flaws in their own code. In addition, as evidenced in the survey carried out, most SMEs are limited in the assignment of profiles dedicated exclusively to testing tasks.

In general and for most types of projects, it is recommended that at least one of the test levels should be handled by independent testers. Developers must participate in testing, especially at lower levels, so that they can exercise control over the quality of their own work.

## 2.5 The Role of Communication

Communication in this process is essential, especially when you want to implement it in an organization that has not incorporated this type of practice. Generating a good atmosphere in the work team is essential for the success of the appropriation of a "new" process in an organization. Communication involves details as small as a description of a finding or its classification. That is why it is recommended that common language forms be established for this type of process to help the success of the project and the quality of the product developed. Throughout the implementation process of this proposal, it is intended that the greatest number of people involved in the project can communicate their needs about the quality of the product. Both from the moment of diagnosis and in the successive cycles of improvement, the intention is to generate conversations about what is expected of the product. Short cycles should incentivize meetings, at least every two weeks, which allow setting objectives and reviewing progress.

Everyone involved in the improvement process must be able to communicate internally and with external groups. Good internal communication is essential to working as a cohesive group. Communication with clients allows testers to know their expectations. Communication with developers is essential to understand the design details of the system.

Clearly defining test objectives and communicating them in the plan has important psychological implications. Most people tend to align their plans and behaviors with the goals set by the team. It is also important that they adhere to these goals with minimal personal bias.

### 2.6 The Full Implementation

What does it mean that all axes reached the top level or with a full implementation? It means that the organization managed to implement a controlled and formalized testing process. There is a formalized plan that covers the different test levels, where the test cases are fully specified and include combined static and dynamic test activities. Testing is carried out in controlled and adequate environments, with complete traceability and supported by tools. There is a testing base that is taken as the starting point of each iteration or phase and increases as the product evolves. The result of this proposal, in addition, is adaptable to a particular project and, as it is executed in different areas, provides feedback on the process on a permanent basis.

By achieving the level of full implementation, the foundations are laid to aspire to achieve an ISO or CMM certification, since the process is recognized, standardized and formalized. The proposed methodology incorporates activities and structures that are common to all the models analyzed, so that the organization in general is in a more favorable and better predisposed situation to carry out these challenges. It is the kick-start in the generation of new tasks, areas and responsibilities with awareness of the effects of the tests. This improvement in the process will inevitably be reflected in the improvement of the quality of the final product.

## 3 Results Obtained

### 3.1 Summary of Expert Evaluation

Due to the global pandemic situation, the proposal could not be put into practice in a real context. Instead, it was submitted to the judgment of experts, both from the knowledge area and from the industry, to evaluate its feasibility and its contribution to the problem.

Regarding the proposal itself, they generally considered that it was clear and well organized. They highlighted its simplicity and practicality, as well as its ability to adapt to different development models. They also valued its possibility of application both in developer companies and in systems areas within any type of organization, mainly for its comprehensive approach, the contribution of support tools and the guide for its implementation. It is considered as a valid, successful alternative, focused on the typical limitations of SMEs and that it has enough information to be put into practice immediately.

With respect to the proposed axes, they considered that these were opportune, balanced and consistent with the improvement activities that they propose. The division into axes and levels was considered clear since it allows visualizing the real and specific state of each one. It was mentioned that the methodology in general allows an exhaustive analysis to be carried out without being dense or complex, and that its progressive approach also increases its level of acceptance when carrying it out. On the other hand, it was recommended to include the integration tests in the partial implementation level of the category "Affected test levels", since they could be carried out in conjunction with the unit tests. Correspondingly, it was recommended to evaluate the possibility of including metrics at this level, such as code coverage, since it is easy to calculate and use.

Finally, regarding how relevant the proposal is for software developing SMEs in the region, all experts recognized the problem as something that often happens in the industry. Based on their experience, the evaluators consider that the main barriers that SMEs have to adopt frameworks or improvement models of this type are the complexity of traditional models and the lack of them adapted to their realities. They highlighted the need for quality processes to comply with service contracts, customer involvement in these processes and the need for continuous improvement in the industry.

### 3.2 Conclusions

Different institutions have developed different models and standards for the software development continuous improvement, mainly focused on the testing stage. However, there are none applicable to software developing SMEs in our region, as could be corroborated from both state of the art and a survey on the region. Then a management proposal was stated that aims to improve the testing process in SMEs, selecting the most relevant parts and simplifying several widely accepted improvement models, and suggesting tools to support the whole process. The proposal is born from the analysis of the needs and limitations of SMEs and is based on those activities and practices that have been shown to add value to the improvement models analyzed. In this sense, an agile process is proposed, without too many requirements for its implementation, with the minimum possible documentation and that generates reusable work products for deployment in the different projects of the organization. Finally, the proposal was submitted to an evaluation by industry experts, the result of which was positive.

### 3.3 Future Work

The main challenge for the future is to test this proposal in a real environment. This will allow new challenges to be posed based on the ability to adapt the proposal to the particular situations of each organization. This will allow its constant evolution depending on the different scenarios in which it is applied.

It is important to emphasize that the proposal exposed tries to be a first step for those organizations defined as immature. A challenge for the future is to achieve the next step.

**Acknowledgments.** This work was founded by the research project SIUTICU0005297TC of the National Technological University, Argentina. The authors also want to extend their gratitude to Tomás Casanova, for proofreading the translation, and the anonymous reviewers of this work for their valuable comments and suggestions.

## References

1. ISO/IEC/IEEE 29119:2013 Software Testing Strandard (2013). https://doi.org/10.1109/IEEESTD.2013.6588543.
2. Pressman, R.S.: Ingeniería Del Software Un Enfoque Práctico. McGraw-Hill (2010). http://zeus.inf.ucv.cl/~bcrawford/Modelado%20UML/Ingenieria%20del%20Software%207ma.%20Ed.%20-%20Ian%20Sommerville.pdf
3. Sommerville, I.: Software Engineering. Addison-Wesley (2011). https://doi.org/10.1111/j.1365-2362.2005.01463.x
4. Paulk, M.C., Curtis, B., Chrissis, M.B., Weber, C.V: Capability Maturity Model for Software, Pittsburgh, Pennsylvania (1993)
5. SEI: CMMI for Development, Version 1.3 (2010)
6. Van Veenendaal, E.: Test Maturity Model Integration (TMMi) Foundation (2018)
7. Pol, M., Teunissen, R., Van Veenendaal, E.: Software Testing: A Guide to the TMap Approach. Addison-Wesley (2002)
8. Koomen, T., Van Der Aalst, L., Broekman, B., Vroon, M.: TMap Next for result-driven testing. UTN Publishers: Netherlands (2006)
9. van der Aalst, L., Broekman, B., Koomen, T., Vroon, M.: TMap Next, the Essentials of TMap. Sogeti Nederland B.V, Vianen (2010)
10. IT Mark. http://it-mark.eu/?&lang=es
11. Competisoft: COMPETISOFT. Mejora de Procesos para Fomentar la Competitividad de la Pequeña y Mediana Industria del Software de Iberoamérica (2008)
12. ISO/IEC 29110 Software Engineering – Lifecycle profiles for Very Small Entities (VSEs) – Management and engineering guide (2011)
13. Ley 27506 - Régimen de Promoción de la Economía del Conocimiento, http://servicios.infoleg.gob.ar/infolegInternet/verNorma.do?id=324101
14. Programa de Recuperación Productiva - Ley 27264, http://servicios.infoleg.gob.ar/infolegInternet/anexos/260000-264999/263953/norma.htm. Accessed 26 Oct 2020
15. OPSSI, CESSI: Reporte anual sobre el Sector de Software y Servicios Informáticos de la República Argentina, Reporte año 2018 (2019)
16. Argüello, M., Casanova, C., Cedaro, K.: Encuesta para Conocer el Estado del Testing en las empresas de la región (2019) V1 (2021). https://doi.org/10.17632/c7cg6bmnwv.1
17. Pino, F.J., et al.: Software process improvement in small and medium software enterprises: a systematic review. Softw. Qual J. **16**, 237–261 (2008). https://doi.org/10.1007/s11219-007-9038-z
18. Pusatli, O.T., Misra, S.: Software Measurement Activities in Small and Medium Enterprises: an Empirical Assessment (2011)
19. Kituyi, G.M., Amulen, C.: A software capability maturity adoption model for small and medium enterprises in developing countries. Electron. J. Inf. Syst. Dev. Ctries. **55**, 1–19 (2012). https://doi.org/10.1002/j.1681-4835.2012.tb00389.x

20. Iqbal, J., Nasir, M.H.N., Khan, M., Awan, I., Farid, S.: Software process improvement implementation issues in small and medium enterprises that develop healthcare applications. J. Med. Imaging Heal. Informatics. **10**, 2393–2403 (2020). https://doi.org/10.1166/jmihi.2020.3187
21. Karlström, D., Runeson, P., Nordén, S.: A minimal test practice framework for emerging software organizations. Softw. Testing, Verif. Reliab. **15**, 145–166 (2005). https://doi.org/10.1002/stvr.317
22. International Software Testing Qualifications Board: Certified Tester Foundation Level Syllabus. 85 (2011)
23. Naik, K., Priyadarshi, T.: Software Testing and Quality Assurance Theory and Practice. John Wiley & Sons Inc, Hoboken, New Jersey (2008)

# Author Index

Printed in the United States
by Baker & Taylor Publisher Services

Printed in the United States
by Baker & Taylor Publisher Services